DEFINED IMMUNOFLUORESCENCE AND RELATED CYTOCHEMICAL METHODS

ANNALS OF THE NEW YORK ACADEMY OF SCIENCES
Volume 420

DEFINED IMMUNOFLUORESCENCE AND RELATED CYTOCHEMICAL METHODS

Edited by Ernst H. Beutner, Russell J. Nisengard, and Boris Albini

The New York Academy of Sciences
New York, New York
1983

Library of Congress Cataloging in Publication Data
Main entry under title:

Defined immunofluorescence and related cytochemical
   methods.

   (Annals of the New York Academy of Sciences, ISSN 0077–8923; v. 420)
      Proceedings of the Seventh International Conference
Defined Immunofluorescence, Immunoenzyme Studies and
Related Labeling Techniques, organized by the Dept. of
Microbiology, State University of New York at Buffalo,
et al., held on June 8–11, 1982 in Niagara Falls, NY.
      Includes bibliographies and index.
      1. Immunofluorescence—Methodology—Congresses.
2. Immunochemistry—Methodology—Congresses. I. Beutner,
Ernst H., 1923–   . II. Nisengard, R. J. III. Albini, B. IV. International Conference
Defined Immunofluorescence, Immunoenzyme Studies and
related Labeling Techniques (7th: 1982: Niagara Falls,
N.Y.) V. State University of New York at Buffalo.
Dept. of Microbiology. VI. Series. [DNLM: 1. Fluorescent
antibody technic—Congresses. W1 AN626YL v.420 / QY 250
D313 1982]
Q11.N5   vol. 420a   [QR187.I48]   500s   [616.07'5]   84-4802
ISBN 0-89766-238-5
ISBN 0-89766-239-3 (pbk.)

Cover: Biopsy of rabbit kidney with immunofluorescent staining for bovine serum albumin. Antigen is localized in basement membrane only. Courtesy of B. Albini.

SP
Printed in the United States of America
ISBN 0-89766-238-5 (cloth)
ISBN 0-89766-239-3 (paper)
ISSN 0077-8923

ANNALS OF THE NEW YORK ACADEMY OF SCIENCES

VOLUME 420

December 30, 1983

DEFINED IMMUNOFLUORESCENCE AND RELATED CYTOCHEMICAL METHODS[a]

Editors

ERNST H. BEUTNER, RUSSELL J. NISENGARD, AND BORIS ALBINI

CONTENTS

I. Background for Defined Immunocytochemistry

II. Standardization of Immunofluorescence

[a]This volume is the result of a conference entitled Seventh International Conference Defined Immunofluorescence, Immunoenzyme Studies and Related Labeling Techniques, organized by the Department of Microbiology of the School of Medicine and the Department of Periodontology of the School of Dentistry of the State University of New York at Buffalo and the International Service for Immunodermatology Laboratories (ISIL) with advice from the IUIS-WHO Subcommittee on Standardization in Immunofluorescence, held on June 8–11, 1982 in Niagara Falls, NY.

III. Immunoenzyme Methods and Their Application

IV. Substrate Processing for Immunofluorescence

VII. Immunochemical Procedures in Virology, Bacteriology, and Parasitology

Recognition of Antibodies
as Labeled Globulins

MICHAEL HEIDELBERGER

Department of Pathology
New York University Medical Center
New York, New York 10016

As early as the end of the nineteenth century it was widely suspected that antibodies were proteins, since they normally were found in the globulin fractions of serum or plasma. Proof of their nature was, however, impossible because of the merely qualitative, comparative, and often misleading dilution titers by which antibodies were presumed to be measured.[4] As a result, the belief persisted that antibodies might be substances of unknown nature merely adsorbed to globulins. Even as late as 1922 or 1923 a "protein-free antibody" could be had for the treatment of pneumococcal pneumonia: O.T. Avery and I found that "antibody-free protein" was more descriptive.

Announcement of the precipitin reaction between antigen and antibody by R. Kraus in 1897 stimulated speculation as to whether antigens rendered antibodies insoluble by denaturing them, and whether the precipitates formed contained antigen. This led to the first recorded use, in 1902, of labeled antigens. Von Dungern raised antibodies in rabbits with crab blood containing the copper-protein hemocyanin.[1] The antigen-antibody precipitate turned blue when shaken with air, proving the presence of hemocyanin. P.T. Müller used the phosphoprotein casein as antigen and reported a much higher content of phosphorus in the immunoprecipitate than in a volume of serum equal to that taken for precipitation.[19] Labeled antibodies entered the picture much later, as we shall see, although antibodies in their native state are already labeled globulins separable from accompanying proteins by their combining groups for homologous antigen.

As a chemist suddenly exposed to immunology in the early 1920s, I marveled at the progress that had been made with the primitive, purely relative methods of titration then available to the immunologist, progress that often depended upon a brilliant idea or a shrewd guess rather than sound scientific evidence. Yet such fundamental questions as whether antibodies were proteins, whether precipitins, agglutinins, and protective antibodies were the same or different, and whether complement was merely a colloidal state of fresh serum (Bordet) or a substance (Ehrlich) could not be approached with methods as subjective and grossly inaccurate as those in use. After fascinatedly reading Bordet's *Traité d'Immunité,* papers by Ehrlich, and Arrhenius and Madsen's *Immunochemistry,* I resolved to try to end the polemics that were plaguing immunology by applying the rigid criteria of microanalytical chemistry. This had been attempted previously by Wu *et al.*[21] who used hemoglobin and iodinated egg albumin as labeled antigens, but the results were equivocal.

The time was ripe for such an effort, for new tools had become available. Avery and I, later with W.F. Goebel, had established that the capsular slimes of types II and III pneumococci were chemically different, nitrogen-free polysaccharides,[5,6] so that any nitrogen found in their precipitates with homologous antisera would necessarily come from the sera. Even partially purified antibodies were available, for Felton had found that when potent antipneumococcal horse sera were poured into water most of the

1

antibodies precipitated with the so-called euglobulins, leaving most of the other proteins in solution.[2] With these materials, Forrest E. Kendall and I were able to measure accurately in units of weight, mg per ml, the actual content of precipitating antibodies in antipneumococcal sera[9] and to develop a quantitative theory of the precipitin reaction that explained much, and from which useful predictions could be made.[10] When a single polysaccharide antigen was involved, precipitins, protective antibodies,[14] and agglutinins[7] were found to be identical and the intractable dispute as to complement was decided in favor of Ehrlich's view[3]: complement was found to have weight.

But most antigens are proteins, and so it became necessary to distinguish between antigen nitrogen and antibody nitrogen. Yellow to orange azoproteins were known, but as quantitative colorimetric estimations were to be made, it seemed desirable to strive for deeper colors. Accordingly, benzidine was tetrazotized and combined on one side with R-salt, the salt of 2-naphthol-3,6-disulphonic acid, known to yield red dyes, and on the other side with crystalline egg albumin (Ea).[18] The resulting dark purplish-red protein was rather drastically purified until it reacted only faintly with anti-Ea, as we did not want it said that any antibodies due to DEa were actually only anti-Ea.

DEa was antigenic in rabbits and yielded precipitates with the dye that were pink to red depending upon the proportions in which dye and antiserum were mixed.[11,15] For analysis, the washed precipitates were dissolved in alkali, quantitatively rinsed into a colorimeter cup, and compared with an alkaline solution of DEa of known content of nitrogen. This had to be done visually, for photoelectric colorimeters had not yet been invented. The solutions of the precipitates were then quantitatively rinsed into micro-Kjeldahl flasks and analyzed for total N from which the colorimetrically determined antigen N was subtracted to give antibody N. Study of the characteristics of this precipitin reaction made it possible to extend our quantitative method to colorless native proteins.[12] Although Ea precipitated anti-DEa, the quantitative reaction curve was like that expected for a cross-precipitation and totally different from that of a homologous reaction.[12,13]

With the polysaccharide-antibody system we showed, with Torsten Teorell, that strong salt solutions shifted the equilibrium toward precipitates containing less antibody.[16] This release of antibody theoretically promised a means of isolating pure antibody and the conditions for accomplishing this were found after some difficulty. Analytically pure antibody, 100% precipitable or agglutinable, proved to be a typical globulin,[8] settling the old dispute once and for all and putting immunology on a firm scientific basis. Once the chemical nature of antibodies was proven, immunologists could begin to ask where and how they were elaborated, starting a tremendous, continuing expansion.

Kendall, Kabat, and I also began tentative attempts to label antibodies with R-salt or with malachite green, but the products tended to become insoluble and were not intensely enough colored to use in a projected study of phagocytosis. It remained for Albert Coons to think of fluorescence.

As labeled proteins, antibodies served well in a collaborative study with Rudolph Schoenheimer, Sarah Ratner, and David Rittenberg[17,20] who were using the uptake and release of ^{15}N in their physiological studies of protein behavior. Because of the rapid shuttling in and out of ^{15}N, Schoenheimer proposed that amino acid linkages were labile, opening and closing readily. It was Henry P. Treffers' idea to test this hypothesis with two different immune labels. We immunized a rabbit with killed type III pneumococci, gave it a diet containing ^{15}N, and also injected antiserum to type I pneumococci from another rabbit. Repeated quantitative type I and type III precipitin analyses showed that only the type III antibody, which the rabbit was synthesizing, had taken up ^{15}N—the type I precipitates contained virtually none. Schoenheimer greeted the data with "O Weh!," for the antibodies with their specific markers had required an alteration of his hypothesis.

It is apparent, then, that the specificity of the combining groups of antibodies has served as an adequate label for many purposes. Antibodies even possess a second marker, idiotypy, which has given rise to a whole literature and may even be awaiting further exploitation. However, the conjugation of antibodies with enzymes or radioactive elements or substances has enormously increased the sensitivity of their detection and the advantages of their use, as the preceding conferences of this series and the present one have demonstrated.

REFERENCES

1. VON DUNGERN. 1903. Bindungsverhältnisse bei der Präzipitinreaktion. Zent. Bakt. Parasitenk. I. Orig. **34:** 355–360.
2. FELTON, L. D. 1926. Studies on the protective substance in antipneumococcus serum. Bull. Johns Hopkins Hosp. **38:** 33–60.
3. HEIDELBERGER, M. 1940. A quantitative absolute method for the estimation of complement (alexin). Science **92:** 534–535. 1941. Quantitative chemical studies on complement or alexin. I. A method. J. Exp. Med. **73:** 681–694.
4. HEIDELBERGER, M. 1946. Immunochemistry. *In* Currents in Biochemical Research. D. E. Green, Ed. Chapter 29: 453–460. Interscience Publishers, Inc. New York.
5. HEIDELBERGER, M. & O. T. AVERY. 1923. The soluble specific substance of pneumococcus. J. Exp. Med. **38:** 73–79. 1924. **40:** 301–316.
6. HEIDELBERGER, M., W. F. GOEBEL & O. T. AVERY. 1925. The soluble specific substance of pneumococcus. III. J. Exp. Med. **42:** 727–745.
7. HEIDELBERGER, M. & E. A. KABAT. 1936. Chemical studies on bacterial agglutination. II. The identity of precipitin and agglutinin. J. Exp. Med. **63:** 737–746.
8. HEIDELBERGER, M. & E. A. KABAT. 1938. Quantitative studies on antibody purification. II. Dissociation of antibody from pneumococcus specific precipitates and specifically agglutinated pneumococci. J. Exp. Med. **67:** 181–199.
9. HEIDELBERGER, M. & F. E. KENDALL. 1929. A quantitative study of the precipitin reaction between type III pneumococcus polysaccharide and purified homologous antibody. J. Exp. Med. **50:** 809–823. 1935. **61:** 559–562.
10. HEIDELBERGER, M. & F. E. KENDALL. 1935. The precipitin reaction between type III pneumococcus polysaccharide and homologous antibody. III. A quantitative study and a theory of the reaction mechanism. J. Exp. Med. **61:** 563–591.
11. HEIDELBERGER, M. & F. E. KENDALL. 1935. A quantitative theory of the precipitin reaction. II. Study of an azoprotein-antibody system. J. Exp. Med. **62:** 467–483.
12. HEIDELBERGER, M. & F. E. KENDALL. 1935. III. The reaction between crystalline egg albumin and homologous antibody. J. Exp. Med. **62:** 697–720.
13. HEIDELBERGER, M. & F. E. KENDALL. 1954. Quantitative studies on the precipitin reaction. The role of multiple reactive groups in antigen-antibody union as illustrated by an instance of cross-precipitation. J. Exp. Med. **59:** 519–528.
14. HEIDELBERGER, M., F. E. KENDALL & R. H. P. SIA. 1930. Specific precipitation and mouse protection in type I antipneumococcus sera. J. Exp. Med. **52:** 477–483.
15. HEIDELBERGER, M., F. E. KENDALL & C. M. SOOHOO. 1933. Quantitative studies on the precipitin reaction. Antibody production in rabbits injected with an azoprotein. J. Exp. Med. **58:** 137–152.
16. HEIDELBERGER, M., F. E. KENDALL & T. TEORELL. 1936. Quantitative studies on the precipitin reaction. Effect of salts on the reaction. J. Exp. Med. **63:** 819–826.
17. HEIDELBERGER, M., H. P. TREFFERS, R. SCHOENHEIMER, S. RATNER & D. RITTENBERG. 1942. Behavior of antibody protein toward dietary nitrogen in active and passive immunity. J. Biol. Chem. **144:** 555–562.
18. KENDALL, F. E. & M. HEIDELBERGER. 1930. Quantitative studies on the precipitin reaction. Data on a protein-antibody system. Science **72:** 252–253.

19. MÜLLER, P. T. 1903. Weitere Studien über das Laktoserum. Zent. Bakt. Parasitenk. I. Orig. **34:** 48–60.
20. SCHOENHEIMER, R., S. RATNER, D. RITTENBERG & M. HEIDELBERGER. 1942. The interaction of antibody protein with dietary nitrogen in actively immunized animals. J. Biol. Chem. **144:** 545–554.
21. WU, H., L. H. CHENG & C. P. LI. 1927/1928. Composition of antigen-precipitin precipitate. Proc. Soc. Exp. Biol. Med. **25:** 853–855; WU, H., P. P. T. SAH & C. P. LI. 1928/1929. **26:** 737–738.

ALBERT H. COONS

Albert H. Coons

In memoriam

ASTRID FAGRAEUS

Department of Immunology
National Bacteriological Laboratory
S 105 21 Stockholm, Sweden

As we are gathered here it is my privilege to take you back some years in remembrance of Albert Coons. Since I am his oldest friend in this society, I have been asked to bring homage to his memory. It is exactly 30 years and 30 days since I met him the first time at Harvard.

As we all know his scientific achievements starting in 1942 with the finding of pneumococci in tissue by fluorescent antibody technique, I am not going to discuss his scientific contributions. I will instead give some personal glimpses. Albert Coons was a Harvard boy, or let us rather call him a Harvard gentleman. He graduated from Harvard in 1937. He was a classmate of another famous research worker, Lewis Thomas. In his book *The Medusa and The Snail,* Thomas has a chapter, "Note on Medical Economics." Albert Coons, like the majority of his class that later went into research and teaching, was quite sure when he graduated he would be a practitioner. Harvard graduates were better paid because they were better physicians than graduates from other schools; that was what they told themselves. Coons also wanted to practice. He chose internal medicine. He wanted to do something socially important and not just earn money like many of the other graduates. We are grateful that he did not set up this practice.

Indeed, Coons did do something essential in research—and perhaps he did not earn so much money, at least not to start with. One example: In June 1952, I visited Albert Coons who demonstrated the brand new sandwich technique on cells forming antibodies to diphtheria toxoid. In 1953, during the Sixth International Congress of Microbiology in Rome, I was asked to arrange a symposium on antibodies. I was glad to succeed in the end to have Albert Coons present, to explain his immunofluorescent technique to a larger international audience. "In the end"—that is where the money comes in again—Albert Coons got no governmental money for travelling. His father's economic support made the participation in the Rome congress possible. Things were more difficult in those days and Albert Coons was a very modest, self-critical man, not very clever in selling himself. But not even his modesty could stop the fantastic success of his technique. He initiated a new field of study in pathology, cell biology, microbiology, and, of course, in immunology.

In 1961, Albert Coons gave a presidential address to the American Association of Immunologists. To read this address is to learn what qualifications make a great pioneer research worker. He finished his address by giving what might be called a scientific will. It concerned the education of young people in laboratory research work and reflects his own life experience. Firstly, the problem you attack, Coons said, should be a principle one; secondly, it is no doubt better and more stimulating to work on a problem of your own than to carry out a set task. Albert Coons found his own research problem and he never gave up the idea he got in 1939 while sitting alone in a room in Berlin: to label antibodies for the study of Aschoff's nodules. He returned in 1940 to Harvard Medical School with this idea hammering away in his mind. At Harvard he had a most stimulating environment. People listened, as for example John Enders, who told him that he did not believe that Coons should solve the rheumatoid disease problem by hooking a fluorescent dye on antibodies but that the idea was very good for many other purposes. Enders also introduced him to chemists working with isocyanate for other purposes. Coons thus was fortunate to carry on his pioneer work in an outstanding environment, but his success depended on his character. He had scientific imagination; he could create enthusiasm in workers in other research fields and he could cooperate. He was keen on giving credit to his co-workers and was absolutely loyal to Harvard until his death. Last, but not least, he never gave up and he was himself a hard worker. He was in the lab most weekends and holidays. One of his maxims was the Greek saying "Strive for grace under pressure." But in spite of being so scientifically involved he had time for his large family, he was interested in music, especially the chamber music of Beethoven, and in poetry. He had many good friends who appreciated his sense of humor and listened to his laughter around a good dinner table. We are grateful to have known him.

Introduction:
The Nature of Defined
Immunocytochemical Studies

ERNST H. BEUTNER

Department of Microbiology
Schools of Medicine and Dentistry
State University of New York at Buffalo
Buffalo, New York 14214

Modern, quantitatively defined immunocytochemical studies historically arose from two distinct branches of immunology, notably from studies on dye-labeled antigens and antibodies for the characterization and quantitation of antibodies and their reactions with antigens,[9,10,13,17] and secondly from the uses of fluorochrome and other dye-labeled antibodies, antigens, and other probes as histochemical markers.[2,3,8,9,22]

Dr. Michael Heidelberger's presentation in this volume[8] affords a uniquely authentic, historical review of the genesis of the first of these two pillars of this field. He credits von Dungern's work in 1903[24] with being the first to use antigens labeled with a dye as markers. Drs. Heidelberger and Kendall[9,10] were not only the first to identify antibodies unequivocally as proteins but also to quantitate them in part with dye-labeled antigens. These methods, and refinements of them, yielded two of the three pillars of defined immunocytochemistry, notably defined tests for specificity and measures of the sensitivity of the labeled reagents. For the third and most important technical advance, the microscopic detection of labeled antibodies, we all owe a debt of gratitude to Albert Coons.[2,3] Dr. Astrid Fagraeus' presentation gives us insights into the human side of Dr. Coons. In 1941, he and his associates[2] reported on the use of the then recently described method of Fieser and Creech[6] for synthesizing β-anthryl conjugates of proteins and in 1942[3] for binding fluorescein covalently to proteins with a carbamido linkage. Among the many important uses of labeled antibodies by Coons and his associates was the confirmation of the findings of Fagraeus that antibodies are synthesized by plasma cells.[4,18] Drs. Nairn,[19–22] Kawamura,[14,15] and others,[7] as well as the proceedings of these conferences,[1,11,12,16] now provide us with scholarly reviews on the uses of Albert Coon's methods and the now widely used modification of Riggs *et al.,*[23] which employs thiocarbamido linkages for fluorochrome, as well as many other useful methods for dye-probe studies of tissues and cells by microscopy.[22]

In these introductory remarks, I would like to stress the critical importance of integrating all three of these well-founded types of studies on labeled antigens and antibodies, i.e. studies for characterization of their specificity, for their quantitation, and for their use as cytochemical probes. The synthesis of these three types of studies, which provides us with a sound foundation for modern, defined immunofluorescence and other quantitatively controlled immunocytochemical studies, stands in sharp contrast to other cytologic studies with markers. This contrast can be expressed succinctly as shown in TABLE 1.

The contributions of our own group[1] have been, and continue to be the introduction and promotion of methods integrating these precepts of defined immunofluorescence with quantitative serology *a là* the classic teachings of the Witebsky school of immunology.[25]

TABLE 1. Differentiation of Defined Immunocytochemical Methods from Other Cytologic Methods

Examples of Types of Morphologic Studies	Criteria for Selection of Markers (e.g., dyes or conjugates)
Histology, classic histochemistry, and other morphologic methods	Visually "optimal" contrast or differentiation
Quantitatively controlled immunocytochemical methods, e.g. "defined immunofluorescence"	Specificity of antibodies or antigens Quantitative measures of their sensitivity (i.e., concentration and degree of labeling) in cytologically detectable ranges.

The current state of the art of the industry that provides reagents for defined immunofluorescence studies serves to illustrate both some of the present prospects and problems of defined immunocytochemistry. Specifically, fluorescein isothiocyanate–labeled anti-human IgG antibodies can now be obtained commercially from a sizable number of companies throughout the developed countries of the world. Many of them now provide such reagents with quantitative data on their most salient characteristics. These data on characteristics can and do provide simple functional guides for the performance of defined immunofluorescence studies, such as the titration of human antibodies to microbial or tissue antigens. TABLE 2 summarizes the most relevant data provided with one such reagent, the recommended assay methods employed by the manufacturer, and the use of these data in the performance of defined immunofluorescence studies with this conjugate.

By the use of such quantitatively characterized anti-human IgG conjugates as the one listed in TABLE 2, at dilutions with properly selected and defined antibody protein

TABLE 2. Characteristics of One FITC-labeled Anti-human IgG Conjugate and Their Use for Defined Immunofluorescence Studies

Factors Tested	Test Methods	Data Given[a]	Conclusions on Conjugate
Specificity: anti-human IgG	Immunoelectrophoresis	Anti-H-IgG	Desired specificity for human IgG
Concentration	Biuret protein	11.6 mg/ml	50 μg/ml for use obtained by 50/1300 or 1:26 dilution[b]
Sensitivity			
Weight F/P[a] ratio molar F/P ratio	NCCLS[a] approved methods	6.1×10^{-3} 2.4	In desired range of F/P ratios
Antibody concentration	Reverse radial immunodiffusion	1.3 mg/ml	1.3/11.6 $\times$ 100 = 11.2%, i.e. sensitivity of 10% is acceptable 50 μg/ml for use obtained by 50/1300 or 1:26 dilution[b]

[a]NCCLS, National Committee for Clinical Laboratory Standards recommended methods for F (fluorescein) and P (protein) assays and F/P ratio calculations.

[b]This needs to be checked by performance tests using chessboard titrations. For details see report of Beutner *et al.* in this volume.

concentrations, the immunocytochemical reaction can be rendered more reproducible and interpretable. That is, interpretations of reaction patterns can, with greater confidence, focus on the properties of the antigen preparations used and the antibodies that combine with them rather than on the means of detecting these reaction patterns. This brings us closer to the ultimate goal of defined immunocytochemistry, to gain reliable information on the relation of structure and function.

The reason for the existence of this conference is clear from these considerations. As long as there is a need for studies of the relation of structure and function, and a desire to check on the reliability of the observations of these relationships, there will be a continued need for further refinements in defined immunocytochemical methods such as defined immunofluorescence.

REFERENCES

1. BEUTNER, E. H., ED. 1971. Defined Immunofluorescent Staining Ann. N.Y. Acad. Sci. Vol. 177.
2. COONS, A. H., H. J. CREECH & R. N. JONES. 1941. Immunological properties of an antibody containing a fluorescent group. Proc. Soc. Exp. Biol. **47:** 200–202.
3. COONS, A. H., H. J. CREECH, R. N. JONES & E. BERLINER. 1942. The demonstration of pneumococcal antigens in tissues by the use of fluorescent antibody. J. Immunol. **45:** 159–170.
4. COONS, A., E. LEDUC & J. M. CONNOLLY. 1955. Studies on antibody production. I. A method for histochemical demonstration of specific antibody and its application to a study of the hyperimmune rabbit. J. Exp. Med. **102:** 49–60.
5. FAGRAEUS, A. 1983. Albert Coons—In memoriam. Ann. N.Y. Acad. Sci. (This volume.)
6. FIESER, L. F. & H. J. CREECH. 1939. The conjugation of amino acids with isocyanates of the anthracene and 1,2-benzanthracene series. J. Am. Chem. Soc. **61:** 3502–3506.
7. GOLDMAN, M., ED. 1968. Fluorescent Antibody Methods. Academic Press. New York.
8. HEIDELBERGER, M. 1983. Recognition of antibodies as labeled globulins. Ann. N.Y. Acad. Sci. (This volume.)
9. HEIDELBERGER, M. & F. E. KENDALL. 1929. A quantitative study of the precipitin reaction between type III pneumococcus polysaccharide and purified homologous antibody. J. Exp. Med. **50:** 809–823.
10. HEIDELBERGER, M. & F. E. KENDALL. 1935. The precipitin reaction between type III pneumococcus polysaccharide and homologous antibody. III. A quantitative study and a theory of the reaction mechanism. J. Exp. Med. **61:** 563–591.
11. HIJMANS, W. & M. SCHAEFFER, EDS. 1975. Fifth International Conference on Immunofluorescence and Related Staining Techniques. Ann. N.Y. Acad. Sci. Vol. 254.
12. HOLBOROW, E. J., ED. 1970. Standardization in Immunofluorescence. Blackwell Scientific Publications. Oxford.
13. KABAT, E. A. 1968. Structural Concepts in Immunology and Immunochemistry. Holt, Rinehart and Winston. New York.
14. KAWAMURA, A., JR., ED. 1969. Fluorescent Antibody Techniques and Their Applications. 1st edit. University of Tokyo Press. Tokyo.
15. KAWAMURA, A., JR., ED. 1977. Fluorescent Antibody Techniques and Their Applications. 2nd edit. University of Tokyo Press. Tokyo.
16. KNAPP, W., K. HOLUBAR & G. WICK, EDS. 1978. Immunofluorescence and Related Staining Techniques. Elsevier/North Holland Biomedical Press. Amsterdam.
17. LANDSTEINER, K. 1945. The Specificity of Serological Reactions. 2nd edit. Harvard University Press. Cambridge, Mass. (Reprinted in paperback, Dover Publications, New York. 1962. The Basic Early Studies on Immunological Specificity and The Introduction of Small Haptenic Groups into Proteins to Modify Antigenicity.)
18. LEDUC, E., A. COONS & J. M. CONNOLLY. 1955. Studies on antibody production. II. The primary and secondary responses in the popliteal lymph node of the rabbit. J. Exp. Med. **102:** 61–72.

19. NAIRN, R. C., ED. 1962. Fluorescent Protein Tracing. 1st edit. F & S Livingstone, Ltd. Edinburgh.
20. NAIRN, R. C., ED. 1964. Fluorescent Protein Tracing. 2nd edit. Williams and Wilkins Co. Baltimore.
21. NAIRN, R. C., ED. 1969. Fluorescent Protein Tracing. 3rd edit. Williams and Wilkins Co. Baltimore.
22. NAIRN, R. C., ED. 1976. Fluorescent Protein Tracing. 4th edit. Churchill Livingstone. Edinburgh.
23. RIGGS, J. L., R. J. SEIWALD, J. H. BURCKHALTER, C. M. DOWNS & T. G. METCALF. 1958. Isothiocyanate compounds as fluorescent labeling agents for immune serum. Am. J. Pathol. 34: 1081–1097.
24. VON DUNGERN. 1903. Bindungsverhältnisse bei der Präzipitinreaktion. Zent. Bakt. Parasitenk. I. Orig. 34: 355–360. (From Heidelberger. This volume. Ref. 1)
25. WITEBSKY, E. 1957. An immunologist's vade mecum. N.Y. State J. Med. 57: 3615–3622.

Measurement of Variation and Significance in Serologic Tests

ROGER N. TAYLOR

Centers for Disease Control
Atlanta, Georgia 30333

It has been traditionally held, and is frequently cited in a wide variety of literature, that the variation in replicate serologic test results on a single specimen is plus or minus one tube (usually a twofold dilution). This idea has become firmly ingrained in immunology despite the fact that it is vague and undocumented. It has also become almost universally accepted that a fourfold rise in titer on sequential specimens is significant serologic evidence of recent infection.

Even though these ideas are commonly accepted, little thought has been given to what they actually mean or what the implications of their acceptance are. In most cases, the plus-or-minus-one-tube rule has been taken to mean that as long as none of the results differ by more than one tube from a central result the variation is acceptable. In some cases, it is interpreted to mean that variation is acceptable as long as replicate titers remain within a twofold range. Sometimes the plus-or-minus-one-tube rule is used when dilution schemes other than twofold are used, or when irregular dilution schemes are used.

These interpretations do not take into account the fact that frequently the true result may be between two tubes. They also ignore the fact that within these limits there can be great differences in the percentage of results that are identical when the same specimen is repeatedly measured. The implicit assumption underlying this rule and these interpretations is that the variation in all serologic tests, in all laboratories, and under all conditions is the same.

FIGURE 1 shows some hypothetical examples of possible frequency distributions of serologic test results having results within plus or minus one twofold dilution of the median. It is obvious that the variation in these three sets of data is different in spite of the fact that in each case all of the results are within one twofold dilution of the median result.

The problem of measurement of variation in serologic test results has been investigated by others at Centers for Disease Control,[1,10] but we prefer the simpler method that has been in use in proficiency testing since 1972.[6-9] Our method consists simply of converting the results to logarithms (to any convenient base), calculating the mean and standard deviation, and converting back to the original form (antilogarithm).

The simplicity of the calculation is demonstrated in the following example:

Result	$\log_2$	$(\log_2)^2$
1	0	0
2	1	1
4	2	4
$N = 3$	$\Sigma X = 3$	$\Sigma X^2 = 5$

<u>Geometric Mean</u>   <u>Geometric Standard Deviation</u>

$$\overline{X}_G = \text{antilog } \overline{X}_{\log} \qquad S_G = \text{antilog } S_{\log}$$

$$\overline{X}_{\log} = \frac{\Sigma X}{n}$$

$$S_{\log} = \sqrt{\frac{\Sigma X^2 - \dfrac{(\Sigma X)^2}{n}}{n-1}}$$

$$\overline{X}_{\log} = \frac{3}{3}$$

$$\overline{X}_{\log} = 1$$

$$S_{\log} = \sqrt{\frac{5 - \dfrac{(3)^2}{3}}{3-1}}$$

$$\overline{X}_G = 2$$

$$S_{\log} = 1$$

$$S_G = 2$$

This method is a simple adaptation (log conversion) of the method that is commonly used in other clinical laboratory disciplines and defines precision in terms of standard deviations (geometric). It avoids the confusion that has resulted from the conflicting definitions of repeatability, replicability, and reproducibility that are used in serology.[2–4] Variation can be measured within or among samples, technologists, tests, or laboratories. The geometric standard deviations (S_Gs) and 95% ranges for the hypothetical examples are shown in FIGURE 1. The geometric standard deviations reflect the differences in variation in these three distributions.

This method of measuring variation in serologic test results is similar to the method used in other laboratory disciplines, but the results must be log-transformed because serologic test results are log-normally distributed.

It is convenient to describe variation in measurements by reference to the normal curve for a number of reasons: (1) The mean value is the best estimate of the true value. (2) The calculations of two simple parameters (mean and standard deviation) completely describe the distribution and will therefore permit calculations of the percentage of results that will be within any defined limits on the curve. (3) Many common statistical tests are based on the normal curve or modifications of it. Therefore, comparison between or among items can be made and the probability of significant differences can be estimated.

The concept that "plus-or-minus-one-tube" is acceptable variation implies that results are symmetrically clustered around the mode, but this is not always the case. In fact, the geometric mean or median is not always one of the possible values in the dilution scheme. Let us consider an example involving a test that has a S_G of 1.725. The three distributions of results from a dilution scheme starting at 1:10 shown in FIGURE 2 could be obtained from samples with geometric means of 40, 33, and 28, respectively. In the top histogram, the results are symmetrically distributed around the median and geometric mean of 40, and 96% of the results are within one tube of the mode. The middle histogram shows a distribution with the mean of 33, slightly lower than an obtainable result. Although 95% of the results are between 20 and 80, the other 5% are lower, and no results are higher than this range. In the bottom histogram the results are again symmetrical, but the geometric mean of 28 and the median are midway between two possible results. Neither of the results, 20 or 40, satisfactorily describes the central analyte level of this sample. All three distributions could be described by "40 plus-or-minus-one-tube," but their differences are better described by stating their geometric means and standard deviations.

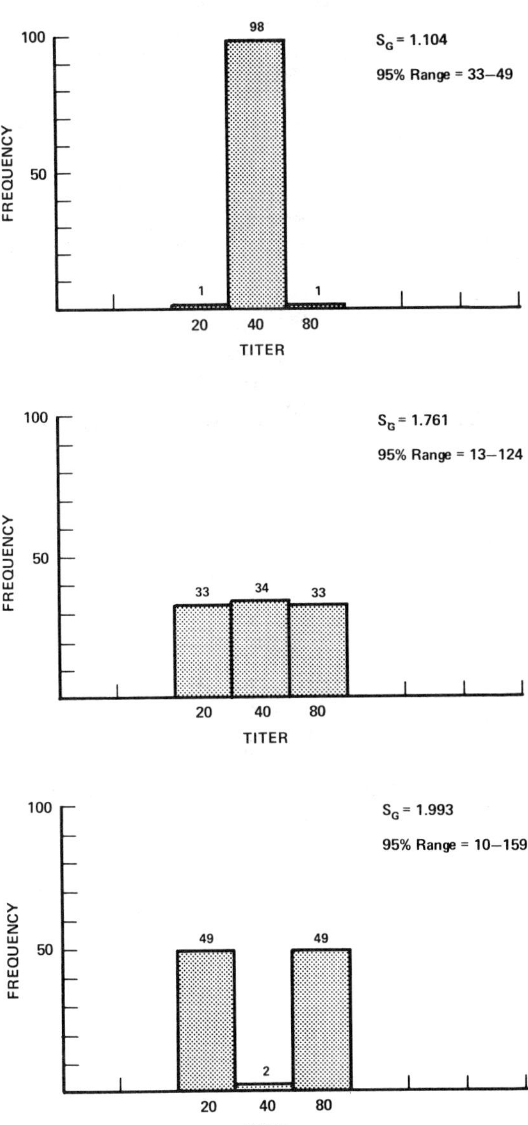

FIGURE 1. Examples of possible frequency distributions having results within plus or minus one twofold dilution of the median.

Any system of measurement of variation in serologic test results must take into consideration the fact that the best estimate of central tendency (analyte concentration) will usually not be a possible result in any particular dilution scheme, and the estimate of variation should not change appreciably when samples with slightly different means are tested. The system we use meets these requirements as can be seen in this example. The S_Gs were all near 1.725.

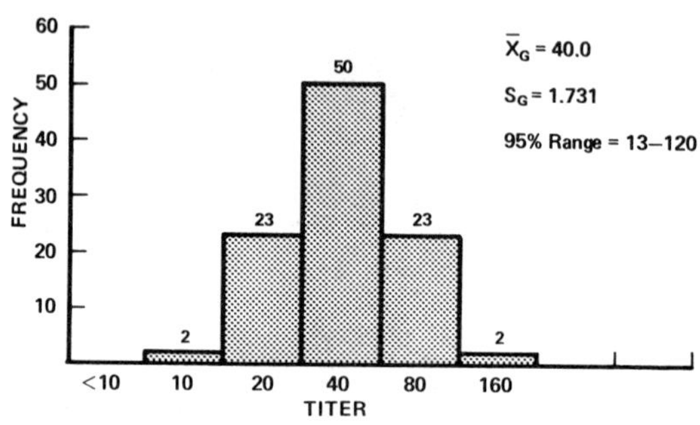

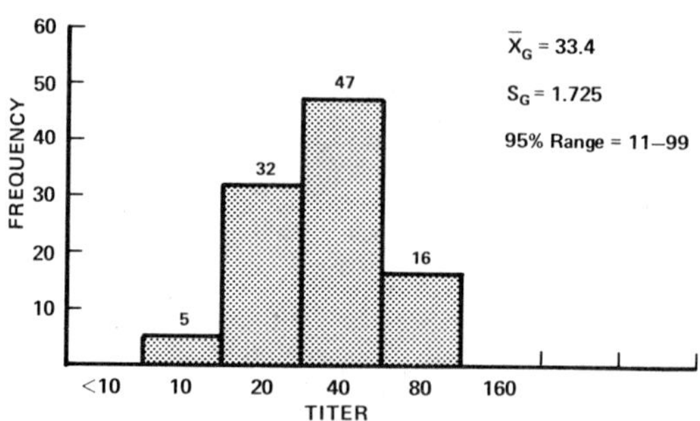

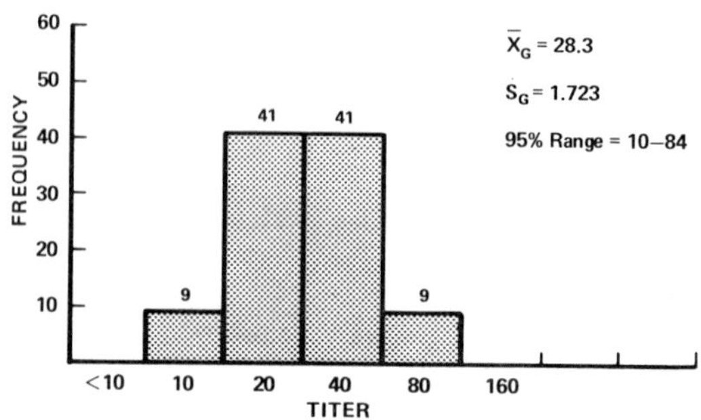

FIGURE 2. Examples of frequency distributions with similar variation, but with different central values.

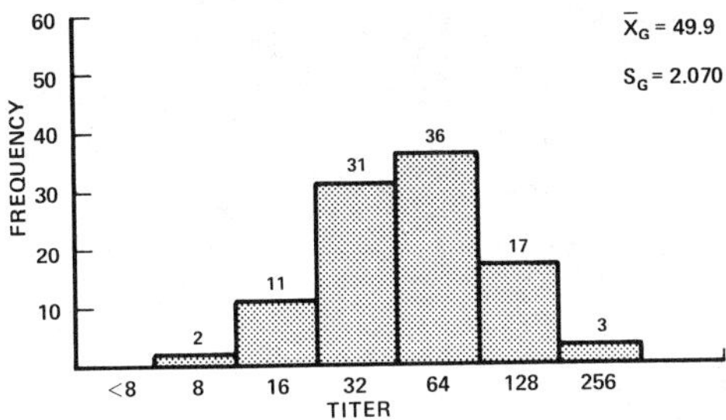

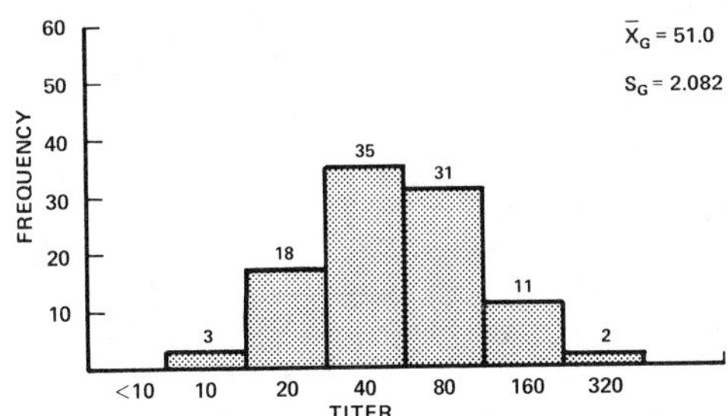

FIGURE 3. Comparison of frequency distributions of test results from the same population using different dilution schemes.

Another prerequisite for a system of measurement of variation in serologic test results, especially if it is to be used for interlaboratory or intermethod comparisons, is that it must not be biased toward any particular dilution scheme. FIGURE 3 shows histograms of results from two different dilution schemes that might be expected from a population with a true analyte titer of 50 and with a geometric standard deviation of about 2.08. With our system, the geometric means and standard deviations were very near the expected values in spite of the fact that the results were taken from two different dilution schemes and the true titer did not occur in either scheme.

TABLE 1 shows comparisons of measures of central tendency and spread of the serologic test results shown in FIGURE 3. Both the median and the mode are obviously biased toward the system from which they were derived. The arithmetic mean is too high in both schemes. By far the best estimate of the true value is the geometric mean, which is also the least biased.

Comparison of measures of spread shows that the central 95% limits are biased and differ substantially between dilution schemes. The two standard deviation limits (arithmetic) frequently extend below zero and are not practically useful. The geometric standard deviation squared limits are not biased (they are almost identical from both dilution schemes) and they are more realistic (i.e., theoretically possible).

Another desirable feature of any measurement system is that it be adaptable to irregular dilution schemes such as the Rantz and Randal system used for antistreptolysin O. The possible titers in this system are 12, 50, 100, 166, 250, 333, 500, 625, 833, 1250, 2500, with ratios between successive titers of 4.17, 2, 1.25, 1.33, 1.51, 1.33, 1.50,

TABLE 1. Comparisons of Measures of Central Tendency and Spread for Distributions from Two Dilution Schemes

Measures	Dilution Scheme	
	Base 8 Starting Dilution 1:8	Base 10 Starting Dilution 1:10
Central Tendency		
Median	64	40
Mode	64	40
Arithmetic mean	64.3	66.7
Geometric mean	49.9	51.0
True value	50	50
Spread		
Central 95% limits	16 to 128	20 to 160
$2\,S$ limits (arithmetic)	-35 to 164	-44 to 178
S_G^2 limits (geometric)	12 to 214	12 to 221

and 2, respectively. In this system, a one-tube variation has substantially different meaning and depends on what level is being measured. The dilution scheme recommended by Centers for Disease Control is much simpler and has uniform ratios between successive tubes.[5] It consists simply of two twofold dilution series with starting dilutions of 60 and 85. The resulting dilutions (60, 85, 120, 170, 240, 340...) all differ by a ratio of 1.41. FIGURE 4 shows that the method we use gives accurate estimates of precision even though the results are from an irregular dilution scheme.

If the maximum allowable variation permitted in a test is plus or minus one twofold difference from the mean and 95% of the results must be within this range, then the midpoint between the first and second dilution from the mean (on a logarithmic scale) will define the $S_G^{1.96}$ limits (95% limits) and they will differ by eightfold. For example, if a twofold dilution scheme is used, then possible values of test results could be 1, 2, 4, 8, 16, 32.... By convention it is accepted that a test result of 8 means that the true value is closer to 8 than it is to either 4 or 16. Since this is a geometric progression, it means that it is closer on a logarithmic scale and the true value is between $\sqrt{4 \times 8}$ and $\sqrt{8 \times 16}$ or, in other words, between 5.66 and 11.31. In this example, results of not only 8 but also 4 and 16 would be considered acceptable. Since a value of 4 means that it is closer to 4 than it is to 2, the lower limit for acceptability would be $\sqrt{4 \times 2}$, or 2.83. Similarly, the upper limit would be $\sqrt{16 \times 32}$, or 22.63. Thus, there is an eightfold difference between the upper and lower limits ($22.63/2.83 = 8$) and the difference between the geometric mean and either the upper or lower limit would be $\sqrt{8}$, or 2.83.

The 95% limits in arithmetic data include all values between the mean minus 1.96 standard deviations and the mean plus 1.96 standard deviations ($X \pm 1.965$), but with geometric data, the 95% limits include all values between the geometric mean divided by the geometric standard deviation to the 1.96 power and the geometric mean multiplied by the geometric standard deviation to the 1.96 power ($X_G \times/\div S_G^{1.96}$). For this example, the 2.83-fold range is equivalent to $S_G^{1.96}$, so $S_G = \sqrt[1.96]{2.83}$ or 1.70. Therefore, a distribution with an S_G of less than 1.70 would have at least 95% of the results within a twofold difference from the mean and would be considered to have acceptable variation by the traditional rule.

With the new test methods that are not based on the traditional dilution technique (e.g., nephelometry, fluoroimmunoassay, and enzyme immunoassay) the results are less discrete, i.e., more nearly continuous, than older serologic tests. Much of the benefit of better precision would be lost if the variation in results were not actually measured. The method we use is applicable to these kinds of results and should be used.

By using the method described, technologists can perform multiple tests on quality control samples, calculate the actual variation, and establish quality control charts. These charts can then be used to more accurately determine when the test is out of control because they are based on actual, measured variation and calculated limits.

The method can also be used to fill other needs for estimates of variation and significance, such as a difference in results between or among tests, laboratories, procedures, or technologists; for measurement of variation in analyte levels in population; and for measurement of significance of differences between samples.

In addition to accepting the idea that serologic tests have a variation of plus or minus one tube, serologists have also accepted the idea that a fourfold rise in titer constitutes a significant change. The idea that a twofold or fourfold titer difference constitutes a significant change again disregards the fact that variations in serologic test results are not the same in all tests. In practice, such change may or may not be significantly different. Examination of FIGURE 1 reveals that the probability of two samples having a fourfold difference varies substantially in these three distributions. In the upper distribution, the probability is .01 times .01, or 0.0001; in the center it is .33 × .33, or 0.11; and in the lower it is .49 × .49, or 0.24. Obviously if two samples from

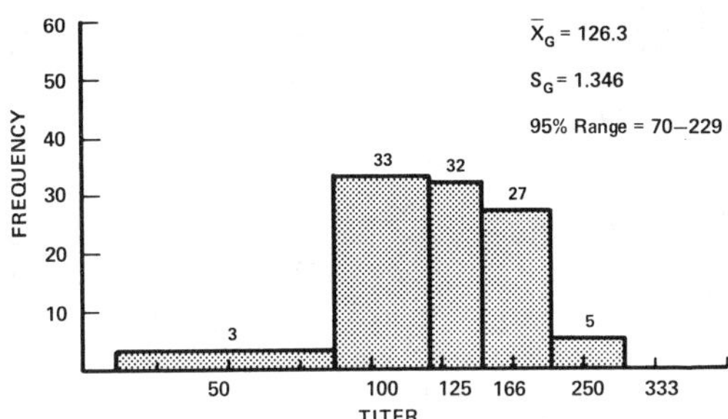

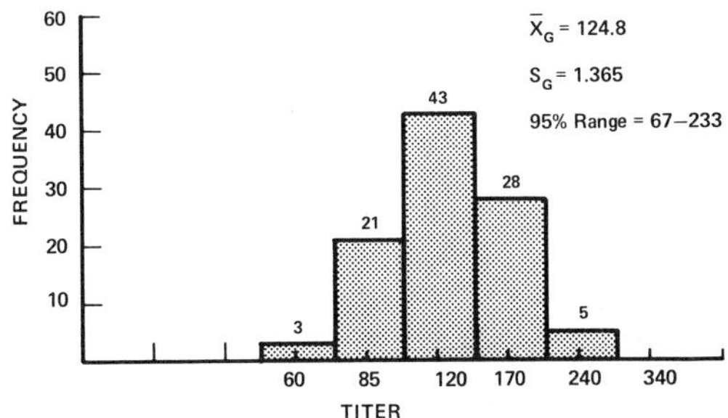

FIGURE 4. Comparison of frequency distributions of test results from the same population using irregular versus uniform dilution schemes.

TABLE 2. Ratio of Results Required Between Two Serologic Test Results for Differences To Be Statistically Significant for Selected Tests

Test (Method)	Intralaboratory Precision (S_G)	Ratio Required for Significant Difference		
		p < .05	p < .01	p < .001
Antinuclear antibodies (Indirect immunofluorescence)	1.72	2.89	4.03	5.99
Cytomegalovirus (Indirect immunofluorescence)	2.152	4.49	7.17	12.54
Herpes simplex virus (Indirect immunofluorescence)	1.880	3.45	5.07	8.03
Rheumatoid factors				
(Latex tube)	1.212	1.46	1.64	1.89
(Latex slide)	1.596	2.50	3.33	4.68
C-reactive protein (Nephelometry)	1.117	1.24	1.33	1.44
Infectious mononucleosis				
(Tube agglutination)	1.378	1.87	2.28	2.88
(Ox cell hemolysin)	1.232	1.51	1.71	1.99
Syphilis				
(Venereal disease research laboratory)	1.273	1.60	1.86	2.22
(Rapid plasma reagin)	1.343	1.78	2.13	2.65

the same distribution can have a fourfold difference 24% of the time from chance alone, a fourfold difference cannot be considered statistically significant.

TABLE 2 shows intralaboratory precision (S_G) selected from proficiency-testing summaries. The values listed under the <.05, <.01, <.001 probabilities are the geometric standard deviations raised to the 1.96, 2.57, and 3.30 powers, respectively (i.e., the powers required to define the respective probabilities). The values show the size of the ratio between two serologic test results that would be required for the differences to be statistically significant at various probability levels. It can be seen

TABLE 3. Ratio of Results Required Between Two Serologic Test Results for Differences To Be Statistically Significant for Selected Geometric Standard Deviations

Precision	Ratio of Results Required for Significant Difference		
	p < .05	p < .01	p < .001
1.000	1.000[a]	1.000[a]	1.000[a]
1.111	1.229	1.311	1.414[a]
1.144	1.302	1.414[a]	1.559
1.193	1.414[a]	1.574	1.790
1.234	1.510	1.717	2.000[a]
1.310	1.698	2.000[a]	2.438
1.424	2.000[a]	2.480	3.211
1.522	2.278	2.943	4.000[a]
1.715	2.878	4.000[a]	5.930
2.028	4.000[a]	6.154	10.312

[a]Ratios occurring in common dilution schemes.

from this table that a fourfold difference in cytomegalovirus indirect-immunofluorescence titer would not be statistically significant at the 0.05 probability level, but that a twofold difference in infectious mononucleosis (ox cell hemolysin) or rheumatoid factor (latex tube) titer is significantly different at the 0.001 probability level.

It is obvious from these results that a fourfold titer change may be highly significant with some tests in some laboratories and may not be significant at all in others. TABLE 3 shows similar ratios for selected values of S_G. This table can be used to determine the significance of differences between results at given S_Gs. For example, if the S_G is less than 1.424, then a twofold difference is significant at the 0.5 probability level. If each laboratory measured the intralaboratory precision for each test, it could very easily report to the physician not only titers on paired sera but also the probability that the differences were statistically significant. On single samples the laboratory could provide titers and confidence intervals.

We use a simple method for measuring variation in serologic test results that is a modification of the procedure commonly used in other laboratory disciplines. It is easy to perform; is not biased toward any dilution scheme; can be used with irregular dilution schemes; can be used for interlaboratory, intermethod, or intertechnologist comparisons; can be used to establish quality control checks; and can be used in conjunction with common statistical tests to determine probability and significance levels.

We recommend that serologists measure the actual variation in the tests they perform, that they establish controls with calculated acceptable limits, and that the significance of differences in results be calculated.

REFERENCES

1. HALL, E. C. & M. B. FELKER. 1970. Reproducibility in the serological laboratory. Health Lab. Sci. **7:** 63–68.
2. MANDEL. J. 1971. Repeatability and reproducibility. Mater. Res. Stand. **11:** 8–16.
3. MANDEL, J. & L. F. NANNI. 1978. Measurement evaluation. *In* Quality Assurance Practices for Health Laboratories. S. L. Inhorn, Ed.: 209–272. America Public Health Assoc. Washington, D.C.
4. PALMER, D. F. & J. J. CAVALLARO. 1980. Some concepts of quality control in immunoserology. *In* Manual of Clinical Immunology. 2nd edit. N. R. Rose & H. Friedman, Eds.: 1078–1082. American Society for Microbiology. Washington, D.C.
5. PALMER, D. F., J. J. CAVALLARO & R. H. GALT. 1973. Laboratory Diagnosis by Serologic Methods. Center for Disease Control. Atlanta, Ga.
6. TAYLOR, R. N. 1974. Derivation of Evaluation Survey Data from Proficiency Testing for Infectious Mononucleosis. Doctoral Dissertation. University of Utah. Salt Lake City, Ut.
7. TAYLOR, R. N. & K. M. FULFORD. 1981. Theory and Application of the CDC Diagnostic Immunology Proficiency Testing Program. Centers for Disease Control. Atlanta, Ga.
8. TAYLOR, R. N., A. Y. HUONG & K. M. FULFORD. 1975. Measurement of Variation in Serologic Tests. Center for Disease Control. Atlanta, Ga.
9. TAYLOR, R. N., A. Y. HUONG, K. M. FULFORD, V. A. PRZYBYSZEWSKI & T. L. HEARN. 1979. Quality Control for Immunologic Tests. DHEW Publication No. (CDC) 79-8376. Center for Disease Control. Atlanta, Ga.
10. WOOD, R. J. & T. M. DURHAM. 1980. Reproducibility of serological titers. J. Clin. Microbiol. **11:** 541–545.

External Quality Control and Standardization in Labeled Immunoassay

ROGER N. TAYLOR AND KAREN M. FULFORD

Centers for Disease Control
Atlanta, Georgia 30333

INTRODUCTION

Proficiency testing programs provide a convenient way for laboratory workers to check the accuracy of their procedures and performance by comparing their results with results from the same samples obtained by other workers. This external quality control should be an integral part of every laboratory's quality-control program. Proficiency testing can also be used to detect problem areas in testing and to provide mechanisms for standardizing procedures or reporting results.

In 1977 and 1978, two proficiency testing studies were conducted by the Proficiency Testing Branch of the Centers for Disease Control to determine the feasibility of standardizing results of rheumatoid factor (RF) tests through the use of a standard.[4,12] Substantial improvement was seen in the agreement of results with each other, and the same degree of agreement has continued in subsequent surveys.[9,11]

The success of the RF studies prompted the suggestion that similar results could be obtained in other areas. As newer tests, such as immunofluorescence tests, enzyme immunoassays, fluoroimmunoassays, and radioimmunoassays, have become more widely used in clinical laboratories, laboratory workers, clinicians, and others are faced with results that are given in various units that, for a number of reasons, are not comparable. Under such conditions, the need for standardization with reliable reference materials is readily apparent.

Proficiency testing services are in a unique position to help identify and possibly lessen the effect of these divergent results by recognizing them and providing a means of interlaboratory comparisons and standardization. This paper provides baseline data for areas where isotopic- and nonisotopic-labeled immunoassays have become popular and discusses an attempt to improve the comparability of results in antinuclear antibody (ANA) testing.

MATERIALS AND METHODS

Most of the sera or plasma used for specimen preparations were purchased from commercial suppliers under government contract. Other sources donated pools of human serum of known reactivity in specific tests.

Surveys are sent routinely on a quarterly or semiannual basis, and special study surveys are provided as necessary. Procedures have been described previously.[8-11]

Results were analyzed by using previously described statistical calculations, notations, and methods.[10,14] Geometric means ($\overline{X}_G$) and standard deviations (S_G) were calculated by converting results to logarithms, calculating the parameters, and converting back to original form (antilog).[14]

RESULTS

Results of indirect immunofluorescence (IIF) tests for ANA showed considerable variation, depending on the source and type of substrate used.[a] Mean results for each of seven samples are given by substrate type in TABLE 1. The relative sensitivities of substrates noted for these samples are not consistent with some previously reported comparisons,[2,3,5–7] but they are consistent with previous proficiency-testing data.[9]

Many false-negative results were reported for all the samples in a special study survey for ANA. Comparison of methodology variables did not reveal any correlations between qualitative results and procedures or equipment. A comparison of qualitative and quantitative results showed that samples with mean titers of greater than 40 were generally reported as positive, but lower titered samples were frequently called negative (TABLE 2). A study using prestained mouse kidney slides was conducted to determine if the inability to read fluorescence was a major reason for reporting negative results. The results of this study are given in TABLE 3. Familiarity with the

TABLE 1. Antinuclear Antibody Results by Substrate Type

Substrate Type	Geometric Mean Titer by Sample							Mean No. Labs	S_G
	A	B	C	D	E	F	G		
HCL	13.4	14.7	9.9	59.3	98.9	72.3	36.1	178	2.53
H O	6.1	6.3	6.5	10.1	–	96.6	56.6	12	2.55
ML	21.6	19.3	16.5	195.0	124.4	70.2	45.5	30	2.92
MK	18.8	17.8	14.4	129.1	115.5	83.2	43.6	79	2.76
MCL	14.4	17.3	10.6	13.3	68.8	68.6	43.6	58	2.80
RL	20.8	21.5	19.8	216.3	135.8	115.7	64.0	92	2.86
RK	15.5	21.9	16.1	57.3	–	65.0	46.7	15	2.32
Ot	16.1	17.1	10.5	61.2	67.2	102.9	40.0	52	2.93
Total	15.8	16.8	12.3	69.9	103.1	83.9	45.3	515	2.89

HCL = human cell line, H O = other human cell, ML = mouse liver, MK = mouse kidney, MCL = mouse cell line, RL = rat liver, RK = rat kidney, Ot = other, and S_G = geometric standard deviation.

substrate was important, and people who did not routinely use tissue sections commented that the slides were difficult to read, even if they did report correct results. In general, if the first samples were reported negative a greater chance existed that the prestained slides would also be reported negative.

Standardization, by use of a standard serum, did not improve the comparability of the results in the special study survey. A serum that had been standardized against the World Health Organization standard for ANA-homogeneous pattern[1] was provided to participants who then were asked to report results in international units per milliliter as well as in raw titers. TABLE 4 gives the variances obtained when results were not standardized or were standardized against various other materials. A significant decrease in variance ($p < .05$) was observed only when similar materials were compared.

Another area in which labeled immunoassays have become increasingly common is the TORCH series, serologic tests for toxoplasma, rubella, cytomegalovirus, and herpes simplex virus antibodies. The standard test for detecting rubella antibodies has

[a]The data from this study are also reviewed by Beutner *et al.* in this volume.

TABLE 2. Comparison of Mean Quantitative Results and Qualitative Results for Antinuclear Antibody Testing

Sample Number	Geometric Mean	Percent Reported Positive
BGO-D40	257	94
BGO-B19	212	98
BGO-A10	117	69
BGO-B20	42	59
BGO-A09	41	63
BGO-C30	28	44
BGO-C29	27	42
BGO-D39	<20	5

been hemagglutination inhibition (HAI), but many laboratories are switching to less complex methods. The passive hemagglutination (PHA) test is used primarily as a qualitative test; the fluorescence immunoassay (FIA, FIAX) and the enzyme immunoassay (EIA) tests are reported quantitatively in units that are not necessarily comparable to each other or to HAI titers, although they may be close (TABLE 5). A standard or reference material is available for rubella, but it is not widely used.[15]

The most common methods for testing for toxoplasma antibodies are IIF and PHA. Some discrepancy in results is apparent between methods and also between sources of antigen slides.[13] Tests for herpes simplex virus (HSV) and cytomegalovirus antibodies are most often either IIF or complement-fixation (CF). The HSV results by antigen source show some differences; however, no distinction was made between herpes type I and type II antibodies and this may have contributed to the variation in these results.

Radioimmunoassay (RIA) and EIA tests are used to test for carcinoembryonic antigen. There appeared to be greater differences in mean values and greater variation between sources of reagents than between methods (TABLE 6). The simpler methods gave more reproducible results.

Hepatitis Bs antigen test results are usually in good agreement when any of the third generation tests are used, with the exception of samples containing low levels of

TABLE 3. Results of Stained-Slide Survey for Antinuclear Antibody Testing

Substrate	Results of Previous Survey	Percent Results Reported as Positive on Well Number					Total Wells
		2	3	4	5	6	
Rat liver	Positive	100	100	100	100	80	96
	Negative	75	80	50	80	40	65
Mouse CL	Positive	80	80	80	80	80	80
	Negative	50	75	75	75	25	50
Mouse kidney	Positive	100	100	80	100	80	92
	Negative	75	75	50	75	50	65
Human CL	Positive	60	60	60	80	25	58
	Negative	83	100	100	83	67	87
Amniotic	Positive	75	100	100	100	50	85
	Negative	75	100	100	50	25	70

CL = cell line.

TABLE 4. Comparison of Variances of Standardized Antinuclear Antibody Test Results by Standard Material

Sample Number	None V	50 IU V	50 IU C	S01 V	S01 C	S09 V	S09 C	S10 V	S10 C
B10-S02	2.37	3.70	+56	0.75	−68	ND		ND	
B10-S05	1.84	2.76	+50	ND		ND		0.69	−62
B10-S06	2.18	3.49	+60	1.01	−54	ND		ND	
B10-S07	2.30	3.54	+54	0.99	−57	3.96	+72	1.30	−43
B10-S08	5.03	6.96	+38	ND		0.94	−81	ND	
B10-S09	4.86	6.87	+41	4.34	−11	ND		4.47	− 8
B10-S10	1.66	2.85	+71	1.24		4.47	+69	ND	

V = variance, C = change from no standard, and ND = not determined.

antigen. For two such samples, RIA appeared to be more sensitive than other methods, including EIA (TABLE 7).

DISCUSSION

Proficiency-testing programs can be used to detect problem areas in testing and provide mechanisms for standardizing testing procedures, as well as assessing performance. When results vary considerably among laboratories, a comparison of methodologies will frequently indicate potential sources of problems.

One of the criteria used to select tests for inclusion in the Centers for Disease Control program is the difficulty in obtaining useful results. Problems in obtaining useful results may arise from the technical difficulty of the test or from the use of nonstandardized reagents or test methods. Consistency of results within or among laboratories certainly increases the usefulness of those results. Standardization of reagents, methods, or reporting of results can often be used to improve the consistency of the results.

TABLE 5. Comparison of Rubella Results by Method

Sample Number	HAI	FIA	EIA
BP1-B17	138	141	84
BP0-D37	103	82	136
BP0-B18	97	63	100
BP1-B18	88	59	63
BP0-A08	82	45	−
BP0-A07	81	37	−
BP0-B17	80	52	77
BP0-C28	75	57	51
BP0-C27[a]	53	7	8
BP0-D38	10	9	6

[a]Contained nonspecific inhibitors.

HAI = hemagglutination inhibition, FIA = fluoroimmunoassay, and EIA = enzyme immunoassay.

TABLE 6. Comparison of Carcinoembryonic Antigen Results by Method and Manufacturer

Method	BI1-A07		BI1-A08		BI1-B17		BI1-B18	
	$\overline{X}_G$	S_G	$\overline{X}_G$	S_G	$\overline{X}_G$	S_G	$\overline{X}_G$	S_G
Roche RIA	17.9	1.61	41.2	1.86	5.1	1.42	9.2	1.26
Roche Col	16.0	1.18	38.7	1.72	5.2	1.36	9.9	1.30
Abbott RIA	21.6	1.11	34.4	1.16	5.8	1.20	11.0	1.13
Abbott EIA	21.3	1.14	31.9	1.16	5.6	1.16	10.3	1.16
Total	18.8	1.43	37.6	1.64	5.4	1.30	10.4	1.26

RIA = radioimmunoassay, Col = column + RIA, EIA = enzyme immunoassay, $\overline{X}_G$ = geometric mean, and S_G = geometric standard deviation.

Standardization through the use of a standardized reference material achieved the desired goals in proficiency-testing studies of RF tests, and commercial standards are now available for these tests. The same approach has not been successful with IIF tests for ANA, probably because of the number of different antinuclear antibodies involved and the more subjective nature of the procedures.

Indirect immunofluorescence tests for ANA are widely used in clinical laboratories and many different reagents are available from commercial or local sources. The number of available substrates, conjugates, and procedural modifications has increased greatly in the last few years. It has been shown that IIF-ANA results are not comparable from procedure to procedure, or from laboratory to laboratory.

It will be difficult to standardize IIF tests for ANA because of the variability of antisera, conjugates, and types of substrate. Conjugated antisera may vary as to animal source, specificity, sensitivity, and degree of labeling. The heterogeneity of antibodies and their reactions with the various substrates probably mean that a single serum standard is not sufficient as a method of standardization for this test. Multiple standards may be helpful, but, standardization of substrate, conjugate, and methodology should also be considered.

As labeled immunoassays are used in more laboratories and for more tests, it would be advantageous to compare and standardize tests so that accurate, consistent results are reported. This is particularly true in premarital or prenatal rubella screening. Many laboratories are switching to tests that are easier to perform than the standard HAI test. These new tests become available from commercial suppliers with few

TABLE 7. Comparison of Hepatitis Bs Antigen Results by Method

Sample Number	% Correct		Sample Number	% Correct	
	RIA	EIA		RIA	EIA
BH0-A01	99.6	100	BH1-A01	98.5	100
BH0-A02	99.0	100	BH1-A02	99.2	100
BH0-A03	99.6	100	BH1-A03	100	100
BH0-A04	100	100	BH1-A04	100	100
BH0-A05	99.6	100	BH1-A05	NA	NA
BH0-B06	87.4	0	BH1-B06	100	100
BH0-B07	100	100	BH1-B07	98.7	100
BH0-B08	89.7	0	BH1-B08	98.6	96.6
BH0-B09	95.6	100	BH1-B09	97.3	93.3
BH0-B10	94.8	100	BH1-B10	100	100

NA = data not available, RIA = radioimmunoassay, and EIA = enzyme immunoassay.

comparative studies. In some cases, the antibodies measured may differ from those detected by standard tests in specificity or in the type of antibody detected (i.e., IgG versus IgM).

Standardization of test results so that interlaboratory and interprocedure comparability is possible is one of the most critical needs of diagnostic immunology. It is especially true at this time when the field of immunology is expanding so rapidly. As technology improves and better tests are developed in which new procedures are used, it is important that laboratories and physicians are able to relate the results of these tests to the results obtained by other procedures. It is essential that government, private, and professional organizations cooperate in the development of physical and procedural standards.

REFERENCES

1. ANDERSON, S. G., I. E. ADDISON & H. G. DIXON. 1971. Antinuclear factor serum (homogeneous); an international collaborative study of the proposed research standard 66/233. Ann. N.Y. Acad. Sci. **177:** 337.
2. BLASZCZYK, M., E. H. BEUTNER, T. ROGOZINSKI, G. RZESA, M. JARZABEK CHORZELSKA, S. JAKLONSKA & T. P. CHORGELSKI. 1977. Substrate specificity of antinuclear antibodies in scleroderma. J. Invest. Dermatol. **68:** 191–193.
3. CLEYMAET, J. E. & R. M. NAKAMURA. 1972. Indirect immunofluorescent antinuclear antibody tests: Comparison of sensitivity and specificity of different substrates. Am. J. Clin. Path. **58:** 388–393.
4. FULFORD, K. M., R. N. TAYLOR & V. A. PRZYBYSZEWSKI. 1978. Reference preparation to standardize results of serological tests for rheumatoid factor. J. Clin. Microbiol. **7:** 434–441.
5. LUBINIECKI, A. S. & C. A. DORSCH. 1976. Comparison of mouse liver sections and cultured mouse fibroblasts as substrates for the detection of antinuclear antibodies. Am. J. Clin. Path. **66:** 892–898.
6. MANDEMA, E., V. E. POLLAK, R. M. KARK & J. REZAIAN. 1961. Quantitative observations on antinuclear factors in systemic lupus erythematosis. J. Lab. Clin. Med. **58:** 337–352.
7. McCARTY, G. A. & J. R. RIVE. 1980. Characterization and comparison of commercially available antinuclear antibody kits using high-titered single pattern index sera. J. Rheumatol. **7:** 339–347.
8. TAYLOR, R. N. & K. M. FULFORD. 1980. Proficiency testing summary analysis. Diagnostic immunology special study—antinuclear antibodies. Centers for Disease Control. Atlanta, Ga.
9. TAYLOR, R. N. & K. M. FULFORD. 1981. Assessment of laboratory improvement by the Center for Disease Control diagnostic immunology proficiency testing program. J. Clin. Microbiol. **13:** 356–368.
10. TAYLOR, R. N. & K. M. FULFORD. 1981. Theory and application of the CDC diagnostic immunology proficiency testing program. Centers for Disease Control. Atlanta, Ga.
11. TAYLOR, R. N. & K. M. FULFORD. 1982. Centers for Disease Control diagnostic immunology proficiency testing program results for 1980. J. Clin. Microbiol. (In press.)
12. TAYLOR, R. N., K. M. FULFORD & W. L. JONES. 1977. Reduction of variation in results of rheumatoid factor tests by use of a serum reference preparation. J. Clin. Microbiol. **5:** 42–45.
13. TAYLOR, R. N., K. M. FULFORD & V. A. PRZYBYSZEWSKI. 1982. Proficiency testing critique—diagnostic immunology 1980. Centers for Disease Control. Atlanta, Ga.
14. TAYLOR, R. N., A. Y. HUONG, K. M. FULFORD, V. A. PRZYBYSZEWSKI & T. L. HEARN. 1979. Quality control for immunologic tests. DHEW Publ. No. CDC 79-8376. Center for Disease Control. Atlanta, Ga.
15. WORLD HEALTH ORGANIZATION EXPERT COMMITTEE ON BIOLOGICAL STANDARDIZATION. 1971. Technical report series 463. p. 18. W.H.O. Geneva.

Prospects and Problems in the Definition and Standardization of Immunofluorescence

I. Present Levels of Reproducibility and Disease Specificity of Antinuclear Antibody Tests[a]

ERNST H. BEUTNER, SUSAN KRASNY, VIJAY KUMAR, ROGER TAYLOR,[b] AND TADEUSZ P. CHORZELSKI[c]

Department of Microbiology
School of Medicine
State University of New York at Buffalo
Buffalo, New York 14214

IF Testing Service
Kenmore, New York 14223

[b] Centers for Disease Control
Public Health Service
U.S. Department of Health and Human Services
Atlanta, Georgia 30333

[c] Department of Dermatology
Warsaw Academy of Medicine
Warsaw, Poland

INTRODUCTION

Much of the earlier work on standardization of antinuclear antibody (ANA) tests centered on the definition of fluorescein-labeled antihuman Ig antibodies in terms of their specificity,[6,13] their effects on titers,[2,5,7–9,16] and material standards of ANA and the conjugates.[14,20,24] The studies of techniques have yielded a now well-established group of methods that are widely used. (See APPENDIX A for a brief review.) Other studies have dealt with the role of optical systems[2,7,9,21] and with the disease specificity of ANA.[3,4,11,12,19]

Some of the more recent studies on ANA standardization deal with the titer variability in different ANA test systems.[23,26,27] These studies serve to shift the focus of our attention from conjugates to the role of differences in antigenic substrates and other factors that contribute to ANA titer variability as well as to the means of analyzing such variabilities.

This report on ANA standardization deals with the identification of current problems and the means of coping with them. It is based, in large part, on comparisons between findings in a survey carried out and reported by the Centers for Disease

[a] Supported by the International Society of Immunology Laboratories, the IF Testing Service, and the Warsaw Academy of Medicine.

Control (CDC) in 1980[26] on a group of reference sera prepared by the International Society for Immunology Laboratories (ISIL)[d] and findings of the ISIL on the same reference sera in two studies, one in 1979 and one in 1982. The latter part of this report includes data relevant to the disease specificity of ANA from another ISIL study, which confirms and extends some of the previous findings on the subject.[11]

The past and present studies on the standardization of ANA tests point to one fundamental tenet on ANA tests standardization and four corollaries or extensions of it. This basic tenet is that patients with ANA titers of 640, for example, do not exist; such patients have ANA titers of 640 on a given antigenic substrate such as mouse liver sections or HEp-2 cells. For example, the results reported in the CDC survey[26] demonstrated that the use of "units of ANA" based on titers obtained with a reference serum[e] in place of the titers as such actually decreased reproducibility of findings. That is, this survey revealed divergent sensitivities of selected antigens for ANA of different specificities. This tenet is even more clearly illustrated by the finding[26] of a 12.5-fold difference in geometric mean titer $(\overline{X}_G)$ of the serum of one patient on two substrates but a difference of only 1.1-fold with another patient's serum on the same two substrates and no difference in a third case in studies by the same 148 laboratories of all three cases. Obviously, we cannot equate for example, VDRL and Wassermann test titers for antibodies to cardiolipin; and even more so, we cannot equate ANA titers on different antigenic substrates since both the antibody specificities of the patients' sera and the antigenic specificities of nuclei in different cell preparations are divergent. This fact is also well illustrated, for example, by studies on anticentromere antibodies[23]; these antibodies react with human cell lines but fail to react with tissue sections or mouse fibroblast cell cultures.

One of the corollaries of the basic tenet of our studies is that while the performance of ANA titrations with presently available commercial ANA kits using specified antigenic substrates hold out the greatest promise for reproducible and thus reliable test systems, the observed variances of many of the commercial kits now available exceed acceptable limits. (The criteria for acceptability are presented in another report[25] in this volume.) This can again be documented by data derived from the findings in the CDC survey[26] as illustrated graphically in FIGURE 1. The interlaboratory variances (which serve as a measure of reproducibility of ANA titers between laboratories in FIGURE 1) of eight commercial kits with four types of antigenic substrates differed at the extremes by a factor of 14-fold. Interestingly, the interlaboratory variances for the four kits that use cell cultures for antigenic substrates correlated significantly with their intralaboratory variances. Although the reason(s) for this correlation cannot be identified on the basis of these data, it clearly lends support to the view that significant differences do exist in the reproducibility of ANA titers obtained by laboratory workers with ANA kits of different manufacturers. Whether this is due to kit variability, differences in the methods employed by the laboratory workers who use them, or both, remains to be determined. The reason there was not a statistically significant correlation between the interlaboratory and intralaboratory variances for the tissue-section test results may be that heterogeneity for tissues used is responsible for much greater variation than intralaboratory variation.

Another corollary of the basic tenet of this study and the one to which we have given primary attention in this report is that more effective control of technical factors can reduce variances in ANA titers obtained with any given test system. A key factor

---

[d]The prime function of the ISIL is to promote the complete standardization of immunologic tests. Further information about the ISIL can be obtained from 219 Sherman Hall, SUNY at Buffalo, Buffalo, NY 14214.

[e]World Health Organization antinuclear factor reference standard (homogeneous).[14]

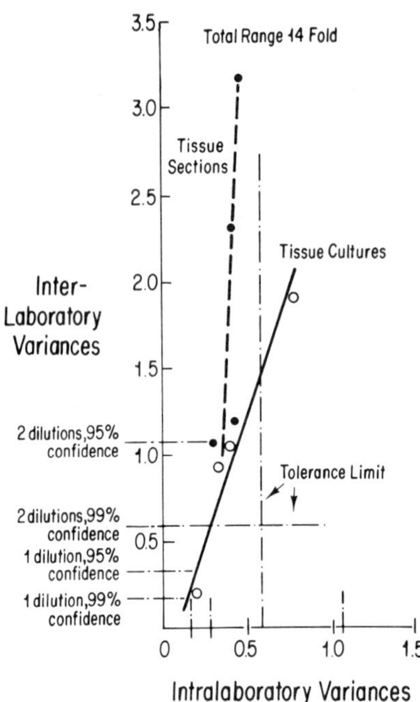

FIGURE 1. Variance is $(\log_2 S_G)^2$ where S_G is the geometric standard deviation of the ANA titers. Variances afford a measure of the reproducibility of a test system. According to R. Taylor[25] a change of two doubling dilutions in titers can be considered significant at the 95% confidence limit only if the geometric standard deviation of the test system, i.e. the S_G, is 2.028 or less, i.e. if the variance is 1.045 or less. The change is significant at the 99% confidence limit only if S_G is 1.72 or less, i.e. if the variance is 0.605. The latter value is shown as the "tolerance limit" in FIGURE 1 for the intra- and interlaboratory variances depicted. The variance limits for one doubling dilution shown in FIGURE 1 were reported in the 1980 CDC survey[26] for the eight most widely used ANA kits. The interlaboratory variances ranged from 0.226 for one of the human cell lines (Palomar) to 3.2 for one of the rat liver section preparations, a 14-fold range. For the four manufacturers of ANA kits with cell culture preparations, a significant correlation was found between the inter- and intralaboratory variances, as determined by the coefficient correlation ($p < 0.05$). On the other hand, for the four manufacturers of ANA kits with tissue sections, the slope of the line for inter- versus intralaboratory variances did not differ significantly from the slope of the y axis. The data base for the values plotted in FIGURE 1 is as follows:

Manufacturer		Interlab. Variance	Intralab. Variance
Bio Dx	Rat liver sections	1.073	0.309
CSI	Rat liver sections	3.192	0.460
Zeus	Rat liver sections	2.268	0.388
Calbio	Mouse kidney sections	1.190	0.424
Ab Inc.	Human cells	0.914	0.354
ENI	Human cells	1.044	0.400
Palomar	Human cells	0.226	0.197
Meloy	Mouse fibroblast	1.885	0.761

revealed by previous studies in this and other laboratories is the role of the anti-IgG (or other) conjugates used in ANA test systems, i.e. their specificity, concentration, dilution, and fluorescein-to-protein ratio.[5–7,9,10] APPENDIX A of this report presents a synopsis of the defining characteristics and an updated listing of the recommended methods for standardizing them. Two factors on which relevant data are given in this report are the effects of optical systems on ANA titers and a method for checking the accuracy of dilutions. Technical details on the latter appear in APPENDIX B of this report. The findings in this report indicate that other technical factors, such as variations in reactivity of substrates, also need to be analyzed further and controlled more effectively to further improve the reproducibility of ANA titers.

The last corollary to the basic tenet of these studies, and the most important one from the standpoint of the diagnostic utilization of ANA test systems (with acceptable low interlaboratory and intralaboratory variances), is that the disease specificity of ANA tests must be evaluated separately for each antigenic substrate.

MATERIALS AND METHODS

Preparation of Reference Sera

Plasmapheresis samples collected for treatment of patients with connective tissue diseases[15] used for these studies were coded as sera of Cases A, B, and C, defibrinated, checked for the presence of hepatitis B antigen, delipidated, passed through bacteriologic (Millipore) filters, and diluted with a mixture of phosphate-buffered saline, 4% bovine serum albumin, and 0.1% sodium azide. Serum of Case A was prediluted 1:100 while sera of Cases B and C were prediluted 1:10. TABLE 1 lists the clinical findings on Cases A, B, and C.

All sera were then aliquoted, frozen at about −20°C and either stored at that temperature or lyophilized. Sera of Cases A to C, which were used in the 1979 study (titrations performed at the ISIL laboratories), were examined for the presence of antibodies to native double-stranded DNA,[1] RNP,[22] and other extractable nuclear antigens. The results of these studies as well as the clinical findings on the patients are summarized in TABLE 1. These three sera as well as one additional serum of a patient with anti-RNP antibodies and a tentative diagnosis of mixed connective tissue disease were used in the CDC survey.[26]

TABLE 1. Clinical Findings on Donors of ANA Positive[a] ISIL Reference Sera and Some of Their Serologic Characteristics

Basic Findings	Case A	Case B	Case C
Clinical diagnoses	Systemic sclerosis[b]	Systemic LE[b]	Systemic LE
ARA preliminary criteria			
Number of positives	not applicable	4	5
Facial erythema		no	yes
Raynaud's		no	yes
Photosensitivity		yes	yes
Arthritis without deformity		yes	yes
Chronic BFP syphilis serology		no	yes
Profuse proteinuria		yes	no
Cellular casts		yes	no
(Other seven criteria)		no	no
Selected serology			
anti-RNP[c]	negative	negative	positive
Other: anti-			
ENA[c] or DNA[c]	negative	negative	negative
HBsAg[c]	negative	negative	negative

[a]ANA titers are given in TABLES 2 to 5

[b]SLE or systemic LE = systemic lupus erythematosus.

[c]Anti-RNP (ribonucleoprotein) or anti-ENA (extractible nuclear antigen) antibodies assayed by gel precipitation according to the methods of Reichlin and Maddison.[22] Anti-double-stranded DNA antibodies detectable by indirect immunofluorescence test[1] with Calbiochem-Behring kit (AFT-II). HBsAg (hepatitis B surface antigen) detectable with radioimmune assay of Abbott Laboratories.

ANA Tests of ISIL Reference Sera

Four types of ANA test systems were used in this ISIL study in 1979 (see below). All titrations were read independently by at least two trained laboratory workers with two types of optical systems. The ANA titers listed in this report include the ranges of titers reported for the multiple ANA tests and readings. The ANA patterns recorded for most of the test sera were "ranges" since there was marked variation in pattern readings. The ANA test system used in the 1979 ISIL studies was as follows.

Human Cell Line

The ANA kit with human KB cell cultures, control reagents, and conjugate (characteristics not specified) supplied by ENI (Electronucleonics, Inc., 4809 Auburn Ave., Bethesda, MD 20014) were used as prescribed in the package insert supplied with the kit in 1979.

Mouse Fibroblasts

The ANA kit with mouse fibroblast cell cultures, ANA positive and negative control reagents, and conjugates (characteristics not provided) were used as prescribed in the package insert supplied with the kit in 1979 by Meloy Laboratories, Inc. (6715 Electronic Dr., Springfield, VA 22151).

Mouse Kidney Sections

The AFT kit components supplied by Calbiochem-Behring, Inc. (10933 Torrey Pines, La Jolla, CA 92037) with mouse kidney sections, ANA positive and negative controls, and an anti-IgG conjugate with a molar F/P ratio of 1.0 to 1.5 prediluted to contain 1/4 unit/ml (see APPENDIX A) were used as prescribed in the package insert supplied with the kit in 1979. For later studies mouse kidney sections were cut in-house.

Rat Liver Sections

Two types were employed, fresh cut and commercial.

Fresh cut. Unfixed 4 μm sections of rat liver were tested with an anti-human conjugate with a molar F/P of 4.0 diluted to contain 1/4 units/ml of antibody, i.e., approximately 50 μg antibody/ml.

CSI kit. (Clinical Sciences, Inc., 30 Troy Rd., Whippany, NJ 07981) with rat liver sections were employed with the reference sera and conjugate (no characteristics given) as specified in the package insert supplied with the kits in 1979.

In the 1982 study, mouse liver sections and HEp-2 cells were also employed. Four samples of each of the four cases, A to C plus another serum of Case D were coded at the CDC, returned for testing and then decoded. (See *Analyses of Data*) The added substrates were prepared in-house. These were tested with an anti-IgG conjugate with a molar F/P ratio of 3.0 to 4.0 prediluted to contain 1/4 unit/ml.

Sera for Studies on Clinical Significance of ANA Titers

Sera of 24 cases of systemic lupus characterized according to American Rheumatism Association (ARA) provisional criteria, 51 cases of systemic sclerosis, 51 "hospital patient controls," and 91 blood donors selected for their age-sex distribution were tested for ANA on mouse kidney and mouse liver sections and on two types of human cell cultures in 1981 and 1982, HEp-2 and KB cells. This report deals only with the frequencies of false-positive and false-negative ANA of these sera in systemic lupus and systemic sclerosis and relative substrate sensitivities.

Optical Systems Employed

All ANA preparations were examined with the following two optical systems in the initial 1979 study. Epi-illumination system employed was an American Optical (AO) Model 2071 H vertical illumination unit with a 500 nm dichroic mirror, a 10× eyepiece and both a 20× (NA50) objective for scanning and a 40× (NA66) objective for reading patterns, a mercury vapor light (HBO-50), BG12+KV418 primary filters, and 0G515 secondary filters. Substage illumination system employed was an AO Model 10 microscope with a toric lens darkfield condenser, the same eyepiece and objective lenses as for epi-illumination, and BV type filters with a Schott BG 12 primary filter and OG-1 secondary filter.

Analyses of Data

Standard methods for determinations of geometric mean titers $(\overline{X}_G)$, geometric standard deviations (S_G), and variances as described by Taylor *et al.*[25,29] were used throughout these studies.

In the comparisons of the data from the ISIL studies and the CDC survey, the mode of listing the data reported in the CDC survey was modified in one respect. Notably, the number of laboratories reporting ANA titers of 10 and <10 were grouped as laboratories reporting titers of <20 because many laboratories start ANA titrations at a 1:20 dilution. This is recommended by manufacturers of some of the kits because ANA titers of 10, on KB cells and tissue sections for example, are of dubious validity as is borne out by some of the data set forth in this report. The mode of analyzing the data was left as specified on the CDC survey.

To evaluate the reproducibility of these (or other serologic) test systems the following standard deviation ranges may be applied according to R. Taylor.[25] A difference of one doubling dilution in paired titrations is statistically significant at the 95% confidence level if the mean and median S_G are 1.424 or less; at the 99% confidence level, if the mean and median S_G are 1.310 or less. A difference of two doubling dilutions in paired titrations is statistically significant at the 95% confidence level if the mean and median S_G are 2.028 or less; at the 99% confidence level if the mean and median S_G are 1.715 or less.

These values and the corresponding variances (V) can be summarized as follows:

Range of Variation (Number of Twofold Dilutions)	Confidence Level			
	95%		99%	
	S_G	Variance	S_G	Variance
1 dilution (2×)	1.424	0.2601	1.310	0.1519
2 dilutions (4×)	2.028	1.0409	1.715	0.6054

The following titers obtained in four replicate tests of ISIL reference standard from Case A on four antigenic substrates in a coded study serve to illustrate the method used to determine S_G values for intralaboratory variances:

Antigenic Substrate	Samples of Case A				Geometric mean $(\overline{X}_G)$	Geometric Standard Deviations (S_G)
	1st	2nd	3rd	4th		
Human cells	128	128	128	128	128	1.000
Mouse fibroblasts	128	128	128	64	107.6	1.414
Mouse kidney	128	64	128	128	107.6	1.414
Mouse liver	256	128	256	128	181	1.492

A total of 20 S_G values were determined with coded samples of five[f] ISIL reference sera on the above-listed four antigenic substrates by comparable analytic methods. The 20 S_G values (determined at the CDC) were: 4 at 1.000, 7 at 1.414, 6 at 1.492, 2 at 1.761, and 1 at 1.942; the mean S_G value was 1.416. Consequently, a twofold titer change between paired samples was deemed to be statistically significant at the 95% confidence level[25] or a fourfold change was deemed significant at the 99% confidence level in the analyses of data obtained in 1982 on the clinical significance of ANA titers.

RESULTS

Reproducibility of ANA

FIGURE 2 depicts some of the typical patterns obtained with sera of Cases A, B, and C on mouse kidney sections; FIGURE 3 illustrates their reactions on rat liver sections. FIGURE 4 shows the ANA patterns of the three reference sera on a human cell line and FIGURE 5 depicts their reactions on a mouse fibroblast cell preparation. Tests of ISIL reference sera from Cases A, B, and C yielded diverse ANA patterns. The Case A serum gave an unequivocal nucleolar pattern particularly on tissue sections (FIGURES 2a and 3a). On some of the cell culture preparations the nucleoli appeared fragmented, as seen in tests of the Case A serum on human KB cells and mouse fibroblasts (FIGURES 4a and 5a). The latter differed clearly from the speckled pattern on cell cultures (FIGURES 4b and 4c). The sera of Cases B and C gave variously speckled, homogeneous, or rim patterns on the four substrates tested.

ANA patterns varied not only between substrates for a given serum but also with serum dilutions and importantly, sometimes even within a given preparation. For example, in one test on a human cell line, the Case C serum gives a speckled pattern in most areas but a rim pattern in some, as shown in FIGURES 4c and 4d. FIGURE 6, which depicts the ANA reactions of a serum from another patient with systemic lupus erythematosus (this serum contained antibodies to double-stranded DNA) on a mouse kidney section, illustrates the fact that multiple ANA patterns may be seen even within a given field. Different ANA patterns, for example rim versus homogeneous or speckled versus homogeneous, were also observed in replicate tests of a given serum at

---

[f] The five sera included four used in the CDC survey; of these four, three were used in the ISIL study of 1979. The results reported here deal primarily with these three.

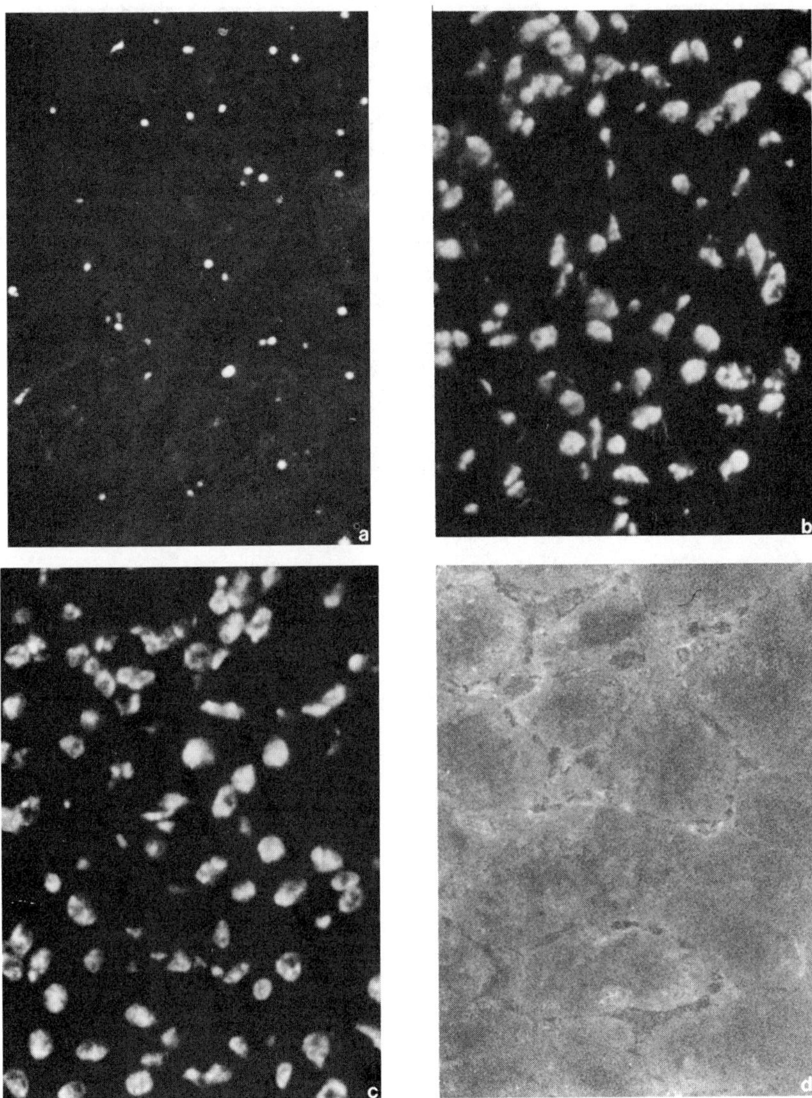

FIGURE 2. IF reactions of human sera on mouse kidney sections. (a) illustrates the nucleolar ANA pattern of the serum of Case A at a real dilution of 1:100 (listed as "undiluted"). Note nucleolar staining can be detected in some, but not all cells. ×1120. (b) shows the ANA of Case B serum and (c) of Case C serum, both at real dilutions of 1:10 ("undiluted"). Note a speckled and homogeneous ANA pattern. ×1120. (d) shows the negative ANA of a normal serum at a 1:10 dilution. ×1120.

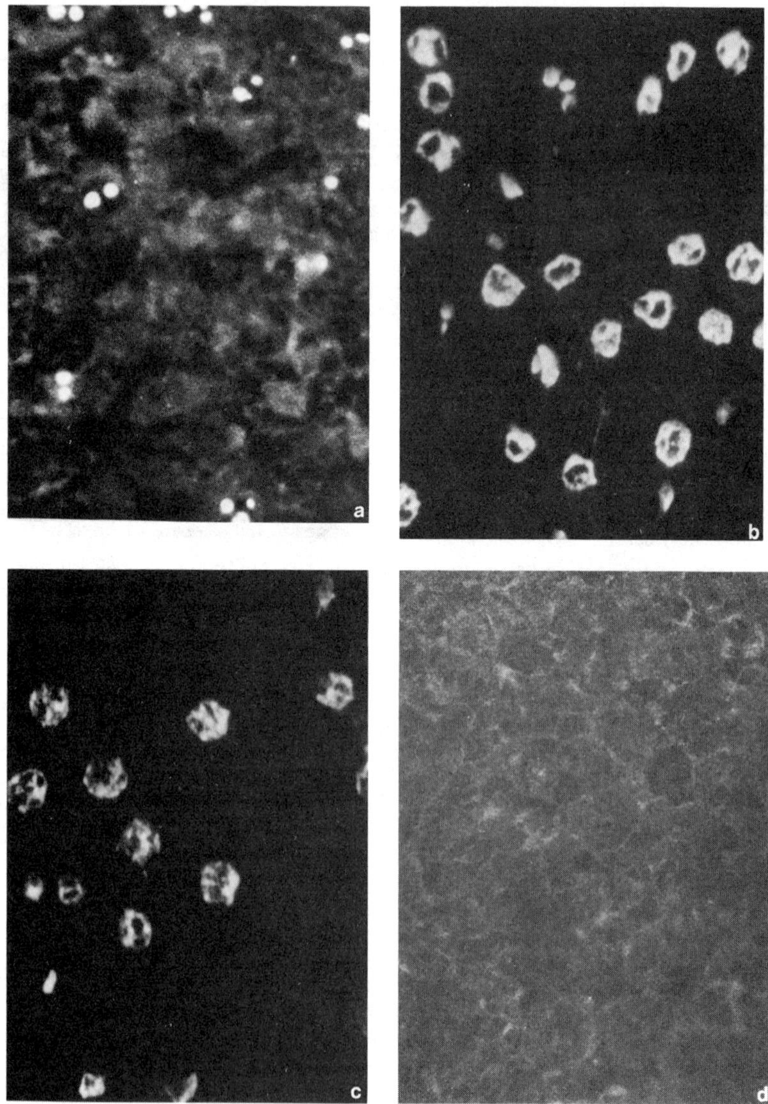

FIGURE 3. IF reactions on rat liver sections. (a) shows the nucleolar ANA pattern of the serum of Case A at a real dilution of 1:100 on a fresh-cut rat liver section. Note nucleolar staining of some, but not all nuclei. ×1120. (b) illustrates the ANA patterns of serum B and (c) of serum C both at real dilutions of 1:10. Note predominantly speckled ANA patterns in both (b) and (c) ×1120. (d) shows a negative ANA reaction of a normal human serum diluted 1:10. ×1120.

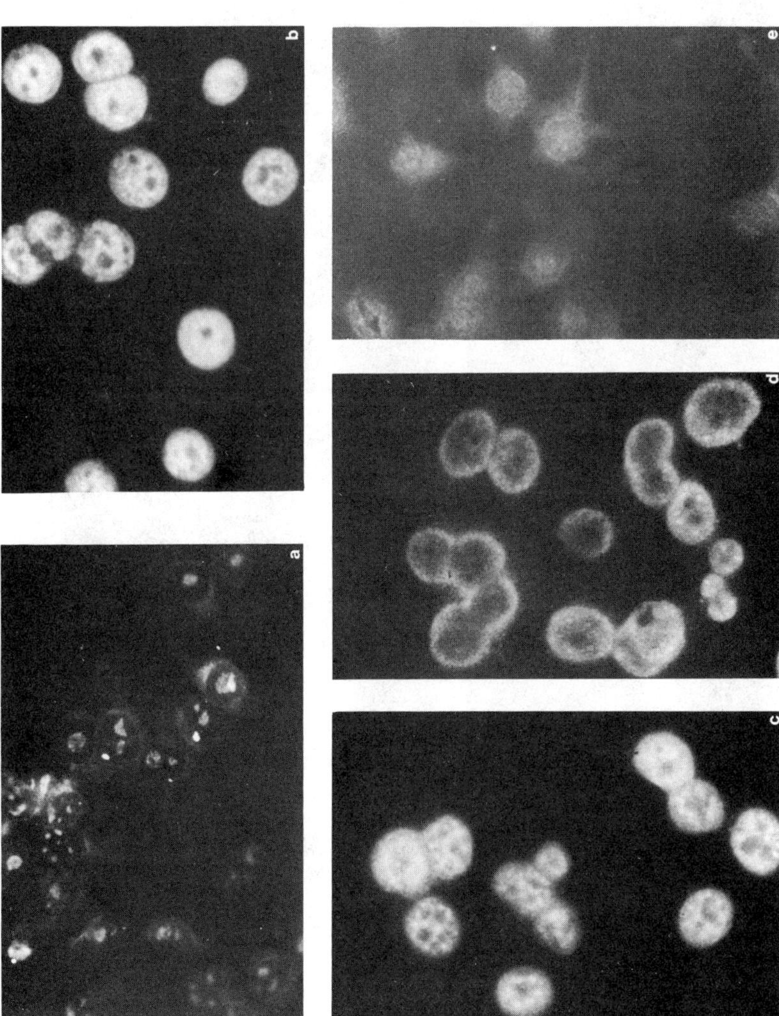

FIGURE 4. IF reactions on human KB cells. (a) shows the nucleolar ANA pattern of the serum of Case A at a real dilution of 1:100 ("undiluted"). Note fragmented nucleoli, which are clearly distinct from the fine speckled ANA pattern. ×960. (b) portrays the fine speckled ANA pattern of serum B at a real dilution of 1:10. ×960. (c) and (d) depict two distinct patterns observed in a single preparation of KB cells treated with the "undiluted" (prediluted 1:10) serum of Case C. Note the speckled ANA pattern that occurred in most field (c) and the rim pattern in some areas (d). ×960. (e) shows a negative ANA of a phosphate-buffered saline (and conjugate) control on a rat liver section. ×960.

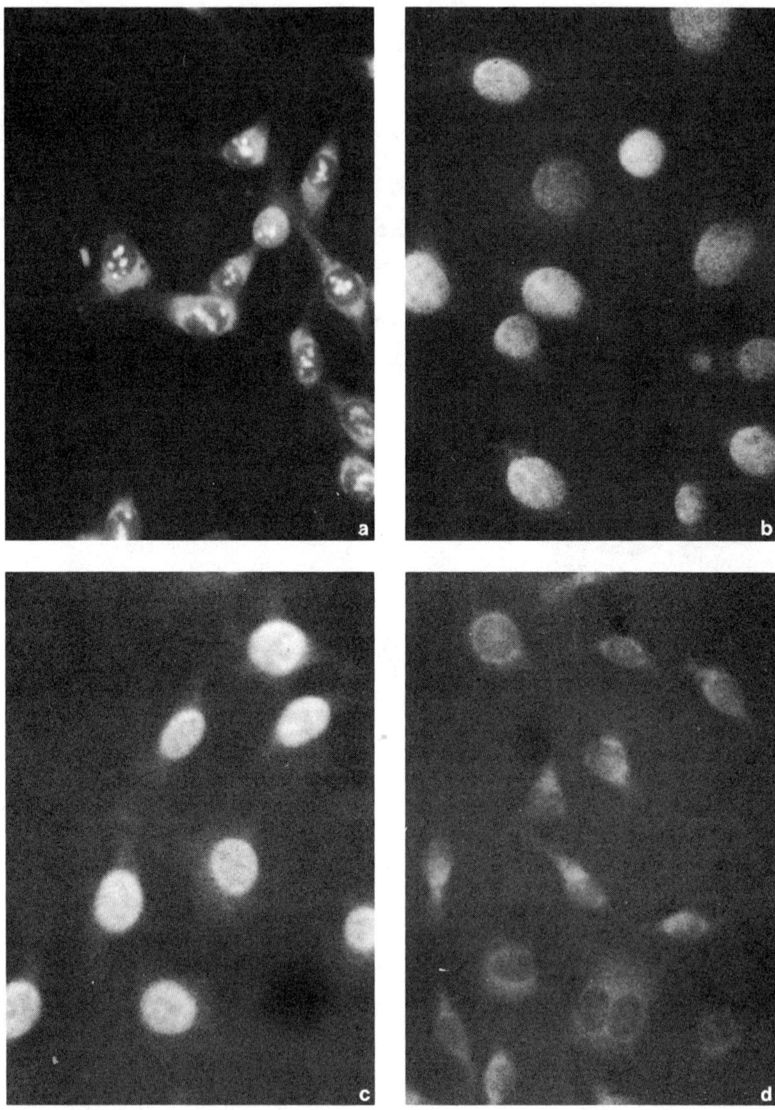

FIGURE 5. IF reactions on mouse fibroblast cell culture preparations. (a) illustrates the nucleolar ANA pattern of the serum of Case A at a real dilution of 1:100. Note fragmentation of nucleoli. ×1120. (b) depicts the ANA pattern of Case B serum and (c) the pattern of the Case C ANA both at real dilutions of 1:10. Note that both sera gave predominantly homogeneous ANA patterns on this substrate. ×1120. (d) shows the negative ANA of a normal serum at a 1:10 dilution. ×1120.

the same dilution on the same substrate. These observations point to the need for recording ranges or spectra of ANA patterns of each serum and to the fact that, in studies of real patient sera, we should not expect to find a single ANA pattern in a given case of, for example, systemic lupus erythematosus.

TABLE 2 provides comparisons between the frequencies with which given ANA titers of the serum of Case A were reported in the 1980 CDC survey[26] and those observed in the ISIL studies of 1979 and 1982. The latter two studies yielded the similar titer ranges on mouse kidney and one to two dilutions higher on human cells and mouse fibroblasts in the 1982 study. A different human cell line was used in 1982

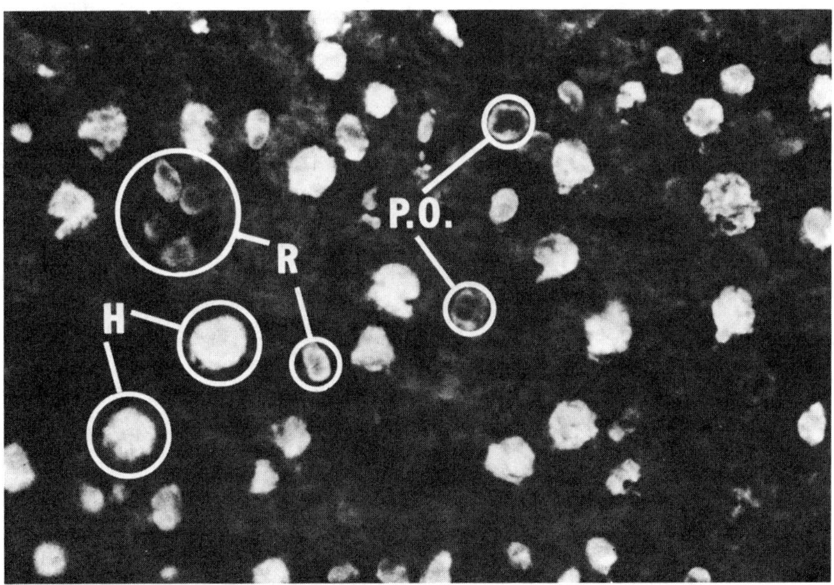

FIGURE 6. The ANA of a serum of a systemic LE patient, which contains anti-native DNA antibodies. (This serum is not included with the ISIL reference sera). Test on a rat liver section (CSI) yields diverse ANA patterns including homogeneous (marked H), peripheral or rim pattern (marked R), and artifactually produced "punched out" nuclei (marked PO). The irregular, sharp, inside edges of the latter are produced by cuts through nuclei (made in sectioning). The rim pattern, in contrast, give the appearance of a strongly reactive nuclear rim with gradual fading of the ANA staining toward the center of the nuclei. Because of the diagnostic significance of the rim pattern it must be clearly distinguished from the artifactually produced punched out nuclei. ×1120.

(HEp-2 versus KB cells) and the mouse fibroblast substrate was modified by the manufacturer in 1982. Mouse liver sections instead of rat liver sections were employed in the tests performed in the 1982 studies. In the 1979 study, reading with epi-illumination tended to yield higher titers than those made with a substage illumination microscope. TABLE 3 lists the analyses of these ANA titers of the Case A serum and provides a summary of the analyses of the ANA titers of Cases B and C reference sera; TABLE 4 provides statistical comparisons of ANA titers obtained with two optical systems.

TABLE 2. Data on Frequencies of ANA Titers Reported on One Serum from a Systemic Sclerosis Case (A)[a] Reported in a CDC Survey[b] and Two ISIL Studies[b] Using Non-commercial and Four Types of Commercial Test Systems

Types of ANA Tests Source and Substrates	No. Mfg.	Optics Used[c]	Examined By	Yr	No. of Tests	Frequency of Reports of ANA Titers of							
						<20 <16	20 16	40 32	80 64	160 128	320 256	640 512	≥1280 ≥1024
I. Non-commercial	?	?	CDC	'80	120	21	5	9	18	32	24	8	3
Uncommon commerical kits	?	?	CDC	'80	40	15	6	9	4	4	1	0	1
Subtotal	?	?	CDC	'80	160	36	11	18	22	36	25	8	4
II. Cell culture commercial													
Human cells[f]	3	?	CDC	'80	228	5	22	46	82	55	14	4	0
Human cells[f]	1[g]	Sub[d]	ISIL	'79	5		1	3	1				
Human cells[f]	1[g]	Epi[e]	ISIL	'79	5			3	2				
Human cells[f]	1[h]	Epi	ISIL	'82	4					4			
Mouse fibroblasts	1	?	CDC	'80	80	44	12	10	7	5	2	0	0
Mouse fibroblasts	1[i]	Sub	ISIL	'79	3		1	2					
Mouse fibroblasts	1[i]	Epi	ISIL	'79	3		1	2					
Mouse fibroblasts	1[i]	Epi	ISIL	'82	4				1	3			
III. Tissue sections commercial													
Mouse kidney[j]	1	?	CDC	'80	68	4	0	4	14	14	13	16	3
Mouse kidney[j]	1	Sub	ISIL	'79	2			2					
Mouse kidney[j]	1	Spi	ISIL	'79	3				2	1			
Mouse kidney[j]	1	Epi	ISIL	'82	4				1	3			
Rat liver[k]	3	?	CDC	'80	90	1	3	4	?	23	26	15	11
Rat liver[k]	2[l]	Sub	ISIL	'79	5				5				

TABLE 2. *Continued*

Rat liver[k]	2	Epi	ISIL	'79	5		3	2	
Mouse liver[m]	1	Epi	ISIL	'82	4			2	2

[a]Case 1 (See TABLE 1).

[b]The Center for Disease Control (CDC) proficiency testing survey[1] entailed ANA tests of duplicate (differently coded) samples of each test serum by the participating laboratories. Two studies were done by the International Society for Immunology Laboratories (ISIL), one in 1979 and one in 1982. In the latter study, the ISIL sera were tested at the IF Testing Service (963 Kenmore Avenue, Kenmore, NY). Each laboratory used one of the antigenic substrates described in footnotes *f–h*.

[c]The two types of optical systems were employed in the ISIL studies (epi-illumination and substage illumination).

[d]Epi-illumination system employed was an American Optical (AO) Model 2071H vertical illumination unit with a 500 nm dichroic mirror, a 10% eyepiece, and both a 20× (NA50) objective for scanning and a 40× (NA66) objective for reading patterns, a mercury vapor light (HBO-50), BG12 + KV418 primary filters, and OG515 secondary filters.

[e]Substage illumination system employed was an AO Model 10 microscope with a toric lens darkfield condenser, the same eyepiece and objective lenses as for epi-illumination and BV type filters with a Schott BG 12 primary filter and OG-1 secondary filter.

[f]Human cell lines employed by 114 laboratories in the CDC survey and in two ISIL studies included those marketed by Ab, Inc. and prepared in-house.

[g]Ab Inc. or Antibodies, Inc. using cultures of HEp-2 cells (20 labs; 40 samples in CDC survey); P.O. Box 442, Davis, CA 95616.

[h]Calbiochem-Behring also put out a trial HEp-2 preparation in 1982 that was used in an ISIL study.

[i]Mouse fibroblast cultures employed by 40 laboratories in tests of 80 samples in the CDC survey and in the 1979 ISIL study were marketed until 1982 by Meloy Laboratories, Inc. (6715 Electronic Drive, Springfield, VA 22151).

[j]Mouse kidney sections employed in the CDC survey were marketed by Calbiochem-Behring Corp. (34 labs; 68 samples in the CDC survey; also used in ISIL studies) (10933 North Torrey Pines Road, La Jolla, CA 92037).

[k]Rat liver sections employed by 90 laboratories in the CDC survey included those marketed by Bio Dex (6 labs; 12 samples in CDC survey; 103 Washington Street, Morristown, NJ 07960).

[l]CSI or Clinical Science, Inc. (24 labs; 48 samples in CDC survey; also used in 1979 ISIL study) (Whippany, NJ 07981). In-house rat liver preparations used in 1979 ISIL study gave essentially the same ANA titers as CSI slides. Zeus (15 labs; 30 samples in CDC survey; Foot of Thompson Street, Raritan, NJ 08869).

[m]Mouse liver sections employed in the 1982 ISIL study were prepared in-house.

TABLE 3. Analyses of Data on ANA Titers[b] of Case A[a] and Summaries of Cases B[a] and C[a] Data Analysis

Types of ANA Tests or Substrates	Study by	No. of Mfg.	No. of Titers	Geometric Mean $\overline{X}_G^p$	Standard Dev. S_G^p	No. of Labs	Range[q]: one S_G^2 above and below $\overline{X}_G$ $\overline{X}/S_G^2$ to $\overline{X} \times S_G^2$	Fold Range of titers $\overline{X} \times S_G^2 \div \overline{X}/S_G^2$	Case B[a] Summary $\overline{X}_G$	S_G	Case C[a] Summary $\overline{X}_G$	S_G
I. Non-commercial	CDC	?	120	86.7	47.7	60	0.0381 to 197300	5.18×10^6	14.0	2.71	17.3	3.21
Non-commercial + uncommon kits	CDC	?	160	129	1.23×10^4	80	$8.53\ 10^{-7}$ to 1.95×10^{10}	2.29×10^{16} 2.29×10^{16}	8.12	4.95	11.8	5.51
II. Commercial kits with cell culture												
Human cells[f]	CDC	3	228	76.4	2.39	114	13.4 to 4364	32.6	7.1	1.92	12.0	2.54
	ISIL '79	1[n]	10	36.8	1.55	1	15.3 to 88.4[n]	5.8	6.2	2.09	35.9	1.33
	ISIL '82	1[h]	4	128	1.00	1	128[h]	1.0	8.0	1.00	11.3	1.49
Mouse fibroblasts	CDC	1[i]	80	14.5	3.53	40	1.2 to 180[i]	155	12.0	2.44	19.3	3.31
	ISIL '79	1[i]	6	25.4	1.43	1	12.4 to 51.9[i]	4.2	4.6	1.89	14.7	2.37
	ISIL '82	1[o]	4	107.6	1.41	1	54.1 to 214[o]	4.0	6.7	1.41	11.3	1.49
III. Commercial kits with tissue sections												
Mouse kidney[j]	CDC	1	68	180.8	3.63	34	13.7 to 2382	17.4	13.7	2.32	19.6	2.37
	ISIL '79	1	5	55.7	1.79	1	17.4 to 179	10.3	8.8	1.82	21.5	1.43
	ISIL '82	1	4	107.6	1.41	1	54.1 to 214	4.0	8.0	1.76	8.0	1.00
Rat liver[k]	CDC	3	90	254.0	2.97	45	28.8 to 2241	77.8	20.9	2.63	21.9	2.87
	ISIL '79	2	10	73.5	1.34	1	40.9 to 132	3.2	14.5	1.58	38.7	1.84
Mouse liver[m]	ISIL '82	1	4	181.0	1.49	1	81.5 to 402	4.9	9.5	1.41	9.5	1.41

[a]See TABLE 2 for footnotes *a* to *m*. For details on Cases A and B, see TABLE 1.

[n]Electronucleonics, Inc. (ENI) using cultures of KB cells (84 labs; 504 samples in CDC survey; also used in 1979 ISIL study). ENI, 4890 Auburn Avenue, Bethesda, MD.

[o]In 1982, a modified formulation of mouse fibroblasts marketed by Meloy was used in the second ISIL study.

[p]Statistical analyses are done according to the method of Taylor *et al.*[20,27] Geometric mean ANA titer ($\overline{X}_G$) = antilog $[(\Sigma \log X)/N]$. Geometric standard deviation (S_G) = antilog $\sqrt{\Sigma (\log X - \log \overline{X})^2/N - 1}$, where X = ANA titer.

[q]Range for 95% confidence limits.

TABLE 4. Comparisons of Geometric Mean Titers and Geometric Standard Deviations of ANA Titers of Sera of Cases A,[a] B,[a] and C[a] with Epi- Versus Substage Illumination in Tests on Four Types of Antigenic Substrates in the 1979 ISIL Study[b]

Antigenic Substrates	Illum. Type[c]	N	Case A[a] $\overline{X}_G^p$	S_G^p	N	Case B[a] $\overline{X}_G$	S_G	N	Case C[a] $\overline{X}_G$	S_G	Mean S_G	Mean Percent Change in S_G sub → epi[d]	total → epi[e]
Cell cultures[f,i]													
Human cells	Sub	5	32.0	1.63	4	4.8	2.83	3	32.0	1.00	1.82	+27.5%	
Human cells	Epi	5	42.2	1.46	4	8.0	1.00	3	40.3	1.49	1.32		
Human cells	Total	10	36.8	1.55	8	6.2	2.09	6	35.9	1.33	1.66		+20.5%
Mouse fibroblasts	Sub	3	25.4	1.49	5	3.5	2.14	4	8.0	1.76	1.80	+9.4%	
Mouse fibroblasts	Epi	3	25.4	1.49	5	6.1	1.46	4	26.9	1.94	1.63		
Mouse fibroblasts	Total	6	25.4	1.43	10	4.5	1.89	8	14.7	2.37	1.90		+14.2%
Tissue sections[j,k]													
Mouse kidney	Sub	3	32.0	1.00	7	6.6	1.69	7	16.0	1.00	1.23	−14.6%	
Mouse kidney	Epi	2	80.6	1.49	7	10.0	1.43	7	29.0	1.30	1.41		
Mouse kidney	Total	5	55.7	1.79	14	8.8	1.82	14	21.5	1.43	1.68		+16.1%
Rat liver	Sub	5	64.0	1.00	7	10.8	1.45	11	28.2	1.68	1.38	−10.9%	
Rat liver	Epi	5	84.4	1.46	7	19.5	1.40	11	53.0	1.73	1.53		
Rat liver	Total	10	73.5	1.34	14	14.5	1.58	22	38.7	1.84	1.59		+3.8%
Mean S_G	Sub			1.28			2.03			1.36	1.56	+5.8%	
Mean S_G	Epi			1.48			1.32			1.62	1.47		
Mean S_G	Total			1.53			1.85			1.74	1.71		+14.0%

[a]See TABLE 1 for details on Cases A, B, and C.

[b]1979 ISIL study data. See TABLE 2 footnote *b*.

[c]See footnote *c* in TABLE 2.

[d]Percent change from S_G of ANA titers with substage to S_G with epi-illumination = $(S_G \text{ sub} - S_G \text{ epi}/S_G \text{ sub}) \times 100$.

[e]Percent change from S_G of ANA titers of total reading to S_G with epi-illumination = $(S_G \text{ total} - S_G \text{ epi}/S_G \text{ total}) \times 100$.

[f,i]See TABLE 2 footnotes for cell cultures used in 1979 ISIL study.

[j,k]See TABLE 2 footnotes for tissue sections used in 1979 ISIL study.

[p]See TABLE 3.

The analyses of the geometric mean $(\overline{X}_G)$[g] ANA titers of the ISIL reference serum of Case A and their geometric standard deviations (S_G)[g] shown in TABLE 3 demonstrate that 95% confidence limits[g] of titers for undefined, noncommercial tests, plus uncommonly used commercial ANA kits used in 80 laboratories, cover a range of about 2.29×10^{16}-fold! On the other hand, the analyses of the 95% confidence limits of interlaboratory variations in titers of the same Case A reference serum in the CDC survey of test performed by 233 laboratories using one of four specified types of antigenic substrates (i.e., either one of two types of cell culture systems or one of two types of tissue section) covered ANA titer ranges of 33-fold to 155-fold. Comparable analyses of intralaboratory variations of Case A serum titers in the uncoded 1979 ISIL study point to ANA titer ranges of 3.2-fold to 10.3-fold (mean 5.9-fold) and in the coded 1982 study to titer ranges of 1-fold to 4.9-fold (mean 3.5-fold) that fall within the 95% confidence limits. Obviously defined test systems offer some hope of attaining reproducible ANA titers, undefined test systems do not.

As indicated by the summaries of analyses of the ANA titers of the sera of Cases B and C, the difference in the S_G values between undefined types of antigenic substrates (section I in TABLE 3) and more clearly defined substrates (sections II and III) were considerably smaller than for the ANA titers of the serum of Case A. Though less dramatically elevated than the S_G values of Case A serum ANA titers, they are important because of the consistency of elevation of S_G values. Also, as indicated by our findings (see *Disease Specificity of ANA*) the substrate specificity of the type displayed by serum of Case A is not unique for nucleolar antibodies.

The substrate specificity of the serum of Case A was also demonstrated by the statistically significant differences between its ANA titers on mouse fibroblasts and rat liver sections in both the CDC survey and the 1979 ISIL study. The modified mouse fibroblast preparations of Meloy Laboratories available for the 1982 ISIL study gave ANA titers with Case A serum that did not differ significantly from the titers obtained with the other three substrates.

The data listed in TABLES 2 and 3 implicate three factors in the variability of ANA titers obtained with given, commercially prepared ANA kits, notably variations in kits and in optical systems, and laboratory errors in processing, possibly due to errors in the preparation of dilutions. Analytic studies were undertaken on the effects of the optical systems and of dilutions on reproducibility of ANA titers.

TABLE 4 summarizes the analyses of the effects of optical systems on the geometric means and standard deviations of ANA titers in the 1979 ISIL study in which tests were read with both substage illumination and epi-illumination types of fluorescence microscopes. Epi-illumination microscope readings yielded higher ANA titers than readings with substage illumination; the difference[h] was 1.76-fold, i.e. over one half of a doubling dilution. As indicated in TABLE 4 the S_G values for ANA titers read by

---

$$^g \log \overline{X}_G = \frac{\Sigma \log x}{N} \, ; \, \log S_G = \sqrt{\frac{\Sigma(\log x - \log \overline{X}_C)^2}{n-1}} \, ; \text{ and}$$

95% confidence limits $= \overline{X}_G / S_G^2 \quad$ to $\quad \overline{X}_G \times S_G^2.$

^{h}Comparisons were made by

$$\frac{\Sigma^{\overline{X}_{sub}} / \overline{X}_{epi}}{N},$$

where N is the number of comparisons of geometric mean $(\overline{X})$ ANA titers observed with substage $(\overline{X}_{sub})$ and epi-illumination $(\overline{X}_{epi})$ fluorescence microscopes.

substage illumination also tended to be somewhat higher on the average (5.8%), though this was not a consistent finding. Importantly, the S_G values were consistently greater for the total of the ANA titer readings in the 1979 study than for the readings made only with an epi-illumination fluorescence microscope. As judged by these analyses, the use of different optical systems can give rise to about a 14% increase in the standard deviations of ANA titers in intralaboratory comparisons. For an S_G value of 2 or a variance of 1 (4 of 8 interlaboratory variances are about 1 as shown in FIGURE 1), this would reduce their variances to 0.61, the maximum value for the 99% confidence level in the significance of a two-dilution difference. (See METHODS for details.) However, there are no assurances that these findings in the 1979 ISIL study are applicable to the 1980 CDC survey.

Since laboratory errors affecting accuracy of serum dilutions may increase the variability of ANA titer and since the mean S_G values for the 1979 study were over

TABLE 5. A Method for Checking Accuracies of Serologic Dilutions with Hydrolyzed Fluorescein Diacetate (FDA)

	Tube Number			
	#1	#2	#3	#4
Volumetric 1:100 dil. of FDA stock sol. $(100\ \mu g/ml)$				
Observed $E_{1\ cm}^{1\ \mu g/ml}$ at 490 nm = 0.218				
Three serial 1:2 serologic dil. of FDA and further dil.	1:8	1:16	1:32	1:64
Initial volume of dilution	0.1 ml	0.1 ml	0.1 ml	0.2 ml
Volume with 0.9 ml (or 0.8 ml) 0.1N NaOH	1.0 ml	1.0 ml	1.0 ml	1.0 ml
Final dilutions	1:80	1:160	1:320	1:320
Expected extinction (EE)[a]	0.273	0.136	0.068	0.068
Range from EE -20%	0.218	0.109	0.054	0.054
to EE $+20\%$	-0.327	-0.164	-0.082	-0.082
Examples of observed extinctions				
First test (Ji.)	0.235	0.130	0.080	0.075
(% error, pipette #2)	(-14%)	(4.4%)	$(+18\%)$	$(+10\%)$
Second test (Am)	0.225	0.110	0.045	0.040
(% error, pipette #2)	(-18%)	(-19%)	(-33%)	(-41%)
(mean % error pipette #2)	(-16%)	(-12%)	(-8.1%)	(15.4%)

[a]EE or expected extinction = Final dil. (eg., 1:80) $\div$ 1:100 $\times$ 0.218 = 0.273.

1.31 (required for ascribing significance at the 99% level of confidence to one dilution difference) or even over 1.42 (required for the 95% confidence level), a method for checking that had been in use in this laboratory earlier was reinstituted. It is based on the use of a fluorescein diacetate reference standard with a known extinction coefficient in place of sera in standard titrations. This method checks serologic dilutions spectrophotometrically. Details appear in TABLE 5 with accompanying text presented in Appendix B. This method revealed, for example, that extension devices used for certain automatic pipettes increase pipetting errors. On the basis of these findings, standard practices for routine use of the abovementioned method for checking accuracy of dilutions have been instituted in this laboratory and are recommended as a simple, functional routine procedure that increases the accuracy of titrations.

TABLE 6. Comparisons of ANA Titers and Patterns of 216 Sera on Four Antigenic Substrates

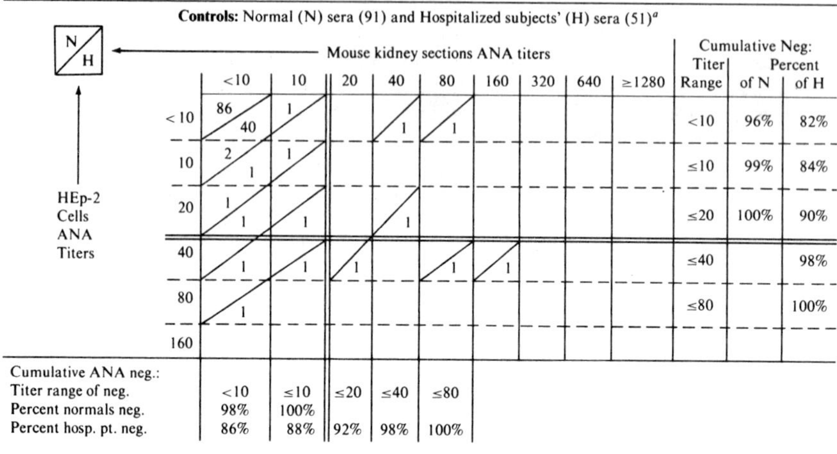

Controls: Normal (N) sera (91) and Hospitalized subjects' (H) sera (51)[a]

Legend box: N (upper) / H (lower). Middle columns = Mouse kidney sections ANA titers. Last three columns = Cumulative Neg: Titer Range, Percent of N, Percent of H.

HEp-2 Cells ANA Titers	<10	10	20	40	80	160	320	640	≥1280	Titer Range	Percent of N	Percent of H
<10	86 / 40	1		1	1					<10	96%	82%
10	2 / 1	1								≤10	99%	84%
20	1 / 1	1		1						≤20	100%	90%
40	1	1	1		1	1				≤40		98%
80	1									≤80		100%
160												

Cumulative ANA neg.:

	<10	10	20	40	80
Titer range of neg.	<10	≤10	≤20	≤40	≤80
Percent normals neg.	98%	100%			
Percent hosp. pt. neg.	86%	88%	92%	98%	100%

Systemic lupus erythematosus (22 cases)

Middle columns = Mouse kidney section ANA titers and patterns[a]. Last two columns = Cumulative Pos: Titer Range, Percent Pos.

HEp-2 Cells ANA Titers and Patterns[a]	<10	10	20	40	80	160	320	640	≥1280	Titer Range	Percent Pos.
<10										normal range	
10		H[b]								normal range	
20										normal range	
40				H	H					≥40	95%
80										≥80	86%
160						S,H				≥160	86%
320			S[b]					H		≥320	77%
640							R\H[b], H			≥640	68%
≥1280						H,H\R		H	6-H[b], 2-S S\H,H\R	≥1280	59%

Cumulative ANA pos.:

	<10	10	20	40	80	160	320	640	≥1280
Titer range	Normal range		≥20	≥40	≥80	≥160	≥320	≥640	≥1280
Percent pos.			95%	95%	86%	82%	64%	55%	45%

TABLE 6. *Continued*

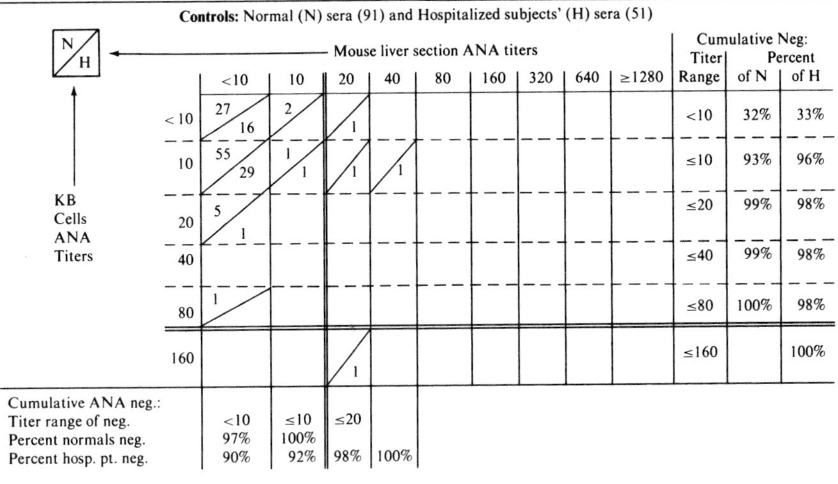

Controls: Normal (N) sera (91) and Hospitalized subjects' (H) sera (51)

The matrix shows **KB Cells ANA Titers** (rows) against **Mouse liver section ANA titers** (columns). Cells split diagonally give N (upper) / H (lower) counts.

KB Cells ANA Titers	<10	10	20	40	80	160	320	640	≥1280	Cumulative Neg: Titer Range	Percent of N	Percent of H
<10	27 / 16	2	1 (H)							<10	32%	33%
10	55 / 29	1 / 1	1 (H)	1 (H)						≤10	93%	96%
20	5 / 1									≤20	99%	98%
40										≤40	99%	98%
80	1 (N)									≤80	100%	98%
160			1 (H)							≤160		100%

Cumulative ANA neg.:

	<10	10	20	40
Titer range of neg.	<10	≤10	≤20	
Percent normals neg.	97%	100%		
Percent hosp. pt. neg.	90%	92%	98%	100%

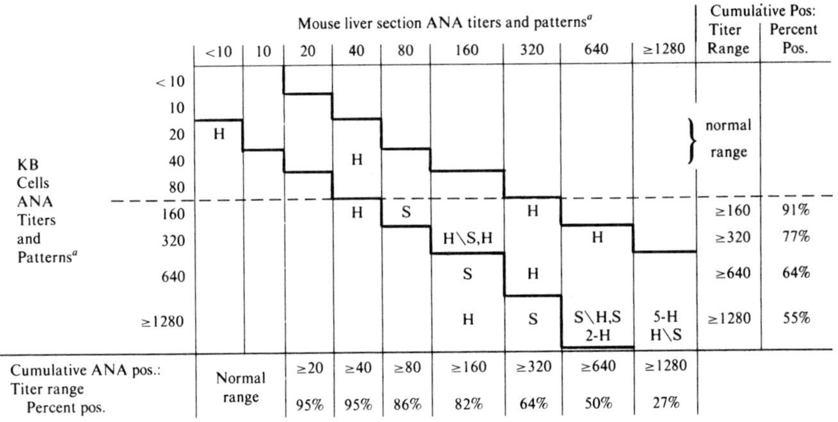

Systemic lupus erythematosus (22 cases)

The matrix shows **KB Cells ANA Titers and Patterns**[a] (rows) against **Mouse liver section ANA titers and patterns**[a] (columns).

KB Cells ANA Titers and Patterns[a]	<10	10	20	40	80	160	320	640	≥1280	Cumulative Pos: Titer Range	Percent Pos.
<10											
10										} normal	
20	H									range	
40				H							
80											
160				H	S		H			≥160	91%
320						H\S,H		H		≥320	77%
640						S	H			≥640	64%
≥1280						H	S	S\H,S 2-H	5-H H\S	≥1280	55%

Cumulative ANA pos.:

	<10	10	20	40	80	160	320	640	≥1280
Titer range	Normal range		≥20	≥40	≥80	≥160	≥320	≥640	≥1280
Percent pos.			95%	95%	86%	82%	64%	50%	27%

TABLE 6. *Continued*

Systemic sclerosis

Diffuse and acrosclerosis (25 cases) C.R.E.S.T. (27 cases)

HEp-2 Cells ANA Titers and Patterns[a]	Mouse kidney section ANA titers and patterns[a]									Cumulative Pos:	
	<10	10	20	40	80	160	320	640	≥1280	Titer Range	Percent Pos.
<10	—	H[b],H	H							normal range	
10											
20				H							
40	2-C[b]	N\S	H				N\H			≥40	90%
80	H	C\H	C\H	H,H	2-H	H,N\H / N\H		N-H\S[b]		≥80	81%
160	C		C\H	H,C\H	N[b],H / N\H					≥160	60%
320	2-C	N\H / C\H	C\H	S	N\H / C\S		2-H			≥320	46%
640	C, N		C\H		N					≥640	25%
≥1280	N	N		H		2-S,H			2-S,S	≥1280	17%
Cumulative ANA pos.:- Titer range	Normal range		≥20	≥40	≥80	≥160	≥320	≥640	≥1280		
Percent pos.			67%	56%	42%	25%	13%	8%	6%		

Systemic sclerosis

Diffuse and acrosclerosis (25 cases) C.R.E.S.T. (27 cases)

KB Cells ANA Titers and Patterns[a]	Mouse liver section ANA titers and patterns[a]									Cumulative Pos:	
	<10	10	20	40	80	160	320	640	≥1280	Titer Range	Percent Pos.
<10										normal range	
10	H	H									
20	H			N\S							
40	C				H	H					
80	3-C,N	C\H / N\H		N							
160	C	C\H,N	2-H / C\H	C\S	2-H		N			≥160	75%
320	2-C		C\H	H,N	N	2-H	2-N	S		≥320	56%
640	H				N\H,H	H,H				≥640	35%
≥1280	C				C\H,N	H,S	S,S,H	S,H	S,2-H	≥1280	25%
Cumulative ANA pos.: Titer range	Normal range		≥20	≥40	≥80	≥160	≥320	≥640	≥1280		
Percent pos.			67%	60%	50%	35%	23%	12%	6%		

[a] The hospital cases which served as controls include patients with the following disorders:- 12 suspected thyroid disorders, 10 kidney and heart diseases, 10 infections, 3 malignancies, 16 other disorders. Biologic false-positive ANA appeared randomly in these subgroups. Of these 51 controls, most (31 cases) were females and most (39 cases) were over 40 years old.

[b] ANA patterns on the predominant or most important patterns are listed. The following notations for given patterns appear at the specified titers on the two substrates in each of the four graphs for SLE and SS: H = homogeneous; S = speckled; R = rim; N = nucleolar; C = centromere.

Other ANA pattern notations:

a) e.g. R\H = R on cell cultures and H on tissue sections; b) e.g. C or H = C or H pattern of ANA in serum from patient with CREST syndrome; c) e.g. 6-H or 2-C = six cases with H pattern on 2 CREST cases with C pattern at specified titers on two substrates; d) e.g. N-H\S = mix N and H pattern on cells but S pattern on sections.

Disease Specificity of ANA

TABLE 6 summarizes some of the key findings in appropriately standardized test systems in a survey of ANA titers and patterns of 143 control sera and sera of 75 connective tissue disease patients on four antigenic substrates: two human cell lines, HEp-2 cells and KB cells, and two types of tissue sections, mouse kidney and mouse liver. Comparisons of the ANA reactions of the 143 control sera on these four antigenic

TABLE 7. Characterization of FITC[a] labeled Anti-Ig Conjugates

Characteristics	Assay Methods
Specificity	IEP[a] gel precipitation (IIF[a] tests with Ig class restricted Ab)[a]
Sensitivity	
F[a] and P[a] conc.	NCCLS methods
F/P ratios	NCCLS methods
Ab[a] concentration ($\geq 10\%$)	Reverse radial immunodiffusion (chessboard titration unitages)
Use dilution	50 μg Ab/ml (25–100 μg/ml) check with chessboard titration[a] for PEP[a] and NSS[a]

[a]FITC = fluorescein isothiocyanate, IEP = immunoelectrophoresis, Ab = antibody, IIF = indirect immunofluorescence, NCCLS = National Committee for Clinical Laboratory Standards, F = fluorescein, P = protein, PEP = plateau endpoint of conjugate, and NSS = nonspecific staining of cytoplasm. Example of a chessboard titration on mouse kidney sections with serial dilutions of an anti-human IgG conjugate and a human serum containing ANA:

Strength of ANA obtained with dilutions of conjugate containing the following antibody concentrations in units/ml (1/2 to 1/64) and in estimated antibody protein (100 to 3.1 mg/ml).

Human Sera Tested	Dilutions Tested	Units/ml mg/ml	1/2 100	1/4 50	1/8 25	1/16 12.5	1/32 6.3	1/64 3.1
Saline Normal	1:10		NSS					
ANA pos.	1:80		+ +	+ +	+ +	+ +	+	+
	1:160		+ +	+ +	+ +	+ +	+	weak
	1:320		+	+	+	+	weak	±
	1:640		+	+	+	weak	±	±
	1:1280		weak	weak	weak	±	±	±
	1:2560		±	±	±	±	±	
Titers			1280	1280	1280	640	320	160

Comments:	ANA plateau titer	
Some	Use	Conj.
NSS	dilution	PEP

substrates reveal the following relationships. (1) The ANA titer range of normal human sera on HEp-2 cells, mouse kidney, and mouse liver was from <10 to 10, but on KB cells it was from <10 to 40. (2) Sera of hospitalized control subjects who do not have connective tissue diseases have higher ANA titers than normal sera. Comparisons of the range of ANA titers of sera from hospitalized control subjects indicate that 98%

(51/52) had titers of ≤40 on both HEp-2 and KB. Thus, their difference in normal ANA titer ranges may not be significant in clinical laboratory studies. Further studies on selected disease controls are needed to resolve this question. (3) The appearance of ANA titers on these four antigenic substrates that are beyond the range of the normal control sera but within the range of the titers of sera from hospital control subjects is of doubtful significance.

Comparisons of ANA titers of sera from 22 patients with systemic lupus erythematosus (SLE) and 52 patients with systemic sclerosis (SS) on the same four antigenic substrates (TABLE 6) reveal the following relationships. (1) ANA titers of SLE sera tended to be somewhat higher on mouse kidney sections than on HEp-2 cells; KB cells and mouse liver sections yielded comparable ANA titers with the same group of SLE sera. (2) Significant ANA titers of SS sera appeared more frequently in tests on HEp-2 and KB cells than on the tissue sections included in these studies. This was caused primarily by the presence of anticentromere antibodies in the ANA of sera from SS patients with the CREST syndrome. (3) These two findings suggest that screening of sera to rule out SLE or SS can be done with ANA tests on two substrates such as rodent tissue sections and a human cell line.

DISCUSSION

The basic tenet of these studies and its extensions and corollaries afford the best prospect for standardization of ANA test systems. It can be restated as: *reproducible and interpretable ANA titers exist only on given antigenic substrates.*

Since marked differences in variances, that is, in the reproducibility of ANA titers on given substrates from different sources have been observed (FIGURE 1), the basic tenet should be modified to specify that: the reproducibility and interpretation of ANA titers on a given antigenic substrate must be established by each producer of an IF test system, regardless of whether the producer is a manufacturer of a commercial ANA kit or the test system is produced only for in-house use. As indicated in the introduction, commercially manufactured ANA kits afford better prospects for standardization; this will become increasingly evident because of the many factors entailed in complete standardization of an ANA test. On the basis of the previous and the present findings, these factors include the following. (1) Antigen reactivity is obviously a key factor in achieving reproducible ANA tests as illustrated by the data summarized in FIGURE 1. This factor, which is largely the responsibility of producers, clearly merits detailed analytic studies. (2) Conjugate specificity and sensitivity, which have been the prime focus of attention in standardization studies in the past, have been shown to affect the reproducibility in a number of studies[2,5-7,9,10] and can be controlled with standard methods. (See APPENDIX A.) (3) Optical system sensitivity plays a significant role in interlaboratory variances, as shown in this study. This factor can in part be controlled with optical standard (OS) slides.[9] Attention to this factor may bring the interlaboratory variance of several of the currently used ANA test systems into an acceptable range. (4) Statistical analyses of interlaboratory and intralaboratory variances by established analytic procedures[25,27,28] need to be carried out at regular intervals, preferably four times per year. If titers vary beyond statistically established tolerance ranges, the cause of variability needs to be established and corrected. (5) Biologic false positive (BFP) and biologic false negative (BFN) frequencies need to be evaluated separately for each ANA test system. The use of the term biologic in the expressions BFP and BFN is indicated, first because they are due to biologic phenomena, not to experimental errors, and secondly, this term calls attention to the nature of the phenomena that cause the appearance of BFP and BFN ANA. The results of this study clearly indicate that BFP and BFN ANA frequencies vary independently on different substrates. Clearly, the normal human sera of selected ages and sexes, which have been used in most studies to date, need to be supplemented with appropriate disease controls.

These five factors and any others that affect the reproducibility and/or interpretability of ANA tests must be dealt with effectively as determined by established statistical methods to achieve completely standardized test systems. The promotion of the development of complete standardization programs (for example, Beutner *et al.*[9]) based on the use of both material and methods standards and on evaluation by established statistical methods[25,27] is the primary objective of the ISIL.

ACKNOWLEDGMENTS

This report is dedicated to the memory of William L. Hale, M.A., Clinical Assistant Professor of Microbiology at SUNY at Buffalo, Schools of Medicine and Dentistry who passed away in April 1980.

The writers gratefully acknowledge the help of Professor Stefania Jablonska, Chairman of the Department of Dermatology of the Warsaw Medical School in Warsaw, Poland in characterizing the clinical features and selection of the patients whose sera serve as ISIL reference standards for these studies, and of the following individuals of SUNY at Buffalo, Department of Microbiology: Mr. Jay Huang who programmed a calculator for the statistical analyses reported here, Mrs. Patricia Greenlee who did the technical work, and Mrs. Gloria Griffin who prepared this report for publication.

The use of trade names is for identification purposes only and does not constitute endorsement by the ISIL or by the Centers for Disease Control, Public Health Service, U.S. Department of Health and Human Services.

REFERENCES

1. AARDEN. L. A., E. R. DE GROOT & T. E. W. FELTKAMP. 1975. Immunology of DNA. III. *Crithidia luciliae*, a simple substrate for the determination of anti-ds-DNA with the immunofluorescence technique. Ann. N.Y. Acad. Sci. **254:** 505–515.
2. ALBINI, B., F. HERZOG & G. WICK. 1972. Photometric evaluation of indirect immunofluorescence chessboard titrations for the characterization of FITC-labelled antibodies. Immunology **23:** 203–208.
3. BERGQUIST, R. N. 1973. Antinuclear antibodies: Significance of titers. *In* Immunopathology of the Skin: Labeled Antibody Studies. E. H. Beutner, T. P. Chorzelski, S. F. Bean & R. E. Jordon, Eds. **1:** 394–401. Dowden, Hutchinson and Ross. Stroudsburg, Pa.
4. BERGQUIST, N. & D. DANIELSSEN. 1973. Evaluation of two different antigen preparations in the fluorescent antibody test for antinuclear antibodies (ANA). A comparison between two laboratories. Acta Pathol. Microbiol. Scand. **81:** 446–452.
5. BEUTNER, E. H. 1971. Defined immunofluorescent staining: past progress, present status and future prospects for defined conjugates. Ann. N.Y. Acad. Sci. **177:** 506–526.
6. BEUTNER, E. H., W. L. BINDER & V. KUMAR. 1980. State of the art of immunofluorescence techniques in tissues. *In* Diagnostic Immunology: Current and Future Trends. P. W. Keitges & R. M. Nakamura, Eds.: 89–99. College of American Pathologists. Skokie, Ill.
7. BEUTNER, E. H., J. S. DENG, S. SHU, P. ANDERSEN & F. PEETOM. 1975. Studies on defined immunofluorescence in clinical immunopathology. III. Past progress, some new methods and prospects for the future. Ann. N.Y. Acad. Sci. **254:** 573–591.

8. BEUTNER, E. H., E. J. HOLBOROW & G. D. JOHNSON. 1965. A new fluorescent antibody method. Mixed antiglobulin immunofluorescence or labelled antigen indirect immunofluorescence staining. Nature **208:** 353–355.

9. BEUTNER, E. H., V. KUMAR & P. GREENLEE. 1983. Basics in standardization and practical applications of immunofluorescent microscopy. Standardization of antinuclear antibody tests. *In* Immunodiagnostics. J. Hyun & R. M. Aloisi, Eds.: 49–73. Alan R. Liss. New York.

10. BEUTNER, E. H., G. WICK, M. SEPULVEDA & K. MONIN. 1970. A reverse immunodiffusion assay for antibody protein concentration in antisera or conjugates to human IgG. *In* Standardization in Immunofluorescence. E. J. Holborow, Ed.: 165–169. Blackwell Scientific Publications. Oxford, England.

11. BLASZCZYK, M., S. JABLONSKA, E. H. BEUTNER, T. ROGOZINSKI & M. JARZABEK-CHORZELSKA. 1977. Substrate specificity of antinuclear antibodies (ANA) in scleroderma. J. Invest. Dermatol. **68:** 191–193.

12. CLAYMOET, J. & R. NAKAMURA. 1982. Indirect immunofluorescent antinuclear antibody test: Comparison of sensitivity and specificity of different substrates. Am. J. Clin. Pathol. **58:** 388.

13. HIJMANS, W., B. E. R. SCHUIT, A. P. M. JONGSMA & J. S. PLOEM. 1970. Performance testing of fluorescent antisera against human immunoglobulins. *In* Standardization of Immunofluorescence. E. J. Holborow, Ed.: 193–202. Blackwell Scientific Publications. Oxford, England.

14. HOLBOROW, E. J. 1983. Standardization in immunofluorescence. Ann. N.Y. Acad. Sci. (This volume.)

15. JABLONSKA, S., T. P. CHORZELSKI, M. BLASZCZYK, J. DASZYNSKI & E. H. BEUTNER. 1982. Plasma exchange in dermatology. *In* Current Dermatologic Therapy. S. Maddin, Ed.: 510A–510B. W. B. Saunders Co. Philadelphia, Pa.

16. McCARTY, G. A. & J. R. RICE. 1980. Characterization and comparison of commercially available antinuclear antibody kits using single pattern index sera. J. Rheumatol. **7:** 339–347.

17. McKINNEY, R. M., G. W. PEARCE & J. T. SPILLANE. 1964. Fluorescein diacetate as a reference color standard in fluorescent antibody studies. Anal. Biochem. **9:** 474–476.

18. NATIONAL COMMITTEE FOR CLINICAL LABORATORY STANDARDS. August 1975. Determinants of fluorescein/protein ratios in fluorescein isothiocyanate labeled protein.

19. NISENGARD, R. J. 1979. Antinuclear antibodies: Significance of titers. *In* Immunopathology of the Skin. 2nd edit. E. H. Beutner, T. P. Chorzelski & S. F. Bean, Eds.: 387–398. John Wiley & Sons, Inc. New York.

20. NORBERG, R. 1983. Future development of immunologic reference preparations. Ann. N.Y. Acad. Sci. (This volume.)

21. PLOEM, J. S. 1973. Immunofluorescence microscopy. Comparisons of conventional systems with interference filters and epi-illumination. *In* Immunopathology of the Skin: Labeled Antibody Studies. E. H. Beutner, T. P. Chorzelski, S. F. Bean & R. E. Jordon, Eds.: 248–270. Dowden, Hutchinson and Ross. Stroudsburg, Pa.

22. REICHLIN, M. & P. J. MADDISON. 1979. Soluble tissue autoantigens which precipitate with sera of patients with connective tissue diseases. *In* Immunopathology of the Skin. 2nd edit. E. H. Beutner, T. P. Chorzelski & S. F. Bean, Eds.: 351–362. John Wiley & Sons, Inc. New York.

23. TAN, E. M., G. P. RODNAN, I. GARCIA, Y. MARIO, M. FRITZLER & C. PEEBLES. 1980. Diversity of antinuclear antibodies in progressive systemic sclerosis. Anticentromere antibody and its relationship to CREST syndrome. Arthr. Rheum. **23:** 617–625.

24. TEN VEEN, J. H. & T. E. W. FELTKAMP. 1971. Studies on the standardization of antinuclear antibody (ANA) determination. Ann. N.Y. Acad. Sci. **177:** 346–353.

25. TAYLOR, R. N. 1983. Measurement of variation and significance in serologic tests. Ann. N.Y. Acad. Sci. (This volume.)

26. TAYLOR, R. N. & K. M. FULFORD. 1980. Proficiency testing summary analysis. Diagnostic immunology special study—antinuclear antibodies. Centers for Disease Control. U.S. Department of Health and Human Services.

27. TAYLOR, R. N. & K. M. FULFORD. 1981. Assessment of laboratory improvement by the

Centers for Disease Control, Diagnostic Immunology Proficiency Testing program. J. Clin. Micro. **13:** 356–368.

28. TAYLOR, R. N. & K. M. FULFORD. 1983. External quality control and standardization in labeled immunoassay. Ann. N.Y. Acad. Sci. (This volume.)

29. TAYLOR, R. N., A. Y. HUONG, K. M. FULFORD, V. A. PRZYBYSZEWSKI & T. L. HEARN. 1979. Quality control for immunologic tests. HHS Publication No. (CDC) 82-8376. Centers for Disease Control. Atlanta, Ga.

APPENDIX A

Criteria for Characterization, Selection, and Dilution of Anti-IgG Antibodies Conjugated with Fluorescein Isothiocyanate

TABLE 7 summarizes the key factors in the characterization of FITC conjugates of globulins with anti-Ig antibodies. The major assays are as follows.

Specificity Assays

For IgG-specific conjugates, immunoelectrophoresis with whole serum should give only the IgG lines and no lines for IgM and IgA. This test itself may give misleading results and should be supplemented with gel diffusion tests. Specificity tests can also be done with indirect immunofluorescence using class-restricted antibodies.[6] For ANA tests, conjugates should detect IgG class antibodies.

Basic Assays for Sensitivity (Relative and Absolute)

1. Total protein (P) assays by the biuret method give reliable values if done with appropriate controls.[5,18] Standard conjugates should be adjusted to 10 mg protein/ml.[5]

2. Total fluorescein (F) assays according to the method of McKinney *et al.*[17,18] are based on 490 nm extinctions of conjugate dilutions made in 0.1 N NaOH and checked with a hydrolyzed fluorescein diacetate standard.

3. Fluorescein to protein (F/P) ratios calculated by standard methods[5,18] must be restricted to a constant range for reproducible indirect IF tests for ANA or other antibodies. Low molar F/P ratios (e.g., 1.0 to 1.5) give "clean" ANA tests with low background staining but somewhat lower intensities and titers of ANA. Higher molar F/P ratios (e.g., 2.0 to 3.0) give stronger intensities of IF staining and higher ANA titers[5] but give stronger background staining.

4. Antibody protein assays performed by reverse radial immunodiffusion tests for anti-Ig antibody protein concentrations[10] afford the method of choice.[7] Antibody protein content should be at least 10% of the total protein,[5] i.e., for an FITC conjugate with 10 mg protein/ml, the antibody content should be at least 1 mg antibody/ml.

Selection and Dilution of Conjugates

Conjugates that detect only IgG class antibodies, contain 10% or more antibody protein, and fall within a constant, predetermined F/P ratio range should be selected for ANA tests. Dilutions of selected conjugates for ANA tests should be adjusted to contain 25, 50, or 100 μg/ml of antibody protein. These should give the same titers of ANA. For optimal function of the indirect IF tests, 50 μg antibody/ml should be used. The dilution used to obtain this antibody concentration should be based on the abovementioned antibody protein assay.[7,10]

The selected dilution should be checked with a shortened indirect IF chessboard titration using conjugate dilutions containing e.g. 50, 25, 12.5, 6.3, 3.1, and 1.6 μg/ml of labeled antibody. For example, ANA titrations with conjugate dilutions containing 50 and 25 μg/ml should yield the same ANA titers; those containing 12.5, 6.3, and 3.1 μg/ml yield variable titers and those with 1.6 μg/ml or less of antibody should give significantly lower ANA titers with present-day optics. If major deviations from such expected results occur, this chessboard titration and possibly also the antibody protein assay need to be rechecked. Chessboard titrations such as the one shown in the footnote of TABLE 7 should, strictly speaking, be performed on all conjugates used for ANA tests or other indirect IF tests.

APPENDIX B

A Method for Checking Accuracies of Serologic Dilutions

The hydrolyzed fluorescein diacetate (FDA) reference standard used for F assays[17,18] gives a highly reproducible extinction coefficient at its absorption maximum. The latter falls in the 490 nm to 495 nm range and should be determined for each spectrophotometer. It may be, e.g. 492 nm. Specifically, for FDA in 0.1 N NaOH:

$$E_{1cm}^{1\,\mu g/ml}, \text{ e.g. } 492 \text{ nm} = 0.208\ (\pm 10\%).$$

A standard stock solution of 100 μg FDA/ml[18] is used for checking accuracies of serologic dilution as shown in TABLE 5. A volumetric 1:100 dilution of this stock solution serves as a check on the method; i.e. if the extinction coefficient is within 10% of the expected value, the method is operative. If not, check for error(s). The observed extinction coefficient is then used for calculating the extinctions expected (EE) in the serologic dilution. Starting with 100 μg FDA/ml, twofold serial dilutions of 0.1 ml are prepared for 1:2, 1:4, 1:8, 1:16, 1:32, and 1:64. Discard the first two (1:2 and 1:4) and add 0.9 ml of 0.1 N NaOH to the 1:8 (Tube #1), 1:16 (Tube #2), and 1:32 (Tube #3) dilutions and, if desired, 0.8 ml to the 0.2 ml from which the 1:64 dilution can be prepared. Calculate the EE for Tube #1 in TABLE 5, for example.

$$EE = 1/80 \div 1/100 \times E_{1cm}^{1\,\mu g/ml} = 100/80 \times 0.218 = 0.273.$$

Calculate the EE $\pm$ 20% range as shown in TABLE 5.

Duplicate check of each new pipette should be made. The mean % error should be less than $\pm 20\%$ as in the example of pipette #2 shown in TABLE 5. If the error is greater, repeat the test. Obviously only those automatic pipetting systems that deliver within the tolerance limits of this test system are acceptable for routine use. This method revealed, for example, that extension tubes provided with certain automatic pipettes introduce such pipetting errors that their use is contraindicated.

Prospects and Problems in the Definition and Standardization of Immunofluorescence

II. A Quantitative Assay for Antibody Protein of FITC-labeled Anti-IgG Conjugates

JAY G. HUANG, ERNST H. BEUTNER, AND
VIJAY KUMAR

Department of Microbiology
Schools of Medicine and Dentistry
State University of New York at Buffalo
Buffalo, New York 14214

INTRODUCTION

Reproducibility in the immunofluorescence system depends to a great extent on the use of well-characterized conjugates. The concentration of specific antibodies (Ab) in a conjugate represents one of the main parameters to be determined in order to produce a well-characterized reagent. For routine use in the laboratory, a gel precipitation titration for a unitage assay[3] and the plateau endpoint determination[5] provide practical semiquantitative measures of conjugate Ab levels. For more quantitative determinations, two methods of Ab assay have been recommended for use in immunofluorescence, the reverse radial immunodiffusion (RRID) technique,[6] and a radioimmunoassay.[1] The RRID method employs hyperimmune sera, globulins, or conjugates that have a known Ab concentration as standards by which unknown reagents can be assayed for Ab content by single radial immunodiffusion. The simplicity of this technique renders it the method of choice when a reliably assayed reference standard is available. Bergquist and Norberg[1] have recommended a second method for conjugate Ab quantitation that obviates the requirement for a reference standard. A radiolabel attached to the conjugated protein enables counts of the Ab portion, adsorbed onto antigen-coupled Sepharose, to be compared with counts of the conjugate as a whole. This information, together with the protein content of the conjugate, allows a calculation of the Ab concentration. While this radioimmunoabsorption method is clearly advantageous in situations where an Ab reference standard is unavailable, its application is limited to those facilities having the equipment and expertise for handling radioactive material.

This report describes a method of quantitating the specific Ab in fluorescein isothiocyanate (FITC)-labeled antihuman IgG conjugates that requires only a spectrophotometer and no biological reference standards. It is based solely on standard assays for fluorescein (F) and protein (P) and two types of affinity chromatography.

TABLE 1. Characteristics of Anti-IgG Conjugates Used in This Study

Conjugate	Source	F/P (μg/mg)[a]	F/P (molar)	Units/ml Ab
1	UB goat 005	4.10	1.68	8
2	UB goat 005	3.63	1.48	16
3	UB goat 005	17.06	6.80	16
4	UB goat 005	2.64	0.88	8
5	UB goat 005	5.30	1.87	8
6	UB goat 005	12.19	4.24	8
Kallestad	Commercial[b]	5.65	2.32	16
NBL	Field trial study[c]	3.4	1.4	8–16
Fuji	Field trial study	3.0	1.2	4
Eiken	Field trial study	–	–	4

[a] Calculated according to methods advocated by the National Committee for Clinical Laboratory Standards;[8] P determinations by biuret assay and F determinations using fluorescein diacetate as a reference standard.[7]

[b] Kallestad Laboratories, Inc. (Chaska, MN) Lot F204N016-3. Values are those supplied on the package insert.

[c] These conjugates have been described in detail in a previous study.[2]

MATERIALS AND METHODS

Conjugates

TABLE 1 gives details of the nine conjugates used in this study. Six antihuman IgG conjugates were prepared by methods described previously.[4,5] A single large batch of ammonium sulfate–precipitated goat globulin provided the source for conjugates 4 through 6, coupled at three different F/P ratios. A portion of this same batch of globulin, affinity-purified on a human fraction II column and then coupled at two F/P ratios, yielded conjugates 2 and 3. Conjugate 1 was prepared in the same manner as 4 through 6, but from a separate ammonium sulfate precipitate. FITC-conjugated goat antihuman IgG was purchased from a commercial source (Kallestad Laboratories, Inc., Chaska, MN). Three conjugates (designated Fuji, Eiken, and NBL) were reconstituted from previously unopened, lyophilized material that has been described in detail in a previous study.[2]

F/P Ratio Determination

Human Cohn fraction II (Miles Laboratories, Inc., Elkhart, IN) covalently coupled to agarose gel at 15 mg per ml of packed gel and poured into pasteur pipettes (2 ml packed gel each), served as the ligand in a micro scale affinity chromatography purification of conjugates 2 through 6. To each column 1.5 ml of conjugate was applied, followed by a wash with 5.0 ml of 1.4 M NaCl in phosphate-buffered saline (PBS; .01 M phosphate, .15 M NaCl, pH 7.2). The bound antibody was eluted with 5.0 ml of 3 M NaSCN in PBS. The eluates were dialyzed for 24 hr in 4 L PBS (four changes), and analyzed for fluorescein and protein using standard methods[7,8] for antibody (Ab) F/P ratio determinations.

Preparation of Adsorbent

Forty-two milliliters (packed) of CM-BioGel-A (Bio-Rad Laboratories, Richmond, CA) was covalently coupled with 737 mg of human fraction II according to manufacturer's instructions. Briefly, after washing and equilibration in coupling buffer (3.0 mM PBS, pH 6.3), the gel was mixed for 30 minutes with 20 ml antigen solution (45 mg/ml in coupling buffer). Addition of 400 mg 1-ethyl-3-(3-dimethylaminopropyl)carbodiimide (EDAC) with constant adjustment of pH to 6.3 for one hour, followed by gentle overnight stirring, resulted in 98% coupling efficiency. The adsorbent was distributed into 1.5 ml polyethylene conical-bottom microcentrifuge tubes at 0.3 ml packed gel per tube.

Conjugate Ab Assay

FIGURE 1 schematically outlines the method of Ab determination. The conjugate to be assayed is applied to adsorbent gel contained both in a pasteur pipette column, as described above, and in microcentrifuge tubes (1.5 ml of conjugate in the former instance, and 0.2 ml of undiluted, 1/2 and 1/4 diluted samples in the latter). The pipette column–purified material, eluted from the beads as described above, affords a measure of the Ab F/P ratio. Adsorption of the specific Ab in the microcentrifuge tubes occurs over a period of three hours at room temperature with constant agitation. Examination of the supernatants by a capillary-tube precipitation test for residual Ab indicates complete adsorption under these conditions. Washing the contents of each tube with 4.0 ml PBS removes unbound material. Overnight agitation at room temperature with 1.0 N NaOH (1.2 ml per tube) results in complete extraction of the bound F from the adsorbent, as revealed by non-fluorescence of the beads by UV

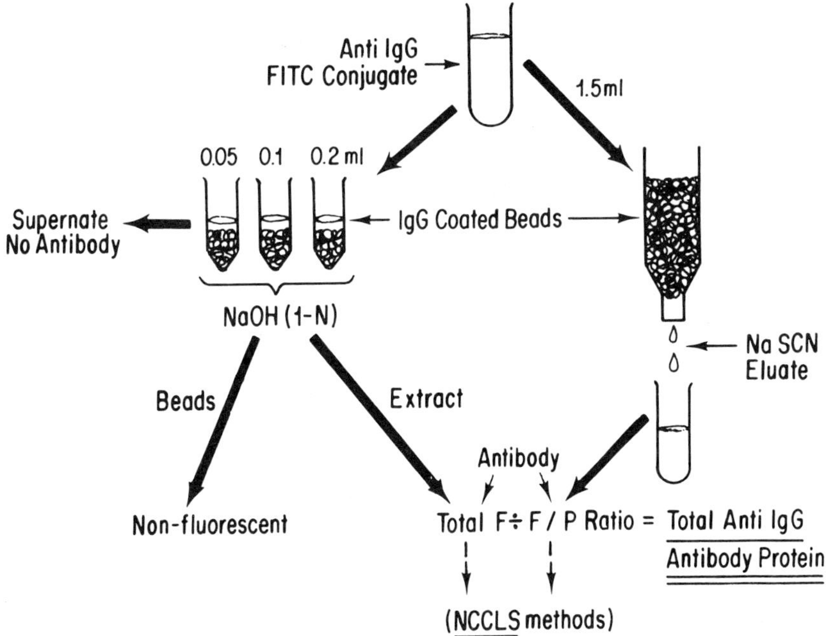

FIGURE 1. Schematic representation of fluorescein assay method of conjugate antibody quantitation.

microscopy. The supernatant extracts are assayed for F by the National Committee for Clinical Laboratory Standards (NCCLS) recommended method of McKinney *et al.*[7] The F content of each tube, when divided by the Ab F/P ratio, provides a measure of Ab P per added volume of conjugate.

RRID determinations of Ab concentration were performed according to previously published methods[6] for comparative purposes.

RESULTS

TABLE 2 summarizes data on the reproducibility of the test system over a number of trials with different conjugates. Analysis of the two conjugates, which were assayed frequently enough for statistical treatment, indicates an acceptable degree of precision

TABLE 2. Replicate Ab Assays of Five Conjugates by Fluorescein Assay Method

Experiment Number	Conjugate (mg/ml)				
	1	4	2	Kal	NBL
9	1.93				
10	2.16				
11	2.04				
12	1.66				
14	1.91				
16		2.65			
16		2.40			
17		2.49	5.44		
18		2.82	5.44		
21		2.52			
21		2.43			
23		2.24		1.54	
23		2.62			
24		2.56		1.59	
24		2.38			
25	2.00	2.64			2.53
25	1.93	2.79			2.55
$\overline{X} \pm$ s.d.	1.95 ± .15	2.55 ± .17	5.44	1.57	2.54
c.v.	7.9%	6.7%	—	—	—

as indicated by coefficient of variation of 7.9% and 6.7% for conjugates 1 and 4, respectively.

Antibody content values obtained by this method correlated well with those determined by the RRID technique (TABLE 3). As a further confirmation of the accuracy of the F assay method of conjugate Ab quantitation, conjugates that had been assayed by this technique were used as standards in a RRID assay of three previously studied conjugates. Independent Ab content values of these three and other conjugates, obtained in 1969 by seven different laboratories, have been reported elsewhere.[2] The mean values in the present study appeared to be within 7% of the 1969 values for two of the conjugates (TABLE 4). The third appeared to have deteriorated over time, as revealed by its turbidity upon reconstitution. This probably accounts for the 46% lower assay value in this study.

TABLE 5 affords comparisons of whole conjugate F/P ratio with the specific Ab F/P ratio. The purified Ab fractions have significantly lower F/P ratios than

TABLE 3. Comparison of Mean Ab Protein Values by RRID and F Assay Methods

Conjugate	F Assay (mg/ml)	RRID (mg/ml)
1	1.95 (7)[a]	2.10 (3)
4	2.55 (12)	2.54 (6)
Kal	1.57 (2)	1.51 (4)
NBL	2.54 (2)	2.74 (1)

[a]Numbers in parentheses indicate the number of assays of which the value is the mean.

conjugates prepared from ammonium sulfate precipitates, whereas additional affinity purification of the precipitated globulin reduces the difference to an insignificant level. Conjugates 4, 5, and 6, prepared from the same batch of ammonium sulfate fraction at different F/P ratios, have similar differences in Ab and whole conjugate F/P ratios, thus indicating that the percentage change in F/P is independent of degree of labeling. Conjugate 1, prepared from the same goat serum, using the same method as 4, 5, and 6, but from a different batch of ammonium sulfate precipitate, shows a difference between the Ab F/P and the whole conjugate F/P of 28%. This percent change, when compared to that of conjugates 4 through 6, is quite different.

DISCUSSION

Determinations of the Ab content in conjugates that are not based on immunoprecipitation generally make use of a label to trace the specific Ab. The radiolabel I-125 serves this purpose in the radioimmunoabsorption method.[1] The Ab assay described in this paper takes advantage of the FITC label already coupled to the conjugate protein. The elimination of further labeling of the conjugate circumvents potential problems that could occur as a result of adding another label (e.g. denaturation of some of the Ab, with a subsequent false low Ab concentration). An additional advantage of the use of fluorescein as the Ab tracer is the fact that a routine and ubiquitous methodology exists for its measurement, which includes satisfactory provisions for standardization.[7]

Quantitative recovery of adsorbed Ab could not be reliably attained by NaSCN elution. For this reason, NaOH extraction was employed for the total removal and measurement of the bound Ab. However, NaSCN elution was satisfactory for the determination of Ab F/P since it was only the relative amounts of F and P that need to

TABLE 4. Comparison of RRID Values Obtained in 1982[a] and 1969[b]

Conjugate	1969 Value (mg/ml)	1982 Value (mg/ml)
NBL	2.6	2.74
Eiken	0.56	0.52
Fuji	0.84	0.45[c]

[a]Determinations based on standards calibrated by fluorescein assay technique. Conjugates reconstituted from lyophilized material.
[b]Mean of values obtained in seven different laboratories.[2]
[c]Lyophilized conjugate showed signs of deterioration.

be determined, quantitative recovery being unnecessary. The NaOH extraction was unsuitable for this purpose because significant amounts of antigen protein were removed from the beads along with the Ab.

The fact that the F/P ratio of the conjugate may not be the same as that of the specific Ab in the conjugate necessitates the inclusion of a micro scale affinity purification step for Ab F/P determination. However, the data in TABLE 5 indicate that in affinity-prepared conjugates little difference exists between the F/P ratios of a whole conjugate and its Ab fraction. Thus, the micro scale Ab purification step can be eliminated in Ab assays of conjugates prepared by affinity chromatography.

In our hands, the F assay method of conjugate Ab quantitation is as reproducible as the RRID method. The two methods give comparable results for the several conjugates that have been assayed repeatedly by each method. Ab determination by F assay exhibits several advantages over the existing methods. There is no additional labeling of the conjugate required, and the reagents and equipment are readily available and simple to use. The F determination is a technique that many laboratories perform on a

TABLE 5. Comparison of Fluorescein to Protein Weight Ratios of Unfractionated Versus Specific Antibody Fraction of Six Conjugates

Conjugate	F/P Whole Conjugate (μg/mg)[a]	F/P Specific Antibody (μg/ml)[a]	% Change	Globulin Preparation Method
1	4.10	2.94	28.2	Conventional[b]
2	3.63	3.60	0.8	Affinity[c]
3	17.06	16.55	3.0	Affinity
4	2.64	2.13	19.3	Conventional
5	5.30	4.56	14.0	Conventional
6	12.19	10.31	15.4	Conventional

[a]Calculated according to methods advocated by the National Committee for Clinical Laboratory Standards;[8] P determinations by biuret assay and F determinations using fluorescein diacetate as a reference standard.[7]

[b]Ammonium sulfate–precipitated globulin.[4]

[c]Affinity purification of anti-FII antibody from ammonium sulfate–precipitated globulin.

routine basis and has been standardized by the NCCLS. Finally, it is not necessary to have a previously calibrated reference conjugate for this method.

The volume of sample required (2.05 ml) to perform this assay precludes its application at the user level for most situations. RRID, with its minimal volume requirements and simplicity of performance continues to be the method of choice here. The F assay for conjugate Ab is indicated at the manufacturer's level, since such sample volumes are not prohibitive. A further application of this method is as an independent means of determining the Ab concentration in conjugates to be used as reference standards for RRID.

REFERENCES

1. BERGQUIST, N. R. & R. NORBERG. 1974. Determination of specific antibody content of fluorescein conjugates by radioimmunoabsorption. J. Immunol. Methods 5: 275–282.

2. BEUTNER, E. H. 1971. Field trials by ten laboratories of six commercial conjugates prepared from antisera to human IgG. Ann. N.Y. Acad. Sci. **177:** 361–403.
3. BEUTNER, E. H., E. J. HOLBOROW & G. D. JOHNSON. 1967. Quantitative studies of immunofluorescence. I. Analysis of mixed immunofluorescence. Immunology **12:** 327–337.
4. BEUTNER, E. H., R. J. NISENGARD & V. KUMAR. 1979. Defined immunofluorescence: Basic concepts and their application to clinical immunodermatology. *In* Immunopathology of the Skin., 2nd edit. E. H. Beutner, T. P. Chorzelski & S. F. Bean, Eds.: 29–75. John Wiley & Sons, Inc. New York.
5. BEUTNER, E. H., M. R. SEPULVEDA & E. V. BARNETT. 1968. Quantitative studies of immunofluorescence staining. II. Relationships of characteristics of unabsorbed antihuman IgG conjugates to their specific and nonspecific staining properties in an indirect test for antinuclear factors. Bull. World Health Org. **39:** 587–606.
6. BEUTNER, E. H., G. WICK, M. SEPULVEDA & K. MONIN. 1970. A reverse immunodiffusion assay for antibody protein concentration in antisera or conjugates to human IgG. *In* Standardization in Immunofluorescence. E. J. Holborow, Ed.: 165–169. Blackwell Scientific Publications. Oxford.
7. MCKINNEY, R. M., J. T. SPILLANE & G. W. PEARCE. 1964. Fluorescein diacetate as a reference color standard in fluorescent antibody studies. Anal. Biochem. **9:** 474–476.
8. NATIONAL COMMITTEE FOR CLINICAL LABORATORY STANDARDS. August, 1975. Determinants of fluorescein/protein ratios in fluorescein isothiocyanate labeled protein.

Standardization in Immunofluorescence

E. J. HOLBOROW

Bone and Joint Research Unit
The London Hospital Medical College
University of London
London E1 2AD England

G. D. JOHNSON

Department of Immunology
The Medical School
University of Birmingham
Birmingham B15 2TJ England

Immunological reagents used in clinical chemistry or clinical immunology laboratories are for the most part biological materials that, at present at any rate, cannot be characterized completely by chemical or physical means alone. The increase in use of immunofluorescent techniques in routine laboratories, which took place during the 1960s, focussed attention on the problems encountered in different laboratories using immunofluorescent reagents from different sources. The consequent variation in results not only between laboratories using similar procedures, but also within individual laboratories performing tests on different occasions posed a challenge to effective quality assurance. This challenge was, and is still, being met by a World Health Organization/IUIS (WHO/IUIS) Subcommittee on immunofluorescence standardization, which has long implemented a policy of providing common reference materials (standards) for use in comparative assay according to established principles of biological standardization (TABLE 1). The assumptions upon which such comparative assays are based in physiology and pharmacology are that like is being compared with like, that the variable being measured is relevant to some characteristic (e.g. binding affinity) of the entity under test, that there is true randomness in the treatment of variables, and that the assay provides internal evidence of its own validity.

Thus, standards should ideally be qualitatively identical to test materials, plots of their respective dose-responses should be parallel and show linearity, and they should have long-term stability. If these conditions are met the ratio of the potency of the test material to that of the standard is independent of the method used and should remain constant. The assignment of an arbitrary unitage to the standard or reference material and the expression of results in relative units minimize interlaboratory differences caused by technical variation. Unfortunately, biological materials rarely satisfy all the above requirements. Their performance is affected by qualitative and quantitative differences related to antibody avidity and specificity and, in the case of fluorochrome-labeled reagents, additional factors related to the method of preparation. Many of these diverse factors affecting performance are well documented so that their adverse effects can be minimized by careful preparation, characterization, and selection of materials that satisfy defined minimal requirements. However, although the suitability of labeled reagents for use as standards can be predicted from a knowledge of selected physicochemical characteristics, their practical adequacy can only be established by their performance in a definitive system. The prime function of international immunofluorescence standards and reference preparations is thus to enable reactions detectable by immunofluorescence to be quantified in uniform terms throughout the world and also to provide a check on specificity. The WHO nomenclature used for immunological comparison materials of this sort is given in TABLE 1.

INTERNATIONAL COLLABORATIVE ASSAYS

The suitability of immunofluorescence standards has been assessed by a sequence of collaborative assays carried out on an international basis with the aim of answering the question "Are the proposed materials useful to ensure uniformity of results?" 'Results,' here, means attempts to measure the amount or activity of reagents in various immunofluorescence reactions. It might be thought that uniformity of results would be best achieved by specifying a standard technique to be used in each laboratory routinely performing a given test. However, the variables in immunofluorescence methodology are numerous and it would not be practicable to impose technical restrictions on them all. They include the optical features and illumination

TABLE 1. Immunological Comparison Materials

Material	Intended Use	Available for Immunofluorescence
WHO International standards	Calibration of local materials for control of quality and for potency	FITC anti-human Ig conjugates:[b] —Sheep anti-human Ig —Sheep anti-human IgM (anti-μ-chain) —Sheep anti-human IgG (anti-γ-chain)
WHO International reference preparations	Provided as a service reagent; also suitable for calibrating local material	—Anti-nuclear factor (homogeneous) 66/233[a] (100 IU/ampule) —IgM class anti-nuclear antibody HL[b]
WHO International reference	Qualitative use in identification (high specificity)	(In preparation)

[a] Anti-nuclear factor 66/233 available from: The Director, WHO International Laboratory for Biological Standards, Statens Seruminstitut, 80 Amager Boulevard, DK 2300 Copenhagen S, Denmark; or The Director, National Institute for Biological Standards and Control, Holly Hill, Hampstead, London, NW3 6RB, England.

[b] ANA IgM (HL), FITC-labeled sheep anti-human IgM, FITC-labeled sheep anti-human Ig, and FITC-labeled sheep anti-human IgG available from: World Health Organization, Immunology Unit, 121, Geneva 27, Switzerland.

systems used in the microscopy, the properties of the fluorochrome-conjugated antibody, in indirect immunofluorescence, the quality and quantity of antibody in the patient's serum, the antigen-containing substrate, and the observer. Thus, the accepted approach to promoting uniformity of reporting immunofluorescence test results is to use collaborative studies to establish a standard material of uniform and stable potency suitable for monitoring local reference preparations for potency and specificity.

As to the relevance of optical features and illumination systems to standardization, we may note that by the beginning of the 1970s, most of the essential techniques for immunofluorescence had been established. A general realization of their potential in diagnosis has stimulated an increasing demand for improved and simplified fluorescence microscopy, and optimally labeled reagents of defined specificity and potency. An additional spur to improved microscopy was the rapid growth of interest in membrane determinants of lymphocytes and tissue culture cells and their functional behavior. This encouraged the production of a range of moderate and higher power objectives of high numerical aperture (NA 1.30–1.40) in combination with wide-angle low-power eyepieces, allowing detection of much smaller quantities of fluorochrome than conventional systems. With the introduction of vertical illuminators capable of providing epi-illumination, reproducible conditions of illumination were achieved, the use of oil was avoided, and the advantage of increasing intensity with increasing objective power attained. Another development in microscopy was the production of a wide range of interference filters of different short-pass and narrow-band transmittances, capable of providing filter combinations to deal with staining by more than one fluorochrome, in combination with phase-contrast microscopy.[a] Modern microscopes with epi-illumination have easily interchangeable blocks containing complete sets of primary and secondary filters so that the required illumination may be instantly assembled by the user.

All scientific observations are subject to variance and immunological tests are no exception. Variance in its correct literal sense means discord and discrepancy. It is common experience that unacceptable levels of inter- and intra-laboratory comparability are found when attempts are made to observe and measure this variance. External reference preparations come to our aid in providing accredited material whose potency and specificity have been established on the basis of multicenter international evaluation and enable us to calibrate our systems according to agreed yardsticks. In autoimmune serology, standard antibody preparations provide the means of assessing the efficiency of the test system (and its operator) in the detection of specific antibodies by the characteristic staining patterns. Reference conjugates find their application in manufacture rather than in the clinical laboratory although the user may appreciate their value in confirming specificity when unexpected results are obtained with a regular conjugate. Thus, reference materials—both labeled and unlabeled—indeed have a useful place in the diagnostic laboratory where they have the potential to improve the quality of the service provided.

With the present emergence of the second generation of applications of immunofluorescent-labeling techniques in which monoclonal antibodies are used in combination with the cell analysis and sorting potential of flow cytometry and scanning fluorescence microscopy, and with image analysis of cells becoming a practical procedure, it is clear that the need for standardization of relevant reagents will not decrease. It may also turn out that the role of standardization in immunofluorescent methodology in these new contexts may require re-evaluation.

[a] PLOEM, J.C. 1971. A study of filters and light sources in immunofluorescence microscopy. Ann. N.Y. Acad. Sci. **177**: 414–429.

Future Development of Immunologic Reference Preparations

RENÉE NORBERG

Department of Immunology
National Bacteriological Laboratory
S - 105 21 Stockholm, Sweden

The World Health Organization (WHO) has hitherto accepted three different FITC-conjugated anti-human immunoglobulin standards, a polyvalent anti-immuno-globulin, a specific anti-γ-chain conjugate, and the third one is specific against μ chains. There has been a great demand for these standards, which has encouraged the IUIS subcommittee for standardization of immunofluorescence to continue.

The production of a standard preparation, whether FITC-labeled or not, is not only a very laborious procedure but also connected with large expenses. Thus, it is not possible to produce standards for every conceivable conjugate to be used in immuno-fluorescence techniques. In this context it should be mentioned that IUIS has a subcommittee working on the production of enzyme-labeled anti-immunoglobulin conjugates. To a certain extent the requirements of FITC- and enzyme-labeled conjugates are the same. This applies, for example, to the specificity and content of antibodies. Therefore, it seems rational that the evaluation of antisera is done in collaboration between the two committees.

The next standard will presumably be an FITC-labeled anti-mouse immunoglobu-lin conjugate. Because of the extensive use of mouse monoclonal antibodies both in research and in clinical diagnostic work, the need for such a standard is urgent. There are, however, certain requirements that must be fulfilled by the anti-mouse immuno-globulin conjugate and that have apparently not been so essential for the polyspecific anti-human immunoglobulin conjugates. Anti-mouse immunoglobulin conjugates will without doubt mainly be used to detect reactions elicited by monoclonal antibodies. There are at least six different mouse immunoglobulin classes, IgM, IgG1, IgG2a, IgG2b, IgG3, and IgA, which are actual as monoclonal reagents and the anti-mouse conjugate must contain antibodies reacting with all these antibody classes. Theoreti-cally, the presence of antibodies against light chains should be sufficient to detect all immunoglobulin types, but the sensitivity of the test system increases many times when antibodies to heavy chains are included. This might depend not only on the increased number of epitopes covered but, at least in some antibody complexes, also on the better availability of antigenic determinants on the heavy chains. Therefore, the tests designed to evaluate conceivable anti-mouse conjugates must include a sufficient number of monoclonal antibodies of different immunoglobulin classes and different specificities. The anti-mouse immunoglobulin preparation found suitable to be used as a fluorescein-labeled standard will presumably also fulfill the criteria of an enzyme-labeled standard.

The second standard to be established might be a FITC-labeled conjugate against rabbit immunoglobulins. Although the use of rabbit antisera has diminished very much after the introduction of the mouse monoclonal antibodies, rabbit antibodies are still commonly used, for example, for rapid viral diagnoses. The evaluation of the anti-rabbit immunoglobulin standard must therefore include test systems utilizing various rabbit antiviral antibodies.

Hitherto all WHO anti-immunoglobulin standards have been FITC-labeled sheep IgG preparations. Many cells, however, have receptors for IgG-Fc, which might increase the possibility of nonspecific immunoglobulin reactions, especially when investigating antigens on the surface of live cells. As large a field of application, at least in clinical immunology and immunopathology, is classification of various cells with mouse monoclonal antibodies. It might be discussed whether the anti-mouse immunoglobulin standard should instead be an $F(ab)_2$ preparation.

For many years a serum pool, 66/233, containing human antinuclear antibodies (ANA) has been available at WHO. These antibodies can be used as middle-layer in IFL standard tests. Moreover, a serum, H.L., containing human IgM antibodies exhibiting a speckled nuclear pattern can be acquired from WHO. An IUIS subcommittee, chaired by Dr. T.E.W. Feltkamp in Amsterdam, is working on the standardization of DNA antibodies. But so far, the IUIS subcommittee on standardization of immunofluorescence has taken limited interest in the establishment of reference sera to be used as middle-layers in clinical immunologic tests.

However, at a meeting in Vienna (February 1981), our subcommittee decided to analyze the need of such sera, which in the first place should be used as references in IFL tests of auto-antibodies. This work is rather preliminary and it is for practical reasons impossible to produce standards or references for every possible test. The following preparations together with the already established ANA references, representing the most common auto-antibodies of diagnostic importance have been preferentially considered: (1) Antibodies to RNase-sensitive, extractable nuclear antigens, (sRNP) giving speckled nuclear fluorescence. (2) Antibodies to actin (smooth muscle antibodies, SMA). Actin is a constituent of all cells, but the auto-antibodies are usually demonstrated by their staining of smooth muscle fibers. (3) Antibodies to mitochondria, type M2. These auto-antibodies are mainly seen in patients with primary biliary cirrhosis. (4) Antibodies reacting with the intercellular substance of skin and mucosa ("pemphigus antibodies"). (5) Antibodies reacting with the basal membrane of skin and mucosa ("pemphigoid antibodies").

These auto-antibodies give characteristic IFL staining patterns. Moreover, the specificity of at least the first three antibodies can be documented also by other techniques.

Antibodies against RNase-sensitive, extractable nuclear antigen (sRNP) are common in various rheumatic diseases.[5,6] The reacting antigen has been characterized and its reaction with the corresponding antibodies has been demonstrated by various test systems.[4-6]

Actin, which is a constituent of all eukaryotic cells, can be isolated from various tissues but preferentially from skeletal muscle. Actins of all origins are chemically and immunologically almost identical. Absorption with any actin will extinguish the reaction of any anti-actin antibodies.[2]

Antibodies specific to mitochondria constitute a heterogeneous group, although they give a rather similar IFL staining pattern on tissue sections. The antigens, however, can be biochemically defined and the antibodies connected with primary biliary cirrhosis, the type M 2 antibodies, are shown to react with a protein present in the chloroform-released ATPase preparations of submitochondrial particle membranes.[3]

The skin antibodies give very characteristic IFL patterns and are highly disease-specific.[1] Therefore, they can be used as references without further evaluation.

The collection and evaluation of suitable starting sera for production either of FITC-labeled anti-immunoglobulin standards or for preparations to be used as references in clinical diagnostic tests, is a time-consuming procedure. The subcommittee will have enough work with the standards and reference preparations mentioned during the next years.

REFERENCES

1. BEUTNER, E. H., T. P. CHORZELSKI & S. F. BEAN. 1979. Immunopathology of the Skin. 2nd edit. John Wiley and Sons, Inc. New York.
2. FAGRAEUS, A. & R. NORBERG. 1978. Anti-actin antibodies. Curr. Topics Microbiol. Immunol. **82:** 1–13.
3. FAYEMEEK, K., D. DONIACH & H. BAUM. 1980. Mitochondrial antibodies in chronic liver diseases and connective tissue disorders: Further characterization of the auto-antigens. Clin. Exp. Immunol. **41:** 43–54.
4. JONSSON, J. & R. NORBERG. 1980. Some functional properties of the RNP nucleoplasmic antigen. Scand. J. Immunol. **12:** 119–127.
5. NAKAMURA, R. M. & E. M. TAN. 1978. Recent progress in the study of autoantibodies to nuclear antigens. Human Pathol. **9:** 85–91.
6. SHARP, G. C., W. S. IRVIN, C. M. MAY, H. R. HOLMAN, F. C. MCDUFFIE, E. V. HERS & F. R. SCHMID. 1976. Association of antibodies to ribonucleoprotein and SM antigen with mixed connective tissue disease, systemic lupus erythematosus and other rheumatic diseases. New Engl. J. Med. **295:** 1149–1153.

Quality Control of Fluorescein Isothiocyanate-labeled Reagents

SHIREEN CHANTLER AND IRENE BATTY

Wellcome Research Laboratories
Beckenham, Kent BR3 3BS
England

INTRODUCTION

The principle of combining the unique specificity of antibody interaction with the sensitivity of fluorescent-light detection for localizing the site of antigen-antibody reactions was first described by Coons and his colleagues.[9,10] Its diagnostic potential however was not exploited until the late 1950s following the introduction of the more stable thiocyanate derivatives of fluorescein.[19] The subsequent expansion of immunofluorescence technology into routine laboratories for detecting and identifying autoantibodies, microbial antibodies, and antigens in clinical materials emphasized the existence of considerable interlaboratory variation in test results and a clear need for commercial reagents of consistent quality. The purpose of this paper is to emphasize the major factors known to influence reagent quality and the measures adopted by reputable manufacturers to ensure effective quality control.

The principle and technical steps involved in immunofluorescence procedures are simple yet test results are influenced by multiple factors (TABLE 1). Two of these variables, namely instruments and reagent quality, have been extensively investigated by commercial and research institutions: the current manufacture and assessment of reagents is based largely on this extensive published work.[1,3,5,6,8,15,23,24]

QUALITY CONTROL OF PREPARATIVE PROCESS

Production and Selection of Antisera for Conjugation

Antiserum used for the preparation of fluorescein isothiocyanate–labeled reagents must be highly potent in order to reduce nonspecific reactivity in the final reagent, and of the required specificity. Rabbits, guinea pigs, sheep, or goats may be immunized but it is imperative that highly purified antigens containing all representative antigen epitopes are used as immunogen and that multiple injections are given to ensure the production of potent antisera with high ratios of antibody to non-antibody globulin.

Those experienced in the art of raising antisera will know that different preparations of similar antisera, although prepared in an identical manner, seldom exhibit exactly the same qualitative and quantitative properties. Polyclonal sera are heterogeneous with regard to immunoglobulin class, antibody avidity, and specificity; hence their properties may vary in different test methods. Preliminary evaluation of specificity and potency of antisera is best performed therefore by comparative evaluation with a serum of proven suitability in the same defined system. Anti-immunoglobulin antisera can be selected readily on potency from the results of titration in Ouchterlony gel diffusion tests against the appropriate purified immunoglobulin at 1 mg/ml in the central well: sera giving a good precipitation line at a dilution of 1/16 or more are suitable for labeling.[3] Alternatively, a simple enzyme-linked immunosorbent assay (ELISA) may be used in which polystyrene microtiter plates coated with the purified immunoglobulin are incubated first with dilutions of sera followed by the working dilution of enzyme-labeled anti-species conjugate. In our experience end-point titers obtained in this type of assay may vary when tested on different occasions but the inclusion of a reference serum of known potency enables the selection of antisera of an equivalent or greater activity.

If the lack of suitable soluble antigen makes such tests impossible, indirect immunofluorescence tests should be done utilizing a range of dilutions of pre- and post-immunization sera as the intermediary layer followed by the appropriate labeled anti-species immunoglobulin. Test samples, which may be histological preparations or cell films, should be appropriately prepared (some prior knowledge of the system is almost essential) and should represent both 'positive' (antigen-containing) and 'negative' (non-antigen-containing) materials. Antibacterial and antiviral sera for conjugation are frequently selected in this way.

Similar test systems may be used to determine whether the specificity of the antisera is adequate by utilizing a spectrum of appropriate target antigens, the

TABLE 1. Major Variables in Immunofluorescence Procedures

Instrumentation	Filter selection
	Lamp deterioration
	Microscope alignment
Technical	Substrate preparation
	Substrate deterioration
	Sample/reagent application
	Effective washing
	Reagent quality
	Mountant pH
Assessment	Fading of fluorescence
	Subjectivity

selection of which requires a knowledge of the likely cross-reactivities. In defining the specificity criteria for an antiserum it must be recognized that absence of undesirable cross-reactivity in some systems, such as gel diffusion and immunoelectrophoresis, does not preclude reactivity in the more sensitive immunofluorescence procedures.[7] It is mandatory that specificity checks on the final reagent utilize the most demanding and sensitive systems available.

The results of these tests will determine whether the antiserum is accepted, rejected, or subjected to further purification by absorption with appropriate immunoadsorbents or by affinity chromatography. In our experience, repetitive processing can result in preparations with lowered and often inadequate potency and/or stability.

The recent advances in hybridization techniques[17] offer the possibility of using monoclonal sera of defined avidity and epitope specificity for the preparation of conjugates. Whilst this avoids the need for removal of antibodies of unwanted specificity by absorption, it becomes necessary first to determine whether monoclonal antibody of restricted epitope specificity is capable of reacting with all molecules of the designated type and second, to establish the stability of products of each different clone. Our preliminary investigations with selected monoclonal products indicate that the performance of the derived conjugates is comparable to that of polyclonal reagents of apparently similar specificity.

Control of Labeling Parameters

The methods of labeling proteins with fluorescein isothiocyanate (FITC) are well documented[24] but considerable attention must be given to the purity of the reactants, the relative amounts used, and pH control. These parameters influence the degree of labeling, which may affect the suitability of a conjugate in use.

Since the binding efficiency of FITC is dependent upon its purity[20] and the coupling rate varies with different proteins,[21] only FITC of the highest purity and homogeneous protein preparations should be used for conjugation if some measure of control is to be achieved. The purity of the FITC and its binding efficiency can be determined readily by published methods.[22] A variety of methods are available for separating the antibody-containing immunoglobulin fraction from the selected anti-serum[13,26,27] but irrespective of the method used, the homogeneity of the derived fraction must be confirmed. The absence of contaminating serum proteins can be checked by cellulose-acetate electrophoresis[11] or by immunoelectrophoresis against an anti-species whole serum using the test preparation at a concentration of 10 mg/ml in the well. Only in the most demanding systems will it be necessary to prepare immunospecific antibody rather than a homogeneous immunoglobulin preparation for labeling. The reduced yield, varying level of antibody denaturation, preferential elution of less avid antibodies, and decreased stability of immunopurified antibody may offer limited advantages over the whole immunoglobulin fraction, provided suitable sera are selected in the first instance. It is prudent to reassess potency at this stage to ensure that antibody activity has been retained following fractionation.

If the immunoglobulin preparation and FITC are satisfactory, labeling may be performed by conventional procedures using the most appropriate level of fluorochrome. In our experience, ratios of 1 to 2.5 mg FITC:100 mg immunoglobulin are satisfactory for most applications but higher ratios may be desirable for antibacterial reagents[12] and for the detection of cell-surface markers. The pH should be monitored at intervals to ensure that it lies between 9–9.5. Extremes of pH decrease the efficiency of labeling and may cause antibody inactivation. Despite careful control of the conjugation conditions, the labeled product will contain a heterogeneous mixture of immunoglobulin molecules with differing levels of FITC and free unbound dye. Unbound FITC must be removed by dialysis and column chromatography and its absence checked by thin layer chromatography. Ion exchange chromatography may be used to obtain fractions containing immunoglobulins with a restricted range of fluorescein:protein ratios[30] but in practice this is rarely necessary for the majority of applications provided potent antisera are used and gross overlabeling has been avoided. Simple tests should be undertaken initially to ensure adequate labeling at this stage. In our experience a reliable first estimate of labeling efficiency can be made by measuring the absorption at 493 and 280 nm and deriving the optical-density ratio. This is not an absolute measurement but provides an index that can be derived readily. If this optical-density ratio is between 0.6–0.9, the possibility of gross over or underlabeling is unlikely.

After labeling, but before extended evaluation, the conjugate is adjusted to a suitable protein concentration (usually 6–10 mg/ml) by simple calculation from the initial protein concentration and the residual volume after chromatography.

QUALITY CONTROL OF CONJUGATE ASSESSMENT

The essential requirements of a conjugate are that it exhibits the stated immunological specificity, that it gives good specific staining with minimal background
fluorescence at the working dilution, and that it is predictably stable on storage.

A knowledge of the preparative methods and the chemical characteristics of a
particular reagent are of considerable predictive value in determining whether a
reagent will be satisfactory in use. Important measurements include total protein
(before the addition of stabilizing proteins),[18] specific antibody protein,[2,4] and fluores-
cein:protein (F/P) molar ratios.[22,29] Experience has shown that satisfactory staining is
obtained with conjugates in which the antibody protein is at least 10% of total protein
and the F/P molar ratio ranges between 1.5 and 4.5. These measurements fail to
identify the immunological specificity of the final reagent or the dilution range over
which the reagent can be used without interference by nonspecific background
fluorescence.

Specificity Assessment

The criteria used should be clearly defined and take into account possible
interactions due to Fc receptors or shared epitopes on apparently unrelated antigens.
The choice of test for determining the specificity of a reagent will vary but the most
sensitive system is used.

Comparative specificity evaluation of anti-immunoglobulin conjugates is custom-
arily performed in indirect immunofluorescence tests employing intermediary layers
containing antibody of defined immunoglobulin class or in direct tests employing
defined monoclonal bone marrow preparations[14] or purified antigens chemically linked
to an inert substrate.[25,28] In most instances we have found that assessments of this kind
performed in parallel with preparations of proven specificity (e.g., anti-human
immunoglobulin conjugates distributed by WHO) ensure that acceptable specificity
requirements are met and also control the technical parameters of the assay. However,
our recent experience indicates that meticulous assessment of specificity in these tests
may fail to detect low levels of cross-reactivity demonstrable in some applied systems
with sophisticated optical equipment. We have found that ELISA systems employing
purified immunoglobulins passively adsorbed to polystyrene microtiter wells as target
antigens, readily reveal weak cross-reactions. Our data indicate that this procedure
provides an additional objective measure of immunological specificity in labeled
anti-immunoglobulin reagents, provided reference reagents of established performance are included for comparison to exclude effects attributable to minor impurities
in the antigens employed in such a sensitive assay.

Most specificity assessments are designed to exclude the possibility of expected,
undesirable cross-reactions and relatively little attention is paid to confirming whether
the required spectrum of activity occurs. This is particularly relevant for polyspecific
reagents intended for screening purposes. Immunofluorescence evaluation of polyspecific antibacterial conjugates should include organisms containing the major antigen
variants and those for polyspecific anti-species immunoglobulin reagents should utilize
intermediary sera with antibody of defined immunoglobulin class. When such sera are
not available, other tests employing purified immunoglobulins may be substituted but
a sound knowledge of their limitations is mandatory if erroneous results are to be
avoided. The use of whole serum in immunoelectrophoresis for determining the
polyspecificity of anti-human immunoglobulin conjugates commonly gives rise to
misleading results. The failure to obtain lines of precipitation characteristic of IgM
and IgA in this system with anti-immunoglobulin reagents, whose polyspecificity is

reliant on anti–light chain activity, is caused by antibody saturation by IgG rather than absence of reactivity with IgM and IgA.

Potency Assessment

The specificity of a reagent must be established before its effective working dilution range or potency is measured. Evaluation by chessboard titration in a representative immunofluorescence system is invaluable for determining the dilution range that will enable different levels of antibody to be detected without troublesome background staining under laboratory conditions.[16] Comparative evaluation in parallel with a reference preparation excludes the possibility that major technical variables are operative and provides some measure of relative potency thereby ensuring consistency between different production batches of reagents.

Performance Assessment

In the majority of cases the above evaluation of conjugate activity is sufficient for quality assessment confirmation. In others, particularly when the test substrates differ substantially from those used in a clinical laboratory, additional evaluation on clinical specimens is desirable.

SUMMARY

Fluorescein-labeled reagents are not amenable to the same precision of definition as are chemical substances. Nonetheless extensive experimental evidence shows that certain preparative requirements and selected measurable chemical parameters of the product are of considerable predictive value in determining whether a reagent will be suitable in use. Adherence to these criteria and their continuous reappraisal combined with performance assessment and information on stability are all essential to ensure that reagents are of a quality acceptable to laboratory workers and legislative bodies. The need for manufacturers to document each procedural step and record the results is an essential element of good manufacturing practice now imposed by regulatory bodies. All reputable manufacturers adhere to these requirements, which serve not only to maintain standards but also to form a basis for corrective action.

REFERENCES

1. BATTY, I. & G. TORRIGIANI. 1976. Standardization of reagents and methodology in immunology. *In* Manual of Clinical Immunology. N. R. Rose & H. Friedman, Eds. American Society for Microbiology. Washington, D.C.
2. BERGQUIST, N. R. & R. NORBERG. 1974. Determination of specific antibody content of fluorescent conjugates by radioimmunoabsorption. J. Immunol. Methods. **5:** 275–282.
3. BEUTNER, E. H., M. R. SEPULVEDA & E. V. BARNETT. 1968. Quantitative studies in immunofluorescent staining. II. Relationships of characteristics of unabsorbed anti human IgG conjugates to their specific and nonspecific staining properties in an indirect test for anti nuclear factors. Bull. W. H. O. **39:** 587–606.
4. BEUTNER, E. H., G. WICK, M. SEPULVEDA & K. MONIN. 1970. A reverse immunodiffusion assay for antibody protein concentration in antisera or conjugates to human IgG. *In* Standardisation of Immunofluorescence. E. J. Holborow, Ed.: 165. Blackwell. Oxford, England.
5. BEUTNER, E. H. 1971. Field trials by ten laboratories of six commercial conjugates prepared from antisera to human IgG. Ann. N.Y. Acad. Sci. **177:** 361–404.

6. BRIGHTON, W. D., C. E. D. TAYLOR, A. W. TOMLINSON & A. WILKINSON. 1967. Provisional specifications for fluorescein labelled antibody against human IgG. Mon. Bull. Minis. Hlth. Lab. Serv. **26:** 179–181.

7. CHANTLER, S. M. & M. HAIRE. 1972. Evaluation of the immunological specificity of fluorescein labelled anti human IgM conjugates. Immunology **23:** 7–12.

8. CHERRY, W. B. & C. B. REIMER. 1973. Diagnostic immunofluorescence. Bull. W. H. O. **48:** 737–746.

9. COONS, A. H., H. J. CREECH, R. N. JONES & E. BERLINER. 1942. The demonstration of pneumococcal antigen in tissues by the use of fluorescent antibody. J. Immunol. **45:** 159–170.

10. COONS, A. H. & M. H. KAPLAN. 1950. Localization of antigen in tissue cells. II. Improvements in the method for the detection of antigen by means of fluorescent antibody. J. Exp. Med. **91:** 1–13.

11. HEBERT, G. A., B. PITTMAN, R. M. McKINNEY & W. B. CHERRY. 1972. The preparation and physicochemical characterization of fluorescent antibody reagents. U.S. Department of Health, Education and Welfare. Center for Disease Control. Atlanta.

12. HEBERT, G. A., B. PITTMAN & R. M. McKINNEY. 1981. Optimal fluorescein to protein ratios of bacterial direct fluorescent antibody reagents. J. Clin. Microbiol. **13:** 498–502.

13. HEIDE, K. & H. G. SCHWICK. 1978. Salt fractionation of immunoglobulins. *In* Handbook of Experimental Immunology. D. M. Weir, Ed. Blackwell, Oxford, England.

14. HIJMANS, W., H. R. E. SCHUIT & F. KLEIN. 1969. An immunofluorescence procedure for the detection of intra cellular immunoglobulins. Clin. Exp. Immunol. **4:** 457–472.

15. HOLBOROW, E. J., G. D. JOHNSON & E. H. BEUTNER. 1967. Quantitative studies of immunofluorescence. Immunology **12:** 327–337.

16. JOHNSON, G. D., E. J. HOLBOROW & J. DORLING. 1978. Immunofluorescence and immunoenzyme techniques. *In* Handbook of Experimental Immunology. 3rd edit. D. W. Weir, Ed. Blackwell Scientific Publications. Oxford, England.

17. KÖHLER, G. & G. MILSTEIN. 1975. Continuous culture of fused cells secreting antibody of predefined specificity. Nature **256:** 495–497.

18. LOWRY, O. H., N. J. ROSENBERG, A. L. FARR & R. J. RANDALL. 1951. Protein measurement with the Folin phenol reagent. J. Biol. Chem. **193:** 265–275.

19. MARSHALL, J. D., W. C. EVELAND & C. W. SMITH. 1958. Superiority of fluorescein isothiocyanate (Riggs) for fluorescence antibody technique with a modification of its application. Proc. Soc. Exp. Biol. Med. **98:** 898–900.

20. McKINNEY, R. M., J. T. SPILLANE & G. W. PEARCE. 1964. Determination of purity of fluorescein isothiocyanates. Anal. Biochem. **7:** 74–86.

21. McKINNEY, R. M., J. T. SPILLANE & G. W. PEARCE. 1964. Factors affecting the rate of reactions of FITC with serum proteins. J. Immunol. **93:** 232–242.

22. McKINNEY, R. M., J. T. SPILLANE & G. W. PEARCE. 1966. A simple method for determining the labelling efficiency of fluorescein isothiocyanate products. Anal. Biochem. **14:** 421–428.

23. MEDICAL RESEARCH COUNCIL. WORKING PARTY. 1971. Characterisation of antisera as reagents. Immunology **20:** 1–10.

24. NAIRN, R. C. 1976. Fluorescent Protein Tracing. 4th edit. Livingstone. London.

25. RAYBOULD, T. J. G. & S. CHANTLER. 1979. Comparison of activation and blocking procedures in the use of defined antigen substrate spheres (DASS) for quantitative and visual assessment of conjugate activity. J. Immunol. Methods **27:** 309–318.

26. STANWORTH, D. R. 1960. A rapid method of preparing pure serum gammaglobulin. Nature **188:** 156–157.

27. STEINBACH, N. & R. AUDRAN. 1969. The isolation of IgG from mammalian sera with the aid of caprylic acid. Arch. Biochem. Biophys. **134:** 279–284.

28. VAN DALEN, J. P. R., W. KNAPP & J. S. PLOEM. 1973. Microfluorometry on antigen-antibody interaction in immunofluorescence using antigens covalently bound to agarose beads. J. Immunol. Methods **2:** 383–392.

29. WELLS, A. F., C. E. MILLER & M. K. NADEL. 1966. Rapid fluorescein and protein assay method for fluorescent antibody conjugates. Appl. Microbiol. **14:** 271–275.

30. WOOD, B. T., S. H. THOMPSON & E. GOLDSTEIN. 1965. Fluorescent antibody staining. III. Preparation of FITC labelled antibodies. J. Immunol. **95:** 225–229.

Preparation of Monomeric Fab′-Horseradish Peroxidase Conjugate Using Thiol Groups in the Hinge and Its Evaluation in Enzyme Immunoassay and Immunohistochemical Staining[a]

EIJI ISHIKAWA, SHINJI YOSHITAKE,
AND MASAYOSHI IMAGAWA

Department of Biochemistry
Miyazaki Medical College
Kiyotake, Miyazaki 889-16, Japan

AKINOBU SUMIYOSHI

Department of Pathology
Miyazaki Medical College
Kiyotake, Miyazaki 889-16, Japan

INTRODUCTION

We previously conjugated Fab′ with β-D-galactosidase by using N,N'-o-phenylene-dimaleimide[9,15] to perform highly sensitive enzyme immunoassays.[8,9] However, maleimide groups of this compound are not stable at neutral pH, so we synthesized a stable maleimide compound, N-succinimidyl 4-(N-maleimidomethyl) cyclohexane-1-carboxylate,[10,17] which was useful for conjugating antibodies with glucose oxidase.[17] By using the stable maleimide compound, we have recently conjugated Fab′ and horseradish peroxidase in high yields with a minimal polymerization and without impairing the activities of Fab′ and peroxidase.[7,16] A feature of our method is the use of thiol groups in the hinge of Fab′ for conjugation but no random use of amino groups. This paper describes some properties of conjugates prepared by using thiol groups in the hinge and amino groups of Fab′ and evaluates them in sandwich enzyme immunoassay and immunohistochemical staining.

MATERIALS AND METHODS

Materials

Horseradish peroxidase was obtained from Boehringer Mannheim (grade I), Sigma Chemical Co. (type VI), and Toyobo Ltd. (grade I.C.). (There was no difference in elution profiles of conjugates prepared by the maleimide method (I) using peroxidase from these different sources.) N-Succinimidyl 4-(N-maleimidomethyl)-cyclohexane-1-carboxylate was obtained from Zieben Chemical Co. and is available

[a]Supported in part by the Naito Foundation Research Grant for 1982.

from Pierce Chemical Co. *N*-Succinimidyl *m*-maleimidobenzoate was obtained from Sigma Chemical Corp. *N*-Succinimidyl 3-(2-pyridyldithio)propionate, Sephadex G-25, CNBr-activated Sepharose 4B, and concanavalin A-Sepharose 4B were obtained from Pharmacia Fine Chemicals AB. Ultrogel AcA 44 was a product of LKB. Bovine serum albumin (fraction V) was obtained from Armour Pharmaceutical Co. Alkaline phosphatase from calf intestine was obtained from Boehringer Mannheim. Human chorionic gonadotropin (hCG) was obtained from Calbiochem-Behring Corp. Rabbit fibrinogen was obtained from Tokyo Research Labs., Kowa Co., Ltd. Standard human serum containing α-fetoprotein was obtained from Behringwerke AG. Goat anti-rabbit IgG serum, fluorescein-labeled rabbit anti-human IgG, IgG, and rabbit anti-hCG serum were obtained from Miles Labs., Inc. Sheep anti-rabbit fibrinogen IgG was obtained from Cappel Labs., Inc. Goat anti-human α-fetoprotein serum was a generous gift from Medical and Biological Labs., Ltd. Polystyrene balls (3.2 mm in diameter) were obtained from Precision Plastic Ball Co. Trypsin was obtained from Difco Labs. Other chemicals were obtained from Nakarai Chemicals, Ltd.

Calculations of the Amount of Proteins

The amounts of IgG, F(ab')$_2$,[6] and Fab'[6,15] were calculated from absorbance at 280 nm, and the amount of peroxidase was calculated from absorbance at 403 nm.[7,16] The amount of rabbit fibrinogen was calculated from absorbance at 280 nm by taking its molecular weight and extinction coefficient to be 340,000 and 1.5 cm^2/mg, respectively.[12]

Preparation of Protein-Sepharose 4B

Proteins (about 10 mg each) were coupled to CNBr-activated Sepharose 4B (1 g) following the instruction of Pharmacia.

Preparation of IgG, F(ab')$_2$, Fab', and SH-blocked Fab'

IgG and F(ab')$_2$ were prepared as described previously.[6] Fab' was prepared by reducing F(ab')$_2$ with 10 mM 2-mercaptoethylamine at pH 6.0 at 37°C for 1.5 hr. The average number of thiol groups generated per Fab' molecule was about one.[15] For the periodate method, Fab' was prepared by reducing F(ab')$_2$ with 10 mM 2-mercaptoethanol at pH 8.2 at room temperature for 1 hr[14] and the average number of thiol groups generated per Fab' molecule was 1.6. The SH-blocked Fab' was prepared by incubating with 12 mM sodium monoiodoacetate at 4°C for 16 hr.[14]

In order to prepare fluorescein-labeled Fab'-peroxidase conjugate, F(ab')$_2$ was mixed with one fortieth in weight of fluorescein-labeled nonspecific F(ab')$_2$, which was prepared by passing fluorescein-labeled anti-human IgG F(ab')$_2$ through a human IgG-Sepharose 4B column.

Determination of Thiol, Maleimide, and Pyridyl Disulfide Groups

Thiol and maleimide groups were determined using 4,4'-dithiodipyridine as described previously.[6] For measuring pyridyl disulfide groups, a sample in a total volume of 0.5 ml of 0.1 M sodium phosphate buffer, pH 6.0 was incubated with 0.02 ml of 0.1 M dithiothreitol at 30°C for 20 min, and absorbance at 343 or 324 nm was measured. The extinction coefficients of pyridine-2-thione at 343 nm[13] and pyridine-4-thione at 324 nm[5] were taken to be 8.08×10^3 and 1.98×10^4 $M^{-1} \cdot cm^{-1}$, respectively.

Assay of Peroxidase Activity

Peroxidase activity was fluorimetrically determined using *p*-hydroxyphenylacetic acid as described previously.[7]

Assay of Alkaline Phosphatase

A sample in 0.1 ml of 0.1 M glycine-NaOH buffer, pH 10.3 containing 1 mM $MgCl_2$, 0.1 mM $ZnCl_2$, 0.05% NaN_3, and 0.025% egg albumin was incubated with 0.05 ml of 3×10^{-4} M 4-methylumbelliferyl phosphate at 30°C for 10 min. The reaction was stopped by adding 2.5 ml of 0.5 M K_2HPO_4-KOH buffer, pH 10.4, containing 10 mM EDTA,[4] and fluorescence intensity was determined using 1×10^{-7} M 4-methylumbelliferone in the same buffer as standard as described for β-D-galactosidase.[9]

Concentration of Proteins

Proteins such as antibodies, their fragments, and peroxidase were concentrated using a collodion bag in ice throughout.

Maleimide Method (I)

In the maleimide method (I), horseradish peroxidase was treated with *N*-succinimidyl 4-(*N*-maleimidomethyl)cyclohexane-1-carboxylate to introduce maleimide groups, and the maleimide-peroxidase obtained was allowed to react with thiol groups in the hinge of Fab' (FIGURE 1).

Horseradish peroxidase (2 mg in 0.3 ml of 0.1 M sodium phosphate buffer, pH 7.0) was incubated with *N*-succinimidyl 4-(*N*-maleimidomethyl)-cyclohexane-1-carboxylate (1.6 mg, 100-fold molar excess, in 0.02 ml of *N,N*-dimethylformamide) at 30°C for 1 hr with continuous stirring. Precipitates formed were removed by centrifugation and the supernatant was subjected to gel filtration on a Sephadex G-25 column (1.0 × 45 cm) using 0.1 M sodium phosphate buffer, pH 6.0. The average number of maleimide groups introduced into peroxidase molecule was 0.9–1.2.

The maleimide-peroxidase obtained (about 1.8 mg in 0.45 ml of 0.1 M sodium phosphate buffer, pH 6.0) was incubated with Fab' (about 2 mg in 0.45 ml of 0.1 M sodium phosphate buffer, pH 6.0 containing 5 mM EDTA) at 4°C for 20 hr or at 30°C for 1 hr. The final concentration of the maleimide-peroxidase and rabbit Fab' in the reaction mixture was about 0.05 mM.

Maleimide Method (II)

The maleimide method (II) was performed in the same way as in the maleimide method (I) except that *N*-succinimidyl *m*-maleimidobenzoate was used in place of *N*-succinimidyl 4-(*N*-maleimidomethyl)cyclohexane-1-carboxylate.

Peroxidase was treated with N-succinimidyl m-maleimidobenzoate at 25°C, and the maleimide-peroxidase was separated by gel filtration using 0.05 M sodium acetate buffer, pH 5.0, since maleimide groups of m-maleimidobenzoic acid were not stable at higher pH.[10,16] Other conditions were the same as in the maleimide method (I). The average number of maleimide groups introduced per peroxidase molecule was 0.5–0.6.

Maleimide Method (III)

In the maleimide method (III), maleimide groups were introduced into peroxidase in the same way as in the maleimide method (I), and the maleimide-peroxidase obtained was allowed to react with thiol groups that were introduced into the SH-blocked Fab' using amino groups by treatment with S-acetylmercaptosuccinic anhydride.

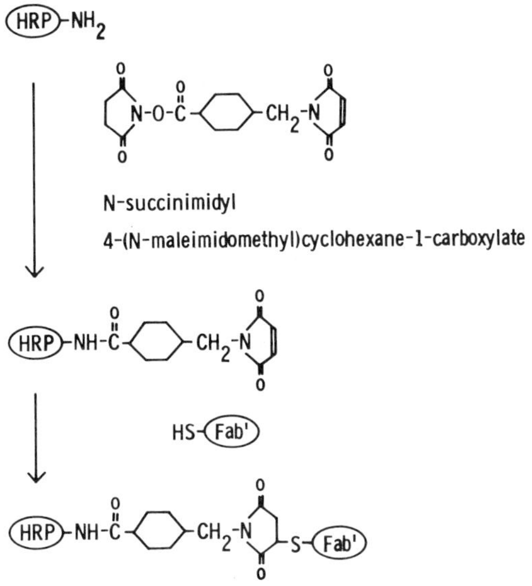

FIGURE 1. Preparation of Fab'-peroxidase conjugate by the maleimide method (I). HRP = horseradish peroxidase.

The SH-blocked Fab' (2.2 mg in 0.44 ml of 0.1 M sodium phosphate buffer, pH 6.5) was incubated with S-acetylmercaptosuccinic anhydride (0.21–0.84 mg, 25–100-fold molar excess, in 0.01 ml of N,N-dimethylformamide) at 25°C for 30 min with continuous stirring. Then, the reaction mixture was mixed with 0.1 ml of 0.1 M Tris-HCl buffer, pH 7.0, 0.01 ml of 0.1 M EDTA, pH 7.0, and 0.1 ml of 1 M hydroxylamine, pH 7.0, incubated at 30°C for 4 min and subjected to gel filtration with a Sephadex G-25 column (1.0 × 45 cm) using 0.1 M sodium phosphate buffer, pH 6.0 containing 5 mM EDTA. The average number of thiol groups introduced per Fab' molecule was 0.99–1.8. The SH-blocked and thus mercaptosuccinylated Fab' was conjugated with the maleimide-peroxidase prepared in the same way as in the maleimide method (I).

Pyridyl Disulfide Method (I)

In the pyridyl disulfide method (I), horseradish peroxidase was treated with
N-succinimidyl 3-(2-pyridyldithio)propionate to introduce pyridyl disulfide groups,
and the pyridyl disulfide-peroxidase obtained was allowed to react with thiol groups in
the hinge of Fab' (FIGURE 2).

Peroxidase (2 mg in 0.3 ml of 0.1 M sodium phosphate buffer, pH 7.5) was
incubated with N-succinimidyl 3-(2-pyridyldithio) propionate[2] (0.78 mg, 50-fold
molar excess, in 0.06 ml of ethanol) at 23–25°C for 30 min. The reaction mixture was
subjected to gel filtration with a Sephadex G-25 column (1 × 45 cm) using 0.1 M
sodium phosphate buffer, pH 6.0. The average number of pyridyl disulfide groups
introduced per peroxidase molecule was 2.5–2.7.

The pyridyl disulfide-peroxidase (about 1.8 mg in 0.22 ml of 0.1 M sodium
phosphate buffer, pH 6.0) was incubated with Fab' (about 2 mg in 0.22 ml of the same
buffer containing 5 mM EDTA) at 30°C for 2.5 hr. The final concentrations of the
pyridyl disulfide-peroxidase and Fab' in the reaction mixture were about 0.1 mM.

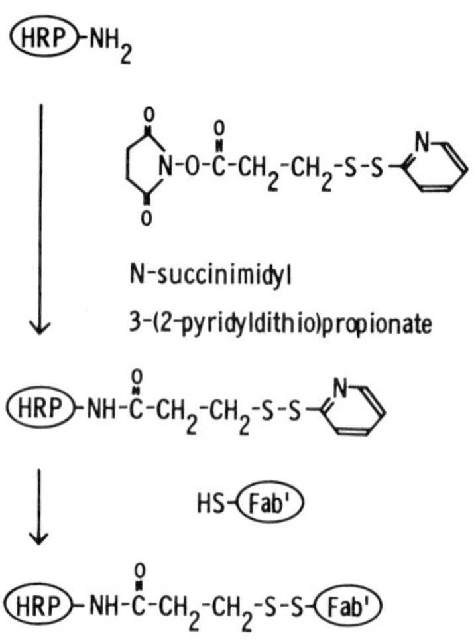

FIGURE 2. Preparation of Fab'-peroxidase conjugate by the pyridyl disulfide method (I). HRP = horseradish peroxidase.

Pyridyl Disulfide Method (II)

In the pyridyl disulfide method (II), Fab' was treated with 4,4'-dithiodipyridine to
introduce pyridyl disulfide groups in the hinge of Fab', and the pyridyl disulfide-Fab'
obtained was allowed to react with thiol groups introduced into peroxidase by
treatment with S-acetylmercaptosuccinic anhydride.

F(ab')₂ (2.2 mg in 0.09 ml of 0.1 M sodium phosphate buffer, pH 6.0) was
incubated with 0.01 ml of 0.1 M 2-mercaptoethylamine, pH 6.0 plus 5 mM EDTA at
37°C for 1.5 hr. After incubation, the reaction mixture was mixed with 0.35 ml of 0.1
M sodium phosphate buffer, pH 6.0 and 0.05 ml of 0.4 M 4,4'-dithiodipyridine in
N,N-dimethylformamide, further incubated at 30°C for 10 min and subjected to gel
filtration with Sephadex G-25 using 0.1 M sodium phosphate buffer, pH 6.0. The
average number of pyridyl disulfide groups introduced per Fab' molecule was 0.84.

The mercaptosuccinylated peroxidase was prepared in the same way as in the preparation of the mercaptosuccinylated Fab′ except that peroxidase was treated with 200-fold molar excess of *S*-acetylmercaptosuccinic anhydride at 30°C for 30 min. The average number of thiol groups introduced per peroxidase molecule was 2.2.

The conjugation was performed in the same way as in the pyridyl disulfide method (I) except that, after the incubation for conjugation, excess of thiol groups of the mercaptosuccinylated peroxidase was blocked by incubating with 1.6 mM *N*-ethylmaleimide at 30°C for 10 min.

Pyridyl Disulfide Method (III)

In the pyridyl disulfide method (III), pyridyl disulfide groups were introduced into peroxidase in the same way as in the pyridyl disulfide method (I), and the pyridyl disulfide-peroxidase obtained was allowed to react with thiol groups that were introduced into the SH-blocked Fab′ in the same way as in the maleimide method (III). The concentrations of the pyridyl disulfide-peroxidase and the SH-blocked, mercaptosuccinylated Fab′ for conjugation were 0.05 mM.

Glutaraldehyde and Periodate Methods

In the glutaraldehyde method, the SH-blocked Fab′ and peroxidase were conjugated by a two-step method as described by Boorsma and Streefkerk.[1] The periodate method was performed as described by Wilson and Nakane[14] except for the following conditions. Peroxidase was oxidized by sodium metaperiodate for 10 min, and the oxidation was stopped by adding ethylene glycol followed by gel filtration with Sephadex G-25.[1]

Gel Filtration of Conjugates

The reaction mixtures for conjugation were subjected to gel filtration with a Ultrogel AcA 44 column (1.5 × 45 cm) using 0.1 M sodium phosphate buffer, pH 6.5. Absorbance at 280 and 403 nm and peroxidase activity of each fraction were measured as described previously.[7]

Conjugates Used for Various Tests

Conjugates used for various tests were taken from fractions 37–39 (FIGURE 3), unless otherwise specified.

Test for the Purity of Conjugates

The proportion of unconjugated peroxidase in the peak fractions of conjugates was assessed by passing samples through a goat (anti-rabbit IgG) IgG-Sepharose 4B column and was expressed in percentage by comparing peroxidase activity in the effluent with that applied.[7] When passed through a normal goat IgG-column as a control, 95–96% of peroxidase activity applied was found in the effluent.

To test the presence of unconjugated F(ab′)₂ in the peak fraction of conjugates, the preparation of fluorescein-labeled Fab′-peroxidase conjugate (45 μg) in 0.1 ml of 0.1

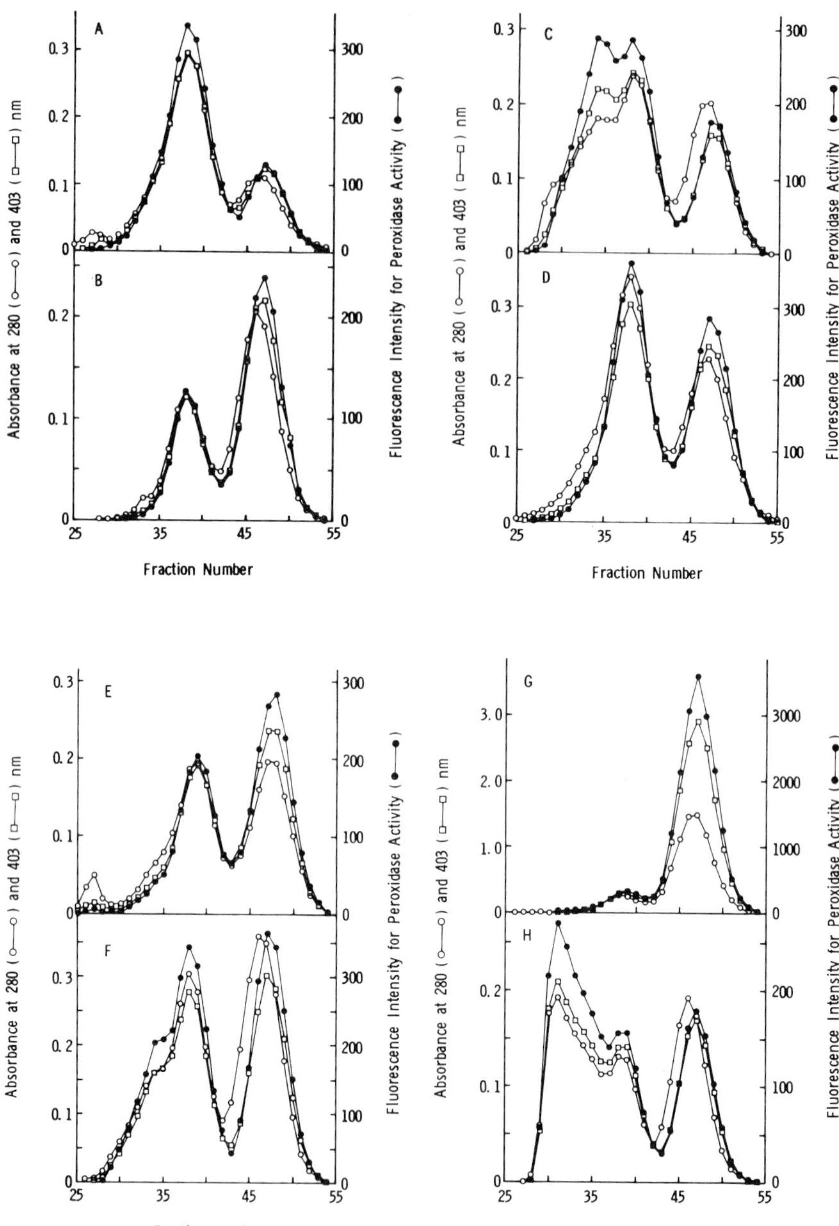

FIGURE 3. Elution profiles from a Ultrogel AcA 44 column of rabbit Fab′-peroxidase conjugates prepared by various methods. Methods tested were: the maleimide methods (I) (A), (II) (B), and (III) (C), pyridyl disulfide methods (I) (D), (II) (E), and (III) (F), glutaraldehyde method (G), and periodate method (H). The concentrations of Fab′ and peroxidase incubated for conjugation were 0.05 mM in the maleimide (I), (II), and (III), pyridyl disulfide (III), and periodate methods, 0.1 mM in the pyridyl disulfide methods (I) and (II), 0.025 mM (Fab′), and 0.12 mM (peroxidase) in the glutaraldehyde method. The fraction volume was 1.0 ml. Peroxidase activity in each fraction was measured by 10-min assay using 0.01 ml samples diluted 2,500-fold with 0.1 M potassium phosphate buffer, pH 7.0 containing 0.1 M NaCl and 0.1% bovine serum albumin.

M sodium acetate buffer, pH 6.0 containing 1 M NaCl, 1 mM $MgCl_2$, 1 mM $CaCl_2$, 1 mM $MnCl_2$, and 0.2% bovine serum albumin was passed through a concanavalin A-Sepharose 4B column (3 × 40 mm) at a flow rate of 1 ml/hr using the same buffer, and fluorescence intensity (490 nm for excitation and 510 nm for emission analysis) in the effluent was compared with that applied. When fluorescein-labeled $F(ab')_2$ (23 μg) was passed through the same column as a control, 95% of fluorescence intensity applied was found in the effluent.

To assess the proportion of peroxidase associated with specific Fab', sheep anti-rabbit fibrinogen Fab'-peroxidase conjugate (200 ng) was passed through a fibrinogen-Sepharose 4B column (3.5 × 20 mm) at a flow rate of 1 ml/min using 0.01 M sodium phosphate buffer, pH 7.0 containing 0.1 M NaCl and 0.1% bovine serum albumin, and peroxidase activity in the effluent was compared with that applied. Normal rabbit Fab'-peroxidase conjugate was applied as a control.

Assessment of the Molecular Weight of Fab'-peroxidase Conjugate

One ml of 0.1 M sodium phosphate buffer, pH 6.5 containing 1 mg of peroxidase, 2 mg of fluorescein-labeled SH-blocked Fab', 0.25 mg of Fab'-peroxidase conjugate, 7 mg of bovine serum albumin, 2 mg of fluorescein-labeled $F(ab')_2$, and 1 mg of alkaline phosphatase was subjected to gel filtration using a Ultrogel AcA 44 column (1.5 × 45 cm) equilibrated with the same buffer. The molecular weights of these proteins were taken to be 40,000,[7,16] 46,000,[15] 66,200,[11] 92,000, and 100,000,[3] respectively. The volume of each fraction was 0.86 ml. Absorbance at 280 and 403 nm, fluorescence intensity (490 nm for excitation and 510 nm for emission analysis), peroxidase activity, and alkaline phosphatase activity were measured. Fab' used was obtained by reducing $F(ab')_2$ at pH 6 and treating with N-ethylmaleimide.

Test for the Stability of Cross-link

Fluorescein-labeled Fab'-peroxidase conjugate stored under an appropriate condition was subjected to gel filtration with Ultrogel AcA 44 to detect release of Fab' from the conjugate by measuring fluorescence intensity (490 nm for excitation and 510 nm for emission analysis).[7]

Affinity Purification of Anti-fibrinogen Fab'-peroxidase Conjugate

Anti-fibrinogen Fab'-peroxidase conjugate prepared by the maleimide method (I) (1.3 mg in 3.2 ml of 0.1 M sodium phosphate buffer, pH 6.5) was adsorbed to a fibrinogen-Sepharose 4B column (5 × 35 mm) and eluted by 0.05 M glycine-HCl buffer, pH 2.9 immediately followed by neutralization with 0.5 M Tris-HCl buffer, pH 8.0.

Sandwich Enzyme Immunoassay

Sandwich enzyme immunoassay was performed as described previously.[7] In brief, a polystyrene ball coated with antibody IgG (0.1 mg/ml) was incubated with antigen at 37°C for 6 hr and at 4°C for 15 hr, washed and incubated with Fab'-peroxidase conjugate at 20°C or 37°C for 6 hr. The specific binding of conjugates was calculated by subtracting peroxidase activity nonspecifically bound in the absence of antigens from that specifically bound in the presence of antigens.

Immunohistochemical Staining

For the localization of fibrinogen and/or fibrin, paraformaldehyde-fixed (3% paraformaldehyde in 0.05 M sodium phosphate buffer, pH 7.4) frozen sections (5 μm thick) were prepared from the rabbit aorta with mural thrombi. The sections prepared were treated with 0.3% H_2O_2 in absolute methanol for 30 min to block endogenous peroxidase activity and washed three times for 5 min each. After washing, the sections were treated for 30 min with normal rabbit serum diluted tenfold with 0.05 M sodium phosphate buffer, pH 7.4, containing 0.15 M NaCl and, after removing the diluted serum, incubated with Fab'-peroxidase conjugates (5 μg/ml in 0.1 M potassium phosphate buffer, pH 7.2, containing 0.1% bovine serum albumin) prepared by various methods in a moist chamber for 1 hr. After washing for 15 min, the sections were stained with 0.05% 3,3'-diaminobenzidine tetrahydrochloride plus 0.01% H_2O_2 in 0.05 M Tris-HCl buffer, pH 7.4, for 5 min. After washing, the sections were counterstained with methyl green, dehydrated, and mounted in Permount. All the washings were performed in the cold with 0.05 M sodium phosphate buffer, pH 7.4 containing 0.15 M NaCl, and processes not specified were performed at room temperature throughout.

For hCG localization, formalin-fixed (10% neutral formalin saline) paraffin sections (4 μm thick) were prepared from the human placenta. The paraffin sections prepared were dewaxed and digested at 37°C for 2 hr with 0.1% trypsin solution containing 0.01% $CaCl_2$, pH of which was adjusted to 7.8 with 0.05 N NaOH. The subsequent processes were performed as described above.

RESULTS

Purity of Conjugates

In the peak fractions of rabbit Fab'-peroxidase conjugates prepared by the maleimide methods (I–III) and pyridyl disulfide methods (I–III), only 1–5% of peroxidase activity was unconjugated. In the peak fractions of the glutaraldehyde and periodate conjugates, larger proportions (19–40% and 6–16%, respectively) were unconjugated, suggesting the presence of dimer of peroxidase (TABLE 1).

In the peak fraction of fluorescein-labeled rabbit Fab'-peroxidase conjugate prepared by the maleimide method (I), 9.5% of fluorescence intensity was not adsorbed to concanavalin A-Sepharose 4B, suggesting the presence of unconjugated F(ab')$_2$ formed by reoxidation of Fab'. The conjugate adsorbed to concanavalin A-Sepharose 4B could be largely (73%) eluted by 0.2 M α-methyl-D-mannoside and was almost completely (98%) adsorbed to anti-rabbit IgG-Sepharose 4B, indicating that the conjugate could be separated from unconjugated F(ab')$_2$, if any.

Size of Conjugate Molecules

The molecular weight of rabbit Fab'-peroxidase conjugate in the peak fraction prepared by the maleimide method (I) was assessed to be 80,000–90,000 by gel filtration, indicating that one molecule of Fab' and peroxidase was cross-linked to form a monomeric conjugate. This was consistent with the fact that the molar ratio of Fab' and peroxidase in the conjugate was calculated to be approximately one.[7,16]

Elution profiles indicated that the conjugates prepared by the maleimide (II), pyridyl disulfide (I) and (II), and glutaraldehyde methods were largely monomeric (FIGURE 3A, B, D, E, and G). However, the conjugates prepared by the maleimide (III) and pyridyl disulfide (III) methods were somewhat widely distributed (FIGURE 3C and F) and heterogeneity increased with increasing number of thiol groups introduced. The periodate conjugates were polymerized to various extents (FIGURE 3H), and highly polymerized conjugates were produced in large proportions by a prolonged oxidation with periodate and eluted in the same fractions as blue dextran.

TABLE 1. Purity of Conjugates, Recovery of Peroxidase Incubated for Conjugation, and Sandwich Enzyme Immunoassay of hCG with Rabbit Anti-hCG Fab'-peroxidase Conjugates

Conjugation Method[a]	Proportion of Unconjugated Peroxidase[b] (%)	Recovery of Peroxidase[c] (%)	Fluorescence Intensity for Peroxidase Activity Bound[d] hCG Added (mU/tube)			
			0	0.1	1	10
Maleimide method						
(I)	1–2	65–74	6	15	66	156
(II)	2	35–49	6	15	53	141
(III)	3	52–74	33	38	65	87
Pyridyl disulfide method						
(I)	2	54–58	6	14	57	152
(II)	3	40–49	6	14	52	145
(III)	5	58	24	28	57	94
Glutaraldehyde method	19–40	4–9	49	50	64	88
Periodate method		58–66				
monomeric	6–16	–	44	47	77	110
polymeric	1	–	98	98	132	188

[a]See FIGURES 1 and 2 and the text for the reaction sequences used for conjugation

[b]The conjugates were applied to goat (anti-rabbit IgG) IgG-Sepharose 4B column, and peroxidase activity in the effluent was expressed in percentages of the enzyme activity applied.

[c]The recovery of peroxidase in the conjugate was calculated from the enzyme activity in the elution profile (FIGURE 3A–H).

[d]The sandwich enzyme immunoassay for hCG was performed by incubation with anti-hCG Fab'-peroxidase conjugates (200 ng/tube) at 37°C for 6 hr, and peroxidase activity bound was determined by a 20-min assay.

Recovery of Peroxidase and Fab' in the Conjugate

Loss of peroxidase and Fab' by gel filtration and concentration before conjugation was less than 10%. The recovery in the conjugate of peroxidase incubated for the conjugation with rabbit Fab' was calculated from absorbance at 403 nm in the elution profile. The recovery of peroxidase by the maleimide method (I) increased up to 60% during the first 5 hr incubation at 4°C for conjugation and reached the maximum (65–74%) by 15–20 hr incubation at 4°C or 1 hr incubation at 30°C (TABLE 1). The recovery of Fab' must have been similar, since the maleimide-peroxidase and Fab' were incubated with a molar ratio of one and converted largely to monomeric conjugate as described above. In the maleimide (II) and (III), pyridyl disulfide (I), (II), and (III), glutaraldehyde and periodate methods, the recoveries of peroxidase were 35–49, 52–74, 54–58, 40–49, 58, 4–9, and 58–66%, respectively (TABLE 1). In the maleimide method (II), the recovery was low, since maleimide groups of the compound used were not stable at neutral pH.[16] The recovery in the maleimide (III) and pyridyl disulfide (III) methods increased with increasing number of thiol groups introduced into Fab'. In the pyridyl disulfide method (I), the recovery reached the maximum during 2.5 hr incubation at 30°C at pH 6.0 and the incubation at 30°C at pH 7.5 resulted in similar recoveries. The incubation at 4°C at pH 6.0 for 20 hr or the use of 50 μM of the pyridyl disulfide peroxidase and Fab' for conjugation resulted in lower recoveries. In the periodate method, the recovery increased up to 90% by a prolonged oxidation with periodate.

Peroxidase Activity in Conjugates

No significant change was observed in peroxidase activity on the basis of absorbance at 403 nm when peroxidase and rabbit Fab' were conjugated by the maleimide methods and pyridyl disulfide methods. However, it tended to be slightly (10% or less) lowered in the glutaraldehyde and periodate methods, and a prolonged oxidation of peroxidase by periodate caused 20–70% loss.

Stability of Cross-link

After rabbit Fab'-peroxidase conjugate prepared by the maleimide method (I) was stored in 0.1 M sodium phosphate buffer, pH 6.5, containing 0.1% bovine serum albumin and 0.002% thimerosal at 4°C for 9 months, no release of Fab' was observed.

Conjugates of Fab' from Sheep, Goat, and Guinea Pig

When peroxidase was conjugated with Fab' from sheep, goat, and guinea pig by the maleimide method (I), no difference was observed in the elution profiles from a Ultrogel AcA 44 column, the yields of conjugates, and peroxidase activity in conjugates as compared with rabbit Fab'-peroxidase conjugate.

TABLE 2. Sandwich Enzyme Immunoassay of Rabbit Fibrinogen with Sheep Anti-rabbit Fibrinogen Fab'-peroxidase Conjugates

| Conjugation Method | Fluorescence Intensity for Peroxidase Activity Bound | | | | | |
| | Rabbit Fibrinogen Added (ng/tube) | | | | | |
	0	1	3	10	100	1000
Maleimide method (I)	1.6	12	41	83	434	823
	(3.0)	(8.6)	(17)	(50)	(189)	(423)
Periodate method						
monomeric	4.7	9.3	20	64	258	464
	(23)	(33)	(37)	(61)	(145)	(287)
polymeric	75	79	109	176	604	1165
	(94)	(106)	(117)	(183)	(520)	(763)

The sandwich enzyme immunoassay for rabbit fibrinogen was performed by incubation with anti-fibrinogen Fab'-peroxidase conjugates (50 ng/tube) at 20°C for 6 hr, and peroxidase activity bound was determined by a 20-min assay. Values in parentheses were obtained by incubation with the conjugates at 30°C for 6 hr.

TABLE 3. Sandwich Enzyme Immunoassay of Human α-Fetoprotein with Goat Anti-human α-Fetoprotein Fab'-peroxidase Conjugates

Conjugation Method	Fluorescence Intensity for Peroxidase Activity Bound				
	α-Fetoprotein Added (pg/tube)				
	0	7	70	700	7000
Maleimide method (I)	1.2	2.3	11.3	136	384
	(2.3)	(3.6)	(13)	(103)	(231)
Periodate method					
monomeric	2.2	2.8	6.7	53	156
	(16)	(14)	(18)	(57)	(117)
polymeric	5.5	5.4	12.8	72	222
	(30)	(31)	(38)	(84)	(186)

The sandwich enzyme immunoassay for α-fetoprotein was performed by incubation with anti-α-fetoprotein Fab'-peroxidase conjugates (100 ng/tube) at 20°C for 6 hr, and peroxidase activity bound was determined by a 20-min assay. Values in parentheses were obtained by incubating with the conjugate at 37°C for 6 hr.

Enzyme Immunoassay

Rabbit anti-hCG Fab'-peroxidase conjugates prepared by various methods were tested in sandwich enzyme immunoassay for hCG (TABLE 1). Nonspecific bindings of the conjugates prepared by the maleimide methods (I) and (II) and pyridyl disulfide methods (I) and (II) were the lowest (0.0040–0.0046% of peroxidase activity added) and specific bindings of these conjugates were the highest. Nonspecific bindings of other conjugates were 4–16-fold higher, and specific bindings of other conjugates were 33–74% lower.

Sheep anti-rabbit fibrinogen and goat anti-human α-fetoprotein Fab'-peroxidase conjugates were also tested and results obtained were similar (TABLES 2 and 3). The maleimide conjugates (I) gave lower nonspecific bindings and higher specific bindings than the periodate conjugates, except that specific binding of highly polymerized periodate anti-fibrinogen conjugate was higher than that of the maleimide conjugate. Incubation with conjugates at 37°C resulted in higher nonspecific and lower specific bindings than that at 20°C.

These results indicated that sandwich enzyme immunoassay could be more easily improved in sensitivity by incubating at 20°C with conjugates prepared using thiol groups in the hinge.

Immunohistochemical Staining

Staining efficiencies of anti-rabbit fibrinogen and anti-hCG Fab'-peroxidase conjugates prepared by various methods were compared using sections from rabbit aorta with mural thrombi and human placenta.

Staining efficiency with the maleimide conjugate (I) was the highest: being similar to or slightly higher than that with the pyridyl disulfide conjugate (I) and much higher than that with the maleimide (III), pyridyl disulfide (III), and periodate conjugates (FIGURES 4 and 5). Stainings with 5 μg/ml of the maleimide (I) and pyridyl disulfide (I) conjugates were similar to those with 40 μg/ml of the monomeric periodate

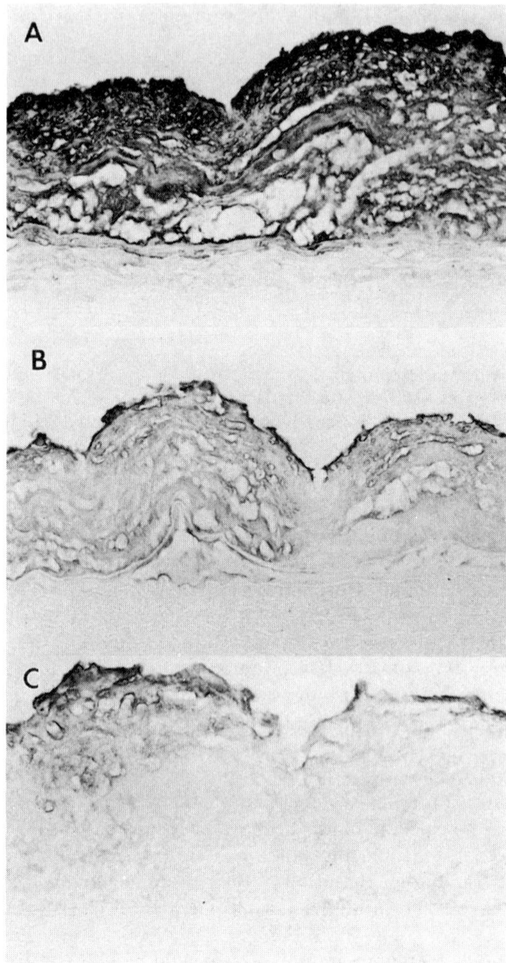

FIGURE 4. Intracytoplasmic localization of hCG in syncytiotrophoblasts of human placenta. Staining was performed using 5 μg/ml of the maleimide conjugate (I) (A), monomeric periodate conjugate (B), and highly polymerized periodate conjugates (C). ×115.

conjugates. The monomeric periodate conjugate was superior to the highly polymerized periodate conjugate. Thus, the order of staining efficiency was maleimide (I) $\geq$ pyridyl disulfide (I) $\gg$ maleimide (III) = pyridyl disulfide (III) $\geq$ monomeric periodate > polymeric periodate > glutaraldehyde.

Background, that is, nonspecific staining, was examined using normal rabbit Fab'-peroxidase conjugates. When sections were incubated with 0.05 mg/ml of various conjugates for 1 hr and with the substrate for 5 min, little background appeared, but the incubation with the substrate for 20 min resulted in different background stainings: little stained with the maleimide conjugate (I) but distinctly stained with the monomeric periodate conjugate and more intensively with the polymeric one.

These results indicated that the conjugates prepared using thiol groups in the hinge were superior in staining efficiency to those prepared using amino groups of Fab'.

Usefulness of Affinity-purified Conjugate

Sheep anti-rabbit fibrinogen Fab'-peroxidase conjugate prepared by the maleimide method (I) was affinity-purified by eluting from a fibrinogen-Sepharose 4B column at pH 2.9 without loss of peroxidase activity, and 95% of the affinity-purified conjugate was adsorbed to fibrinogen-Sepharose 4B, while 36% was adsorbed before affinity-purification and 10% of normal rabbit Fab'-peroxidase conjugate was adsorbed. Staining of fibrinogen and/or fibrin in the sections from rabbit aorta with mural

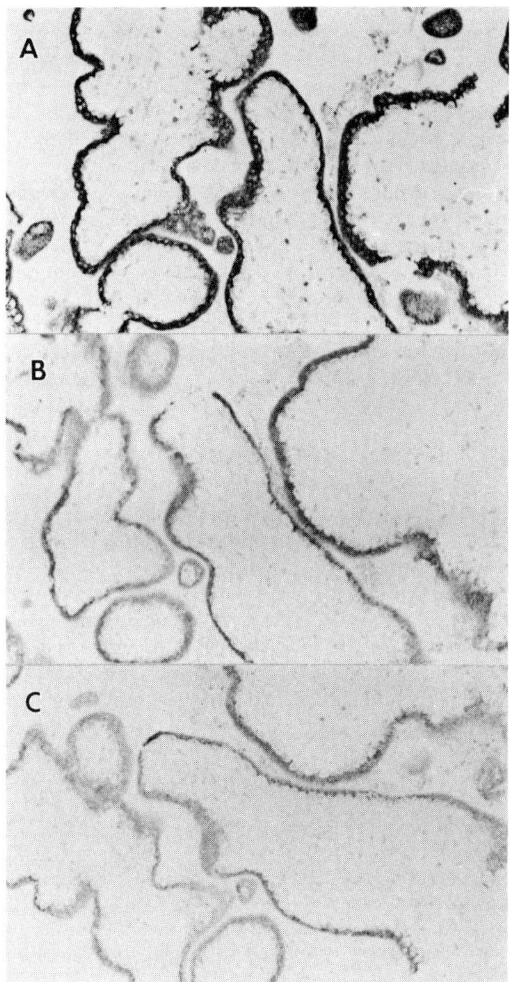

FIGURE 5. Staining pattern for fibrinogen and/or fibrin in the rabbit aorta with mural thrombi. Staining was performed using 5 μg/ml of the maleimide conjugate (I) (A), monomeric periodate conjugate (B), and highly polymerized periodate conjugate (C). ×230.

thrombi with 5 μg/ml of the affinity-purified conjugate was more efficient than with the same concentration before affinity-purification.

DISCUSSION

The conjugates prepared by the maleimide (I) and (II) and pyridyl disulfide (I) and (II) methods were superior in performing both sandwich enzyme immunoassay and immunohistochemical staining to those prepared by the glutaraldehyde and periodate methods. The reason for this may be at least in part based on the fact that thiol groups in the hinge of Fab' molecules, which was remote from the antigen-binding site, were used for the conjugation in the former methods, while amino groups of Fab' used for the conjugation in the latter methods, if not all, may be located in such positions that the antigen-binding activity was sterically impaired by the conjugation. This was supported by the finding that the conjugates prepared by the former methods were superior to those prepared by the maleimide (III) and pyridyl disulfide (III) methods in which amino groups were used for conjugation.

Rabbit Fab-peroxidase conjugates prepared by the glutaraldehyde and periodate methods were similar to the corresponding Fab' conjugates in their elution profiles and sandwich enzyme immunoassay.[7] F(ab')$_2$- and IgG-peroxidase conjugates prepared using amino groups gave less sensitive sandwich enzyme immunoassay than the maleimide conjugate (I).[7] Therefore, Fab-, F(ab')$_2$-, and IgG-peroxidase conjugates prepared using amino groups may also be less effective in immunohistochemical staining than Fab'-peroxidase conjugates prepared using thiol groups in the hinge, although it remains to be tested.

SUMMARY

Horseradish peroxidase and Fab' were conjugated by using thiol groups in the hinge of Fab'. Maleimide or pyridyl disulfide groups were introduced into peroxidase by treatment with *N*-succinimidyl 4-(*N*-maleimidomethyl)cyclohexane-1-carboxylate or *N*-succinimidyl 3-(2-pyridyldithio)propionate and were allowed to react with thiol groups in the hinge of Fab'. The conjugates were obtained in high yields with a minimal polymerization and without impairing the activities of peroxidase and antibodies, and were superior to those prepared using amino groups of Fab' by the glutaraldehyde and periodate methods in performing sandwich enzyme immunoassay and immunohistochemical staining. The conjugate yield was higher in the maleimide method than in the pyridyl disulfide method.

REFERENCES

1. BOORSMA, D. M. & J. G. STREEFKERK. 1979. Periodate or glutaraldehyde for preparing peroxidase conjugates. J. Immunol. Methods **30:** 245–255.
2. CARLSSON, J., H. DREVIN & R. AXEN. 1978. Protein thiolation and reversible protein-protein conjugation. N-Succinimidyl 3-(2-pyridyldithio) propionate, a new heterobifunctional reagent. Biochem. J. **173:** 723–737.
3. ENGSTRÖM, L. 1961. Studies on calf-intestinal alkaline phosphatase. I. Chromatographic purification, microheterogeneity and some other properties of the purified enzyme. Biochim. Biophys. Acta **52:** 36–48.
4. FERNLEY, H. N. & P. G. WALKER. 1967. Studies on alkaline phosphatase. Inhibition by phosphate derivatives and the substrate specificity. Biochem. J. **104:** 1011–1018.

5. GRASSETTI, D. R. & J. F. MURRAY, JR. 1967. Determination of sulfhydryl groups with 2,2'- or 4,4'-dithiodipyridine. Arch. Biochem. Biophys. **119:** 41–49.

6. HAMAGUCHI, Y., S. YOSHITAKE, E. ISHIKAWA, Y. ENDO & S. OHTAKI. 1979. Improved procedure for the conjugation of rabbit IgG and Fab' antibodies with β-D-galactosidase from *Escherichia coli* using N,N'-o-phenylenedimaleimide. J. Biochem. **85:** 1289–1300.

7. IMAGAWA, M., S. YOSHITAKE, Y. HAMAGUCHI, E. ISHIKAWA, Y. NIITSU, I. URUSHIZAKI, R. KANAZAWA, S. TACHIBANA, N. NAKAZAWA & H. OGAWA. 1982. Characteristics and evaluation of antibody-horseradish peroxidase conjugates prepared by using a maleimide compound, glutaraldehyde and periodate. J. Appl. Biochem. **4:** 41–57.

8. IMAGAWA, M., S. YOSHITAKE, E. ISHIKAWA, Y. NIITSU, I. URUSHIZAKI, R. KANAZAWA, S. TACHIBANA, N. NAKAZAWA & H. OGAWA. 1981. Comparison of highly sensitive sandwich enzyme immunoassay and radioimmunoassay for human ferritin. Anal. Lett. **14:** 1679–1692.

9. ISHIKAWA, E. & K. KATO. 1978. Ultrasensitive enzyme immunoassay. Scand. J. Immunol. **8** (Suppl. 7): 43–55.

10. ISHIKAWA, E., Y. YAMADA, Y. HAMAGUCHI, S. YOSHITAKE, K. SHIOMI, T. OTA, Y. YAMAMOTO & K. TANAKA. 1978. Enzyme-labelling with maleimides and its application to the immunoassay of peptide hormones. *In* Enzyme Labelled Immunoassay of Hormones and Drugs. S. B. Pal, Ed.: 43–57. Walter de Gruyer & Co. Berlin.

11. PETERS, T., JR. 1975. *In* The Plasma Proteins. F. W. Putnam, Ed. **1:** 133–181. Academic Press. New York.

12. REGOECZI, E. 1974. Fibrinogen. *In* Structure and Functions of Plasma Proteins. A. C. Allison, Ed. **1:** 133–167. Plenum Press. New York.

13. STUCHBURY, T., M. SHIPTON, R. NORRIS, J. P. G. MALTHOUSE & K. BROCKLEHURST. 1975. A reporter group delivery system with both absolute and selective specificity for thiol groups and an improved fluorescent probe containing the 7-nitrobenzo-2-oxa-1,3-diazole moiety. Biochem. J. **151:** 417–432.

14. WILSON, M. B. & P. K. NAKANE. 1978. Recent developments in the periodate method of conjugating horseradish peroxidase to antibodies. *In* Immunofluorescence and Related Staining Techniques. W. Knapp, K. Holubar & G. Wick, Eds.: 215–224. Elsevier/North-Holland Biomedical Press. Amsterdam.

15. YOSHITAKE, S., Y. HAMAGUCHI & E. ISHIKAWA. 1979. Efficient conjugation of rabbit Fab' with β-D-galactosidase from *Escherichia coli*. Scand. J. Immunol. **10:** 81–86.

16. YOSHITAKE, S., M. IMAGAWA & E. ISHIKAWA. 1982. Efficient preparation of rabbit Fab'-horseradish peroxidase conjugates using maleimide compounds and its use for enzyme immunoassay. Anal. Lett. **15** (B2): 147–160.

17. YOSHITAKE, S., Y. YAMADA, E. ISHIKAWA & R. MASSEYEFF. 1979. Conjugation of glucose oxidase from *Aspergillus niger* and rabbit antibodies using N-hydroxysuccinimide ester of N-(4-carboxycyclohexylmethyl)-maleimide. Eur. J. Biochem. **101:** 395–399.

Neurotypy: The Heterogeneity of Brain Proteins[a]

NANCY H. STERNBERGER AND
LUDWIG A. STERNBERGER

Center for Brain Research
University of Rochester School of Medicine
Rochester, New York 14642

Our work has been stimulated by the concept of Ernst and Berta Scharrer[15,17] that certain principles (now known as peptides) are fundamental to neuronal communication, be this via circulation, neuronal contacts, or, as recently emphasized by Berta Scharrer, "private" and "semiprivate" relay systems.[16] The importance of the Scharrer concept is confirmed by the excitement that neuropeptide research has generated throughout the world. Like all the rare concepts that advance science by leaps and bounds, the Scharrer concept is, indeed, quite simple. However, the evolution of this concept by the Scharrers required exacting histologic techniques and visionary interpretation of the histologic staining observations in mammalian brain and their correlation with neurosecretory organs in widely disparate species, such as insects. At the present time, nearly 40 peptides are characterized and they all share central and peripheral distribution. Most of these peptides have been discovered because they possess strong physiologic activity.

Our working hypothesis, although unproven, was this: The Scharrer concept is fundamental. The number of known peptides has been limited by the chance of their discovery. The complexity of the nervous system is high. If peptides are a form of expression of this complexity, forty peptides would provide an insufficient biochemical basis to explain it. Therefore, there may exist many other neurosecretory peptides whose quantities are insufficient for biochemical detection or whose physiological activity is not dramatic enough to be detected by the pharmacologist. We felt that current biochemical techniques may not be easily applicable to isolation of such unknown principles, present perhaps only in a few neurons within the whole brain and possessing no known physiologic activity to guide their isolation. Nevertheless, such substances may be antigenic and antibodies could be used to detect them. Unfortunately, these substances are not available in pure form. If impure material were used for immunization, it would be necessary to remove by absorption all antibodies against impurities, leaving behind only the minority of antibodies specific to such unknown new substance. Again, this approach is not feasible. However, the advent of monoclonal antibodies may permit detection of such postulated substance. Each monoclonal antibody is specific only to a single antigenic determinant. The approach of searching for a monoclonal antibody specific to a rare, unknown principle within whole brain requires a selection procedure for this principle that eliminates all antibodies against other substances. Most procedures for clone selection in hybridoma technology require, of course, the purified antigen which we do not have and for which an antibody

[a]Supported by the National Institutes of Health grants NS 17652 and HD 12932, a National Multiple Sclerosis Society grant, and the National Science Foundation grant BNS 82-05643.

is sought. However, if immunocytochemistry were substituted for other selection procedures, purified antigen would not be needed. We set out, therefore, to immunize with hypothalamus and select antibody-producing clones by their immunocytochemical staining distribution in sagittal sections of brain. We assumed that most antibodies will react with most neuronal or most glial tissue, and we hoped that perhaps a small proportion of all antibodies, perhaps one in a thousand or one in a hundred, would be specific to only a single group of neurons or a single neuron. We expected to discard all generally reactive antibodies and to concentrate on those reactive only with these single cells. We made this assumption because we felt that although neurons may possess substances specific to their own specialized function, most antigens in brain would be shared among several cells, and, therefore, the chance of finding that one antigen which is specific to only a single cell would be rather low.

We have produced, thus far, 135 antibody-producing hybridomas that react immunocytochemically with brain. Analysis of data with these antibodies has shown that the above assumption probably is incorrect. We found no antibody that reacted with a single neuron or a single defineable group of neurons. Therefore, it appears unlikely to us, at present, that a specific neuronal function is necessarily the result of the expression of a single specific neurosecretory product.

Instead of the expected findings, our observations suggest that neurons, with the assumption of their specificity, have a choice of selecting various combinations of a probably large number of neurosecretory principles. While the large number of such neurosecretory principles may be limited, and may indeed not vary much from species to species (we have no evidence for this latter assumption), the manner of selection of a limited number of these principles in each neuron, that is, the permutation of principles expressed, may be tremendous. What we postulate, and hope to prove or disprove, is that neuronal differentiation and experience is a selective process in which a restricted number of peptides is expressed while other peptides, with which a cell may be genomically endowed, are suppressed. The manner of permutation of these peptides permits variability of individual neuronal function. The total number of basic peptides does not have to be different between mammalian cerebral cortices and insect corpora allata, for instance. However, the ability to express permutations of these peptides is expected to be higher in individual cells of cerebral cortex.

This concept has two requirements. First, any neurosecretory peptide must be the product of a larger, precursor protein that in its breakdown can release a number of bioactive peptides. A putative precursor may have hormonal reactivity and immunologic cross-reactivity with the known, synthesizable peptide. Second, if different cells express different permutations of peptides, they must have different prohormones that may be coded by different mRNAs or that may have been changed posttranslationally. Such different prohormones do not have to be fundamentally different in their total amino acid structure. Changes could be quite subtle. Large sequences may be shared among different prohormones (e.g., the wide distribution of enkephalin compared to the distribution of enkephalin-containing beta-endorphin).[1] Conceivably, the sequence of all neuroendocrine principles may be encoded in a single chromosome, and different prohormones may be obtained by splices in different intron regions, thus conserving the structure of individual peptides but changing the combination with other peptides with which they may be expressed.[5,8] We also postulated that the undifferentiated cell can express many peptides at random and that differentiation is the selective expression of few. Alternatively, if changes are posttranslational, the undifferentiated cell would express combinations of highly bioactive amino acid sequences, while the differentiated cell would express more physiologically and immunologically inert sequences.

We are using two approaches to illustrate or disprove these concepts.

EXISTENCE OF PROHORMONES FOR NEUROPEPTIDES

Prohormones are accepted for the ACTH-lipotropin and for the oxytocin-mesotocin-vasopressin series of peptides.[4,10,11] When immunocytochemical staining is done with antisera to peptides, one says that one localizes the peptide-like substance, not knowing whether one localizes the peptide itself or a similar substance such as an analogue or prohormone. We have recently found that in the luteinizing hormone–releasing hormone (LHRH) system, and perhaps also in the vasopressin system (in collaboration with Mona Castel), we may not localize the expected peptide, but a precursor or analogue.[21] This was based on the finding that pretreatment of sections with high concentrations of LHRH (or vasopressin, respectively), followed by extensive washing, abolished the reactivity with antipeptide antibodies. We felt that the immunoreactive material that actually was localized was held in tissue by a low affinity carrier and could be exchanged for it by treatment with the peptide. The peptide itself, possessing still lower affinity for the carrier, was lost during subsequent immunocytochemical processing. On the basis of this, we felt that treatment of hypothalami with LHRH should elute specifically the analogue or precursor. By high performance liquid chromatography, such material indeed was isolated.

HETEROGENEITY OF NEURON-SPECIFIC MACROMOLECULAR ANTIGENS

Among 135 monoclonal antibodies selected after immunization with hypothalamus, 88 were largely, though not entirely, brain specific.[22] The immunocytochemical

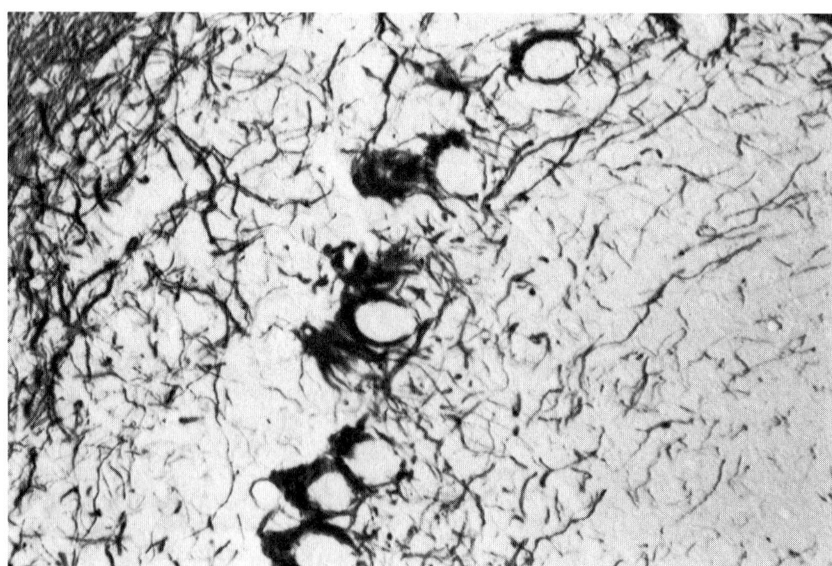

FIGURE 1. Rat cerebellum, sagittal paraffin section, stained with monoclonal antibody 06-17 from the antineurofilamentous group. Nikon-Nomarski optics were used to visualize background. Basket cell fibers stained, Purkinje cells unstained.

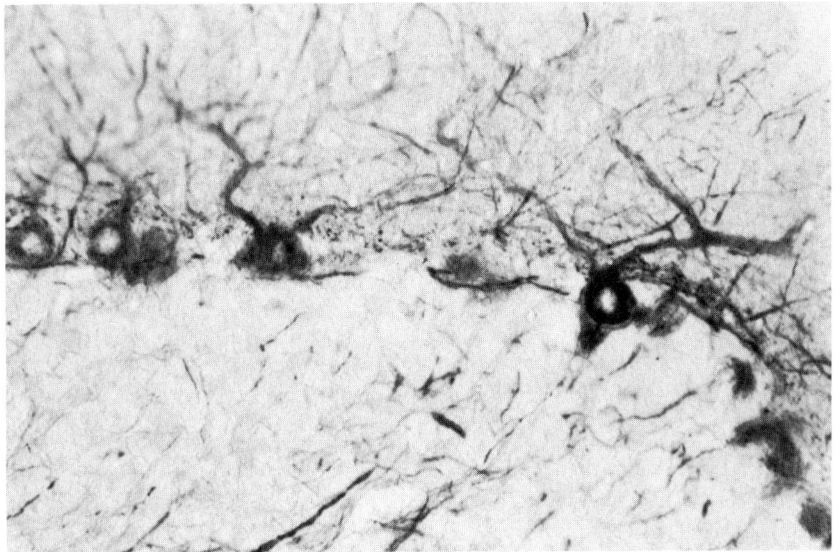

FIGURE 2. Rat cerebellum, sagittal paraffin section, stained with monoclonal antibody 02-135 from the antiperikaryonal-neurofibrillar group. Purkinje cells stained, basket cell fibers unstained.

reactivities of these antibodies have been tabulated previously.[22] In the meantime, many of these clones have been carried as ascites tumors. Their specific immunocytochemical reactivities have been reexamined in the repeated assays necessary to monitor stability during subcloning and before and after ascites production, and the ascites fluids have been used in a number of investigations. With the exception of those clones that completely lost their antibody-producing capabilities, given clones revealed the same characteristic staining patterns at all times, irrespective of passage of the clone and without significant variation among rat brains of equal ages. Staining intensities, in general, were equal for strongly stained structures, whether ascites fluids were used at dilutions of 1:500 or 1:10,000. However, less strongly stained structures revealed reduced staining intensities, generally at dilutions of ascites beyond 1:2,000.

Among the 88, largely brain-specific antibodies, 44 reacted with nonneuronal elements and 37 reacted with neuronal elements. Many of the antibodies to nonneuronal elements reacted with basal lamina of brain capillaries or with pituitary cells. Each single hybridoma, reactive with any one of these components, did reveal an identical reaction pattern. The antineuronal antibodies gave three different overall profiles and were classified, on the basis of immunocytochemical appearance and further analytical studies, into antineurofilamentous (FIGURE 1), antiperikaryonal-neurofilamentous (FIGURE 2), and antisynapse-associated groups (FIGURE 3). However, the antibodies in each of these antineuronal groups provided only a superficial similarity in their staining patterns. In detail, each antibody of a given group yielded a slightly different staining distribution. While all antibodies of the antineurofilamentous group stained basket cell fibers in the cerebellum, there were differences in the fibers stained in the rest of the brain. Thus, antibody 07-5 stained basket cell fibers strongly and white matter fibers weakly (FIGURE 4b). It also stained many fibers in the cerebral cortex. In contrast, antibody 04-7 stained basket cell fibers weakly, but white matter fibers strongly (FIGURE 4a). This antibody stained fibers in the cerebral cortex only weakly. On anatomical grounds alone, these antibodies suggested a microheterogeneity (neurotypy) within the antigens they detected.

In this kind of work, antibodies have been selected and classified merely on the basis of histologic appearance of their staining patterns. One of the major problems arising from this selection method is the biochemical and physiologic identification of the antigens detected by these antibodies. Fortunately, immunocytochemistry can also be used as an analytical tool in biochemical separations with apparently similar sensitivity as that engendered in microscopy. We have separated brain homogenates by electrophoresis in sodium dodecyl sulfate gels and electroblotted the electrophoro-

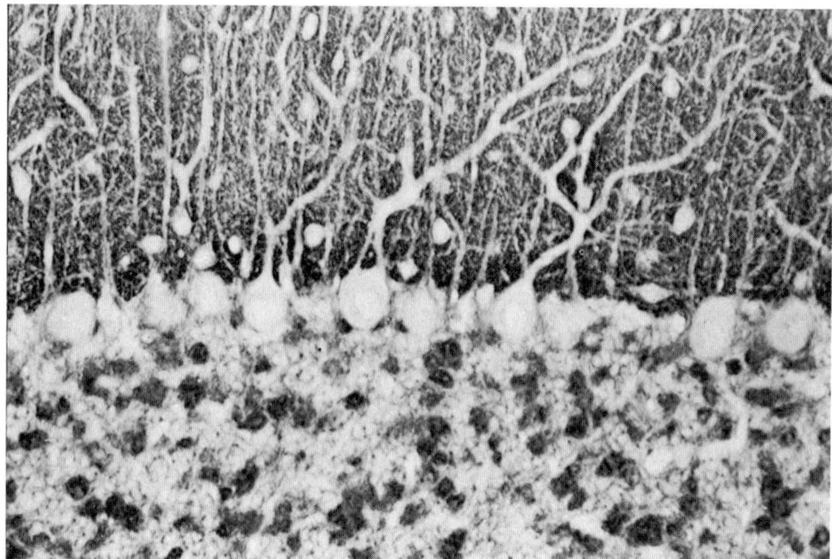

FIGURE 3. Rat cerebellum, sagittal paraffin section stained with monoclonal antibody 02-29 from the antisynapse-associated group. Purkinje cells and their projections outlined in negative images.

grams on cellulose nitrate sheets by the methods of Towbin.[23] Molecular weight standards were stained by Coomassie blue and strips of the cellulose nitrate paper were stained individually[18,19] with monoclonal antibodies, followed by goat antimouse immunoglobulin and peroxidase antiperoxidase (PAP) prepared from monoclonal antiperoxidase. We have used dilutions of 1:1,000 of ascitic fluid for staining of electroblots and of paraffin sections.

Using brain homogenates, we found the 200 K band, and sometimes also the 155 K band, stained with antibodies of the antineurofilamentous group. Bands corresponding to vimentin, glial fibrillary acidic protein, or tubulin on Coomassie blue staining did not stain with the antibodies. Upon use of cytoskeletal preparations obtained in the Triton X-100 insolubility method of Chiu *et al.*[2] or neurofilament preparations obtained by axonal flotation in the method of DeVries *et al.*,[3] the antigens revealed with these

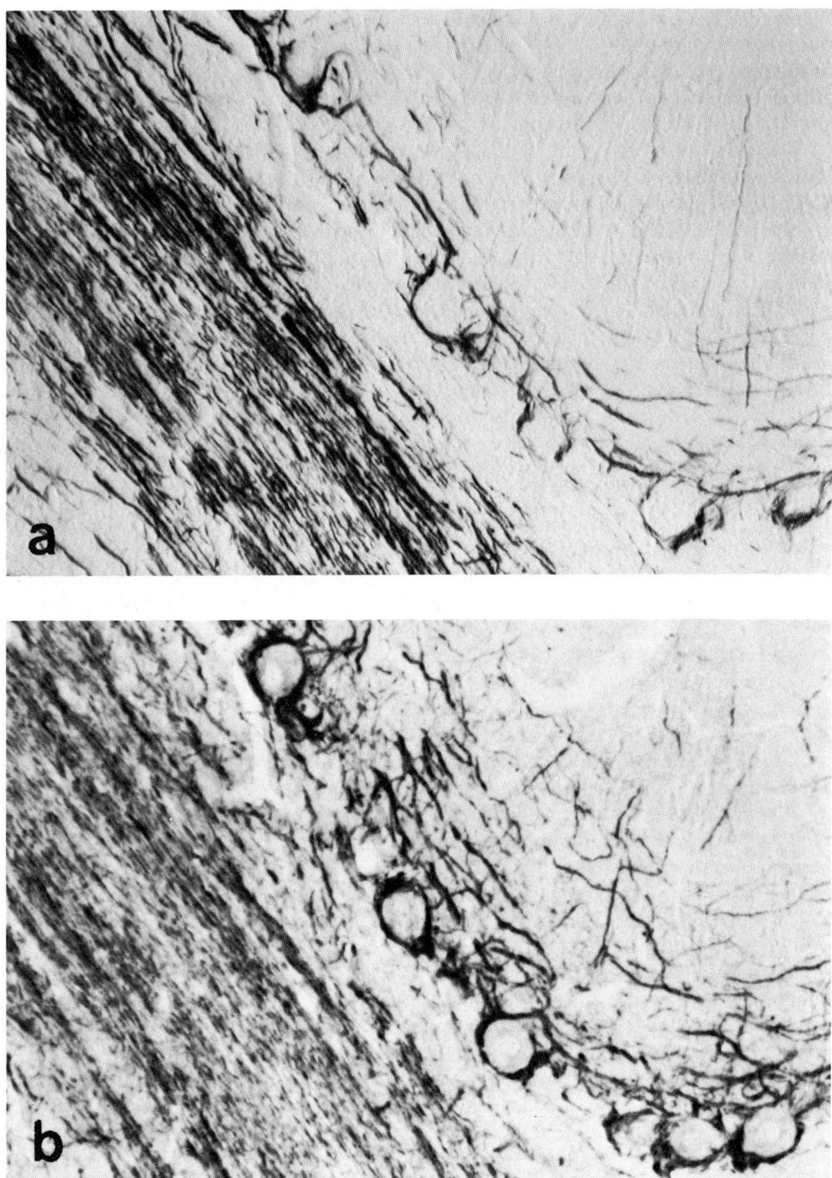

FIGURE 4. Rat cerebellum, adjacent paraffin sections stained with antibodies of the antineurofibrillar group. a, antibody 07-5. White matter fibers stained strongly, basket cell fibers weakly. b, antibody 04-7. Basket cell fibers stained strongly, white matter fibers weakly.

monoclonal antibodies were identified as the 200 K and the 155 K polypeptides of the neurofilament triplet.[7] The 68 K neurofilament band was not stained by any of our antibodies. Both the 200 K and 155 K bands were further resolved into doublets (termed "upper" and "lower" 200 K and 155 K bands). Thus, our antibodies revealed a total of four neurofilament bands.

The identity of neurofilaments as the antigen visualized by antibodies of the "antineurofilamentous" group was not surprising, since as first suggested to us by Sanford Palay,[14] their staining reaction with basket cell fibers gives them a typical appearance of antineurofilament antibodies. However, antibodies of the antiperikaryonal-neurofilamentous group (FIGURE 2), which we originally had termed "perikaryonal" or "neuronal" group, also stained exclusively neurofilament bands in whole brain homogenates, or in cytoskeletal or neurofilament preparations. This was surprising since all these antibodies stained perikarya such as those of Purkinje cells as well as their axons and dendrites, while antibodies of the antineurofilamentous group did not stain perikarya. In addition, most of the former antibodies (FIGURE 2) did not stain basket cell fibers, and indeed, on adjacent sections or by double-staining immunocytochemistry on the same section,[20] there has been no overlap between the staining of antineurofilamentous antibodies (FIGURE 1) and antiperikaryonal-neurofilamentous antibodies (FIGURE 2). An exception was antibody 02-40, which stained both perikarya as well as basket cell fibers. Indeed it was this antibody that had the widest staining distribution throughout the brain.

Electrophoresis confirmed the microheterogeneity of neurofilament polypeptides originally suspected from differences in histologic staining distribution of their antibodies. Thus, the lower band of the 200 K doublet was stained only by antibodies of the antiperikaryonal-antineurofilament group. It, therefore, appears that this lower band represents an antigen present in Purkinje cells but absent in basket cell fibers.

The 155 K doublet was well stained only by antibody 02-40, and poorly by antibody 06-17. These two bands may be characteristic of neurofilaments exemplified by basket cell fibers, but the differences in staining intensities by two antineurofilament group antibodies may suggest additional heterogeneity detectable by biochemical means.

The upper band of the 200 K doublet was stained by all the antibodies examined, whether of the antineurofilamentous group or the antiperikaryonal-neurofilamentous group. Since there is no overlap in staining by all but one of the antibodies of these two groups, the upper 200 K band appears to represent additional microheterogeneity, which, however, is not resolved by one-dimensional SDS gel electrophoresis.

Only antibody 02-40 stained all the four bands of the 200 K and the 155 K doublet. Interestingly, it is this antibody that had the widest histologic staining distribution in the brain among all the antibodies encountered so far.

The microheterogeneity (neurotypy) that appears histologically is wider than that which can be detected by differences in recognition of neurofilament bands. It is expected that additional biochemical separation procedures will establish further degrees of heterogeneity, in consonance with the heterogeneity revealed histologically.

Further evidence for heterogeneity of neurofilament antigens was obtained by developmental studies.[6] Some antibodies do and others do not stain the brains of newborn rats. On the sixth postnatal day, staining is present, although rather sparse, with all antibodies. With many antibodies, especially those that have a less selective distribution in adult brains, staining expands in extent as the brain develops. However, with antibodies that have a more restricted distribution in adults, development of staining patterns proceeds in an opposite direction. Thus, antibody 03-44 stains in the 28-day cerebellum, both basket cell fibers as well as white matter fibers, strongly. In the adult, however, white matter staining largely disappears, and the basket cell fiber staining remains intense. Such observations with neurofilament-reactive antibodies suggest that developing cells express a wider spectrum of heterogeneous neurofilament polypeptides than adult cells, and that differentiation, in contrast to growth, may be accompanied by selection and elimination.

In general, antibodies of the antineurofilamentous group react with neurofibrils in adults as well as in developing animals. However, two antibodies react in adults with neurofibrils and in developing animals with cell nuclei. Conceivably, some of the axons expressed in adults as neurofilament components are expressed in development as nuclear constituents. In this connection, it is of interest that Kornguth et al.[9] observed in 1961 the presence of filamentous structures in nuclei of developing neurons. The possibility of expression of given epitopes in antigens as different as nuclear antigens and neurofibrils may provide a clue for the scrambling of messages necessary to provide for the formation of heterogeneous proteins, such as neurofilaments, that nevertheless share common general properties, such as cytoskeletal structure.

Besides histologic distribution, electrophoretic reactivity, and developmental studies, evaluation of the peripheral immunocytochemistry of monoclonal antibodies yielded a fourth line of evidence for neurofilament heterogeneity.[13] In general, peripheral tissues remained unstained with our monoclonal brain-reactive antibodies. However, some of the antibodies from the antineurofilamentous group reacted with selected nuclei of peripheral cells. Antibody 02-40 also reacted with selected hepatic cells. In general, antibodies reactive with brain neurofilaments also reacted with fibers in peripheral nerves, but here, too, there was a selection: trigeminal nerve, optic nerve, adrenal medullary fibers, and occasional nerves in the thymus were stained by antibodies from the antineurofibrillar and antiperikaryonal-neurofibrillar group, while myenteric fibers were stained only by antibody 06-17. Such peripheral staining variability again attests to heterogeneity of neurofilament polypeptides. This diversity is not merely due to the fact that different monoclonal antibodies react with different epitopes on single polypeptides. The antibodies react with neurofilaments that are present in some cells and absent from others. Hence, they detect a heterogeneity among the polypeptides themselves, and not merely the presence of divergent epitopes in single polypeptides.

The third group of brain-reactive antibodies, besides the antineurofilamentous and antiperikaryonal neurofilamentous groups, was termed antisynapse-associated group, at first merely because of the resemblance of its immunocytochemical staining patterns with those of the monoclonal antibodies produced by Matthew et al.[12] upon immunization of mice with isolated synaptic junctions. Although our antibodies had a general resemblance in distribution with those selected by Matthew et al., details of staining patterns differ from them as well as from each other. Many of these antibodies reacted not only with what appear to be synaptic regions in the brain, but also with macrophages. They did not react with peripheral nerves, except nerve endings in the adrenal medulla. Synaptosome and cerebellar glomeruli preparations were separated on SDS gels into five distinctive bands identified by Coomassie blue. Upon immunocytochemistry of electroblot-transferred preparations, the staining of only one of our antisynapse-associated antibody corresponded to a single band visualized by Coomassie blue. The other antibodies of the antisynapse-associated group also stained one or two bands on the synaptosomal or glomerular preparations but these bands were not identical with those revealed by Coomassie blue. The immunocytochemistry here reveals an expected greater number of bands in SDS-electrophoretically separated preparations than can be detected by the less sensitive Coomassie blue staining. Significantly, each antisynapse-associated antibody revealed a different staining pattern. None of these antibodies reacted with cytoskeletal or neurofilament preparations.

REFERENCES

1. Aronin, N., M. Difiglia, A. S. Liotta & J. B. Martin. 1981. Ultrastructural localization and biochemical features of immunoreactive leu-enkephalin in monkey dorsal horn. J. Neurosci. **1:** 561–677.

2. Chiu, F.-C., W. T. Norton & K. L. Fields. 1981. Intermediate filaments from bovine, rat and human CNS: Mapping analysis of the major proteins. J. Neurochem. **34:** 1149–1159.

3. De Vries, G. H., W. T. Norton & C. S. Raine. 1972. Axons: Isolation from mammalian CNS. Science **175:** 1370–1372.

4. Gainer, H., Y. Sarne & M. J. Brownstein. 1977. Neurophyin biosynthins: Conversion of a putative precursor during axonal transport. Science **195:** 1354–1356.

5. Gilbert, W. 1978. Why genes in pieces. Nature **271:** 501.

6. Goldstein, M. E., N. H. Sternberger & L. A. Sternberger. 1982. Developmental expression of neurotypy revealed by immunocytochemistry with monoclonal antibodies. J. Neuroimmunol. **3:** 203–217.

7. Goldstein, M. E., L. A. Sternberger & N. H. Sternberger. 1983. Microheterogeneity (neurotypy) of neurofilament proteins. Proc. Natl. Acad. Sci. USA **80:** 3101–3105.

8. Growse, L. D., B. C. Schrier, C. H. Letendre & P. G. Nelson. RNA sequence complexity in central nervous system development and plasticity. Curr. Topics Dev. Biol. **16:** 381–397.

9. Kornguth, S., J. Anders & G. Scott. 1967. Observations on the ultrastructures of developing cerebellum of Macacca. J. Comp. Neurol. **130:** 1–24.

10. Lauber, M., P. Nicolas, H. Boussetta, C. Fahy, P. Beguin, M. Camier, H. Vaudry & P. Cohen. 1981. The M_r 80,000 common forms of neurophysin and vasopressin from bovine neurohypophysin have corticotropin- and beta-endorphin-like sequences and liberate by proteolysis biologically active corticotropin. Proc. Natl. Acad. Sci. USA **78:** 6086–6090.

11. Mains, R. E., B. Eipper & N. Ling. 1977. Common precursors to corticotropins and endorphin. Proc. Natl. Acad. Sci. USA **74:** 3014–3018.

12. Matthew, W. P., L. F. Reichardt & L. Tsavaler. 1981. Monoclonal antibodies to synaptic membranes and vesicles. Cold Spring Harbor Reps. Neurosci. **2:** 163–180.

13. Östermann, E., N. H. Sternberger & L. A. Sternberger. 1983. Immunocytochemistry of brain-reactive monoclonal antibodies in peripheral tissues. Cell Tiss. Res. **228:** 459–473.

14. Palay, S. Discussion to Sternberger, L. A. 1982. Peptide heterogeneity and neurotypy. *In* Cytochemical Methods in Neuroanatomy. S. Palay & V. Chan-Palay, Eds.: 241–253. Alan R. Liss. New York.

15. Scharrer, B. 1978. Peptidergic neurons: Facts and trends. Gen. Comp. Endocrinol. **34:** 50–62.

16. Scharrer, B. 1981. Neuroendocrinology and histochemistry. *In* Histochemistry: The Widening Horizons. P. J. Stoward & J. M. Polak, Eds.: 11–20. John Wiley & Sons. New York.

17. Scharrer, E. & B. Scharrer. 1939. Secretory cells within the hypothalamus. Res. Publ. Am. Res. Nervous Mental Diseases **20:** 170–193.

18. Sternberger, L. A. 1979. Immunocytochemistry. 2nd edit. John Wiley & Sons. New York.

19. Sternberger, L. A., P. H. Hardy, J. J. Cuculis & H. G. Meyer. 1970. The unlabeled antibody enzyme method of immunohistochemistry: Preparation and properties of soluble antigen-antibody complex (horseradish peroxidase-antihorseradish peroxidase) and its use in identification of spirochetes. J. Histochem. Cytochem. **18:** 315–333.

20. Sternberger, L. A. & S. A. Joseph. 1979. The unlabeled antibody method. Contrasting color staining of paired pituitary hormones without antibody removal. J. Histochem. Cytochem. **27:** 1424–1429.

21. Sternberger, L. A., J. L. Greenwald, D. Hock, K.-H. Elger & W. G. Forssmann. 1981. A new hypothalamic substance, and not luteinizing hormone-releasing hormone, is

detected immunocytochemically by antibody to luteinizing hormone-releasing hormone. Proc. Natl. Acad. Sci. USA **78:** 5216–5270.

22. STERNBERGER, L. A., L. W. HARWELL & N. H. STERNBERGER. 1982. Neurotypy: Regional individuality in rat brain detected by immunocytochemistry with monoclonal antibodies. Proc. Natl. Acad. Sci. USA **79:** 1326–1330.

23. TOWBIN, H., T. STAEHELIN & J. GORDON. 1979. Electrophoretic transfer of proteins from polyacrylamide gels to nitrocellulose sheets: Procedure and some application. Proc. Natl. Acad. Sci. USA **76:** 4350–4354.

Description of Two Differently Distributed Central Nervous System Antigens with Single Monoclonal Antibody and Different Methods of Fixation[a]

B. ZIPSER AND C. SCHLEY

Cold Spring Harbor Laboratory
Cold Spring Harbor, New York 11724

INTRODUCTION

In this report we address one of the problems that occur with the interpretation of immunocytochemical data—a given monoclonal antibody can react with brain tissue to elicit more than one histological staining pattern. The particular staining pattern generated is under experimental control and the method of fixation used determines the type of tissue component visualized by the monoclonal antibody.

Histological work with antisera has shown that the method of fixation determines the intensity of immunocytochemical staining. The fact that glutaraldehyde impairs immunocytochemical staining is one of the key problems in combining immunocyto-chemistry with electron microscopy. Often, immunocytochemical staining is carried out on 4% paraformaldehyde-fixed tissue. However, other methods of fixation might enhance visualization of particular tissue antigens. Bouin's is a good fixative to obtain optimal staining with some somatostatin antisera.[1] By adding periodate and lysine to paraformaldehyde, the resulting PLP fixative[5] led to greater staining with enkephalin antisera in the hippocampus.[2] In this report we illustrate that a given monoclonal antibody can stain different tissue components as the method of fixation is changed. This is an issue that becomes of particular importance as monoclonal antibodies are increasingly used as cytochemical probes.

METHODS

The leech central nervous system, consisting of a thirty-three–ganglia nerve cord, was dissected out of alcohol-anesthetized leeches of the species *Haemopis marmorata*. The nerve cord was pinned out in a Sylgard dish and was fixed with one of three different types of fixatives: (1) 4% paraformaldehyde in 0.1 M phosphate buffer (pH 7.4), 20 min; (2) PLP fixative, which also contains 4% paraformaldehyde fixative to which lysine (0.1 M) and periodate (0.01 M) have been added, 45 min; and (3) formalin-Ringer (3.7% formaldehyde in leech Ringer, pH 5.1), 7 min. The tissue was processed as a whole mount for immunocytochemistry.[8,9] To ensure antibody penetration through the 250 μm section of tissue, the connective tissue capsule surrounding each ganglion was mechanically opened with forceps. Furthermore, the tissue was

[a]Supported by the National Institutes of Health grant NS 17984-01 and a Whitehall grant.

chemically permeabilized. The 4% paraformaldehyde and PLP-fixed nerve cords were permeabilized by passing the tissue rapidly through an alcohol series into xylene and then returning it through the alcohols into buffer. The formalin-fixed tissue was briefly rinsed with buffer before treatment with acetone (20°C) for 7 min. Furthermore, all antibody incubations and washing steps were done in the presence of 2% Triton X-100. Incubation with the monoclonal antibody took place overnight at room temperature. The next morning, after a brief wash, the tissue was incubated with an HRP (horseradish peroxidase)-conjugated goat anti-mouse IgG (Cappel; 1/30) for 2 hr. The tissue was then reacted with 3',3'-diaminobenzidine (0.03%) and H_2O_2, rinsed, dehydrated, and embedded in Permount.

RESULTS

We isolated the monoclonal antibody, Laz1-130, which was of interest to us because of its initial, apparently neuron-specific, staining. This antibody highlights a small set of neurons. However, it soon became evident that this same antibody also binds to a more generally distributed tissue antigen. One explanation for the two types of staining could be that Laz1-130 is not a monoclonal antibody. However, this possibility was ruled out by extensively recloning the hybridoma line that secretes Laz1-130 and showing that the staining pattern did not change. There is additional evidence suggesting that Laz1-130 is a monoclonal antibody. In the same fusion in which Laz1-130 was generated, a second hybridoma line was isolated that secretes an antibody named Laz1-117. Laz1-117 reacts with the leech nerve cord giving the same staining pattern as Laz1-130 (and the same biochemical data on immunoblots (Hogg and Zipser, unpublished)). (The two different hybridoma lines were isolated from different wells in the same fusion.) The most likely explanation for this phenomenon is that the different hybridoma lines secreting antibodies with similar antigen specificity derive from myelomas that fuse with sibs from a common spleen cell. Thus, two factors—recloning the hybridoma line and the independent isolation of two lines that secrete antibodies eliciting the same immunocytochemical staining patterns—strongly suggest that Laz1-130 is a monoclonal antibody.

We carried out a series of immunocytochemical tests to obtain a definite idea of the tissue distribution of antigens and to optimize neuron-specific staining.

In these histological experiments, immunocytochemical staining was tested in tissues fixed in several different ways. It was found that different fixatives gave different types of antibody staining. Originally, the antibody was isolated by its neuron-specific staining in 4% paraformaldehyde-fixed tissue. Since the intensity of the neuron-specific staining is generally higher in PLP fixatives, this is now the fixative of choice. As seen in FIGURE 1A, the monoclonal antibody highlights two types of cell bodies on the ventral surface in the subesophageal ganglion (FIGURES 1A and 2). They occur as adjacent bilateral pairs with 60 μm and 20 μm cell bodies. The twenty-one unfused midbody ganglia (FIGURE 2) do not show additional stained cell bodies.

Neuron-specific staining also includes a small set of axons and about 3 μm large varicosities resembling typical leech synaptic terminal processes.[6] These antigenically related processes and varicosities are seen not just in the head ganglia. The so-called connective, the thick fiber bundle that links all ganglia along the nerve cord, carries stained axons while the synaptic neuropile of all segmental ganglia are three-dimensionally permeated with varicosities.

Immunocytochemical staining is notoriously variable.[7] However, the specific neuronal staining of Laz1-130 is quite reliably elicited in PLP-fixed tissue. In a total of

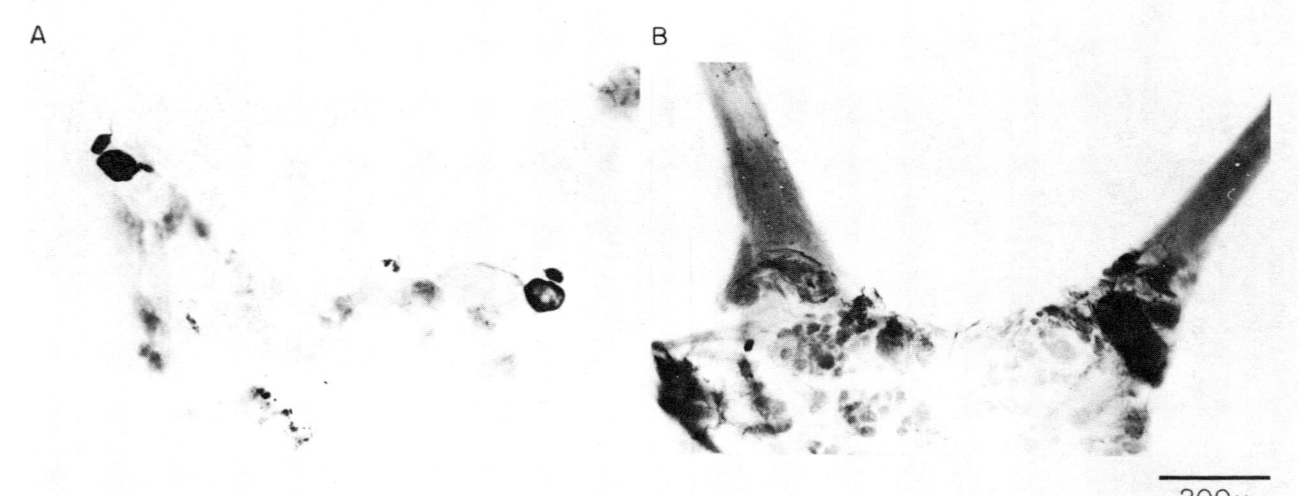

FIGURE 1. Micrographs of two leech head ganglia. In A, a PLP-fixed head ganglion, Laz1-130 strongly stains two pairs of cell bodies at the anterior border of the first subesophageal ganglion. In B, a formalin-fixed ganglion, Laz1-130 stains the connective tissue capsule surrounding the nerve bundles that project anteriorly off the head ganglion. The antibody also gives rise to a generalized cell body stain.

FIGURE 2. A diagrammatic representation of the 33-ganglia leech nerve cord. The Laz1-130–stained cell bodies from the micrograph in FIGURE 1A are indicated in the first head subganglion.

thirty head ganglia fixed with 4% paraformaldehyde or PLP, the ventral cell bodies are distinctly stained in twenty-six cases (FIGURES 1A and 2) while the varicosities show up in all preparations. (In 10% of the head ganglia, a faint pair of dorsal cell bodies are stained. These cell bodies are not entered into the diagrammatic representation of the Laz1-130 cell body map.)

In addition to labeling these specific neuronal cell bodies and processes, Laz1-130 also binds to a tissue antigen that is associated with the connective tissue capsule

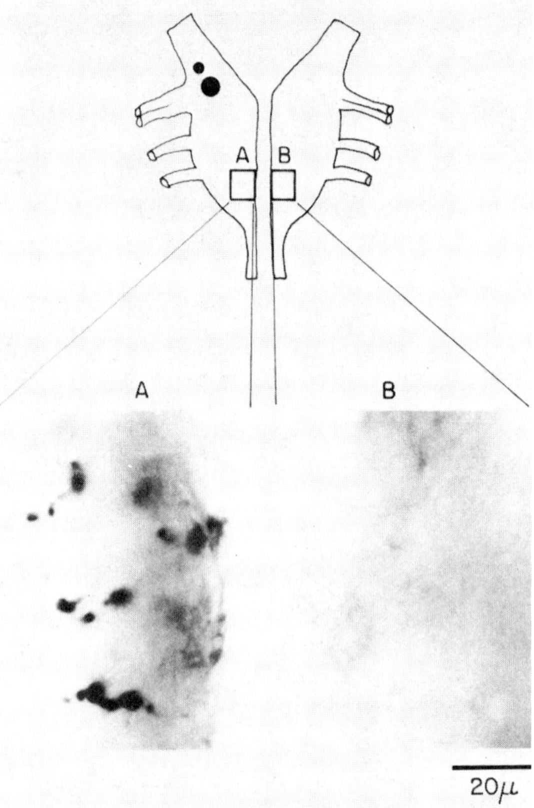

FIGURE 3. A head ganglion cut into halves. The left half was fixed with PLP and the right half with formalin-Ringer. The rectangles in the drawing indicate the location of the synaptic neuropile from which the high-power micrographs in A and B are taken. In A, the PLP-fixed hemiganglion shows deeply stained varicosities resembling typical leech synaptic terminals. In B, no such varicosities are evident in the formalin-fixed synaptic neuropile.

surrounding the nerve cord. To what extent Laz1-130 is a general stain for neuronal cell bodies or material surrounding the cell bodies is not clear. In contrast to other monoclonal antibodies, Laz1-130 tends to stain whole mounts of leech ganglia in such a way as to give a kind of "dirty appearance," which could be indicative of general cell body staining. The relative degree of specific neuronal staining over general tissue staining varies. FIGURE 1A shows the maximal intensity of specific cell body staining while the general tissue staining is rather light. In formalin-Ringer–fixed tissue, the antibody-staining pattern changes (FIGURE 1B). Laz1-130 still stains the connective tissue capsule but it no longer highlights the small set of neuronal cell bodies and varicosities (FIGURES 1A and 2). Formalin-Ringer, which here does not fix the tissue in such a way as to allow specific neuronal staining by Laz1-130, is the choice fixative for several other antibodies. For example, in ganglia fixed with formalin-Ringer, the monoclonal antibodies Laz2-369 and Lan3-8[3] stain both superficial cell bodies and processes deep in the ganglionic neuropile. However in all ten head ganglia fixed with formalin-Ringer, Laz1-130 did not elicit its specific neuronal staining seen in PLP-fixed material.

The effects of formalin-Ringer and PLP fixations were also contrasted on axons and varicosities in head ganglia of the same leech. In five experiments, a live head ganglion was cut into symmetric halves, as illustrated in FIGURE 3. The left half was fixed with PLP and the right half with formalin-Ringer. In FIGURE 1 the cell bodies on the surface of the leech ganglion are in focus; FIGURE 3A and B show high-power views of the central ganglionic core, the synaptic neuropile. The varicosities are prominent in PLP-fixed tissue throughout the entire dorso-ventral extent of the neuropile. They are absent from the same half-head ganglion fixed in formalin-Ringer.

DISCUSSION

We have demonstrated that a single monoclonal antibody (Laz1-130) can give different central nervous system staining patterns. In 4% paraformaldehyde or preferably PLP-fixed tissue, the antibody generally highlights two types of neuronal cells and a small subset of axons and varicosities. Superimposed on this neuron-specific staining by Laz1-130 is a more generalized tissue staining of the connective tissue capsule surrounding the nerve cord and maybe even a weak diffuse staining of the neuronal cell bodies or material associated with them. In formalin-Ringer–fixed tissue, the connective tissue staining persists but the specific neuronal staining is lost.

The antibody Laz1-130 is of particular interest in an electrophysiological study focused on specialized neurons in the head ganglion. Another one of our monoclonal antibodies (Lan 3-1,[7,9]) either binds to the identical pairs of large and small cell bodies visualized by Laz1-130, or our two monoclonal antibodies have discovered a repeated pattern consisting of antigenically related large and small cell bodies. Since Laz1-130 sheds light on novel aspects of neuroanatomical organization in the leech, it became important to verify that there was both a neuron-specific and a diffuse staining whose relative intensities could vary independently. This led to the experiments discussed where different features of antibody binding could be differentiated with different fixatives.

The two observations, that the monoclonal antibody binds to antigens with different tissue distribution and that the differently distributed antigens behave differently under different methods of fixations, suggest that Laz1-130 cross-reacts with at least two different molecules that share a common antigenic determinant. Such cross-reactivities are not uncommon.[4] This cross-reactivity may be of little interest if it is just a coincidental affinity to a similar amino acid sequence on two different and functionally unrelated macromolecules. However, as illustrated in this study, a monoclonal antibody such as Laz1-130, that very likely cross-reacts with different types of molecular moieties, is still useful to neuroanatomical studies and will contribute to our electrophysiological analysis of higher-order neurons in the leech.

The use of different fixatives may be valuable in detecting different antigens that bind the same monoclonal antibody.

ACKNOWLEDGMENT

We would like to thank Murray Flaster for his helpful comments about the manuscript.

REFERENCES

1. FELDMANN, S. C. 1980. Anat. Rec. **196:** A55.
2. GALL, C., H. J. KARTEN & K.-J. CHANG. 1981. J. Comp. Neurol. **198:** 335–350.
3. HOGG, N., M. FLASTER & B. ZIPSER. 1983. J. Neurosci. Res. **9:** 445–457.
4. LANE, D. & A. KOPROWSKI. 1982. Nature **296:** 200–202.
5. MCLEAN, I. & P. NAKANE. 1974. J. Histochem. Cytochem. **22:** 1077–1083.
6. MULLER, K. 1979. Biol. Rev. **54:** 99–134.
7. ZIPSER, B. 1982. J. Neurosci. **2:** 1453–1464.
8. ZIPSER, B. 1980. Nature **283:** 857–858.
9. ZIPSER, B. & R. MCKAY. 1981. Nature **289:** 549–554.

Paucity of HLA-A,B,C Molecules on Human Cells of Neuronal Origin: Microscopic Analysis of Neuroblastoma Cell Lines and Tumor[a]

LOIS A. LAMPSON AND JAMES P. WHELAN

Department of Anatomy
University of Pennsylvania School of Medicine
Philadelphia, Pennsylvania 19104

INTRODUCTION

Monoclonal antibodies are widely recognized as important tools for molecular analysis in the nervous system. They may be used as well-characterized probes for single cell types, molecules, or parts of molecules. We have sought to establish an experimental system in which single antibodies could be used for biochemical and genetic studies in homogeneous tissue, as represented by cloned neuronal cell lines, and for complementary anatomical studies in complex tissue. While this approach has been very fruitful in analyzing the immune system, it has been taken less frequently in studies of the mammalian nervous system.

Here, we review our work with the HLA-A,B,C molecules on human cell lines of neuronal origin. We have previously used monoclonal antibodies in direct binding and direct cytotoxicity assays to show that human neuroblastoma-derived cell lines display a striking lack of these molecules, as compared to glial or lymphoid lines.[7,9] More recently, we have extended these studies to include microscopic analysis of the molecules on sectioned cells. The microscopic assay was then used to show that the paucity of HLA-A,B,C is not limited to the cell lines, but is also a property of neuroblastoma tumor.

We interpret these findings in two contexts. First, we note a possible clinical use. Second, these results add to an increasing appreciation of the variation of expression of HLA-A,B,C (or its murine analogue, H-2) in various cell types.[1,4,17] In more general terms, our results illustrate the feasibility, and the value, of establishing an interplay between complex tissue and tumor-derived cell lines for the detailed study of human neuronal proteins.

MATERIALS AND METHODS

Monoclonal Antibodies

The HLA-A, HLA-B, and HLA-C molecules, which bear the conventional major human transplantation antigens, are each composed of two polypeptide chains. In each case, the heavier chain (HLA chain) bears the polymorphic determinants. In each

[a]Supported by National Institutes of Health grants NS16552 and CA14489 and National Science Foundation grant PCM79 26757.

case, the HLA chain is linked to an invariant light chain, which is β2-microglobulin (β2-m). Monoclonal antibody L368 is specific for β2-m.[10,16] Monoclonal antibody W6/32 recognizes a non-polymorphic determinant on the native two-chain molecule.[2]

Human Cell Lines

IMR-5, NMB, CHP-134A, and SK-N-SH are derived from neuroblastoma, a tumor of the sympathetic neuroblast. CW1-TG1 is derived from an oligodendroglioma. Raji is a B cell line, derived from a Burkitt's lymphoma. Daudi is also derived from a Burkitt's lymphoma, but lacks HLA-A,B,C molecules. The origin and culture of these cell lines is given elsewhere.[16]

Microscopic Assays

Freshly harvested cell lines were washed in phosphate-buffered saline (PBS), fixed with 0.1% glutaraldehyde for 5 min at room temperature, and embedded in agarose.[14] The agarose blocks were then cut into 50 μm vibratome sections for assay. The assays were performed in 96-well Linbro plates, with each section allowed to remain free in one well. For the unlabeled antibody peroxidase-antiperoxidase (PAP) assay, sections were exposed to normal serum, monoclonal mouse antibody, rabbit anti-mouse Ig "bridge," and PAP in sequence, followed by hydrogen peroxide and diaminobenzidine, following a variation of the Sternberger method,[11] as described in Lampson et al.[14] The stained sections were mounted on glass slides in glycerine and photographed under Nomarski optics. Bone marrow was handled in a similar way, except that in some cases red cells were removed with Ficoll-Paque before the cells were fixed and embedded in agarose.

Binding Inhibition

Detergent (Nonidet P-40) extracts of neuroblastoma, lymphoid and glial cells were used to inhibit the binding of each monoclonal antibody to a single B cell target. The extracts were tested at serial two-fold dilutions, and the amount of extract needed to obtain 50% inhibition was expressed in terms of μg of extracted protein.[8]

RESULTS

Microscopic Analysis of Cloned Cell Lines

We had previously used direct binding and direct cytotoxicity assays to show weak expression of HLA-A,B,C molecules on human neuroblastoma-derived cell lines.[7,9] Both of these assays detect cell-surface determinants. In order to study cells that had been cut open and to establish a procedure for later microscopic analysis of tumors and normal tissue, we extended these studies to microscopic analysis of sectioned material. In this case, fixed cells were first embedded in agarose, and then cut into 50 μm vibratome sections. Monoclonal antibodies L368 and W6/32 were each tested at several dilutions.

The results of these assays are illustrated in FIGURE 1. Lymphoid and glial cells

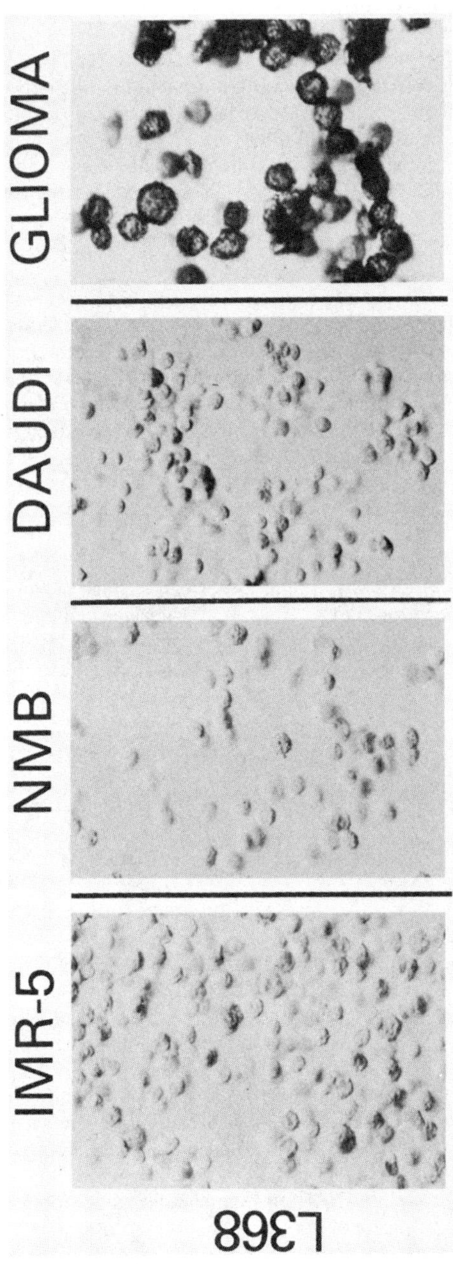

FIGURE 1. Binding of L368 to human cell lines. Cells were fixed with 0.1% glutaraldehyde, embedded in agarose, cut into 50 μm vibratome sections, and exposed to L368 in the PAP assay, as described. All sections shown had been stained with L368. Controls are described in text. (Nomarski optics.)

exhibited a characteristic staining pattern, with a strong ring of stain around the cell, and a less strongly stained cytoplasm. (The glial cell staining with L368 is illustrated.) This staining agrees with the subcellular localization of these antigens that had been reported in other microscopic or biochemical studies.

A negative control for these assays was provided by the cell line Daudi. This B cell line is known to lack HLA-A,B,C molecules, and would not be expected to stain with either of the monoclonal antibodies that were used. The lack of staining seen with L368 is illustrated; the same results were obtained with W6/32.

For each antibody, the staining with four different neuroblastoma-derived lines, IMR-5, NMB, SK-N-SH, and CHP-134A, was indistinguishable from that seen with Daudi. No specific staining was seen with either of the two monoclonal antibodies that were used, L368 and W6/32. The staining seen with L368 and two of the neuroblastoma lines is illustrated in FIGURE 1.

Additional positive and negative controls were also used to evaluate this procedure for later use with complex tissue. Monoclonal antibodies to Ia-like molecules would not be expected to bind to either glial or neuronal cell lines.[9] In fact, anti-Ia antibodies showed no specific staining against glial or lymphoid cell lines in these assays. At the same time, the anti-Ia antibodies did stain the B cell line Daudi, as expected. Anti-neuronal monoclonal antibodies (which we had characterized previously) did, as expected, stain the neuroblastoma-derived cell lines. With different antibodies, both internal and cell surface antigens were stained. Thus, our failure to stain the neuroblastoma cells with anti-HLA-A,B,C reagents did not reflect a general failure to fix internal or external antigens in these cells.

In summary, the microscopic assay extended the results of the binding and cytotoxicity assays. No specific staining of either cell surface or internal HLA-A,B,C molecules was seen, confirming that neuroblastoma-derived cell lines express relatively few of these molecules as compared to glial or lymphoid cells. Moreover, the behavior of the antibodies and controls suggested that this assay could also be applied to more complex tissue.

HLA-A,B,C Expression on Metastatic Neuroblastoma in Bone Marrow

We wished to know whether the lack of HLA-A,B,C expression seen with the neuroblastoma cell lines was also a property of the tumor. To test this, the PAP assay was used to assess HLA-A,B,C expression on metastatic neuroblastoma in bone marrow. Here, tumor cells may appear as readily identifiable aggregates (as well as single cells). Normal bone marrow was used as a control. The results obtained with monoclonal antibody L368 are illustrated in FIGURE 2.

In these assays, L368 showed clear binding to the majority of the nucleated cells in normal bone marrow ($\geq 96\%$). The cells had the characteristic appearance described above, with a strong ring of stain surrounding a less darkly stained cytoplasm (FIGURE 2C and D). The neuroblastoma bone marrow provided a clear contrast. Many fewer ($<50\%$) of the nucleated cells were stained. More strikingly, aggregates of cells with the characteristic appearance of neuroblastoma were completely unstained (FIGURE 2A and B).

Thus, the striking paucity of HLA-A,B,C molecules on neuroblastoma-derived cell lines appears to reflect a property of the neuroblastoma tumor itself. It also suggests a possible clinical use for these antibodies: Since most cells do express the HLA-A,B,C molecules strongly, the apparent absence of the molecules in microscopic assays may be one parameter for distinguishing between single metastatic neuroblastoma cells or residual tumor and surrounding normal tissue. This is discussed more fully elsewhere.[16,17]

Normal Neural Tissue

Previous absorption studies have shown that normal adult human brain expresses little, if any, HLA-A,B,C activity.[3,15] We have confirmed these results in our own

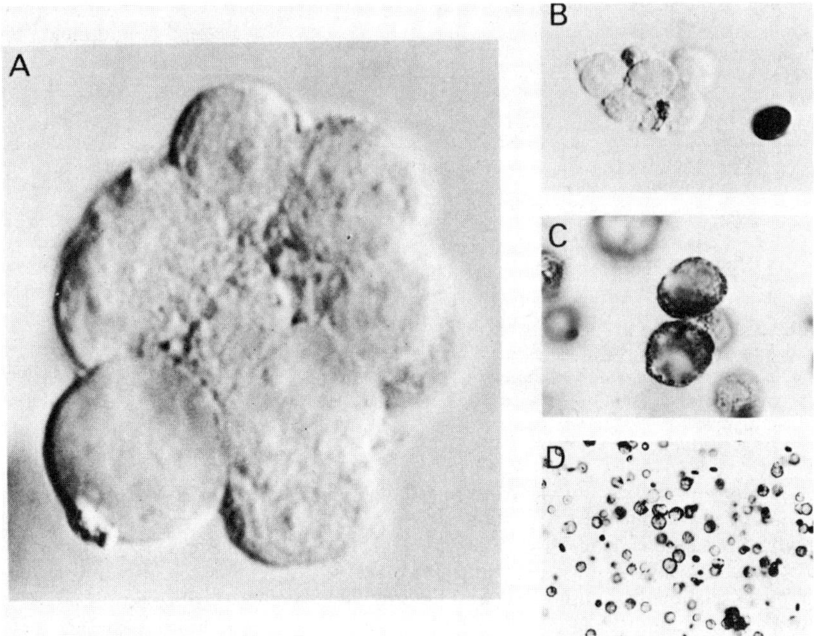

FIGURE 2. Binding of L368 to metastatic neuroblastoma in bone marrow and normal bone marrow. Procedure was as in FIGURE 1, except that red cells were first removed with Ficoll-Paque. All sections shown had been stained with L368. (A) Characteristic aggregate of neuroblastoma cells in bone marrow of patient with stage IV neuroblastoma; (B) a second aggregate, at lower magnification; (C) normal bone marrow cells, at same magnification as in B; (D) normal bone marrow at still lower magnification. Note characteristic ring of stain surrounding less darkly stained cytoplasm in D. (Dark cells in B and D are erythrocytes.) Positive and negative controls were as described in text referring to FIGURE 1. (Nomarski optics.)

laboratory, using both direct radioimmunoassays and binding inhibition assays. We hypothesize that the small amount of HLA-A,B,C activity that is seen may be contributed by the glial cells in the brain, since our findings (above) agree with others that glial lines are rich in HLA-A,B,C molecules.[13] We are now using the microscopic assays described above to test this directly. The important point here is that available evidence suggests that the paucity of HLA-A,B,C on neuroblastoma-derived cell lines may reflect a property of normal adult neurons as well as neuroblastoma tumors.[b]

Biochemical Analysis

Having established that the paucity of HLA-A,B,C on neuronal cell lines is not a culture artifact, we have returned to these lines as a source of homogeneous material for biochemical analysis. Binding inhibition assays have allowed us to quantify the HLA-A,B,C expression. In these assays, cell extracts were used to inhibit the binding of L368 and W6/32 to B cell lines. In ten determinations, we found that each of four neuroblastoma-derived cell lines did contain some inhibitory activity, but the lines were less than 1% as efficient, per μg of extract protein, as were glial or lymphoid lines (TABLE 1).

TABLE 1. Efficiency of Inhibition Relative to Raji

Cell Extract Used as Inhibitor	Inhibition of L368	Inhibition of W6/32
Raji	=1	=1
IMR-5	647, 1770	362
NMB	294, 574	231
SK-N-SH	294, 836	ND
134A	353, 934	ND

Binding inhibition assays were performed as described in the text. For each extract, μg of protein needed to obtain 50% inhibition was determined, and then expressed relative to μg of Raji extract that was needed in the same assay. Thus, it took 362 times as much IMR-5 extract as Raji extract to inhibit W6/32 in one assay.

This low level of activity leads us to ask whether bona fide HLA-A,B,C molecules or cross-reactive molecules are being recognized on the neuronal cell lines. Immuno-precipitation studies from biosynthetically labeled extracts have confirmed that the neuroblastoma-derived lines do synthesize a polypeptide chain that co-migrates with the β2-microglobulin synthesized by glial and lymphoid lines.[16] Studies designed to visualize the chain associated with β2-microglobulin and to determine whether it is a conventional HLA chain are now in progress.

DISCUSSION

The HLA-A,B,C molecules, which bear the conventional human transplantation antigens, are of interest for several reasons. These highly polymorphic molecules provide one major barrier to organ transplantation, they are involved in cellular interactions in the immune response, and a growing list of human diseases have been linked to particular phenotypes. Although these molecules are believed to be present on most nucleated cells, there is increasing evidence for variation in their expression in some cell types.[1,4,5,12,17]

[b]**Note added in proof.** Microscopic analysis of human brain shows that HLA-A,B,C and β2-m are concentrated at the blood vessels, and are expressed weakly, at best, on both neurons and glia (in preparation).

The monoclonal antibodies we have used here are well defined, and very general probes for these molecules. L368 recognizes the invariant β2-m chain of the molecules and W6/32 recognizes an invariant determinant on the native two-chain molecule. With both of these antibodies, we find a striking paucity of the molecules in several types of assays: direct binding (radioimmunoassay), direct C'-mediated cytotoxicity, and, as described here, microscopic analysis and binding inhibition assays.

We wish to stress the potential of establishing this type of interplay between cloned cell lines and complex tissue. By extending our microscopic studies, we can learn whether the paucity of HLA-A,B,C molecules is a property of other neuronal tumors, or of normal fetal or adult neurons (J. Whelan and L. Lampson, in preparation). At the same time, by extending the biochemical studies of the cloned cell lines, we can probe the biochemical basis of these observations: Are conventional HLA-A,B,C molecules absent, present in small quantities, or present in altered or analogous forms? These questions become particularly interesting as we consider the developing genetic picture of these molecules.

The HLA-A,B,C molecules, and their murine counterparts, the H-2 molecules, belong to a larger group of homologous proteins, which includes the lymphocyte differentiation antigens, Qa and TL. Genetic studies indicate that there may be many other analogues that have not yet been characterized.[4,6] Having established a striking paucity of HLA-A,B,C molecules on neuroblastoma tumor and cell lines, we are led to ask whether these cells (and perhaps normal neurons) may express an alternative or analogous form of the molecules. By combining biochemical analysis of homogeneous cell lines, and anatomical studies of complex tissue, we are now able to directly approach this question. In more general terms, if monoclonal antibodies and cloned cell lines are selected with care, the approach we have taken should be useful for the analysis of many other neuronal proteins.

ACKNOWLEDGMENTS

We thank Dr. Jane Chatten, Dr. William Elkins, and Dr. Audrey Evans for human tissue, and Mr. A. Bravo for preparing the manuscript.

REFERENCES

1. ARTZT, K. & D. BENNETT. 1975. Analogies between embryonic (T/t) antigens and adult major histocompatibility (H-2) antigens. Nature **256:** 545–547.
2. BARNSTABLE, C., W. BODMER, G. BROWN, G. GALFRE, C. MILSTEIN, A. WILLIAMS & A. ZIEGLER. 1978. Production of monoclonal antibodies to group A erythrocytes, HLA and other human cell surface antigens—new tools for genetic analysis. Cell **14:** 9–21.
3. BERAH, M., J. HORS & J. DAUSSET. 1970. A study of HL-A antigens in human organs. Transplantation **9:** 185–192.
4. BODMER, W. F. 1981. HLA structure and function: A contemporary view. Tissue Antigens **17:** 9–20.
5. BROWN, G., P. BIBERFELD, B. CHIRSTENSSON & D. Y. MASON. 1979. The distribution of HLA on human lymphoid, bone marrow and peripheral blood cells. Eur. J. Immunol. **9:** 272–275.
6. HOOD, L., M. STEINMETZ & R. GOODENOW. 1982. Genes of the major histocompatibility complex. Cell **28:** 685–687.
7. KENNETT, R. H., Z. L. JONAK, M. MOMOI, M. C. GLICK & L. A. LAMPSON. 1982. Analysis of cell surface molecules on human neuroblastoma cells and leukemia cells. *In* Monoclonal Antibodies in Drug Development. pp. 91–107. American Society for Pharmacology and Experimental Therapeutics.

8. LAMPSON, L. A. 1984. Binding inhibition of monoclonal antibodies by detergent cell extracts. *In* Biotechnology and Bioscience. I. Monoclonal Antibodies and Functional Cell Lines: Progress and Applications. R. Kennett, K. Bechtol & T. McKearn, Eds. Plenum Publishing Corp. New York. (In press.)

9. LAMPSON, L. A. 1981. Expression of the major histocompatibility antigens in the human nervous system. *In* Monoclonal Antibodies to Neural Antigens. R. McKay, M. Raff & L. Reichardt, Eds.: 69–71. Cold Spring Harbor Laboratory. Cold Spring Harbor, N.Y.

10. LAMPSON, L. A. & R. LEVY. 1980. Two populations of Ia-like molecules on a human B cell line. J. Immunol. **125:** 293–299.

11. STERNBERGER, L. 1979. Immunocytochemistry. 2nd edit. John Wiley & Sons, Inc. New York.

12. TROWSDALE, J., P. TRAVERS, W. F. BODMER &·R. A. PATILLO. 1980. Expression of HLA-A, -B and -C and b2-microglobulin antigens in human choriocarcinoma cell lines. J. Exp. Med. **152:** 11s–17s.

13. WELSH, K., G. DORVAL, K. NILSSON, G. CLEMENTS & H. WIGZELL. 1977. Quantitation of b2-microglobulin and HLA on the surface of human cells. II. In vitro cell lines and their hybrids. Scand. J. Immunol. **6:** 265–271.

14. LAMPSON, L. A., J. P. WHELAN & C. J. LAWTON. 1984. Microscopic analysis of monoclonal antibody binding to agarose-embedded cell lines by the unlabeled antibody peroxidase antiperoxidase (PAP) technique. *In* Biotechnology and Bioscience. I. Monoclonal Antibodies and Functional Cell Lines: Progress and Applications. R. Kennett, K. Bechtol & T. McKearn, Eds. Plenum Publishing Corp. New York. (In press.)

15. WILLIAMS, K., D. HART, J. FABRE & P. MORRIS. 1980. Distribution and quantitation of HLA-ABC and DR(Ia) antigens on human kidney and other tissues. Transplantation **29:** 274–279.

16. LAMPSON, L. A., C. A. FISHER & J. P. WHELAN. 1983. Striking paucity of HLA-A,B,C and b2-microglobulin on human neuroblastoma cell lines. J. Immunol. **130:** 2471–2478.

17. LAMPSON, L. A. 1984. Molecular bases of neuronal individuality: Lessons from anatomical and biochemical studies with monoclonal antibodies. *In* Biotechnology and Bioscience. I. Monoclonal Antibodies and Functional Cell Lines: Progress and Applications. R. Kennett, K. Bechtol & T. McKearn, Eds. Plenum Publishing Corp. (In press.)

Immunoenzyme Techniques and Their Application to Diagnostic Studies[a]

CLIVE R. TAYLOR

Department of Pathology
University of Southern California
Los Angeles, California 90033

Immunohistology represents a fortunate marriage, bringing together the specificity of immunologic reactions with the color and morphologic finesse of histochemical techniques. Charles Culling, author of one of the more popular histochemical texts,[5] writes that an essay by Raspail in 1830, "Essai de Chimie Microscopique Applique a la Physiologie," is generally regarded as the beginning of histochemistry. Culling[5] goes on to state that "the object of staining tissue sections is to impart color and therefore contrast to specific tissue constituents." Lillie[9] has it that "the purpose of staining is to make more evident various tissue and cell constituents and extrinsic materials," while Mann, in 1902,[10] is reported[5] to have said that, "it is not sufficient to content ourselves with using acid and basic dyes speculating on the acid or basic nature of the tissues. . . . We should find staining reactions which will indicate the presence of certain elements. . . ." Histochemical stains attempt to exploit predictable chemical reactions between the components of a stain and the various molecules that are to be found in tissues. Histochemical stains have the potential for specific demonstration of tissue elements, but in practice specific recognition of particular cell types is limited by the fact that many enzymes and chemical reactions are common to many different cells.

In contrast, immunohistologic methods offer the potential for specific demonstration of an almost limitless range of tissue components, according to the presence of small three-dimensional "charge/shape" moieties that constitute distinct antigenic determinants, against which it is possible to prepare specific antibodies. These antibodies can then be used as specific probes, or labels, to localize the corresponding antigenic determinants within tissue or cell preparations.

IMMUNOFLUORESCENCE VERSUS IMMUNOPEROXIDASE

Labeled antibody methods utilize some form of label as a means of visualizing the localization of antibody within tissue sections. The immunofluorescence technique was the first of the labeled antibody methods to achieve scientific credibility; its impact upon pathology and immunopathology has been enormous and its applications today are manifold. However, the very success of immunofluorescence techniques led to recognition of a need for development of methods applicable to examination by orthodox light microscopy, methods of direct utility and value to the surgical pathologist.

Immunofluorescence techniques do not permit study of the detailed morphologic features of positively stained cells or of unstained cells; immunoenzyme techniques do,

[a]Supported in part by National Institutes of Health grants 5P01 CA19449 and 5RO1 CA29211 and by a gift from Hybritech Inc.

and this is one of the principal advantages ascribed to the immunoenzyme method. The realization that immunoenzyme techniques could be utilized to demonstrate a wide range of antigens in "routinely processed" tissues,[21] the very tissues upon which the diagnostic pathologist normally practices his art, also proved important, greatly increasing the potential usefulness of the method as a "special stain" in diagnostic surgical pathology.

HISTOPATHOLOGY—ART OR SCIENCE?

Histopathology, although part of the medical sciences curriculum, is viewed with some measure of scorn by practitioners of the "pure" sciences, the micromolecular chemists, the cell physiologists, even the cellular immunologists and immunochemists, who regard histopathology as more akin to the collecting of butterflies than the splitting of the atom:

> [it is] fashionable in some quarters to consider morphology little more than an outmoded
> (if agreeable) pastime, at a level of scientific interest about that of a butterfly collection.
>
> Marcel Bessis, 1976[2]

It is certainly true that, since its inception as a diagnostic method more than one hundred years ago, the practice of diagnostic histopathology has depended as much upon accumulated experience and subjective opinion as upon the application of scientific principles. Nonetheless, it should not be forgotten that for many diseases, particularly the malignant neoplasms, the final diagnosis continues to rest upon examination of a tissue biopsy by a histopathologist; whether it be art or science, it is the final word.

The excitement engendered by the availability of immunoenzyme techniques applied to routinely processed tissues relates to the fact that, for the first time, a truly scientific means of identifying tissue components and cell types had become available to the surgical pathologist.[19] An objective measurement of the interest that this technique has produced is to be found by recourse to the medical literature; in 1974 there were less than ten papers relating to the application of immunoenzyme techniques in diagnostic pathology, in 1981 there were almost 1,000; in 1974 there was grudging recognition that perhaps it was possible to detect one or two antigens in "routinely" processed tissues, in 1982 there was documentation attesting to the demonstration of more than 100 different antigens in paraffin sections (TABLE 1).

IMMUNOENZYME TECHNIQUES APPLIED TO
DIAGNOSTIC SURGICAL PATHOLOGY

The principles of immunoenzyme techniques are quite straight forward, namely the utilization of an antibody for localization of antigen within a tissue section and the attachment of some form of label to the antibody in order to visualize the sites of antigen/antibody reaction. Many different approaches have developed, all with the ultimate goal of attaching the maximum amount of label to the antigen (FIGURES 1–7).[20] Many of the initial studies utilized antibody directly labeled with horseradish peroxidase (conjugated antibody—direct and indirect techniques). Subsequently other means of labeling were sought, to avoid some of the difficulties inherent in chemical conjugation of antibody with enzyme. This was the principal motivation for the

TABLE 1. Antigens Demonstrable in "Routinely" Processed Paraffin Sections

Hormones	Other Cellular Components	Infectious Agents
ACTH	Lysozyme (muramidase)	Herpes I & II
GH	Lactoferrin	Hepatitis B surface Ag
LH	Transferrin	Mouse mammary tumor virus Ag
LTH	Ferritin	Rubella
FSH	Hemoglobin A	Baboon endogenous virus
TSH	Hemoglobin F	Measles Ag
Parathormone	Myoglobin	Respiratory syncytial virus
Calcitonin	Actin	Buffalo pox
Thyroxine	Myosin	Legionella
Thyroglobulin	Keratin	Klebsiella
hCG	α-1-antitrypsin	Group B strep
βhCG	α-1-chymotrypsin	Influenza
Testosterone	α-fetoprotein	Polio
Estradiol	Carcinoembryonic Ag	Varicella-zoster
Progesterone	Mammary epithelial membrane Ag	CMV
Insulin	Hepato-renal Ag	Parainfluenza
Glucagon	Pancreatic Ag	Lymphocytic choriomeningitis virus
Somatostatin	Melanoma Ag	Moloney virus
VIP	Prostatic acid phosphatase	Friend virus
Gastrin	Prostate specific Ag	Shope fibroma virus
Secretin	Glial fibrillary acidic protein	SV 40
Motilin	Tyrosine hydroxylase	Human papilloma virus
Neurotensin	Myelin basic protein	Distemper virus
Cholecystokinin	Enkephalin (like) Ag	Rotavirus
Renin	Substance P	Polyoma virus
Vasopressin	Enolases	Chlamydia
Oxytocin	Adenosine desminase	Mycoplasma
Neurophysin	Terminal transferase	Trichomonas
	Carbonic anhydrase	Toxoplasma
Receptors	Cathepsin D	Trichophyton
Transferrin receptor protein	Converting enzyme	
Estrogen receptor protein	Factor VIII-related Ag	Immunoglobulin Components
	Creatine kinase	κ & λ light chains
Other Tissue Components	Intestinal mucin Ags	γ, α, μ, δ & ϵ heavy chains
Laminin	HLA Ags	J chain
	Blood group Ags	Secretor piece
Fibronectin	Surfactant apoprotein	
Collagens	S100 Ag	Complement Components
Amyloid A protein		C_1, C_3

From the literature, particularly References 6, 12, 20, 24.

development of the enzyme-bridge and PAP methods.[20] Biotin-avidin techniques sought to exploit the high affinity of avidin for biotin, usually employing a biotinylated antibody followed by avidin conjugated with horseradish peroxidase.[7] The PAP method was claimed by many to show a greatly enhanced sensitivity compared to early conjugate methods;[20] similarly the biotin-avidin technique was modified so that, with the application of a large preformed avidin-biotin complex (the ABC method), this technique also achieved a very high order of sensitivity.[7] The use of protein A, either as

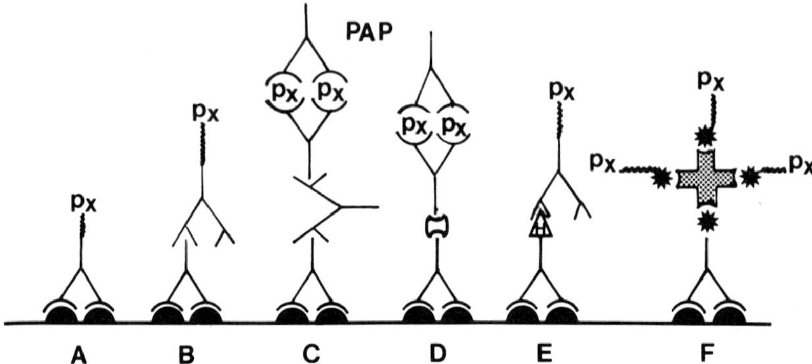

FIGURE 1. Immunoenzyme techniques. A and B: direct and indirect conjugate methods. C and D: PAP methods—in C using specific antibody to the primary antibody and to the PAP complex (PAP) as the linking reagent, in D using Protein A ( ) as the linking reagent (in this latter instance the primary antibody and the PAP reagent could be prepared in different species). E: the hapten-antibody method—the primary antibody is conjugated to a hapten ( ); the secondary antibody, labeled with peroxidase, has specificity against the hapten. F: the ABC method—the primary antibody is biotinylated ( ); to this is added a complex of avidin/bound to biotin/peroxidase. P_x = horseradish peroxidase; = specific antibody.

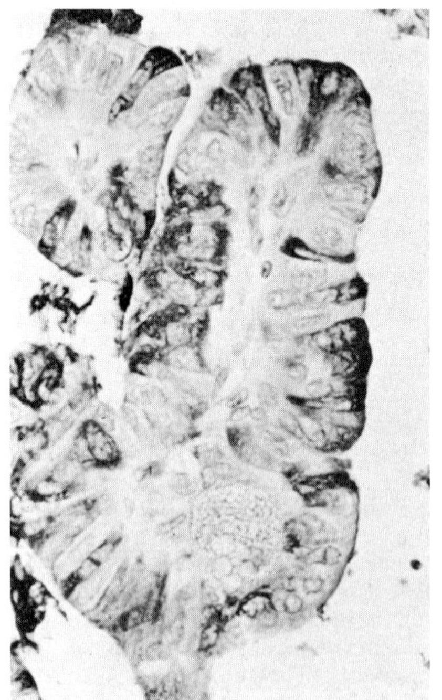

FIGURE 2. Paraffin section: papillary carcinoma of thyroid stained for thyroglobulin (positive cells appear black). PAP method (Immulok Histoset Kits were a gift from Ortho Inc., 1019 Mark Avenue, Carpinteria, CA 93013.) counterstain hematoxylin, × 400.

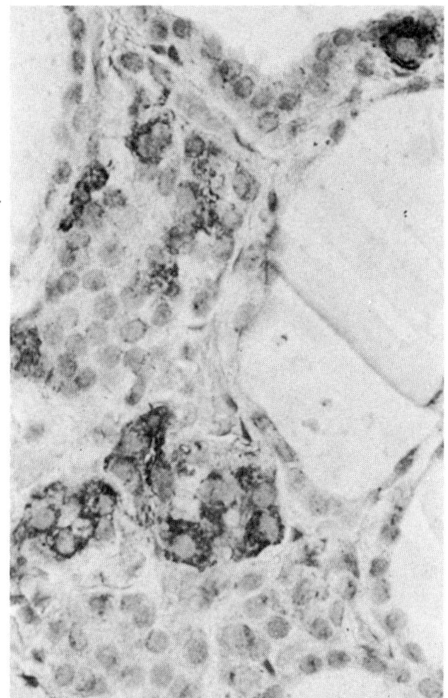

FIGURE 3. Paraffin section: thyroid gland. Staining of "C" cells for calcitonin (positive appears black). PAP method (Immulok Histoset) counterstain hematoxylin, × 400.

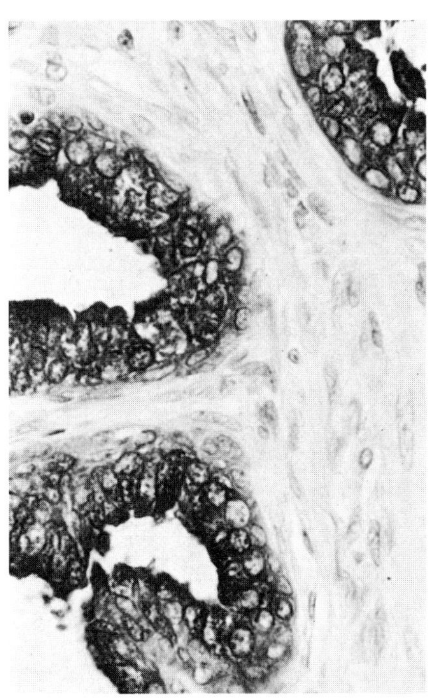

FIGURE 4. Paraffin section: prostate stained for prostatic acid phosphatase using a monoclonal antibody (These monoclonal antibodies were a gift from Hybritech Inc., 11085 Torreyana Road, La Jolla, CA 92121.) Positive staining appears black; absorbed conventional antisera give comparable results. ABC method, counterstain hematoxylin, × 400.

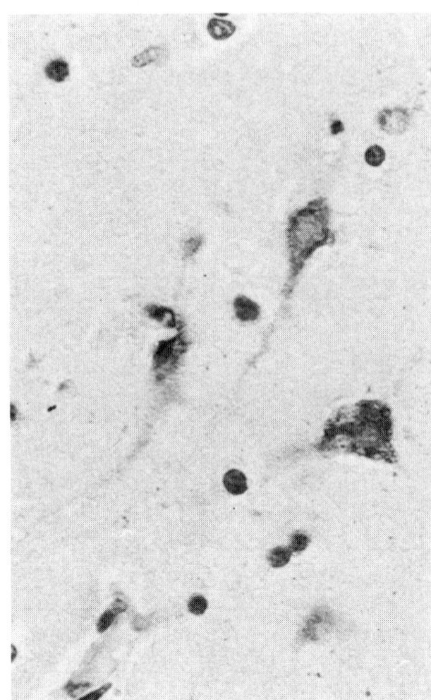

FIGURE 5. Paraffin section of brain from a case of encephalitis, stained for herpes virus antigen. PAP method (Immulok Histoset), counterstain hematoxylin, × 400.

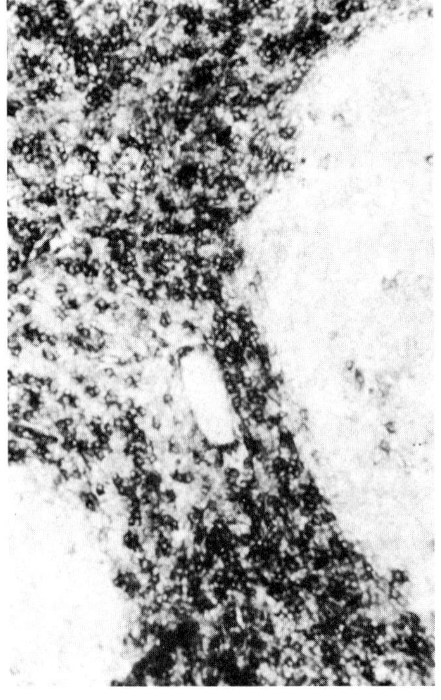

FIGURE 6. Frozen section: tonsil stained with T101, a monoclonal antibody against T lymphocytes (Hybritech). Positive T cells (black) are distributed mainly in the interfollicular area. Parts of two follicles are seen; they contain only rare positive cells. ABC method, counterstain hematoxylin, × 250.

a linking reagent or directly labeled with peroxidase, has been advocated for two principal reasons: first, the extreme rapidity of the protein A-antibody binding makes the method swift; second, that protein A will link to IgG antibodies from many different species makes the method very versatile.[15] This is particularly useful where available primary antibodies are from relatively unusual sources and secondary or linking reagents of high specificity are not readily available (e.g., armadillo antiserum against the leprosy bacillus). Finally, proponents of the hapten-antibody method claim that this variation of the technique may serve to produce a further increase in sensitivity, while at the same time enhancing specificity by reduction of nonspecific binding and cross-reactions of secondary or linking reagents.[3]

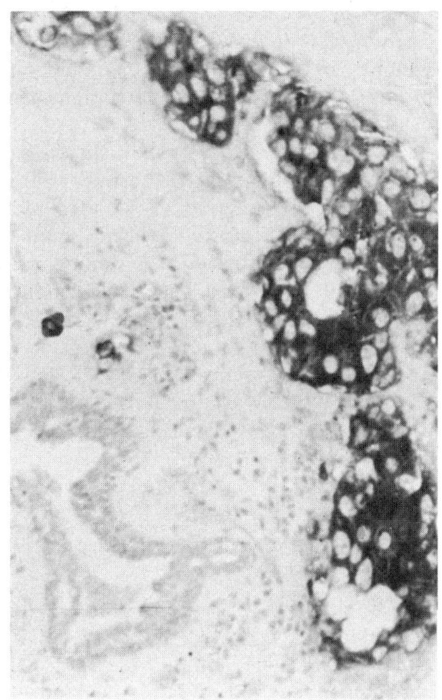

FIGURE 7. Paraffin section of breast cancer stained with a "human" monoclonal antibody (positive cells appear black). This antibody, prepared by Dr. Ashraf Imam at the University of Southern California), is of interest in that the carcinoma cells (upper right) stain intensely, the normal cells (lower left) do not. Four stage PAP method, counterstain hematoxylin, × 250.

Horseradish peroxidase remains the most widely used enzymatic label, though alkaline phosphatase and glucose oxidase have been used successfully by a number of investigators. Glucose oxidase, as a label, avoids one of the principal disadvantages inherent in the use of horseradish peroxidase, namely the occurrence of an endogenous enzyme within human tissues; however, this problem can be circumvented in most circumstances by the "blocking" of endogenous peroxidase activity,[20] and immunoperoxidase techniques still remain the most widely used.

Of the various "chromogens" used in conjunction with horseradish peroxidase, diaminobenzidine continues to enjoy widespread use, although in powder form it is suspected as a carcinogen and should be handled with all due precaution. Alternative chromogens, producing a variety of different colored reaction products, include alpha-1-chloronaphthol, Hanker-Yates reagent, and amino-ethyl-carbazole.[6,20]

SURVIVAL OF IMMUNOREACTIVITY WITHIN FIXED PARAFFIN-EMBEDDED TISSUES

Clearly, labeled antibody methods, however sensitive, will detect only those antigenic determinants that survive fixation and processing substantially intact. To date it has not proved possible to formulate a general rule describing the effects of fixation upon antigenicity; in general terms, the longer the period of fixation the less of the antigen remains in an immunoreactive state. Good results for a wide variety of antigens have been described with Zenker's fixative, B-5 fixative, Bouin's fixative, and neutral buffered formalin,[1,20] with or without prior treatment by proteolytic enzymes.[11] Some antigens appear to be very resistant to the adverse effects of fixation and/or processing, while others are extremely labile and at present appear only to be detectable, with any degree of reliability, in cryostat sections subjected to minimal fixation. Labile antigens include those present at low density within cell surface membranes. Resistant antigens particularly include the smaller polypeptides (e.g., pituitary hormones) present at a high concentration within the cell cytoplasm or within cytoplasmic organelles.

A final general observation should be recorded: an antibody that appears specifically to detect an antigen in serum or in tissue fluids will not necessarily show the same specificity, or the same ability to detect that antigen, within tissue sections. This reservation particularly applies to monoclonal antibodies consisting, as they do, of a single molecular species of immunoglobulin, with a single affinity and specificity for one binding site; excellent specific high affinity binding by monoclonal antibody of antigen in serum shows little correlation with detection of the same antigen in tissue.[14]

For diagnostic purposes the antigens detectable will be considered in two major groups: the first group containing those antigens that are demonstrable within routinely processed tissues, and the second group those antigens that can be usefully detected only in cryostat sections.

ANTIGENS DEMONSTRABLE IN ROUTINELY PROCESSED TISSUES

Of the antigens listed in TABLE 1, experience with some is very limited, and not all have established diagnostic values. Many of the reports in the literature, however, allude to the value, actual or potential, of immunohistologic stains in histopathology, both for conducting investigative studies upon stored paraffin blocks and as special stains in routine diagnostic surgical pathology (TABLE 2).

ANTIGENS DEMONSTRABLE ONLY IN CRYOSTAT SECTIONS

Labile antigens, particularly susceptible to the adverse effects of fixation or processing, and antigens present only in small amounts in tissues, are best demonstrated in frozen sections subject to minimal fixation. Curiously, many of the antigens that fall into these categories are those recognized by use of monoclonal antibodies. The nature of this problem is well illustrated by a study performed in our laboratory:[14] of nine different monoclonal antibodies against prostatic acid phosphatase, only three gave positive staining in paraffin sections, whereas all nine gave intense staining in frozen sections, a finding possibly related to the loss, following fixation, of some of the antigenic determinants of the prostatic acid phosphatase molecule. Large numbers of

TABLE 2. Some Applications of Immunoperoxidase Stains in Diagnostic Surgical Pathology

Antigen	Investigation and Diagnosis of
Immunoglobulin components, J chain	Monoclonal (monotypic) pattern in B cell neoplasms, including myeloma, plasmacytoma, heavy chain disease, plasmacytoid lymphocytic lymphoma, some B cell lymphomas (paraffin and/or frozen sections)
Lysozyme (muramidase) Alpha-1-chymotrypsin Alpha-1-antitrypsin Anti-monocyte antibodies	Monocyte/histiocyte proliferations (specific monocyte antibodies, frozen sections only)
Lysozyme	Lacrimal and salivary gland neoplasms, Paneth cells
HBA, HBF	Erythroid precursors, physiologic erythroid hyperplasia
Alpha fetoprotein Alpha-1-antitrypsin	Liver cancer, germ cell tumors with yolk sac differentiation; liver in alpha-1-antitrypsin deficiency
hCG SPl Placental lactogen	Trophoblast and trophoblastic tumors, some breast cancers
ACTH, LH, GH, TSH, FSH, LTH	Hyperplasia and adenoma of pituitary
Calcitonin	Medullary C cell tumor, thyroid C cell hyperplasia in familial cases
Thyroglobulin, thyroxine	Primary or metastatic thyroid tumors, thyroid function
Prostatic acid phosphatase Prostate epithelial antigen	Primary or metastatic prostatic tumors, normal prostate
Factor VIII-related antigen	Kaposi sarcoma, angiosarcoma
Estrogen, testosterone (estrogen receptor protein)	Hormone producing tumors; possibly estrogen receptor
Insulin, glucagon, gastrin, VIP, PTH, calcitonin, ACTH	Production by endocrine tumors: ectopic hormone production
Myoglobin, actin, myosin	Muscle tumors
Keratin	Squamous carcinoma from other malignancies
CEA, mammary antigen, Pancreatic Ag, hepato-renal Ag	Various forms of adenocarcinoma
Melanoma antigens	Melanoma (chiefly in frozen sections)
Lymphocyte surface antigens (Pan T, helper and suppressor phenotypes, OKT6, OKT10, OKT11, etc); Common leukocyte Ag, Common ALL Ag	Lymphocytic neoplasms (chiefly frozen sections)
GFA, myelin basic protein, Tyrosine hydroxylase, enolases	CNS cells, GFA in astrocytomas
Herpes I & II, Hepatitis B S-Ag, Measles Ag, Toxoplasma, Group B Strep (etc., TABLE 1)	Specific infectious diseases in surgical and autopsy sections; Herpes in cervical smears; Hepatitis B carriers in liver biopsies

From the literature, particularly References 6, 12, 13, 16, 18, 20, 22–25. Studies may be performed upon paraffin sections, except where noted.

monoclonal antibodies are now available; new reports appear daily. That some of these[4,8,17] appear to have tissue specificity, or even specificity against tumor antigens, presents a tantalizing prospect for the pathologist, particularly with the possibility that some may eventually be applicable to fixed paraffin sections as well as to frozen sections.

TABLE 3. Currently Available Immunostaining Kits, Containing Complete Pre-Titered Reagents

Enzymes	Immunoglobulins
Lysozyme	IgA
Prostatic acid phosphatase	IgG
	IgM
	J chain
Oncofetal antigens	Kappa
Alpha fetoprotein (AFP)	Lambda
Carcinoembryonic antigen (CEA)	
	Hormones
Viruses	Adrenocorticotrophic hormone (ACTH)
	Calcitonin
Hepatitis B surface antigen	Chorionic gonadotropin (hCG)
Herpes simplex virus I & II	Chorionic gonadotropin beta subunit (hCG)
	Estradiol
	Follicle stimulating hormone (FSH)
Tissue-specific markers	Gastrin
and other cellular antigens	Glucagon
	Growth hormone (GH)
Alpha-1-antitrypsin	Insulin
Factor VIII-related antigen	Luteinizing hormone (LH)
Ferritin	Parathyroid hormone (PTH)
Fibrinogen	Prolactin
Glial fibrillary acidic protein (GFA)	Somatostatin
Hemoglobin A	Testosterone
Keratin	Thyroglobulin
Myoglobin	Thyroxine
Transferrin	Thyroid stimulating hormone (TSH)
Prostate specific antigen (PSA)	Vasoactive intestinal peptide (VIP)
Pregnancy specific beta-1-glycoprotein (SPl)	

List courtesy of Dr. G. Kledzik, Director of Research, ORTHO Inc., 1019 Mark Avenue, Carpinteria, CA 93013.

THE IMMUNOPEROXIDASE METHOD AS A ROUTINE SPECIAL STAIN

For the practicing surgical pathologist, the question remains as to how immunoenzyme methods, so attractive in theory and so full of potential, can be integrated into the routine surgical pathology laboratory; time is at a premium, as is morphology; neither the expertise nor the knowledge exists for acquiring all of the necessary reagents and tissue controls or for performing the numerous checkerboard titration studies in setting up these staining procedures. The availability of "staining kits" (TABLE 3), containing all of the necessary reagents mutually titered to optimum proportions, goes some way towards meeting this difficulty, producing a range of special stains that are available to all, readily performed, and consistently reproduced, and that the pathologist can easily learn to interpret and use in conjunction with more traditional staining procedures for reaching a histologic diagnosis.

ACKNOWLEDGMENTS

More than sixty of the antigens described herein have been successfully demonstrated in the University of Southern California laboratories. This work has involved many colleagues to whom credit is due: Doctors J.W. Parker, R.J. Lukes, P.K. Pattengale, E. Yanagihara, P. Meyer, A. Cramer, N. Levy, M. Forni, N. Tuzuner, T. Rea, R. Modlin, and F. Hofman; with the excellent technical services of Barbara Felder, Barbara Byrne, Ray Russell, Mary Drushella, Mary Russell, Tilda Cvitanic, Marcellino Pantangco, Cathleen Cooper, Ann Farley, and Wesley Naritoku. The manuscript was prepared by Betty Redmon.

REFERENCES

1. BANKS, P. M. 1979. Diagnostic applications of an immunoperoxidase method in hematopathology. J. Histochem. Cytochem. **27:** 1192–1194.
2. BESSIS, M. 1977. Blood Smears Reinterpreted: xiii. Springer. Berlin.
3. CAMMISULI, S. 1981. Hapten-modified antibodies specific for cell surface antigens as a tool in cellular immunology. *In* Immunological Methods. I. Lefkovits & B. Pernis, Eds. **2:** 139–162. Academic Press. New York.
4. COLCHER, D., P. H. HAND, M. NUTI & J. SCHLOM. 1981. A spectrum of monoclonal antibodies reactive with human mammary tumor cells. Proc. Natl. Acad. Sci. USA **78:** 3199–3203.
5. CULLING, C. F. A. 1974. Histopathological and Histochemical Techniques: 1 and 151. Butterworths. London.
6. DeLELLIS, R. A. 1981. Diagnostic Immunohistochemistry. Masson. New York.
7. HSU, S. M., L. RAINE & M. FANGER. 1981. Use of avidin-biotin peroxidase complex (ABC) in immunoperoxidase techniques: a comparison of ABC and unlabeled antibody (PAP) procedures. J. Histochem. Cytochem. **29:** 577–580.
8. IMAM, A. & Z. A. TOKES. 1981. Immunoperoxidase localization of a glycoprotein on plasma membrane of secretory epithelium from human breast. J. Histochem. Cytochem. **29:** 581–584.
9. LILLIE, R. D. 1965. Histopathologic Technic and Practical Histochemistry: 107. McGraw Hill. London.
10. MANN, G. 1902. Physiologic Histology. Oxford.
11. MEPHAM, B. L., W. FRATER & B. S. MITCHELL. 1979. The use of proteolytic enzymes to improve immunoglobulin staining by the PAP technique. Histochem. J. **11:** 345–357.
12. MUKAI, K. & J. ROSAI. 1980. Applications of immunoperoxidase techniques in surgical pathology. *In* Progress in Surgical Pathology. C. M. Fenoglio & M. Wolff, Eds. **1:** 15–49. Masson. New York.
13. NADJI, M., A. R. MORALES, J. ZIEGLES-WEISSMAN & N. S. PENNEYS. 1981. Kaposi's sarcoma: immunohistologic evidence for an endothelial origin. Arch. Pathol. Lab. Med. **105:** 274–275.
14. NARITOKU, W. Y. & C. R. TAYLOR. 1982. A comparative study of the use of monoclonal antibodies using three different immunohistochemical methods: an evaluation of monoclonal and polyclonal antibodies against human prostatic acid phosphatase. J. Histochem. Cytochem. **30:** 253–260.
15. NOTANI, G. W., J. A. PARSONS & S. L. ERLANDSEN. 1979. Versatility of Staphylococcus aureus protein A in immunocytochemistry. J. Histochem. Cytochem. **27:** 1438–1444.

16. POPPEMA, S., A. K. BHAN, E. L. REINHERZ, R. T. MCCLUSKEY & S. F. SCHLOSSMAN. 1981. Distribution of T cell subsets in human lymph nodes. J. Exp. Med. **153:** 30–41.
17. SHIMANO, T., R. M. LOOR, L. D. PAPSIDERO, M. KURIYAMA, R. G. VINCENT, T. NEMOTO, E. D. HOLYOKE, R. BERJIAN, H. O. DOUGLASS & T. M. CHU. 1981. Isolation, characterization and clinical evaluation of a pancreas cancer-associated antigen. Cancer **47:** 1602–1613.
18. STEIN, H., A. BONK, G. TOLKSDORF, K. LENNERT, H. RODT & J. GERDES. 1980. Immunohistologic analysis of the organization of normal lymphoid tissue and non-Hodgkin's lymphomas. J. Histochem. Cytochem. **28:** 746–760.
19. TAYLOR, C. R. 1978. Immunohistological approach to tumor diagnosis. Oncology **35:** 189–197.
20. TAYLOR, C. R. 1978. Immunoperoxidase techniques: theoretical and practical aspects. Arch. Pathol. Lab. Med. **102:** 113–121.
21. TAYLOR, C. R. & J. BURNS. 1974. The demonstration of plasma cells and other immunoglobulin containing cells in formalin-fixed, paraffin-embedded tissues using peroxidase labelled antibody. J. Clin. Pathol. **27:** 14–20.
22. TAYLOR, C. R., C. L. COOPER, R. J. KURMAN, U. GOEBELSMANN & F. S. MARKLAND, JR. 1981. Cancer **47:** 2634–2640.
23. TAYLOR, C. R., F. M. HOFMAN, E. YANAGIHARA, P. R. MEYER, R. J. LUKES & J. W. PARKER. 1982. Enzyme immunohistochemical studies in human lymphoproliferative diseases. *In* New Directions in Clinical Laboratory Assays. Masson. New York. (In press.)
24. TAYLOR, C. R. & G. KLEDZIK. 1981. Immunohistologic techniques in surgical pathology—a spectrum of "new" special stains. Human Pathol. **12:** 590–596.
25. TUBBS, R. R., K. SHEIBANI, R. A. WEISS, B. A. SEBEK & S. D. DEODHAR. 1981. Tissue immunomicroscopic evaluation of monoclonality of B-cell lymphomas: comparison with cell suspension studies. Am. J. Clin. Pathol. **76:** 24–28.

Single and Double Immunoenzymatic Techniques for Labeling Tissue Sections with Monoclonal Antibodies[a]

D. Y. MASON,[b] Z. ABDULAZIZ, AND B. FALINI

Nuffield Department of Pathology
University of Oxford
Oxford, United Kingdom

H. STEIN

Institute of Pathology
Kiel University
Kiel, West Germany

INTRODUCTION

The purpose of this paper is to review the immunoenzymatic techniques that are used in the authors' laboratories for the single and double immunoenzymatic labeling of antigens in tissue sections.[7,14] The majority of these antigens are detected using monoclonal antibodies (rather than conventional antisera) and in consequence, the techniques discussed below are primarily designed for use in conjunction with monoclonal primary reagents.

SINGLE IMMUNOENZYMATIC STAINING

Immunoperoxidase Techniques

Indirect Immunoperoxidase Labeling

This two-stage technique involves the use of a commercial peroxidase-conjugated anti-mouse Ig antibody (Dakopatts). In the authors' experience, this technique gives labeling reactions that are comparable in intensity to those obtained using the peroxidase–antiperoxidase (PAP) technique (see below). This may at first seem surprising, since it has been claimed that the PAP technique is substantially more sensitive than indirect immunoperoxidase procedures. However, the quality of commercially available peroxidase-conjugated anti-Ig reagents has improved dramatically in recent years as a result of the use of efficient coupling procedures. Consequently, the authors use the two-stage method extensively for both the primary screening of new hybridomas[10] and the detection of tissue antigens using previously characterized monoclonal antibodies.

Three technical points may be made concerning the indirect immunoperoxidase procedure. Firstly, it is usually necessary, when staining antigens in cryostat sections of

[a]Supported by the Leukaemia Research Fund.
[b]Address correspondence to: D. Y. M., Haematology Department, John Radcliffe Hospital, Oxford OX3 9DU.

human tissue, to block the cross-reactivity of the peroxidase conjugate against human Ig. This may be achieved by adding normal human serum to the conjugate or by preliminary exposure to a solid-phase immunoabsorbent. Secondly, new commercial conjugates (and also new batches of conjugate from an established supplier) should be screened to ensure that they function satisfactorily. Thirdly, if weak results are obtained using the two-stage method it may be possible to increase the intensity of labeling by adding a third stage, consisting of peroxidase-conjugated antibody against Ig from the species in which the second-stage conjugate is raised (i.e. when using the Dakopatts conjugate the third stage would consist of peroxidase-conjugated anti-rabbit Ig).

Peroxidase–Anti-Peroxidase (PAP) Technique

Use of the PAP technique for the labeling of monoclonal antibodies necessitates the preparation of PAP complexes containing anti-peroxidase raised in the mouse or rat. Because of the difficulties inherent in producing large volumes of polyclonal antibody in these species we have recently prepared a monoclonal mouse antibody specific for horseradish peroxidase, which can be used to prepare "monoclonal PAP" suitable for labeling a wide range of monoclonal antibodies.[8] However, as noted before, this technique is not of markedly greater sensitivity than the two-stage procedure. Furthermore, occasional monoclonal antibodies are encountered (e.g. OKT8) that give better results by the two-stage technique. Therefore, the PAP technique is used in the authors' laboratory principally for double-labeling techniques (see below) and also to economize on the use of the commercial peroxidase conjugate.

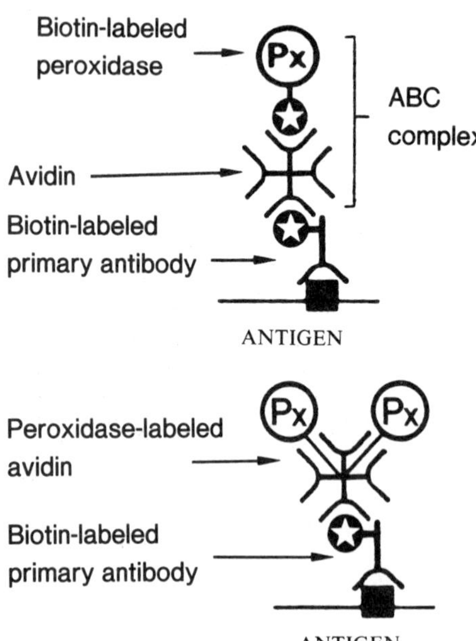

FIGURE 1. Schematic representation of two avidin-biotin procedures that have been found suitable for labeling tissue sections with monoclonal antibodies. In both techniques the primary monoclonal antibody is biotin labeled. In the ABC technique (Top) freshly prepared complexes of avidin- and biotin-labeled peroxidase are used as the second-stage reagent. In (Bottom) a covalent avidin–peroxidase conjugate is used as second-stage reagent.

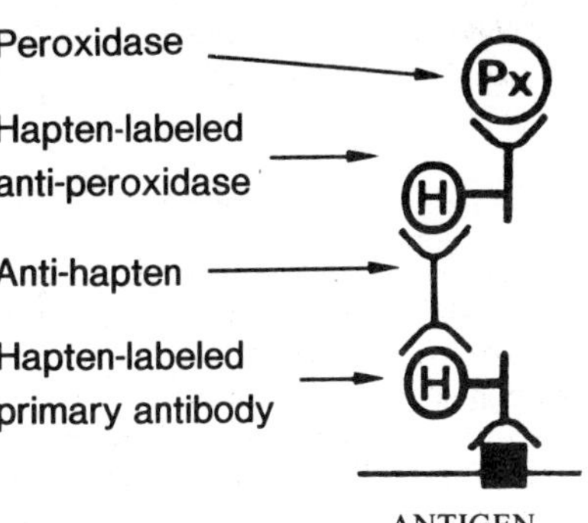

FIGURE 2. Schematic illustration of the hapten sandwich technique as used by the authors for immunoenzymatic labeling of monoclonal antibodies. Since this technique avoids the use of an anti-mouse Ig antibody as a second-stage reagent it is suitable for use in simultaneous double labeling immunoenzymatic techniques (see text).

Avidin-Biotin Systems

There has recently been great interest in the use of immunoenzymatic labeling procedures based upon the interaction between biotin and avidin (FIGURE 1).[2,3,11,17] We have evaluated these techniques in a system involving the use of biotin-labeled monoclonal anti-factor VIII, and found that avidin–biotin complexes (ABC), as originally described by Hsu *et al.* (FIGURE 1a), and a two-stage technique involving the use of an avidin-peroxidase conjugate (FIGURE 1b) give satisfactory results.

It is not apparent from our own results that biotin-avidin systems are clearly superior in terms of sensitivity to conventional procedures nor is there obvious unanimity on this point in the literature. Their main advantage appears to lie in the fact that they can be used for double labeling of monoclonal antibodies (see below). Furthermore, when attempting to detect an antigen in mouse tissues by a mouse monoclonal antibody, it would be possible to label the antibody with biotin, thereby avoiding the background staining caused by endogenous tissue Ig that is likely to occur if anti-mouse Ig is used as the second-stage reagent. It is also probable that biotin-labeled monoclonal antibodies will prove of value for analyzing the competition between pairs of monoclonal antibodies, e.g. by incubating sections with one monoclonal antibody, followed by a biotin-labeled second antibody (of suspected similar specificity) followed by an avidin-based revealing system.

Hapten Sandwich Techniques

This technique is illustrated schematically in FIGURE 2. In the authors' laboratory the hapten most widely used has been arsanilic acid, since this antigenic group can be coupled efficiently to mouse monoclonal immunoglobulin (without loss of antibody activity) using *p*-hydroxybenzimidate as described by Wofsy and colleagues.[16,18] The technique has also been used in the authors' laboratories to conjugate glutamic acid to monoclonal antibodies. A third hapten, of which the authors have no personal experience, is DNP; the original technique for coupling this antigenic group to antibody molecules[4] has recently been improved (B. Jasani, personal communication).

One rationale for using hapten sandwich techniques lies in their applicability to the double labeling of monoclonal antibodies (see below). In addition, haptenated monoclonal antibodies may also offer the advantages noted above for biotin-labeled antibodies, i.e. when staining mouse tissues with mouse antibodies or when analyzing competition between monoclonal antibodies.

Immuno-alkaline Phosphatase Labeling

The use of alkaline phosphatase as an enzymatic marker for monoclonal antibodies is justified by two considerations. Firstly, it is frequently difficult to block endogenous peroxidase activity in cryostat sections of tissues, such as bone marrow or spleen, without denaturing tissue antigens. In contrast, endogenous alkaline phosphatase in the tissues can be readily inhibited by the addition of levamisole to the enzyme substrate,[12] allowing specific reactions to be visualized without background due to endogenous enzyme. Secondly, as discussed below, alkaline phosphatase may be used in conjunction with peroxidase for double immunoenzymatic labeling of tissue antigens.[1,5,6,9]

Alkaline Phosphatase—Anti-alkaline Phosphatase (APAAP) Technique

This procedure (FIGURE 3) is closely analogous to the PAP procedure. Initially the authors employed rabbit antibodies against alkaline phosphatase to prepare APAAP complexes that could be used for the labeling of rabbit primary antisera.[5,6,9] More recently a series of monoclonal antibodies against calf intestine alkaline phosphatase have been prepared and give excellent results when used for the labeling of monoclonal antibodies.

Indirect Immuno-alkaline Phosphatase Technique

Two-stage indirect immuno-alkaline phosphatase techniques have been used by Ormerod and colleagues for labeling tissue antigens.[12,13] The principle disadvantage of this technique is the necessity of preparing covalent conjugates of antibody with alkaline phosphatase. The techniques for this type of conjugation are relatively inefficient and, in the authors' experience, commercial conjugates are frequently unsatisfactory. At present, therefore, the APAAP technique described above is used as the method of preference, although it is possible that in the future more efficient conjugates will increase the attraction of the indirect procedure.

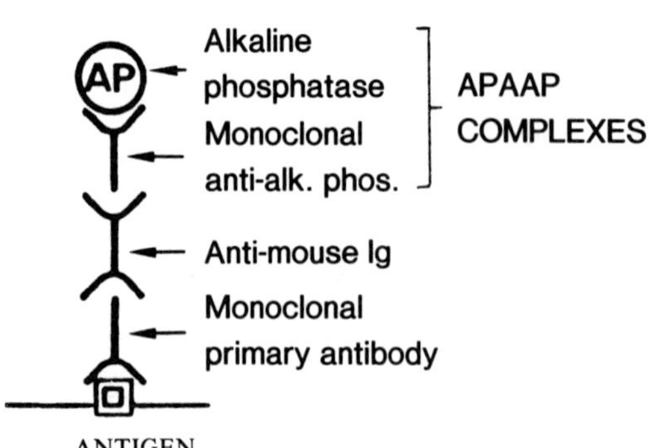

FIGURE 3. Schematic illustration of the APAAP immuno-alkaline phosphatase procedure, which gives immunohistological labeling fully comparable in terms of intensity and absence of background to that obtained with immunoperoxidase procedures. The APAAP procedure offers the advantage of avoiding background staining due to endogenous tissue peroxidase (e.g. when staining cryostat sections of human bone marrow or spleen).

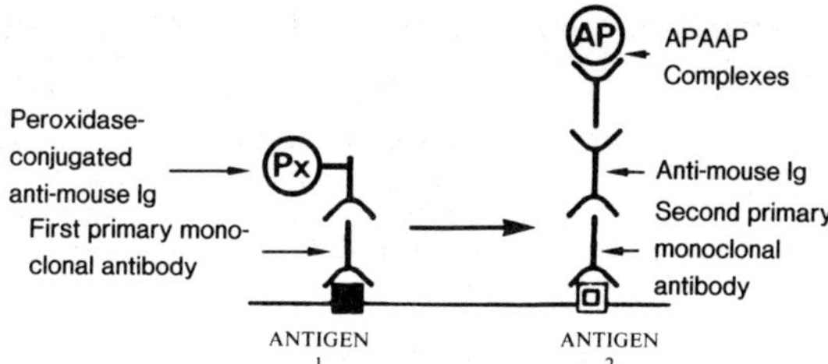

FIGURE 4. Double immunoenzymatic labeling of two monoclonal antibodies based upon the sequential application of an immunoperoxidase sandwich followed (after development of the peroxidase reaction) by application of an immuno-alkaline phosphatase sandwich.

DOUBLE IMMUNOENZYMATIC STAINING TECHNIQUES

The New "Chinese Restaurant Syndrome"

The production in the next few years of an increasingly large number of monoclonal antibodies specific for human tissue constituents is likely to give rise to a new form of the "Chinese Restaurant Syndrome," i.e. severe headache and confusion caused by attempts to distinguish between numerous different antibodies of apparently closely similar specificity. For this reason it will become increasingly important to be able to label simultaneously pairs of monoclonal antibodies and thereby discern their relative distribution patterns.

A major problem in carrying out double immunoenzymatic labeling using monoclonal antibodies is posed by the fact that, since these antibodies are normally raised in the same species (e.g. mouse or rat), conventional indirect labeling procedures involving the use of a second-stage anti-mouse or anti-rat Ig antibody will not be able to distinguish between the two primary antibodies. For this reason the authors have investigated techniques that will allow pairs of primary monoclonal antibodies to be labeled individually without cross-reactivity.

Sequential Immuno-staining Procedure

This technique (FIGURE 4) is a modification of the method proposed by Sternberger and Joseph.[15] It involves the application of an indirect immunoperoxidase sandwich followed by an unlabeled antibody immuno-alkaline phosphatase sandwich (APAAP technique). The rationale of this method is that the development of the first antibody label with diaminobenzidine effectively masks antigenic sites on the first primary antibody. Consequently the anti-mouse Ig antibody in the second stage of the APAAP sandwich fails to recognize the first monoclonal antibody.

The advantages and disadvantages of this method relative to alternative double labeling techniques are set out in TABLE 1. The major drawback of the technique is the time required for its completion.

Hapten Sandwich and Biotin-Avidin Techniques

The practical aspects of these techniques are detailed in TABLE 2 and FIGURES 1 and 2. In the author's experience the principal advantage of these techniques relative to

TABLE 1. Double Immunoenzymatic Staining with Monoclonal Antibodies

Technique	Advantages	Disadvantages
Sequential procedure (FIGURE 4)	It is not necessary to modify the two primary monoclonal antibodies (i.e. by hapten or biotin labeling) before use.	Time-consuming Entails risk that the anti-mouse Ig in the second sandwich will react with the monoclonal antibody in the first sandwich.
Double staining using a hapten sandwich procedure (FIGURE 2) combined with a second hapten sandwich or with a biotin-avidin technique (FIGURE 1)	Both sandwiches can be applied simultaneously	Each primary monoclonal antibody requires biotin or hapten labeling before use

the sequential procedure (TABLE 1) lies in the fact that reagents can be applied simultaneously rather than sequentially, thereby reducing the time required for completion of the labeling procedure. The major restriction to their wider use stems from the necessity to label each antibody individually (with biotin or hapten). Whilst this can readily be performed by published procedures, the quantities of antibodies required (i.e. at least 100 μg) is not available in many laboratories. It is hoped that an increasing number of commercial suppliers of monoclonal antibodies will make their products available in the future in biotin- and/or hapten-labeled form.

ACKNOWLEDGMENTS

We are grateful to Mrs. J. L. Cordell, Mrs. P. M. Knight, and Miss R.-E. Woolston for skillful technical assistance.

TABLE 2. Double Labeling of Human Tissue Cryostat Sections with Two Monoclonal Antibodies[a]

Fixation: Air-dried cryostat sections are fixed in acetone at room temperature for 10 min and then air-dried.

Hapten-labeled first monoclonal antibody: Incubation for 30 minutes.

Anti-hapten antiserum plus biotin-labeled second monoclonal antibody: Incubation for 30 minutes.

Immune complexes of alkaline phosphatase and haptenated–anti-alkaline phosphatase (haptenated APAAP) plus complexes of avidin and biotin–peroxidase (ABC): Incubation for 30 minutes.

Development of peroxidase reaction: Sections are incubated with diaminobenzidine/H_2O_2 for 5–10 minutes.

Development of alkaline phosphatase reaction: Sections are incubated with naphthol AS-MX plus Fast Blue BBN for 10–15 minutes.

Sections are washed in a conventional manner after the application of each reagent. The slides are mounted in an aqueous mountant at the completion of the double staining procedure.

[a]The two component sandwiches in this double labeling technique are illustrated schematically in FIGURES 1a and 2.

REFERENCES

1. FALINI, B., I. de SOLAS, C. HALVERSON, J. W. PARKER & C. R. TAYLOR. 1982. Double labelled antigen method for demonstration of intracellular antigens in paraffin-embedded tissues. J. Histochem. Cytochem. **30:** 21–26.
2. GUESDON, J. -L., T. TERNYNCK & S. AVRAMEAS. 1979. The use of avidin-biotin interaction in immunoenzymatic techniques. J. Histochem. Cytochem. **27:** 771–776.
3. HSU, S. M., L. RAINE & H. FANGER. 1981. Use of avidin-biotin peroxidase complex (ABC) in immunoperoxidase techniques: a comparison between ABC and unlabelled antibody (PAP) procedures. J. Histochem. Cytochem. **29:** 577–580.
4. JASANI, B., D. WYNFORD-THOMAS & E. D. WILLIAMS. 1981. Use of monoclonal anti-hapten antibodies for immunolocalisation of tissue antigens. J. Clin. Pathol. **34:** 1000–1002.
5. MASON, D. Y. & R. E. SAMMONS. 1978. Alkaline phosphatase and peroxidase for double immunoenzymatic labelling of cellular constituents. J. Clin. Pathol. **31:** 454–462.
6. MASON, D. Y., H. STEIN, M. NAIEM & Z. ABDULAZIZ. 1981. Immunohistological analysis of human lymphoid tissue by double immunoenzymatic labelling. J. Cancer Res. Clin. Oncol. **101:** 13–22.
7. MASON, D. Y., M. NAIEM, Z. ABDULAZIZ, J. R. G. NASH, K. C. GATTER & H. STEIN. 1982. Immunohistological applications of monoclonal antibodies. *In* Monoclonal Antibodies in Clinical Medicine. A. McMichael & J. Fabre, Eds.: 585–635. Academic Press. New York.
8. MASON, D. Y., J. L. CORDELL, Z. ABDULAZIZ, M. NAIEM & G. BORDENAVE. 1982. Preparation of PAP complexes for immunohistological labelling of monoclonal antibodies. J. Histochem. Cytochem. **30:** 1114–1122.
9. MASON, D. Y., Z. ABDULAZIZ, B. FALINI & H. STEIN. 1982. Double immunoenzymatic labelling. *In* Immunocytochemistry: Practical Applications in Pathology and Biology. J. M. Polak & S. Van Noorden, Eds.: 113–128. John Wright and Sons.
10. NAIEM, M., J. GERDES, Z. ABDULAZIZ, C. A. SUNDERLAND, M. J. ALLINGTON, H. STEIN & D. Y. MASON. 1982. The value of immunohistological screening in the production of monoclonal antibodies. J. Immunol. Meth. **50:** 145–160.
11. NARITOKU, W. Y. & C. R. TAYLOR. 1982. A comparative study of the use of monoclonal antibodies using three different immunohistochemical methods: An evaluation of monoclonal and polyclonal antibodies against human prostatic acid phosphatase. J. Histochem. Cytochem. **30:** 253–260.
12. PONDER, B. A. & M. M. WILKINSON. 1981. Inhibition of endogenous tissue alkaline phosphatase with the use of conjugates in immunohistochemistry. J. Histochem. Cytochem. **29:** 981–984.
13. SLOANE, J. P., M. G. ORMEROD, S. F. IMRIE & R. C. COOMBES. 1980. The use of epithelial membrane antigens in detecting micrometastases in histological sections. Brit. J. Cancer **42:** 392–398.
14. STEIN, H., A. BONK, G. TOLKSDORF, K. LENNERT, H. ROHT & J. GERDES. 1980. Immunohistologic analysis of the organisation of normal lymphoid tissue and non-Hodgkin's lymphomas. J. Histochem. Cytochem. **28:** 746–760.
15. STERNBERGER, L. A. & F. A. JOSEPH. 1979. The unlabelled antibody method. Contrasting colour staining of paired pituitary hormones without antibody removal. J. Histochem. Cytochem. **27:** 1424–1429.
16. WALLACE, E. F. & L. WOFSY. 1979. Hapten-sandwich labelling. IV. Improved procedures and non-cross-reacting hapten reagents for double-labelling cell surface antigens. J. Immunol. Meth. **25:** 283–289.
17. WARNKE, R. & R. LEVY. 1980. Detection of T and B cell antigens with hybridoma monoclonal antibodies: A biotin-avidin-horseradish method. J. Histochem. Cytochem. **28:** 771–776.
18. WOLFSY, Y., C. HENRY & S. CAMMISULI. 1978. Hapten-sandwich labelling of cell-surface antigens. Contemp. Topics. Molec. Immunol. **7:** 215–237.

Immunohistochemical Markers for Prostatic Cancer

MEHRDAD NADJI[a] AND AZORIDES R. MORALES

Department of Pathology
University of Miami/Jackson Memorial Hospital
Miami, Florida 33101

INTRODUCTION

Although Gomori devised a histochemical test for acid phosphatase in 1941,[12] it was not until 1947 that it became possible to differentiate between prostatic and nonprostatic acid phosphatases by inhibiting the enzymatic activity of the former by tartrate and the latter by formaldehyde.[1,2] Campbell and Cummins in 1951 used acid phosphatase histochemistry to differentiate between prostatic and nonprostatic carcinomas;[7] their method, however, proved to be unreliable in other hands,[9,11,23] where, even after formaldehyde inhibition, a number of normal and neoplastic nonprostatic tissues demonstrated acid phosphatase activity. The lack of organ specificity and the requirement for fresh or formalin-fixed tissue are important drawbacks of acid phosphatase histochemistry and have limited clinical use of these stains.

Prostate-specific acid phosphatase (PSAP) has been isolated in recent years and shown to be immunologically distinct from acid phosphatases from other sources.[3,4,8,10,15,22] Antibodies to PSAP have been developed, well characterized, and used in the immunohistologic investigation of prostatic origin of tumors. Immunofluorescence methods have been employed to detect neoplastic prostatic epithelium, but this technique requires fresh or frozen tissue and complex ultraviolet microscopy. Since the question of the possible prostatic nature of a tumor usually arises after the evaluation of paraffin-embedded material, both immunofluorescence and histochemical techniques are of limited value in the diagnosis of prostatic cancer. These difficulties are not encountered with the immunoperoxidase procedure since it is applicable to routinely fixed paraffin-embedded histologic material. Furthermore, the immunoperoxidase technique does not require ultraviolet microscopy and combines ideal morphology with higher sensitivity. The stained sections are also permanent and can be kept in the files for many years without any change in the intensity of the stain. These and many other advantages of the immunoperoxidase technique are responsible for its rapidly growing popularity in diagnostic pathology particularly in the histogenetic classification of various tumors. Prostatic tumors have recently been studied extensively by the immunoperoxidase method using antibodies to PSAP.[5,6,13,16–20]

In 1979 Wang and associates isolated and purified a prostate-specific antigen (PA) that did not represent PSAP.[24] The value of this antigen as a new immunohistochemical marker for prostatic cancer was shown recently by demonstration of its presence in all known prostatic tumors and its absence in nonprostatic malignancies.[21] Both PA and PSAP serve as specific immunohistochemical markers for prostatic cancer. Nevertheless, knowledge of the potential pitfalls of the technique, particularly those

[a]Address correspondence to: M.N., Department of Pathology (R-1), University of Miami, P. O. Box 016960, Miami, Florida 33101.

associated with nonspecificity of various antibodies, is essential for the correct interpretation of results and valid diagnostic conclusions.

TECHNICAL CONSIDERATIONS

The unlabeled antibody, peroxidase-antiperoxidase technique is the procedure of choice in our laboratory.[20,21] In this procedure, following deparaffinization, tissue sections are sequentially treated with optimal dilutions of primary antibody (anti-PA or anti-PSAP), linking antibody, and peroxidase-antiperoxidase. The results are then visualized by adding diaminobenzidine in the presence of hydrogen peroxide.

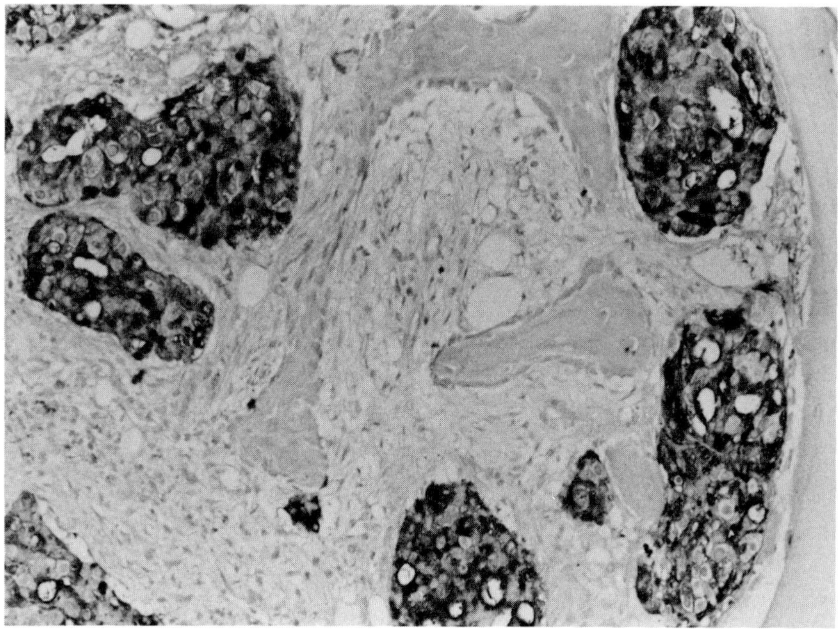

FIGURE 1. Positive reaction for PSAP in metastatic prostatic carcinoma in bone marrow. × 290.

The quality of tissue fixation is an important factor in preservation of various antigens including PA and PSAP. Rapid fixation of tissue and adequate time for penetration of the fixative into the sections are more important than the type of fixative used. Both buffered formalin and Bouin's solution are good for prostate immunohisto-chemistry. Decalcifying solutions do not interfere with the immunoperoxidase reaction, provided that adequate tissue fixation has taken place prior to decalcification.

A wide variety of conventional and monoclonal antibodies to PA and PSAP are now available from commercial sources or through individual investigators. It has been shown that the sensitivity and the specificities of these antibodies vary from one source to another.[18] It is therefore advisable that the specificity of the antibody be determined by using an array of positive and negative control tissues, before any diagnostic conclusions are drawn. Since some of the currently available antibodies to PSAP may exhibit cross-reactivity with the acid phosphatases of nonprostatic tissue,[14,16] we prefer to use anti-PA to demonstrate the prostatic origin of tumors. Thus far, PA has not been shown to be present in nonprostatic tissues.[21,24]

APPLICATION

Deciding on the histogenesis of tumors involving the prostate and adjacent tissues, including the prostatic urethra, urinary bladder, and rectum is not an uncommon diagnostic dilemma. Immunostaining for prostatic markers in tumors that involve both urinary bladder and prostate gland is particularly valuable because these tumors often demonstrate similar histomorphology. Obviously, the differential diagnosis is of vital importance to the patient, since the treatment and prognosis of bladder and prostatic carcinomas are entirely different. Carcinomas of the urinary bladder, however, are uniformly negative for PA or PSAP.[20,21]

A number of histologic variants of carcinoma of prostate gland may also present difficulties in histogenetic classification. However, we have found that all morphologic variants of prostatic cancer are positive for both PA and PSAP. These include: the endometrioid type, undifferentiated small cell variant, adenoid cystic like, carcinoid like, signet ring cell type, and the epithelial component of prostatic carcinosarcoma. We have found no correlation between the intensity of immunostaining for PA or PSAP and the degree of differentiation of prostatic tumors. It appears, however, that

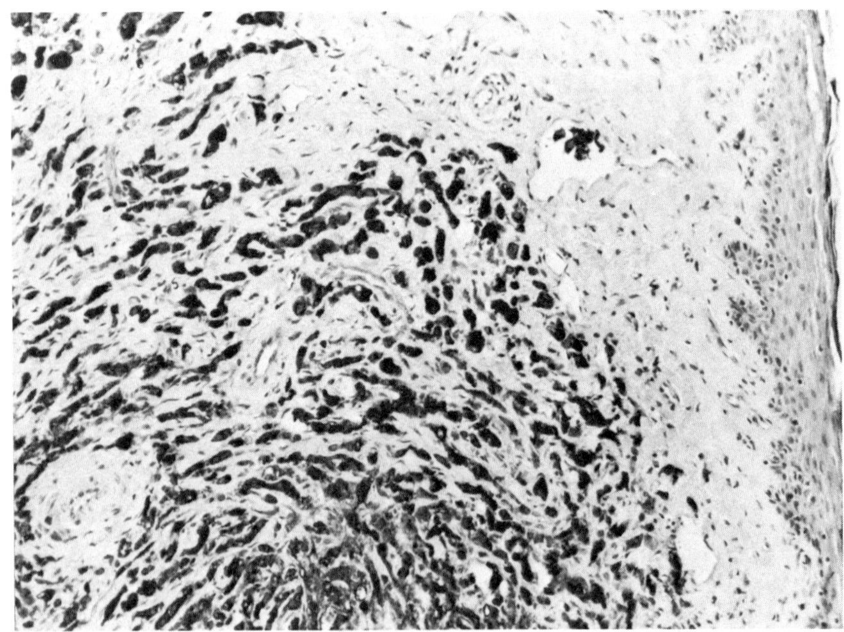

FIGURE 2. Positive reaction for PA in metastatic prostatic carcinoma in breast during estrogen therapy. × 186.

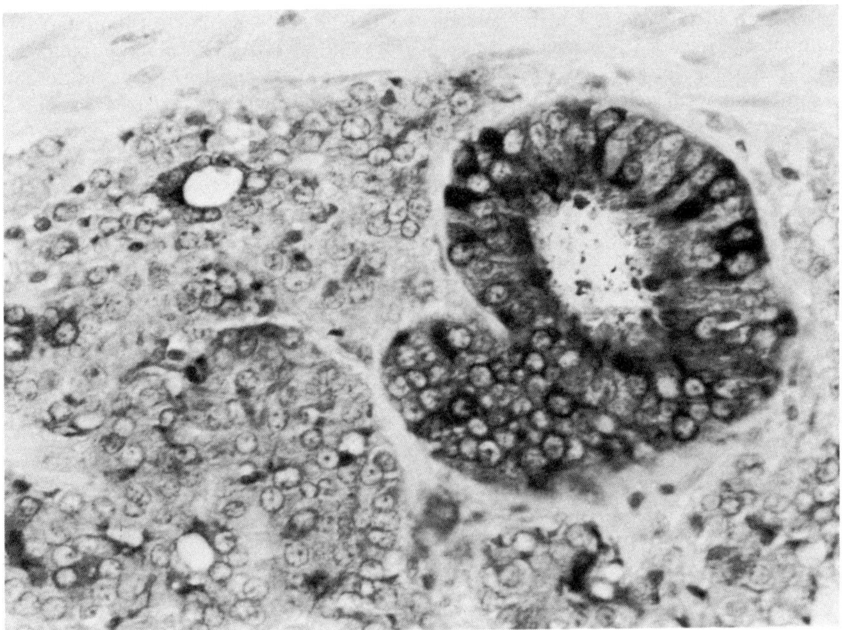

FIGURE 3. Positive, but heterogeneous staining of tumor cells for PA. Primary prostatic carcinoma. × 465.

the intensity of staining in these tumors is closely related to the quality of tissue fixation.

The greatest value of prostate marker immunohistochemistry is in its clear demonstration of the prostatic origin of tumors in patients with or without prior history of prostatic cancer, particularly since serum determination of PSAP is helpful in only 75% of patients with stage IV disease. A simple immunoperoxidase stain for prostatic markers may save a patient from a multitude of time-consuming and expensive biochemical tests, various roentgenologic procedures, and even unnecessary exploratory surgery, in search of a primary malignancy. Often, lymph node or bone marrow metastases are the first clinical manifestation of a prostatic malignancy, and occasionally these tumors may mestastasize to less common sites such as kidney, testis, liver, lung, intestine, central nervous system, skin, and breast (FIGURE 1). The last is of particular interest because to determine whether a given breast cancer, developing during estrogen therapy for prostatic carcinoma, is metastatic from the prostate or a hormone-induced primary mammary tumor may be difficult by histomorphology alone (FIGURE 2).

Immunohistochemistry for prostatic markers may also aid in differentiating metastatic tumors in the prostate from primary cancer of the gland. In our experience, four such tumors, two from the lung, one from colon, and one malignant melanoma were all negative for PA and PSAP. To this date, we have examined more than 300 cases of known primary and metastatic prostatic carcinomas, all of which reacted positively for both PA and PSAP. Conversely, none of the more than 160 examples of nonprostatic malignancies stained positively for either of the markers. The intensity of the reaction in positive tumors varies from cell to cell, from area to area, and from primary to metastatic lesion in the same patient (FIGURE 3). Also, the pattern of immunostaining for PA is usually different from that of PSAP in the same tumor. In double staining for PA and PSAP in the same tissue section utilizing two different chromogens, we have found that in a given tumor island there are few cells that stain for PA only, and few that react with PSAP only; the majority of tumor cells stain for both antigens. In well-differentiated prostatic carcinomas, both markers tend to localize in the supranuclear area and in the secretory pole of the cell.

SUMMARY

In a patient with an unclassifiable primary or metastatic neoplasm, with or without a history of prostatic cancer, immunostaining for PA or PSAP may prove invaluable. The procedure is simple, rapid, inexpensive, and extremely accurate in demonstrating the prostatic origin of tumors. It should be noted however, that the specificity of results is entirely dependent upon the specificity of the primary antibody, which should be meticulously defined before the procedure is used for diagnostic purposes.

ACKNOWLEDGMENT

The secretarial expertise of Lee Ann Moffett is gratefully acknowledged.

REFERENCES

1. ABDUL-FADL, M. A. M. & E. J. KING. 1947. Inhibition of acid phosphatases by formaldehyde. Biochem. J. **41:** 32.
2. ABDUL-FADL, M. A. M. & E. J. KING. 1948. Inhibition of acid phosphatase by D-tartrate. Biochem. J. **42:** 28–29.
3. ABLIN, R. J. 1973. Immunochemical identification of prostatic tissue-specific acid phosphatase. Clin. Chem. **19:** 786.
4. ABLIN, R. J. 1981. Prostatic histogenesis of metastases. Am. J. Clin. Pathol. **75:** 129–130.
5. ANSARI, M. A., R. L. PINTOZZI, Y. S. CHOI & R. F. LADOVE. 1981. Diagnosis of carcinoid-like metastatic prostatic carcinoma by an immunoperoxidase method. Am. J. Clin. Pathol. **76:** 94–98.
6. BURNS, J. 1977. Prostatic acid phosphatase in tissue sections revealed by the unlabeled antibody peroxidase-antiperoxidase method. Biomedicine **27:** 7–10.
7. CAMPBELL, J. H. & S. D. CUMMINS. 1951. Metastases, simulating mammary cancer, in prostatic carcinoma under estrogenic therapy. Cancer **4:** 303–311.
8. CHU, T. M., M. C. WANG, W. W. SCOTT, R. P. GIBBONS, D. E. JOHNSON, J. D. SCHMIDET, S. A. LOENING, G. R. POUT & G. P. MURPHY. 1978. Immunochemical detection of serum prostatic acid phosphatase: Methodology and clinical evaluation. Invest. Urol. **15:** 319–323.
9. FANGER, H. & B. E. BAKER. 1959. Histochemistry of breast diseases. I. Phosphatases. Arch. Pathol. **67:** 293–305.
10. FOTI, A. G., H. HERSCHMAN & J. P. COOPER. 1975. A solid-phase radioimmunoassay for human prostatic acid phosphatase. Cancer Res. **35:** 2446–2452.
11. GLUCK, E. 1952. Les phosphomonoestherases des neoformations mammaires. Etude histochemique. Acta Pathol. Microbiol. Scan. Suppl. **93:** 106–120.
12. GOMORI, G. 1941. Distribution of acid phosphatase in the tissues under normal and under pathologic conditions. Arch. Pathol. **32:** 189–199.
13. JOBSIS, A. C., G. P. DE VRIES, R. R. H. ANHOLT & G. T. B. SANDERS. 1978. Demonstration

of the prostatic origin of metastases. An immunohistochemical method for formalin-fixed embedded tissue. Cancer **41**: 1788–1793.

14. LEATHEM, A. & E. DINSDALE. 1979. Acid phosphatase as marker for carcinoma of prostate. Lancet **I**: 1029.

15. LEE, C. L., M. C. WANG, G. P. MURPHY & T. M. CHU. 1978. A solid-phase fluorescent immunoassay for human prostatic acid phosphatase. Cancer Res. **38**: 2871–2878.

16. LI, C. Y., W. K. W. LAM & L. T. YAM. 1980. Immunohistochemical diagnosis of prostatic cancer with metastasis. Cancer **46**: 706–712.

17. MAHAN, D. E., A. W. BRUCE, P. N. MANLEY & L. FRANCHI. 1980. Immunohistochemical evaluation of prostatic carcinoma before and after radiotherapy. J. Urol. **124**: 488–491.

18. NADJI, M. & A. R. MORALES. 1982. Immunohistiochemistry of prostatic acid phosphatase. Ann. N.Y. Acad. Sci. **390**: 133–141.

19. NADJI, M., S. Z. TABEI, A. CASTRO & A. R. MORALES. 1979. Immunohistological demonstration of prostatic origin of malignant neoplasms. Lancet **1**: 671–672.

20. NADJI, M., S. Z. TABEI, A. CASTRO, T. M. CHU & A. R. MORALES. 1980. Prostatic origin of tumors. An immunohistochemical study. Am. J. Clin. Pathol. **73**: 735–739.

21. NADJI, M., S. Z. TABEI, A. CASTRO, T. M. CHU, G. P. MURPHY, M. C. WANG & A. R. MORALES. 1981. Prostatic-specific antigen: An immunohistologic marker for prostatic neoplasms. Cancer **48**: 1229–1232.

22. PONTES, J. E., B. CHOE, N. ROSE & J. M. PIERCE, JR. 1977. Indirect immunofluorescence for identification of prostatic epithelial cells. J. Urol. **117**: 459–463.

23. REINER, L., A. M. RUTENBURG & A. M. SELIGMAN. 1957. Acid-phosphatase activity in human neoplasms. Cancer **10**: 563–576.

24. WANG, M. C., L. A. VALENZUELA, G. P. MURPHY & T. M. CHU. 1979. Purification of a human prostate specific antigen. Invest. Urol. **17**: 159–163.

Monoclonal Antibodies in Immunoenzyme Studies of Breast Cancer[a]

ROBERT D. CARDIFF,[b] CLIVE R. TAYLOR,[c]
SEFTON R. WELLINGS,[b] DAVID COLCHER,[d]
AND JEFFREY SCHLOM[d]

[b]*Department of Pathology*
University of California Medical School
Davis, California 95616

[c]*Department of Pathology*
University of Southern California Medical School
Los Angeles, California 90033

[d]*Experimental Oncology Section*
Building 37, Room 1A-07,
National Cancer Institute
Bethesda, Maryland 20205

INTRODUCTION

The development and use of various immunoperoxidase techniques has revolutionized surgical pathology.[15,19] Many antigens can be detected in paraffin sections of surgical specimens using immunoperoxidase. These antigens serve as functional markers that can help to identify and classify many morphologically undifferentiated tumors.[15] Thus, the surgical pathologist has a sound scientific basis for diagnosis that supplants authority and opinion of previous eras with objective data.[19]

Pathological diagnosis of breast cancer is generally not difficult for the experienced morphologist. However, some borderline cases of mammary dysplasia and some distant metastases defy the skills of even the most experienced individual. A number of antibodies have been applied using immunoperoxidase techniques for the detection of antigens in breast cancer. Casein and SEMA-70 appear to be breast-specific antigens, but are distributed unevenly and frequently not detectable in breast cancer cells.[1,12,17,18] Other antigens, such as CEA,[6,22] mouse mammary tumor virus,[14] SP-1,[9] hCG,[21] T antigen,[10] transferrin, and lactoferrin[13] have been detected in breast cancer cells, but are not sufficiently specific for diagnostic purposes.

An alternative approach to the problem has been to develop monoclonal antibodies (MCA) to human breast antigens.[2–5,8,20] By careful selection, MCA might be identified that are useful to the surgical pathologist. Of the several strategies available, we have used MCA that were developed against a metastatic human breast cancer.[2]

We have evaluated four monoclonal antibodies. We have been particularly concerned with technical and observer reproducibility from surgical pathologist to surgical pathologist. We report here a collaborative study between three institutions using the same reagents.

[a]Supported by National Cancer Institute grant no. 5R01CA 29211 and contract No1 CP 01008.

MATERIALS AND METHODS

The development and characterization of the monoclonal antibodies used in these studies have been described elsewhere.[2] Briefly, BALB/c mice were immunized with extracts of a human breast cancer metastatic to liver.[2] A standard fusion was performed and hybridomas were screened for antibody production using a solid-phase radioimmunoassay. Clones were selected for their reactivity with extracts of breast cancer metastases and in immunoperoxidase tests.[2,19] The clones were grown in ascites form in BALB/c mice. Ascites fluid was harvested and diluted 1:100.

The diluted ascites fluid was shipped to the two participating Pathology Departments (University of California, School of Medicine, Davis, CA and University of Southern California, School of Medicine, Los Angeles, CA).

The ascites fluids were stored at 4°C until used. The ascites fluids were applied at $1:10^4$ dilution to paraffin sections as previously described.[2] The peroxidase-antiperoxidase (PAP) procedure was used for part of the study.[19] However, the avidin-biotin complex technique gave a higher signal-to-noise ratio and was used through most of the study.[11]

TABLE 1. Positive Staining Breast Cancer Specimens

Study	Monoclonal Antibody (cases positive/total cases)			
Group	B6.2	B50.4	B72.3	B38.1
USC[a]	18/25	12/25	5/25	3/25
UCD[b]	18/26	11/24	7/24	2/24
Totals	36/51	23/49	12/49	5/49
Percentage positive	71%	47%	24%	11%

[a]USC, Univ. of So. California Medical School, Los Angeles.
[b]UCD, Univ. of California, Davis.

All sections were studied independently by two experienced individuals. They were scored for positive or negative reaction products and for the proportion of cells staining. The type of cells staining and the cytoplasmic distribution was also noted.

Blocks of tissues and primary and secondary reagents were exchanged by the participating institutions.

RESULTS

Proportion of Breast Cancer Staining

The proportion of positively staining breast cancers recorded for each MCA in both institutions is tabulated in TABLE 1. Eighty-one percent of all the breast cancers, run under standard conditions, stained with at least one of the MCA in the panel. B6.2 was positive in 71% of the cancers and B50.4 (47%), B72.3 (24%), and B38.1 (11%) were positive in a lower percentage of tumors. When individual blocks were exchanged between institutions, identical results were obtained indicating technical and observer reproducibility. The percentage of positive tumors from the two institutions were similar.

Distribution of Antigens Within Tumors

One of the most remarkable observations in this study was that most of the tumors were antigenically heterogeneous. Very few tumors had positive staining of 100% of the cells with any MCA under the standard conditions. The majority of tumors had less than 50% of the cells staining with any given MCA.

The distribution of stain was affected by the fixation. Formalin fixation was adequate when used properly. Several examples are given in TABLE 2. In these two examples, B6.2 and B50.4 stained the tumor cells in all blocks. However, staining with B72.3 and B38.1 varied in the same tumor from block to block with even the same fixative.

From the viewpoint of the pathologist, these variations between blocks and fixatives are best explained as variations in processing the specimen and not by any capriciousness of the technique. When the specimen is collected, fixed, and processed by a conscientious pathologist the variations are minimized. The difficulty in interpretating the results is compounded by the natural heterogeneity of the antigens in the tissues. Before arriving at a definitive interpretation, the pathologist is well advised to stain more than one block of each specimen.

Ten cases were studied in which both primary tumor and metastases were available. None of the ten metastases had staining patterns identical to the primary tumors. All metastases were stained with at least one of the four MCA. Six of the metastases had antigens that were not detectable in the primary tumor. Four of the metastases did not have antigens detectable in the primary tumor. Six metastases were positive with the same MCA as the primary tumor. Three cases with multiple metastases were of interest since the different metastases from each primary tumor had different sets of detectable antigens.

Distribution of Antigens Within Tumor Cells

The cytological distribution of the antigens was also affected by the fixative and processing. In the majority of tissues fixed in formalin, the staining tended to have a diffuse cytoplasmic distribution. Fixation with B-5 gave the best cytological detail. We were able to distinguish between the cytological distribution of the antigens detected with different MCA best when the tissue was fixed with B-5.

TABLE 2. Effect of Fixation on Reactivity of Cancer Cells with Monoclonal Antibodies

Case	Fixative (Block No.)	Monoclonal Antibodies			
		B6.2	B50.4	B72.3	B38.1
S81-1159	Formalin (1)	+	+	−	−
	Formalin (2)	+	±	±	−
	Bouin's (1)	+	+	±	−
	Bouin's (2)	+	+	−	−
	Bouin's (3)	+	+	ND	ND
	B-5 (1)	+	+	−	+
	B-5 (2)	+	+	−	+
S81-1226	Formalin	+	+	+	+
	Bouin's	+	+	−	±
	B-5	+	+	−	−

TABLE 3. Reactivity of Selected Tumors with Monoclonal Antibodies

Tumors	Monoclonal Antibodies			
	B6.2	B50.4	B72.3	B38.1
Adenocystic carcinoma	+	0	0	0
Bronchogenic carcinoma	0	0	0	0
Gastric carcinoma	+	+	+	0
Medullary carcinoma, thyroid	+	0	+	0
Colonic carcinoma	+	+	+	+
Endometrial carcinoma	+	+	+	0
Ovarian carcinoma	+	+	+	0
Sweat gland carcinoma	±	0	0	0
Lymphoma	+	0	0	0
Granulocytic leukemia	+	0	0	0
Extra mammary Paget's disease	+	+	+	0
Melanoma	±	0	0	0
Carcinoma *in situ* vulva	±	0	0	0
Normal colon	+	+	+	+
Normal duodenum	0	0	0	0
Squamous cell carcinoma, cervix	±	0	0	0

The MCA B6.2 and B50.4 primary stained with cytoplasm of tumor cells. B6.2 generally gave a diffuse cytoplasmic stain, but there was a slight concentration in the Golgi region. B50.4 gave a granular cytoplasmic stain in the best-fixed tissues. The MCA B72.3 and B38.1 tend to stain the apical or basal membranes of cells.

Distribution of Antigens in Other Cells in the Breast

The antigens detected by the MCA panel were clearly differentially expressed in the breast cancer cells. The staining was consistently more intense in the tumor cells than any other cell at any dilution used. However, the staining was not limited to the tumor cells.

B6.2 consistently stained granulocytes in any tissue in the body. Histiocytes surrounding the tumor cells frequently contained cytoplasmic antigens that were identical to the tumor antigens. In some cases tumor antigens were detected in the histiocytes that were not readily detected in the primary tumor. This implied that the macrophages were concentrating the antigen.

Normal quiescent ductal structures were generally negative for the MCA. However, occasionally, individual cells in normal ductal structures adjacent to the tumor cells were positive.

B72.3 and B38.1 stained normal epithelial cells in a breast sample from a 19-year old woman who was five months pregnant. B6.2 and B50.4 did not stain the same breast cells.

Approximately 10–20% of mammary dysplasias stained with the four MCA. We have examples of all types of mammary dysplasia that are stained. Thus far, we have been unable to detect a pattern of reactivity with MCA that corresponds to reputed risk of malignancy for a given tissue lesion.

Distribution of Antigens in Non-Breast Tissue

Although we have concentrated our study on breast cancer, we have had the opportunity to study a variety of other tissues. While not a systematic study, these observations provide some insight into the natural distribution of these antigens. Some staining has been observed in all gastric and colonic carcinomas. B6.2 can be identified in granulocytes, bone marrow myeloid series, and myeloid leukemias. The normal colon contains surface antigen identified by B6.2 and B72.3, and B38.1 can be identified in normal epithelial components of the duodenum.

DISCUSSION

The major purposes of the studies reported here were to determine the technical and observer reproducibility and the diagnostic utility of a panel of four monoclonal antibodies against human breast cancer antigens.

The antigens detected by these antibodies were expressed in human breast cancer at higher levels than in any other normal tissues. In this sense, they were differentially expressed in breast cancer. The technical and observer reproducibility was excellent.

From the outset, the experimental design called for the use of a single concentration of each MCA. With the given concentration, 80% of all breast cancers tested stained with at least one of the MCA panel. It should be noted that independent studies using varying concentrations of MCA demonstrate that an even higher percentage of tumors will give a positive stain, so the data reported here are valid for a single dilution, but the reaction is clearly dose dependent.

None of the MCA were either disease or tissue specific. For example, all four MCA stained colonic carcinomas. B6.2 reacted with normal sweat duct and normal granulocytes. B72.3 and B38.1 reacted with normal components of the duodenum. These observations limit the diagnostic usefulness of this panel in diagnostic surgical pathology.

It must be emphasized, however, that these immunoperoxidase-based studies have not revealed data that preclude the use of these MCA for other scientific and clinical purposes. In fact, these observations contain several insights that are important in consideration of the biology and the treatment of human breast cancer.

First, the antigens identified by this MCA panel are widely dispersed in the body, but are found in the highest concentrations in human breast cancer cells. The widespread distribution serves to emphasize that these are not tumor-specific antigens. Their presence in other tumors underline the possibility that similar mechanisms may be operating in different tumors.

The differential expression in breast cancer of the MCA-detected antigens may be of paramount clinical importance. For example, imaging in nuclear medicine does not require an absolute concentration of the isotope, but merely a differential concentration of antigen. One should also be mindful that these antigens appear on the cell surface of breast cancer cells.[2] The immunoperoxidase technique primarily detects cytoplasmic antigens. The presence of an antigen in the cytoplasm of a normal cell does not necessarily mean that it will also be on the cell surface. The clinical usefulness of these MCA will be primarily based on their affinity for cell surface antigens. The ultimate test of these MCA will be empirical clinical trials.

The second issue that needs emphasis is the apparent antigenic heterogeneity of the tumors. None of the tumors were uniformly positive for any MCA. The variation in staining serves as a warning to those clinicians who plan to treat a given patient with a single monoclonal antibody. Such studies may require panels of MCA to cover the range of heterogeneous antigens expressed in a given tumor.

Finally, these MCA mark the beginning of an important era in our understanding of the biology of the breast. The human breast has been inaccessible to studies of biological differentiation. For example, we have almost no idea of the steps that a stem cell must go through to become a functional epithelial cell. With the utilization of MCA panels, such as presented here, the biology of breast differentiation will become accessible to study.

ACKNOWLEDGMENT

We thank Ms. Cathleen Cooper, Ms. Mary Durshella, Ms. Barbara Felder, and Ms. Teri S. Pratt for their excellent technical assistance.

REFERENCES

1. ARKLIE, J. J., W. TAYLOR-PAPADIMITRIOU, BODMER, M. EGAN & R. MILLIS. 1981. Differentiation antigens expressed by epithelial cells in the lactating breast are also detectable in breast cancers. Int. J. Cancer **28:** 23–29.
2. COLCHER, D., P. HORAN-HAND, M. NUTI & J. SCHLOM. 1981. A spectrum of monoclonal antibodies reactive with human mammary tumor cells. Proc. Natl. Acad. Sci. USA **78**(5): 3199–3203.
3. DAAR, A. S. & J. W. FABRE. 1981. Demonstration with monoclonal antibodies of an unusual mononuclear cell infiltrate and loss of normal epithelial membrane antigens in human breast carcinomas. Lancet **2:** 434–438.
4. FOSTER, C. S., P. A. W. EDWARDS, E. A. DINSDALE & A. M. NEVILLE. 1982. Monoclonal antibodies to the human mammary gland. I. Distribution of determinants in non-neoplastic mammary and extra mammary tissues. Virchows Arch. (Pathol. Anat.) **394:** 279–293.
5. FOSTER, C. S., P. A. W. EDWARDS, E. A. DINSDALE & A. M. NEVILLE. 1982. Monoclonal antibodies to the human mammary gland. II. Distribution of determinants in breast carcinomas. Virchows Arch. (Pathol. Anat.) **394:** 295–305.
6. GOLDENBERG, D. M., R. M. SHARKEY & J. PRIMUS. 1976. Carcinoembryonic antigen in histopathology: Immunoperoxidase staining of conventional tissue sections. J. Natl. Cancer. Inst. **57:** 11–22.
7. HEYDERMAN, E., K. STEELE & M. G. ORMEROD. 1979. A new antigen on the epithelial membrane: Its immunoperoxidase localization in normal and neoplastic tissue. J. Clin. Pathol. **32:** 35–39.
8. HILKENS, J., F. BIUJS, J. HILGERS, P. HAGEMAN, A. SONNENBERG, U. KOLDOVSKY, K. KARANDE, R. P. VAN HOEVEN, C. FELTKAMP & J. M. VAN DE RIJN. 1981. Monoclonal antibodies against human milkfat globule membranes detecting differentiation antigens of the mammary gland. Proc. 29th Annual Colloquim. Protides of the Biological Fluids (Brussels, Belgium). May, 1981.
9. HORNE, C. H. W., I. N. REID & G. D. MILNE. 1976. Prognostic significance of inappropriate production of pregnancy proteins by breast cancers. Lancet 2:279–282.
10. HOWARD, D. R. & C. R. TAYLOR. 1979. A method for distinguishing benign from malignant breast lesions utilizing antibody present in normal human sera. Cancer **43:** 2279–2287.
11. HSU, S. M., L. RAINE & H. FANGER. 1981. Use of avidin-biotin-peroxidase complex (ABC) in immunoperoxidase techniques: A comparison between ABC and unlabeled antibody (PAP) procedures. J. Histochem. Cytochem. **29:** 577–580.
12. IMAM, A. & Z. A. TOKES. 1981. Immunoperoxidase localization of a glycoprotein on plasma membrane of secretory epithelium from human breast. J. Histochem. Cytochem. **29**(4): 581–584.
13. MASON, D. Y. & C. R. TAYLOR. 1978. Distribution of transferrin, ferritin, and lactoferrin in human tissues. J. Clin. Pathol. **31:** 316–327.

14. MESA-TEJADA, R., I. KEYDAR, M. RAMANARAYANAN, T. OHNO, C. FENOGLIO & S. SPIEGELMAN. 1978. Detection in human breast carcinomas of an antigen immunologically related to a group-specific antigen of mouse mammary tumor virus. Proc. Natl. Acad. Sci. USA 75(3): 1529–1533.
15. MUKAI, K. & J. ROSAI. 1980. Applications of immunoperoxidase techniques in surgical pathology. Prog. Surg. Pathol. 1: 15–49.
16. PATTENGALE, P. K., C. R. TAYLOR, C. R. ENGVALL & E. RUOSLAHTI. 1980. Direct tissue visualization of normal cross-reacting antigen in neoplastic granulocytes. Am. J. Clin. Pathol. 73: 351–355.
17. PICH, A., G. BUSSOLATI & A. CARBONARA. 1976. Immunocytochemical detection of casein and casein-like proteins in human tissues. J. Histochem. Cytochem. 24: 940–947.
18. SHOUSHA, S., T. LYSSIOTIS, V. M. GODFREY & P. J. SCHEUER. 1979. Carcinoembryonic antigen in breast cancer tissue: A useful prognostic indicator. Br. Med. J. 1: 777–779.
19. TAYLOR, C. R. & G. KLEDZIK. 1981. Immunohistologic techniques in surgical pathology— a spectrum of "new" special stains. Human Pathol. 12: 590–596.
20. TAYLOR-PAPADIMITRIOU, J., J. A. PETERSON, J. ARKLIE, J. BURCHELL & R. L. CERIANI. 1981. Monoclonal antibodies to epithelium-specific components of the human milk fat globule membrane: Production and reaction with cells in culture. Int. J. Cancer 28: 17–21.
21. WALKER, R. A. 1978. Significance of alpha-subunit HCG demonstrated in breast carcinomas by the immunoperoxidase technique. J. Clin. Pathol. 31: 245–249.
22. WELLS, S. & P. S. HASLETON. 1981. A dilutional immunoperoxidase study of proliferative ductal lesions and carcinomata of the breast. Histopathology 5: 517–526.

Clinical and Experimental Study of the Immune Complex (Herpes Simplex Virus Type 1-IgM) in Herpes Encephalitis Brain

KOZABURO HAYASHI

Laboratory of Viral Immunology
Institute of Medical Immunology
Koriyama, Fukushima, Japan 963

SHUICHIRO TAKAGI

Department of Psychiatry
School of Medicine
Keio University
Tokyo, Japan

NORIYOSHI SEKINE

Department of Virology
Institute of Medical Science
University of Tokyo
Tokyo, Japan

IZUMI KURIHARA

Department of Ophthalmology
Tokyo Women's Medical College
Tokyo, Japan

IMMUNE COMPLEXES (IgM-HERPES SIMPLEX VIRUS) IN PROLONGED HERPES ENCEPHALITIS PATIENT'S BRAIN

A 56-year-old male herpes encephalitis patient died five months after the disease onset. FIGURE 1 illustrates the complement-fixing (CF) antibody titers in the patient's serum and spinal fluid. CF antibody in the serum was first detected 15 days after the onset of the disease and the titer reached a maximum (1:512) at two months of the disease. When he died, the serum CF titer was 1:128. In the spinal fluid, CF antibody was detectable from three weeks through five weeks and reached a titer of 1:32. At his death, almost no CF antibody was detectable in the spinal fluid.

At autopsy, samples of temporal and occipital lobe, caudate nucleus, trigeminal ganglion, and blood were obtained aseptically and carried to the laboratory in sealed plastic Petri dishes in an iced container. Several parts of the autopsied brain were minced finely by scissors for coculture with green monkey kidney cells. TABLE 1 summarizes the results. Only one flask containing minced temporal lobe showed cytopathic effect four weeks after the initiation of the cocultivation. All other flasks were negative for cytopathic effect up to eight weeks of cocultivation.

The isolated virus was identified as herpes simplex virus (HSV) by immunofluores-

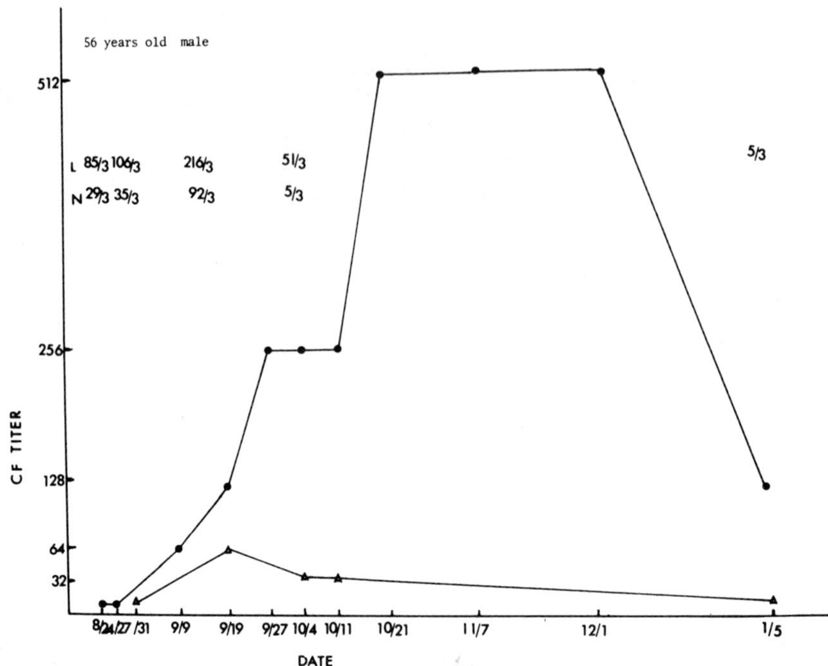

FIGURE 1. CF-antibody in the patient's serum and spinal fluid. L: Number of lymphocytes in the spinal fluid. N: Number of neutrophils in the spinal fluid. ●——●: CF antibody titer in the serum. △——△: CF antibody titer in the spinal fluid.

cence and neutralization with specific antibody. Typing of the isolated HSV was done by Yang's method.[3] Because chick embryonic fibroblasts were not susceptible to this virus, while Vero cells supported viral growth up to 10^{-6} dilution, it was identified as Type 1.

A portion of the brain material was frozen in precooled n-hexane at $-70°C$ to prepare 4 μm thick cryostat sections. They were fixed with acetone for ten minutes at room temperature and stained with FITC-labeled anti-HSV rabbit antibody. Specific fluorescence of HSV antigen was not observed in the preparations. These specimens were stained with either FITC-labeled anti-human IgG, IgA, or IgM antibody. When stained with anti-human IgM, intensive fluorescence was seen in the capillary endothelia and vascular walls of temporal and occipital lobes (FIGURE 2). Occasionally, accumulation of IgM was also seen in the brain parenchyma. Reaction with

TABLE 1. Isolation of HSV by Cocultivation

Occipital lobe	$0^a/5^b$
Temporal lobe	1/5
Caudate nucleus	0/5
Trigeminal ganglion	0/5

[a]Number of flasks showing cytopathic effect.
[b]Number of flasks in which minced brains were cocultured with green monkey kidney cells.

anti-human IgG was totally negative and staining with anti-human IgA revealed sporadic weak fluorescence in the capillary endothelia. The unfixed frozen sections were then treated with 3M NaSCN for 20 minutes at room temperature, washed with phosphate-buffered saline (PBS), and stained with FITC-labeled anti-human IgM or anti-HSV antibody. After this treatment, IgM was no longer detectable and HSV antigen was seen on the vascular walls corresponding to the loci of IgM deposition, as shown in FIGURE 3.

Histopathologically, multiple severe necrotic foci of various sizes were distributed in the brain cortex especially in temporal and occipital lobes, basal nuclei, and pons. Demyelination was seen in these areas. Infiltrations of polymorphonuclear and mononuclear leukocytes were seen around the necrotic foci, and prominent cuffings surrounding the blood vessels were observed (FIGURE 4). When stained with FITC-labeled anti-human T antibody and FITC-labeled anti-human immunoglobulin,

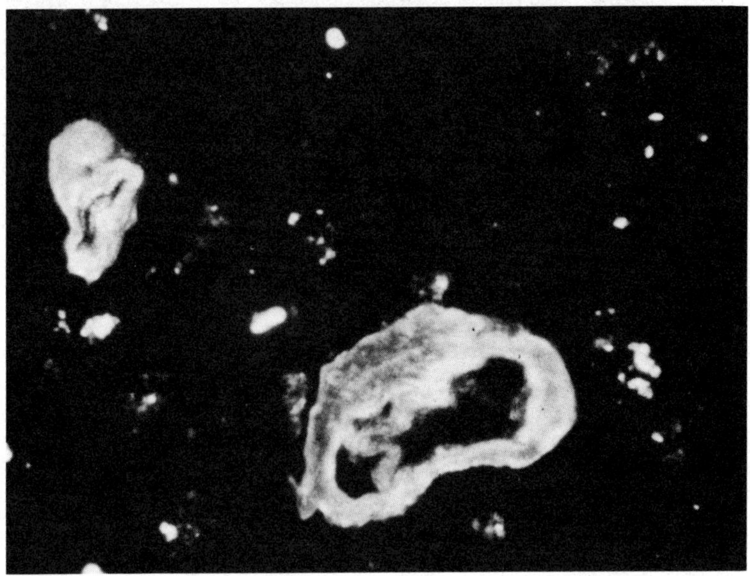

FIGURE 2. IgM deposition on the vascular walls of temporal lobe. Direct immunofluorescence.

approximately 80% of the cuffed cells were identified as T cells and 5% thereof were surface-immunoglobulin–positive B cell lineages.

Serum obtained at autopsy was titrated for its antiviral activity. Results are summarized in TABLE 2. The virus-neutralizing titer of whole serum was 1:640. After 2-mercaptoethanol (2-me) treatment, it showed no reduction of the titer. When it was absorbed with staphylococcal protein A, it reduced to 1:160. When protein A absorption was followed by 2-me treatment, its neutralizing activity was again 1:160. These results might indicate that most of the virus-neutralizing activity of serum at autopsy resided in IgG and some in IgA while almost none in IgM fraction. Thus the dominant serum immunoglobulin class showing the virus-neutralizing activity was clearly different from that of the immune complex found in the brain. Whether IgM in the brain had been produced locally or infiltrated from the circulation at the early

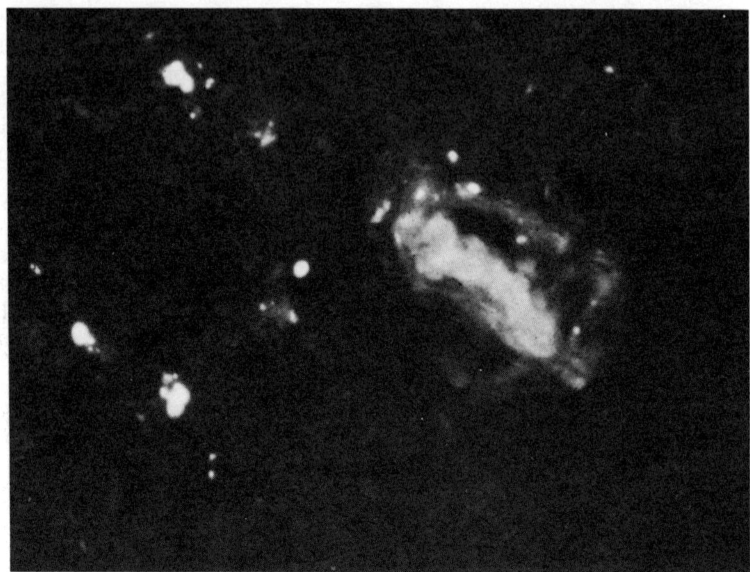

FIGURE 3. HSV antigen on the vascular wall corresponding to the site of IgM deposition.

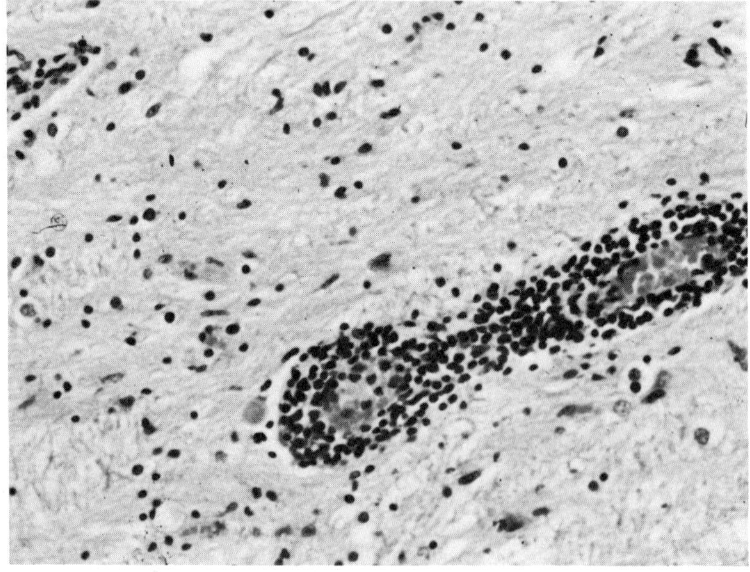

FIGURE 4. Prominent cuffing surrounding blood vessels in the temporal lobe.

phase of infection to deposit as the immune complex is not clear. Another possibility is that the IgM deposited in the brain might be a rheumatoid factor, but this was rendered unlikely by the absence of rheumatoid factor in the patient's serum.

A similar IgM-virus immune complex was also found in a post-vaccine patient's brain.[2]

EXPERIMENTAL STUDY OF PROLONGED HERPES ENCEPHALITIS

To elucidate further details of the immune-complex formation and its role in the pathogenesis of prolonged herpes encephalitis, 4–6 week-old icr (Institute for Cancer Research, Philadelphia) mice were inoculated intracerebrally with 0.025 ml of avirulent Ska strain of type 1 HSV ($10^{6.3}$PFU/ml). It was originally attenuated through repeated "zigzag" passages in chorioallantoic membranes and suckling mouse brains by Yoshino and Taniguchi[4] to such an extent that even intracerebral inoculation into suckling mice failed to kill all of them.

Two to four weeks post-infection, the animals were challenged by intracerebral inoculation either with the homologous Ska or with virulent CHR3 strain of type 1 HSV.[1] Two to four weeks after the challenge virus inoculation, the animals were sacrificed and cryostat sections of the brains were examined by immunofluorescence.

TABLE 2. Neutralizing Activity of Patient's Serum Against HSV-1

Treatment	Neutralizing Antibody Titer
None	640
Protein A	160
Protein A—2-me[a]	160
2-me[a]	640

[a]2-Mercaptoethanol treatment.

Another group of mice were inoculated with 0.3 ml of Ska virus intraperitoneally and these animals were challenged with either Ska or CHR3 virus following the same schedule as the above-described group. One group of control mice was inoculated with Ska virus intracerebrally without challenge virus infection and another group of mice received culture medium in place of virus.

Results are summarized in TABLE 3. Most of the mice that had been inoculated intracerebrally with Ska virus developed marked hydrocephali with thinning of the brain cortex regardless of whether they were challenged with Ska or CHR3 virus. When these brains were studied by immunofluorescence, IgM was noted in the vascular walls or capillary endothelia and occasionally in the brain parenchyma (FIGURE 5).

After treatment with 3 M NaSCN for 30 minutes at room temperature, some of them revealed viral antigen in the areas identical to the place of deposition (FIGURE 6). Histopathologically, marked widening of the bilateral ventricles and accompanying atrophy of the brain cortex were noted (FIGURE 7). Scattered small necrotic foci, mononuclear cell infiltration, and perivascular cuffing and accumulation of lipofuscin in the brain cortex were observed.

On the other hand, only one mouse, which had been inoculated intraperitoneally with Ska virus and received subsequent intracerebral infection of CHR3 virus, developed mild hydrocephalus. Two mice, which had been inoculated with Ska virus

TABLE 3. Hydrocephalus Developing After Prolonged Herpes Encephalitis

Route of Infection		Number of Mice	Virus strain		Immunofluorescence				Virus Isolation		Hydrocephalus		Histopathology
			Initial	Challenge	Viral antigen +	−	IgM +	−	+	−	+	−	
I.C. ⟶ 2–4 weeks	I.C.[a] ⟶ 2–4 weeks	23	Ska	CHR3	5	10	13	4	2	7	19	4	Atrophy of cortex necrosis, cuffing demyelination
I.C. ⟶ 2–4 weeks	I.C. ⟶ 2–4 weeks	15	Ska	Ska	2	9	7	4	1	3	12	3	Same as the above
I.P.[b] ⟶ 2–4 weeks	I.C. ⟶ 2–4 weeks	16	Ska	CHR3	8	0	0	8	0	4	1	15	Small necrotic foci
I.C. ⟶ 2–4 weeks		11	Ska	−	1	6	1	6	0	3	2	9	Small necrotic foci
I.C. ⟶ 2–4 weeks	I.C. ⟶ 2–4 weeks	10	Culture media		0	10	0	10	0	10	0	10	None

[a]Intracerebral inoculation, [b]Intraperitoneal inoculation

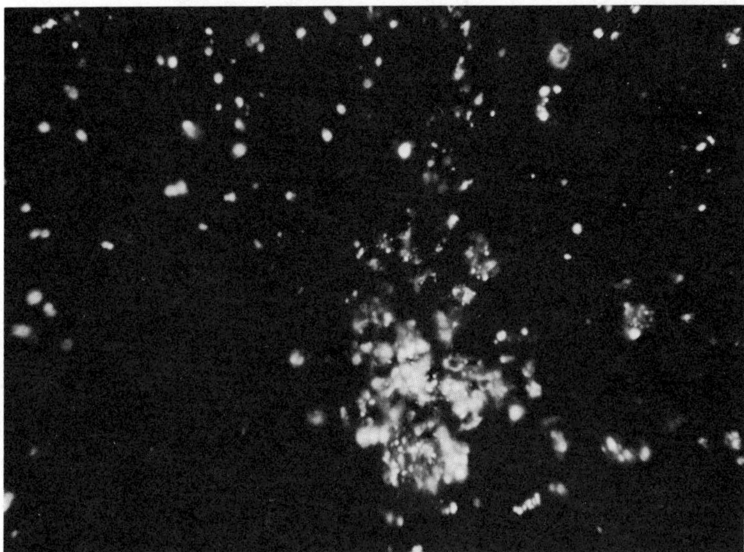

FIGURE 5. IgM deposition in the brain parenchyma. The mouse had been inoculated intracerebrally with Ska virus and then challenged with CHR3 virus.

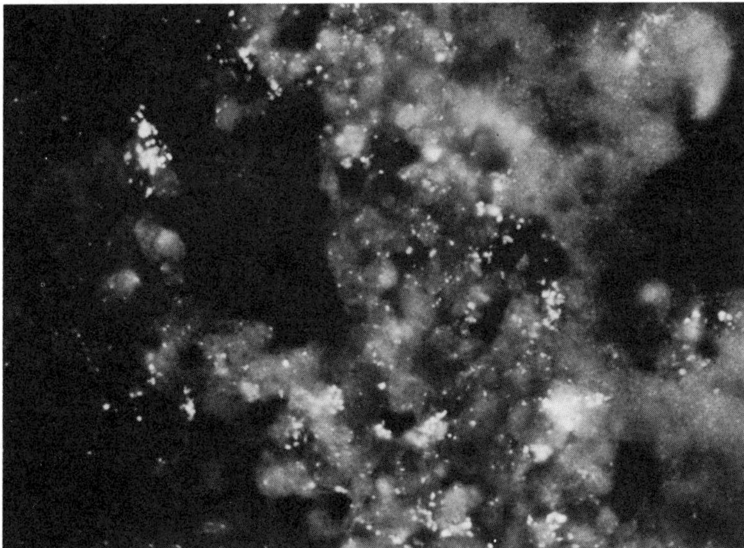

FIGURE 6. HSV antigen in the mouse brain. HSV-IgM immune complex was dissociated by the treatment with 3 M NaSCN for 30 minutes at room temperature.

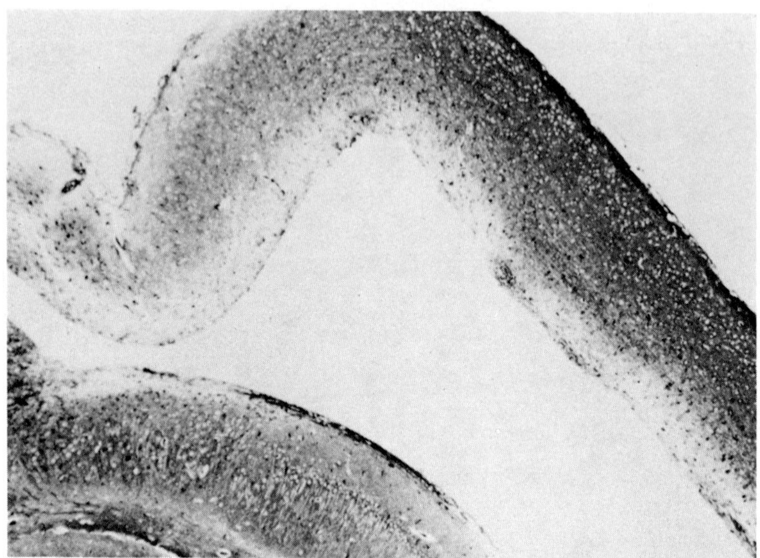

FIGURE 7. Thinning of the brain cortex in hydrocephalic mouse.

intracerebrally and sacrificed 4–8 weeks post-infection, showed mild hydrocephalus while those mice that had received culture medium only did not develop any pathological changes.

These results clearly indicate the similarity between the human case and the animal model with respect to the formation of IgM-HSV immune complex and histopathology in the brain.

The mechanisms of the development of hydrocephalus and the role played by the immune complexes are now under investigation.

TABLE 4. Effect of Varied Concentrations of NaSCN for the Dissociation of Immune Complexes[a]

Immune Complex	FITC-Antibody	1 M	2 M	3 M	4 M	Untreated
HSV-1-Anti-HSV(IgM)	FITC-Anti-HSV	+	+	+ + +	+ + +	+
	FITC-Anti-HuIgM	+ +	+	+	+	+ + +
Uninfected control-Anti-HSV(IgM)	FITC-Anti-HSV	–	–	–	–	–
	FITC-Anti-HuIgM	–	–	–	–	+
HSV-1-Anti-HSV(IgG)	FITC-Anti-HSV	+	+ +	+ + +	+ + +	–
	FITC-Anti-HuIgG	+ +	+ +	+	+	+ + +
Uninfected Control-Anti-HSV (IgG)	FITC-Anti-HSV	–	–	–	–	–
	FITC-Anti-HuIgG	±	–	–	–	–
HSV-1-Anti-HSV(IgA)	FITC-Anti-HSV	+ +	+ +	+ + +	+ + +	±
	FITC-Anti-HuIgA	+ +	+	±	–	+ + +
Uninfected Control-Anti-HSV (IgA)	FITC-Anti-HSV	–	–	–	–	–
	FITC-Anti-HuIgA	±	–	–	–	±

[a]Incubated for 30 minutes at room temperature.

TABLE 5. Effect of Different pH for the Dissociation of Immune Complexes[a]

Immune Complex	FITC-Antibody	pH 2	3	4	Untreated
HSV-1-Anti-HSV(IgM)	FITC-Anti-HSV	+	+	+	+
	FITC-Anti-HuIgM	+ +	+ +	+ +	+ + +
Uninfected control-Anti-HSV(IgM)	FITC-Anti-HSV	−	−	−	−
	FITC-Anti-HuIgM	+	+	+	+
HSV-1-Anti-HSV(IgG)	FITC-Anti-HSV	+ + +	+ + +	+	±
	FITC-Anti-HuIgG	+	+	+	+ + +
Uninfected control-Anti-HSV(IgG)	FITC-Anti-HSV	−	−	−	−
	FITC-Anti-HuIgG	−	−	−	−
HSV-1-Anti-HSV(IgA)	FITC-Anti-HSV	+ + +	+ + +	+	±
	FITC-Anti-HuIgA	±	±	+ +	+ + +
Uninfected control-Anti-HSV(IgA)	FITC-Anti-HSV	−	−	−	−
	FITC-Anti-HuIgA	±	±	±	±

[a]Incubation for two hours at room temperature.

CONDITIONS FOR THE DISSOCIATION OF IMMUNE COMPLEX

Vero cells were infected with CHR3 virus at a multiplicity of infection of 1.0. Sixteen hours after infection, cells were reacted with anit-HSV human antibody of various immunoglobulin classes at 4°C overnight and then washed with PBS. These

TABLE 6. Effect of 3 M NaSCN and Glycine-HCl Buffer (pH 3.0) Treatment on the Dissociation of HSV-1-Anti-HSV(IgM) Complexes

Experiment 1	Treatment with 3 M NaSCN					
Immune Complex	FITC-Antibody	0 min	15 min	30 min	60 min	120 min	180 min
HSV-1-Anti-HSV(IgM)	FITC-Anti-HSV	±	+ + +	+ + +	+	+	±
	FITC-Anti-Human IgM	+ + +	+	+	±	±	±
Uninfected control(Vero)-Anti-HSV(IgM)	FITC-Anti-HSV	−	−	−	−	−	−
	FITC-Anti-Human IgM	+	−	−	−	−	−

Experiment 2	Treatment with Glycine-HCl Buffer (pH 3.0)					
Immune Complex	FITC-Antibody	0 min	30 min	60 min	120 min	180 min	24 hours
HSV-1-Anti-HSV(IgM)	FITC-Anti-HSV	±	±	±	±	±	−
	FITC-Anti-Human IgM	+ + +	+ +	+ +	+ +	+ +	+ +
Uninfected control(Vero)-Anti-HSV(IgM)	FITC-Anti-HSV	−	−	−	−	−	−
	FITC-Anti-Human IgM	+	+	+	±	±	±

TABLE 7. Effect of 3 M NaSCN and Glycine-HCl Buffer (pH 3.0) Treatment on the Dissociation of HSV-1-Anti-HSV(IgG) Complexes

Experiment 1	Treatment with 3 M NaSCN					
Immune Complex	FITC-Antibody	0 min	15 min	30 min	45 min	60 min	90 min
HSV-1-Anti-HSV(IgG)	FITC-Anti-HSV	−	+	+ +	+ + +	+ + +	+ + +
	FITC-Anti-Human IgG	+ + +	+ +	+	±	−	−
Uninfected control(Vero)-Anti-HSV(IgG)	FITC-Anti-HSV	−	−	−	−	−	−
	FITC-Anti-Human IgG	−	−	−	−	−	−

Experiment 2	Treatment with Glycine-HCl Buffer (pH 3.0)			
Immune Complex	FITC-Antibody	0 min	60 min	120 min	16 hours
HSV-1-Anti-HSV(IgG)	FITC-Anti-HSV	−	+ +	+ + +	+ +
	FITC-Anti-Human IgG	+ + +	+	+	±
Uninfected control(Vero)-Anti-HSV(IgG)	FITC-Anti-HSV	−	−	−	−
	FITC-Anti-Human IgG	−	−	−	−

preformed immune complexes were subjected to NaSCN and acid buffer treatments. They were then stained either with FITC-labeled anti-HSV, anti-human IgG, IgA, or IgM antibody.

TABLES 4–8 summarize the results. Treatment with 3 to 4 M NaSCN for 30 minutes at room temperature can effectively dissociate immune complexes of various immunoglobulin classes (TABLE 4). Acid treatment (glycine-HCL buffer with pH 2–3)

TABLE 8. Effect of 3 M NaSCN and Glycine-HCl Buffer (pH 3.0) Treatment on the Dissociation of HSV-1-Anti-HSV (IgA) Complexes

Experiment 1	Treatment with 3 M NaSCN					
Immune Complex	FITC-Antibody	0 min	15 min	30 min	45 min	60 min	90 min
HSV-1-Anti-HSV(IgA)	FITC-Anti-HSV	−	+ +	+ +	+	+	+
	FITC-Anti-Human IgA	+ + +	+	±	−	−	−
Uninfected control(Vero)-Anti-HSV(IgA)	FITC-Anti-HSV	−	−	−	−	−	−
	FITC-Anti-Human IgA	±	−	−	−	−	−

Experiment 2	Treatment with Glycine-HCl Buffer (pH 3.0)				
Immune Complex	FITC-Antibody	0 min	60 min	120 min	180 min	16 hours
HSV-1-Anti-HSV(IgA)	FITC-Anti-HSV	−	+	+ + +	+ +	+ +
	FITC-Anti-Human IgA	+ + +	+ +	+ +	+	±
Uninfected control(Vero)-Anti-HSV(IgA)	FITC-Anti-HSV	−	−	−	−	−
	FITC-Anti-Human IgA	−	−	−	−	−

TABLE 9. Detection of Formalin- or Acetone-Fixed Immune Complexes

Immune Complex	FITC-Antibody	Treatment with 3 M NaSCN				
		15 min	30 min	45 min	60 min	Untreated
HSV-1-Anti-HSV(IgG)	FITC-Anti-HSV	±	±	±	±	±
Formalin fixed	FITC-Anti-HuIgG	+ +	+ +	+ +	+ +	+ + +
Uninfected control-Anti-HSV(IgG)	FITC-Anti-HSV	±	±	±	±	±
Formalin fixed	FITC-Anti-HuIgG	±	±	±	±	±

Immune Complex	FITC-Antibody	Untreated	Acetone	Formalin	Acetone + Trypsin	Formalin + Trypsin
HSV-1-Anti-HSV(IgG)	FITC-Anti-HSV	±	±	±	±	−
	FITC-Anti-HuIgG	+ +	+ + +	−	±	−
Uninfected control-Anti-HSV(IgG)	FITC-Anti-HSV	−	−	−	−	−
	FITC-Anti-HuIgG	−	−	−	−	−

was also effective for dissociation of IgG or IgA antibody-HSV complex (TABLE 5). For the dissociation of IgM-HSV, 15–30-minute treatment with 3 M NaSCN was efficient, while pH 3 glycine HCl buffer treatment was not effective even after 24 hours (TABLE 6). On the other hand, IgG-HSV immune complex needed a 30–90-minute treatment with 3 M NaSCN to permit HSV antigen to bind fluorescein-labeled anti-HSV (TABLE 7). Duration of the NaSCN treatment was longer than that of the IgM-HSV. However, for IgG, pH 2–3 treatment was also effective to dissociate the immune complex. For IgA-HSV, both NaSCN and acid pH treatments showed similar results to those of IgM-HSV, e.g. the 3 M NaSCN and pH 2–3 treatment could reveal HSV antigen (TABLE 8).

When preformed immune complexes were fixed with acetone or formalin, dissociation of the immunoglobulin could hardly be achieved even when they were treated with trypsin for 3 hours at 37°C and then subjected to the treatment with 3 M NaSCN (TABLE 9).

Therefore, in analysis of the particular case of immune-complex formation in the brain, it seems that use of the fresh frozen sections benefit further analyses of the antigen in tissues.

ACKNOWLEDGMENT

We thank Dr. Kamesaburo Yoshino for reviewing this manuscript.

REFERENCES

1. HAMPER, B., A. L. NOTKINS, M. MAGE & M. A. KEEHN. 1968. Heterogeneity in the properties of 7s and 19s rabbits neutralizing antibodies of herpes simplex virus. J. Immunol. **100:** 586–593.
2. KURATA, T., Y. AOYAMA & T. KITAMURA. 1977. Demonstration of vaccinia virus antigen in brains of postvaccinal encephalitis patients. Jap. J. Med. Soc. Biol. **30:** 137–147.
3. YANG, J. P. S., W. CHIANG, J. L. GALE & N. S. T. CHEN. 1975. A chick-embryo cell microtest for typing of herpesvirus hominis.
4. YOSHINO, K. & S. TANIGUCHI. 1969. Isolation of a clone of herpes simplex virus highly attenuated for newborn mice and hamsters. Jap. J. Exp. Med. **39:** 223–232.

Localization of LDL in Arteries: Improvements in Immunofluorescence Procedures[a]

HENRY F. HOFF, DAVID L. FELDMAN, AND
ROSS G. GERRITY

Atherosclerosis Section
Research Division
Cleveland Clinic Foundation
Cleveland, Ohio 44106

Determination of the accumulation of low density lipoproteins (LDL) in arteries has received wide interest since this accumulation has been suggested to play a major role in atherogenesis.[1,19] In light of this interest, we have used immunofluorescence techniques to localize LDL in arteries of humans and experimental animals. The aim of this paper is to describe the development of procedures to improve the localization pattern of LDL in arteries.

During the past years we have striven to improve resolution of the localization pattern by turning from conventional, unfixed cryostat sections, first to sections of formalin-fixed, paraffin-embedded arteries, and recently to sections of paraformalde-hyde-fixed, epoxy-embedded arteries. The rationale for initially changing to formalin-fixed and paraffin-embedded sections was several fold. First, better tissue preservation was achieved,[14,17,20] particularly if perfusion fixation could be accomplished, such as in studies on experimental animals. Secondly, fixation of soluble antigens (such as LDL) resulted in their immobilization, thus minimizing diffusion of the antigen during tissue processing. Embedding in paraffin enabled thinner sections to be cut, thus increasing the resolution of the localization pattern. Further advantages of such embedding over cryostat sections was the fact that less wrinkling of sections was obtained as compared to cryostat sections. Finally, easier storage of blocks and potential performance of retrospective studies on routinely processed tissue blocks could be achieved.

Our initial studies on LDL localization were performed on unfixed cryostat sections of human atherosclerotic lesions.[7,9–12] These studies were performed using FITC-conjugated antibody directed at apoB, the major protein of LDL. This antibody was affinity-purified using a Sepharose-LDL column. We used the transmitted light mode using UG-1 excitation and KP-430 barrier filters in which specific fluorescence was apple-green and autofluorescence either blue (collagen) or blue-white (elastica). Since maximum absorption for the fluorescein (FITC) was not achieved with this system, sensitivity was not optimal. Moreover, later studies suggested that delipidation of cryostat sections with acetone unmasked further reactive sites for apoB, the major protein in LDL. This "unmasking" vastly increased the sensitivity of the technique. FIGURES 1a and b illustrate LDL localization in arteries using the procedures described above. Note the relatively diffuse nature of the localization pattern.

[a]Supported in part by the National Institutes of Health grants HL-28450 and HL-21438.

Abbreviations: LDL = plasma low density lipoproteins, apoB = protein portion of LDL, HSA = human serum albumin, BSA = bovine serum albumin, PBS = phosphate-buffered saline, FITC = fluorescein isothiocyanate.

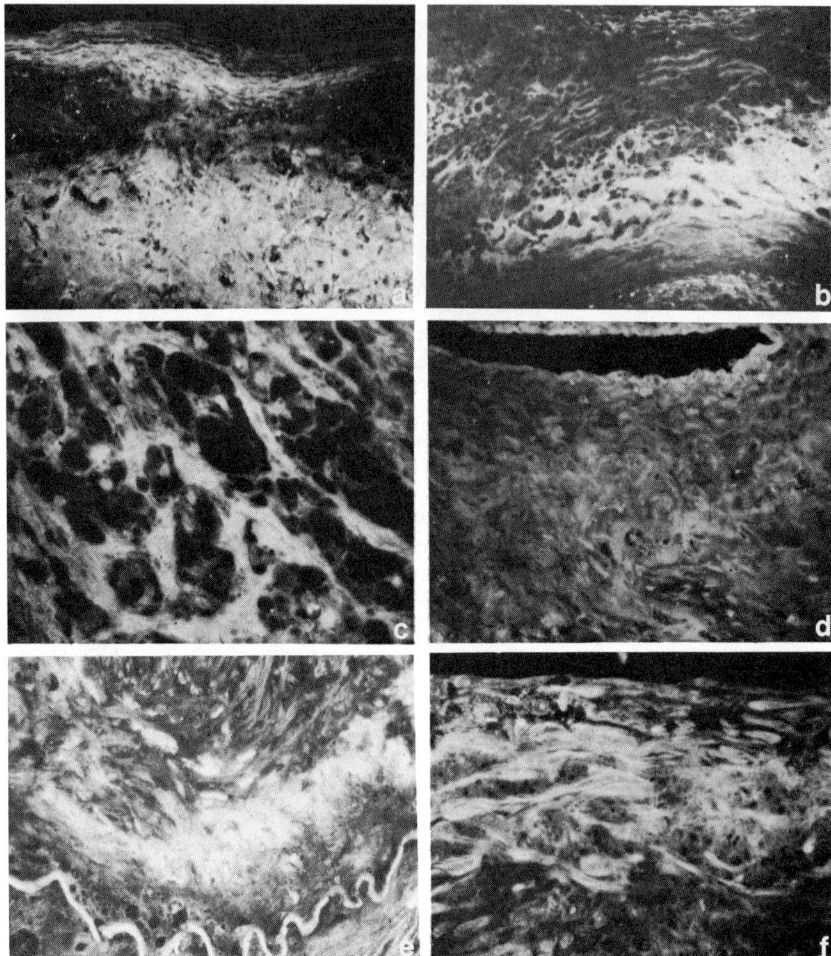

FIGURE 1. (a) Human aortic plaque: LDL (white) localized diffusely in lipid-rich core. Autofluorescence of connective tissue at top. ×60. (b) Human coronary artery (atherosclerotic lesion) showing LDL in lipid-rich core and some LDL close to lumen at top. ×60. (c) Formalin-fixed paraffin-embedded human aortic plaque; LDL localized exclusively in extracellular space between foam cells. ×300. (d)–(f) Formalin-fixed, paraffin-embedded atherosclerotic coronary arteries from hypercholesterolemic cynomolgus monkeys. Spindle-shaped smooth muscle cells and endothelial cells on lumen border also show orange fluorescence due to counterstaining with flazo orange. (d) and (e) ×200; (f) ×300.

The first major improvement was the utilization of formalin fixation and paraffin embedding. In a series of studies we determined that no differences in localization pattern and intensity of immunofluorescence was found when tissues were fixed in 10% neutral formalin or in Bouin's fixative.[8] Moreover, the addition of decalcifying agents to the fixative had no deleterious effects. Tissue could be dehydrated in ethanol, acetone, or dioxane. Warping of sections on slides during incubation could be avoided by adding gelatin and chrom-alum (0.02%) to the water bath used to float the paraffin sections.[16]

Recently, we attempted to improve the intensity of specific immunofluorescence in sections by employing the reflected light mode using a Xenon 150W lamp, in conjunction with oil-immersion objectives, and utilizing a filter system permitting maximum absorption by FITC, namely the Leitz Ploem H_2 filter pack. However, the resultant autofluorescence (yellow) could not be readily distinguished from the specific fluorescence (apple-green). We were obliged to find a counterstain that would change the color of the autofluorescence. This was achieved by any number of counterstains such as methyl green, chelated and non-chelated eriochrome black, and flazo orange.[4,6,13,18] The advantages of one counterstain over the other for specific situations have been described in detail.[8] We have found flazo orange to be most useful for demonstrating the relationship between the cellular elements of the artery and the specific fluorescence. Sections of paraffin-embedded tissue were first deparaffinized in xylene and rehydrated through graded ethanols (95% to 50%) to distilled water. The rehydrated sections were incubated in a moist chamber for one hour at room temperature with the affinity-purified FITC-anti-LDL. One percent human serum albumin (HSA) was added to the FITC-labeled antibody to minimize non-specific binding to the tissue. Following two five-minute washes in phosphate-buffered saline (PBS), slides were immersed in *N,N*-dimethyl formaformamide followed by staining in aluminum chelates of the azo dyes, eriochrome black or flazo orange. These counterstains were diluted 1:10 with chelating reagent as reported[8] or using the non-chelated eriochrome black. Staining times were approximately 30 sec. With methyl green staining time was about 5 min.

Basically, at low magnification little difference can be discerned between the LDL localization pattern from cryostat and paraffin-embedded blocks. FIGURE 1c illustrates the localization of LDL in the human aorta obtained at autopsy and fixed in formalin by immersion. The fatty-fibrous plaque contains numerous foam cells. Note that LDL is confined exclusively in the extracellular space. In black and white photographs the orange-colored cytoplasm is seen as brightly as the specific fluorescence.

We have also studied the localization of LDL in coronary arteries from cynomolgus monkeys. Cross-reactivity between human and monkey LDL permits the use of anti-human LDL to demonstrate monkey LDL. FIGURE 1d shows a partially occluded lumen of an atherosclerotic coronary artery from a hypercholesterolemic cynomolgus monkey. LDL is localized along wavy lines between individual smooth muscle cells. In serial sections these areas positive for LDL stained positive with alcian blue, suggesting the presence of sulfated glycosaminoglycans. FIGURE 1e shows the deposition of LDL as a granular array in the depths of a plaque close to the internal elastic membrane. A similar granular pattern of fluorescence is seen at a higher magnification between the brightly stained (orange) cytoplasm of the spindle-shaped smooth muscle cells (FIGURE 1f).

We have recently studied the possibility of localizing LDL in fixed tissue that is embedded in epoxy resin. This would allow thinner sections to be cut (0.5–1 micron as compared to 4 microns for paraffin-embedded sections), permitting even better resolution of the localization pattern to be achieved. Moreover, subsequent ultrastructural studies of LDL accumulation could be made. Samples of aortas from hypercholesterolemic Yorkshire swine[5] were used for these studies. A more detailed description on the methodology used is being published elsewhere.[3] Briefly, blocks of aorta fixed in 2% paraformaldehyde in phosphate buffer pH 7.3 were embedded in Epon 812-Araldite 6005.[15] Sections 0.5–1 micron thick were deplasticized by immersion for 5 min in a saturated solution of Na ethoxide (aged for at least four days) diluted 1:2 with absolute ethanol.[2] Sections were then rinsed in PBS and PBS containing 1% bovine serum albumin (BSA) and stained for LDL by indirect immunofluorescence. Affinity-purified rabbit anti-swine LDL containing 1% BSA was used as the primary antibody. Incubation was performed for 30 min at room temperature. Following washes in PBS and PBS/BSA, the sections were incubated for 30 min with the secondary antibody (FITC-conjugated goat anti-rabbit IgG). Following further washes, sections were counterstained as described above.

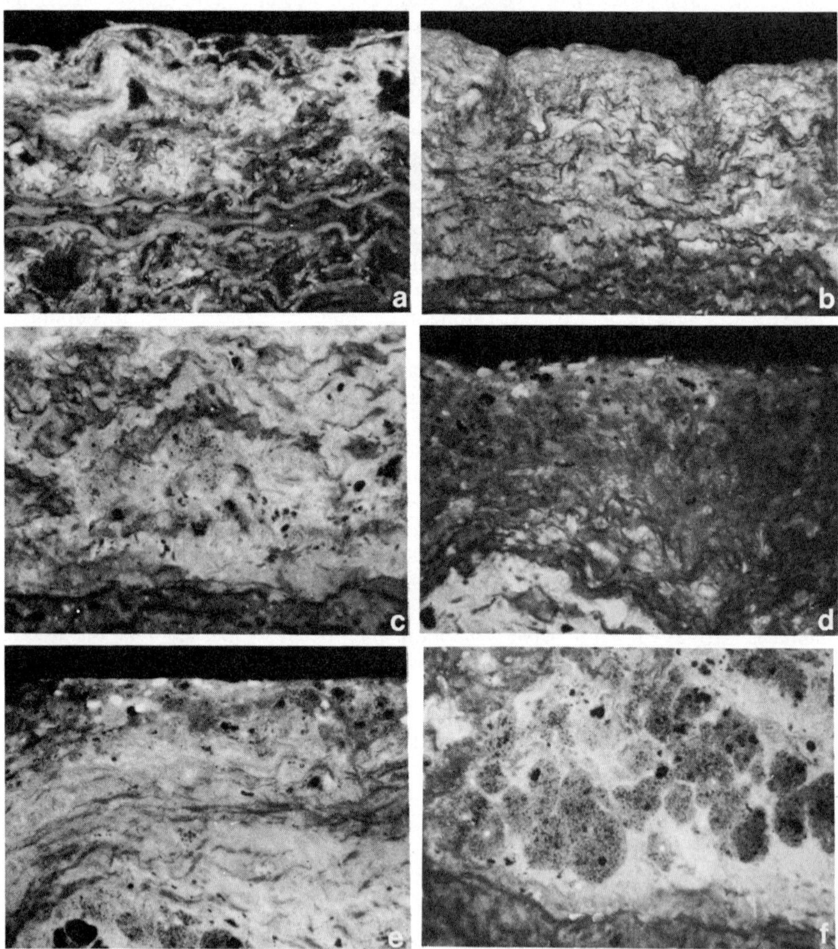

FIGURE 2. Paraformaldehyde-fixed plastic-embedded aorta from hypercholesterolemic swine, flazo orange counterstain. See text for details. All micrographs ×300.

FIGURE 2 gives six examples using the procedures just outlined. At low magnification little difference in LDL localization in paraffin- and plastic-embedded blocks could be discerned. FIGURES 2a and b illustrate the localization of LDL in areas of the aortic arch of swine fed a cholesterol-enriched diet for eight weeks. Immunoreactivity for LDL is seen in the extracellular space between elastic fibers and individual cells. FIGURES 2c through e demonstrate the border between a lipid core and its overlying fibromuscular cap. LDL is present both between the smooth muscle cells of the cap and the foam cells in the core (FIGURE 2f). Also the cytoplasm is devoid of LDL.

SUMMARY

We have described the development of procedures used to localize LDL in arteries with atherosclerotic lesions from humans and experimental animal models. We have first illustrated data obtained from earlier cryostat studies, described the procedures, given examples of results of LDL localization in sections of paraffin-embedded blocks, and finally outlined the procedures used and examples for LDL localization in epoxy-embedded blocks. Future improvements in resolution of LDL localization will probably require performing electron microscopy on ultrathin sections of epoxy-embedded arteries followed by immunocytochemistry.

REFERENCES

1. ADAMS, C. W. M. 1967. General pathology of atherosclerosis. *In* Vascular Histochemistry. p. 35. Year Book Medical Publishers, Inc. Chicago.
2. ERLANDSON, S. L., G. Y. H. PARSONS & C. B. RODNING. 1979. Technical parameters of immunostaining of osmicated tissue in epoxy sections. J. Histochem. Cytochem. **27:** 1286–1289.
3. FELDMAN, D. L., H. F. HOFF & R. G. GERRITY. 1983. Immunofluorescent localization of LDL in aortas from hyperlipemic swine. Preferential accumulation in areas positive for Evans blue dye. (Submitted for publication.)
4. FEY, H. 1972. Eriochrome black, a means for reduction of nonspecificity in immunofluorescence. Pathol. Microbiol. **38:** 271–277.
5. GERRITY, R. G., H. K. NAITO, M. RICHARDSON & C. J. SCHWARTZ. 1979. Dietary induced atherogenesis in swine. Morphology of the intima in pre-lesion stages. Am. J. Pathol. **95:** 775.
6. HALL, C. T. & P. A. HANSEN. 1962. Chelated azo dyes used as counterstains in the fluorescent antibody technic. Zentralbl. Bakteriol. **184:** 548–555.
7. HOFF, H. F. 1976. Apolipoprotein localization in human cranial arteries, coronary arteries and the aorta. Stroke **7:** 390–393.
8. HOFF, H. F., B. M. RUGGLES & M. G. BOND. 1980. A technique for localizing LDL by immunofluorescence in formalin-fixed and paraffin-embedded atherosclerotic lesions. Artery **6:** 328–339.
9. HOFF, H. F., C. L. HEIDEMAN & J. W. GAUBATZ. 1975. Apo-low density lipoprotein localization in intracranial and extracranial atherosclerotic lesions from human normolipoproteinemics and hyperlipoproteinemics. Arch. Neurol. **32:** 600–605.
10. HOFF, H. F., C. L. HEIDEMAN, J. P. NOON & J. S. MEYER. 1975. Localization of apolipoproteins in human carotid artery plaques. Stroke **6:** 531–534.
11. HOFF, H. F., R. L. JACKSON, J. T. MAO & A. M. GOTTO. 1974. Localization of low density lipoproteins in arterial lesions from normolipemics employing a purified fluorescent labeled antibody. Biochim. Biophys. Acta **351:** 407–415.
12. HOFF, H. F., J. T. LIE, J. L. TITUS, R. L. JACKSON, M. E. DEBAKEY, R. BAYARDO & A. M. GOTTO. 1975. Lipoproteins in atherosclerotic lesions. Localization by immunofluorescence of apo- low density lipoproteins in human atherosclerotic arteries from normal and hyperlipoproteinemics. Arch. Pathol. **99:** 253–258.
13. HOKENSON, E. O. & P. A. HANSEN. 1966. Flazo orange for masking of auto and nonspecific fluorescence of leukocytes and tissue cells. Stain Tech. **41:** 9–14.

14. HUANG, S., H. MINCISSIAN & J. D. MOSE. 1976. Application of immunofluorescent staining on paraffin sections improved by trypsin digestion. Lab. Invest. **35:** 383–390.
15. MILLONIG, G. 1976. Laboratory Manual of Biological Electron Microscopy. M. Saviolo, Ed. Mario Saviolo. Vercelli, Italy.
16. PAPPAS, P. W. 1971. The use of a chrome alum-gelatin (subbing) solution as a general adhesive for paraffin sections. Stain Tech. **46:** 121–124.
17. SAINTE-MARIE, G. 1962. A paraffin-embedding technique for studies employing immunofluorescence. J. Histochem. Cytochem. **10:** 250–256.
18. SCHENK, E. Q. & C. J. CHURUKIAN. 1974. Immunofluorescence counterstains. Cytochemistry **22:** 962–966.
19. ST. CLAIR, R. W. 1976. Metabolism of the arterial wall and atherosclerosis. *In* Atherosclerosis Reviews. R. Paoletti & A. M. Gotto, Eds.: 61. Raven Press. New York.
20. TAYLOR, C. R. & J. BURNS. 1974. The demonstration of plasma cells and other immunoglobulin-containing cells in formalin-fixed, paraffin-embedded tissues using peroxidase-labeled antibody. J. Clin. Pathol. **27:** 14–20.

Pre-Embedment Localization of Antigens by Immunofluorescent Methods[a]

RICHARD M. FRANKLIN

Biocenter
University of Basel
CH-4056 Basel, Switzerland

In recent years semi-thin sections of plastic-embedded tissues have contributed to significant improvements in the quality of histological material.[1] Details can be better studied because sections of 0.5 to 2.0 μm involve single-cell layers. Furthermore, many of the embedding media are water miscible so that the tissues do not have to be completely dehydrated. Many monomers can be polymerized at low temperature, another factor resulting in good tissue preservation. Mild and incomplete dehydration and embedding at low temperature permit milder non-coagulative fixation, a further factor leading to improved histology.

A discrepancy between the observed preservation and identification of antigens in histological material using conventional cryosections or paraffin sections and the potential offered by plastic embedding has stimulated several approaches to improve the quality of immunohistochemical preparations. These can be divided into pre- and post-embedding staining methods. The obvious advantage of post-embedding staining is that sections from a single tissue block can be stained with a large number of antibodies whereas a separate embedding has to be carried out for each antibody used in pre-embedding staining. Furthermore a more uniform staining can be achieved with post-embedding staining than with pre-embedding staining. Unfortunately, the water-miscible methacrylate monomers used in the low temperature embedding procedures are virtually impossible to dissolve once polymerized. Therefore the possibility of post-embedding staining after embedding in such plastics is limited to the surface and then only if the monomer has not cross-linked to the antigen of interest. A successful example of post-embedding staining using a methyl methacrylate, 2-hydroxyethyl methacrylate polymer, for embedding whole joints has been described.[7] Plastics that can be etched away from the section, such as various epoxy resins,[9,10] require relatively long curing times at relatively high temperatures (45–60°C). In such cases not all antigens are preserved in a sufficiently native form to allow post-embedding staining. There may also be chemical binding of the monomer to the antigen of interest but this may be alleviated by reaction of the antigen with ethyl acetimidate prior to embedding.[8]

With the present drawbacks of post-embedding immunohistochemistry in mind, we concentrated our efforts on improvements of pre-embedding immunohistochemistry.[5] These improvements were concerned with fixation, staining, and embedding.[3,5] Although we employ a standard procedure, applicable for most tissues and most

[a]Supported in part by a grant (3.392.78) from the Swiss National Fund.

Abbreviations: PBS = phosphate-buffered saline, FITC = fluorescein isothiocyanate, TRITC = tetramethylrhodamine isothiocyanate, HPMA = hydroxypropyl methacrylate, GMA = 2-hydroxyethyl methacrylate, often referred to as glycol methacrylate, PAS = periodic acid-Schiff stain, SOM = refers to a cell containing somatostatin granules, and TLC = thin-layer chromatography.

antigens, preliminary experiments should be carried out by each investigator to determine the most suitable conditions for a particular tissue.

TECHNIQUES

Fixation

The tissues are fixed in 3.5% formaldehyde, which is prepared fresh from paraformaldehyde using Weber-Osborn phosphate-buffered saline (PBS) as buffer.[3] Usually we kill mice by neck dislocation followed by rapid excision of the tissues of interest into the fixative. This has proven satisfactory for most tissues, although preservation of liver chord cells would be superior using perfusion fixation (L. Landman, personal communication). The tissues, once in the fixative, are cleaned, trimmed, and sliced into pieces (about 4–5 mm in one dimension and about 20 mm in the other dimension). After one-hour fixation at room temperature the tissues are sliced into 200 μm slices with the Smith-Farquhar tissue slicer (DuPont-Sorvall Model TC-2) and fixed for another hour. The Oxford Vibratome can also be used for preparing these slices. The fixed tissue is rinsed twice in PBS at room temperature and then permeabilized in absolute methanol at -20 to -30°C (4 × 15 minutes), transferred to absolute ethanol at 4°C and brought to PBS through an alcohol series (100, 94, 80, 70, and 50%), 15 minutes per stage. Permeabilization may not be necessary for all antigens, particularly surface antigens.

We prefer formaldehyde to glutaraldehyde fixation because there is practically no autofluorescence either in the fluorescein isothiocyanate (FITC) or tetramethylrhodamine isothiocyanate (TRITC) bands using formaldehyde fixation but strong autofluorescence, at least in the TRITC band, in the case of glutaraldehyde fixation. Glutaraldehyde fixation also requires a further chemical treatment to block free aldehyde groups,[3] a procedure not necessary with formaldehyde. Nevertheless there may be certain tissues or antigens for which glutaraldehyde fixation is preferable.

Staining

Staining is carried out by placing the tissue slices in microdroplets (10–20 μl) of antibody appropriately diluted in PBS containing 0.01% sodium azide. The micro-

TABLE 1. Hydroxypropyl Methacrylate Embedding Media

	HPMA/2-Butoxyethanol Ratios		
	7:1	9:1	10:1
Solution A			
hydroxypropyl methacrylate	70 ml	90	100
2-butoxyethanol	10 ml	10	10
triethyleneglycol dimethacrylate	4 ml	5	5.6
benzoyl peroxide	0.14 g	0.17 g	0.19 g
Solution B			
polyethylene glycol 400	30 ml		
N,*N*-dimethylaniline	1 ml		

Infiltration in solution A, embedding in solution A + solution B (40:1).

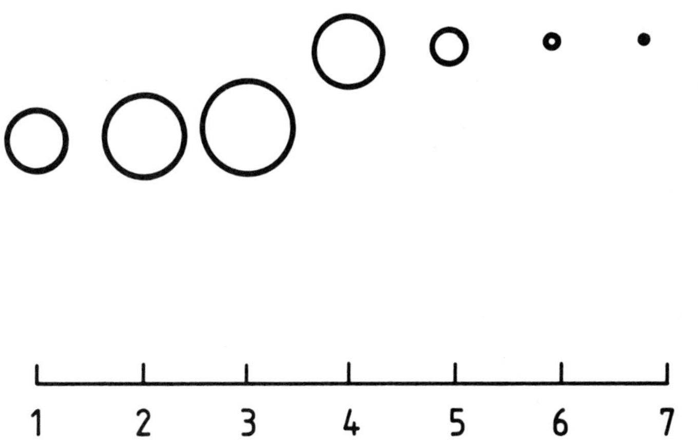

FIGURE 1. Thin-layer chromatography of the hydroquinone stabilizer (hydroquinone monomethylether) present in the hydroxypropyl methacrylate. Sil G200 pre-coated TLC plates, run in benzene-acetone (90:10), developed with Millon's reagent;[10] 1–3 are hydroquinone monomethylether standards (5, 10, and 20 µg), 4–7 are HPMA samples prior to extraction (4), after 2 extractions (5), after 3 extractions (6), and after 4 extractions (7). The hydroquinone monomethylether R_f value in the unknown is influenced by the hydroxypropyl methacrylate.

droplets are deposited on Parafilm® squares in small (30 mm diameter) plastic petri plates, which are in turn placed in a humidified chamber. The time of incubation for both the first and second antibodies should be investigated for a particular tissue. We usually incubated two days for the first antibody and two days for the second antibody, both at 4°C. After the first incubation the slices are washed (3 × 30 minutes in PBS at room temperature) and then incubated in the second antibody. After this incubation the slices are washed again (4 × 30 minutes in PBS at room temperature), dehydrated, infiltrated with monomer, and embedded, all in one day. The long incubation times are necessary to allow the antibodies to diffuse into the tissue slices as deeply as possible.

Unfortunately in the case of compact tissues longer incubation times do not result in better penetration.

Embedding

The tissues are dehydrated at 4°C through the alcohol series (50, 70, 80, 90, and 100%), 15 minutes for each step. Infiltration with hydroxypropyl methacrylate (HPMA) solution A (TABLE 1) is allowed to proceed for two hours at room temperature and one hour at 4°C and then the slices are embedded in HPMA of the appropriate composition as determined by the HPMA/2-butoxyethanol ratios (TABLE 1). These are 7:1 for most soft tissues, 9:1 for lymph nodes, and 10:1 for bone.[4] The tissue slices are carefully arranged so that they lie parallel to the cutting surface. Embedding is carried out at 4°C overnight in Sorvall molding cups with plastic block holders. There is practically no heat generation during the gelation phase at 4°C.[3] The edge of the holder and the hole in the holder are covered with liquid paraffin to limit availability of oxygen, which inhibits polymerization.

The quality of the embedding reagents should be the highest possible. We use commercial hydroxypropyl methacrylate from which we extract methacrylic acid.[6] The resultant acid-free plastic can be used for all types of stains under acidic, neutral, and basic conditions with very little or no background staining. Unfortunately acid-free HPMA is not commercially available; but acid-free 2-hydroxyethyl methacrylate (GMA) can be obtained from SPI Supplies Division, Structure Probe, Inc. (535 E. Gay Street, West Chester, PA 19380). In tests on this material we found a residue of 4.5 μmoles methacrylic acid/ml, as low as that obtained by us[6] and lower than that found for any commercial GMA embedding mixture. Because of the superior sectioning and spreading properties, however, we prefer HPMA-based polymers.

Besides extracting the methacrylic acid from the monomers, we remove the hydroquinone inhibitor by charcoal extraction.[1] In order to reduce the inhibitor to a minimum, thereby permitting good polymerization in a minimal concentration of benzoyl peroxide, we extract with charcoal four times. The extractions are controlled by thin layer chromatography (FIGURE 1). In this example there is less than 0.001% hydroquinone monomethylether left after four charcoal extractions. This control is recommended since different manufacturers use different hydroquinone derivatives at different concentrations to stabilize different methacrylate monomers.

With these appropriate extractions and quality controls for the major methacrylate monomer it is possible to prepare embedding media that yield constant blocks with reproducible sectioning characteristics, permitting reproducible staining.

Sectioning

The tissues are sectioned at 1 μm with Ralph knives.[1] Special attention must be made with respect to knife profile, knife angle, and speed of cutting. All of these factors must be varied with the hardness of the plastic in order to get shatter-free sections. The sections are cut dry, spread on a water surface, and then taken up on coverslips. After drying, the sections are counterstained with Hoechst 33258,[5] dried again, and permanently mounted in Entellan (E. Merck, Darmstadt, FRG). Both the blocks and the sections can be stored for very long periods of time. We routinely store them in the dark at 4°C.

RESULTS

To date a wide variety of antigens have been localized by pre-embedding immunofluorescence. These include proteins at the cell surface, cytoplasmic immunoglobulins, immune complex deposits, viral structural proteins, cytoskeletal proteins, basal membrane proteins, and neuropeptides (R. M. Franklin and R. Timpl, work in

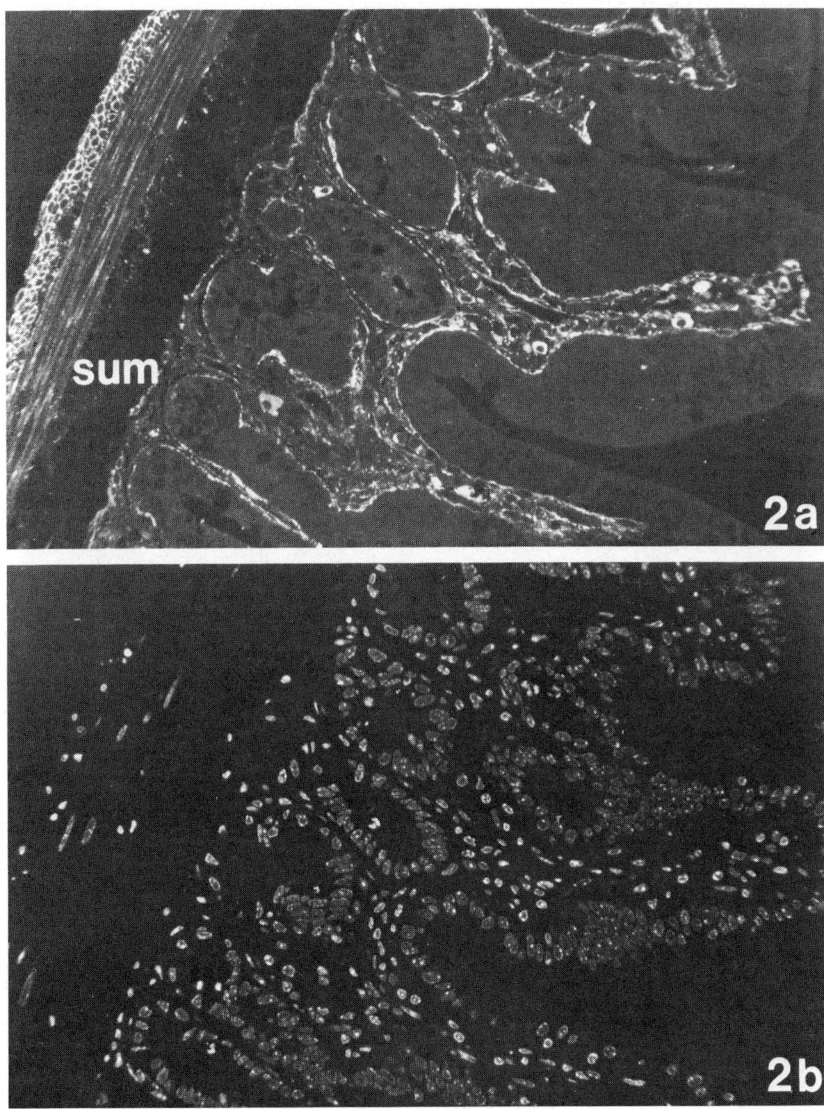

FIGURE 2. Mouse (BALB/c) duodenum, 1 μm section, low power overview, magnification 224×. (a) Anti-fibronectin (FITC) and (b) Hoechst 33258.

progress).[3,5] Limitations in the method are a consequence of the very poor penetration of the immunoglobulins into the tissue rather than as a consequence of any alterations in the antigens. Therefore the tissues that can be readily studied are those of highest penetrability—thymus, spleen, lymph nodes, stomach, intestine, lung, liver, brain, and spinal cord. Muscle, cartilage, and bone[4] are far more difficult to stain uniformly.

Some results are illustrated in FIGURES 2–7. In these sections (all 1 μm thick) the antibody staining was by the indirect method with affinity-purified rabbit anti-fibronectin and anti-tubulin,[5] commercial rabbit anti-somatostatin (Immuno Nuclear Corp., Stillwater, MN 55082), and commercial FITC- or TRITC-labeled antibodies to rabbit IgG. The FITC-labeled wheat germ agglutinin was also commercial (Réactifs

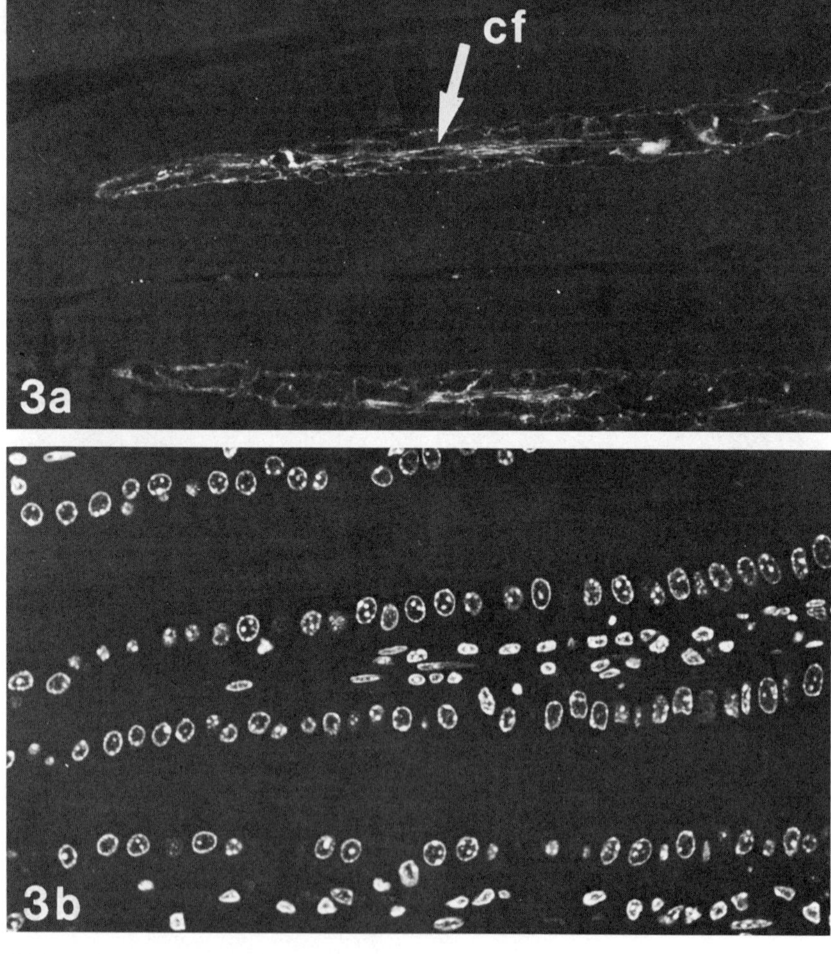

FIGURE 3. Mouse (BALB/c) duodenum, 1 μm section, magnification 400×. (a) Anti-fibronectin (FITC) and (b) Hoechst 33258.

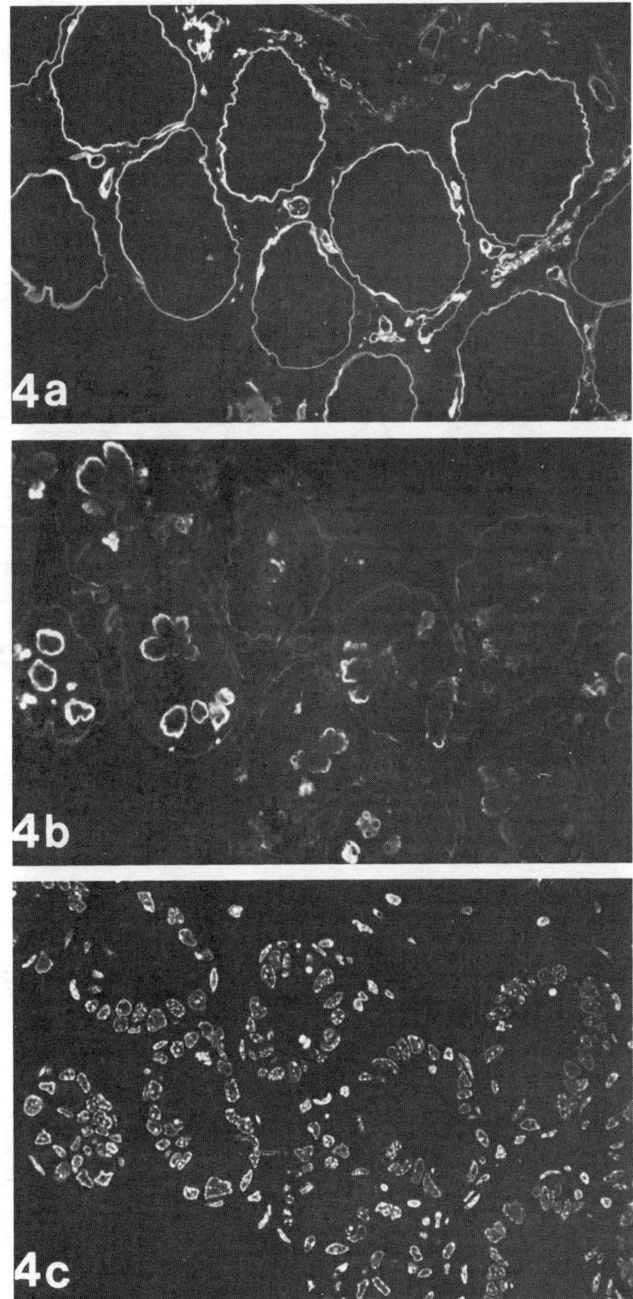

FIGURE 4. Mouse (BALB/c) duodenum, 1 μm section, magnification 338×. (a) Anti-fibronectin (TRITC), (b) FITC wheat germ agglutinin, and (c) Hoechst 33258.

IBF, 92390 Villeneuve-La-Garenne, France). Using appropriate epifluorescence filter systems the FITC and TRITC fluorescence was completely isolated from each other and from the fluorescence of Hoechst 33258 used to stain the nucleus. These filter systems were Leitz filter packs K2, N, and A, respectively.

FIGURE 2 is a mouse duodenum cross-section stained with anti-fibronectin. The smooth muscle cells of the muscularis externa layer (circular and longitudinal

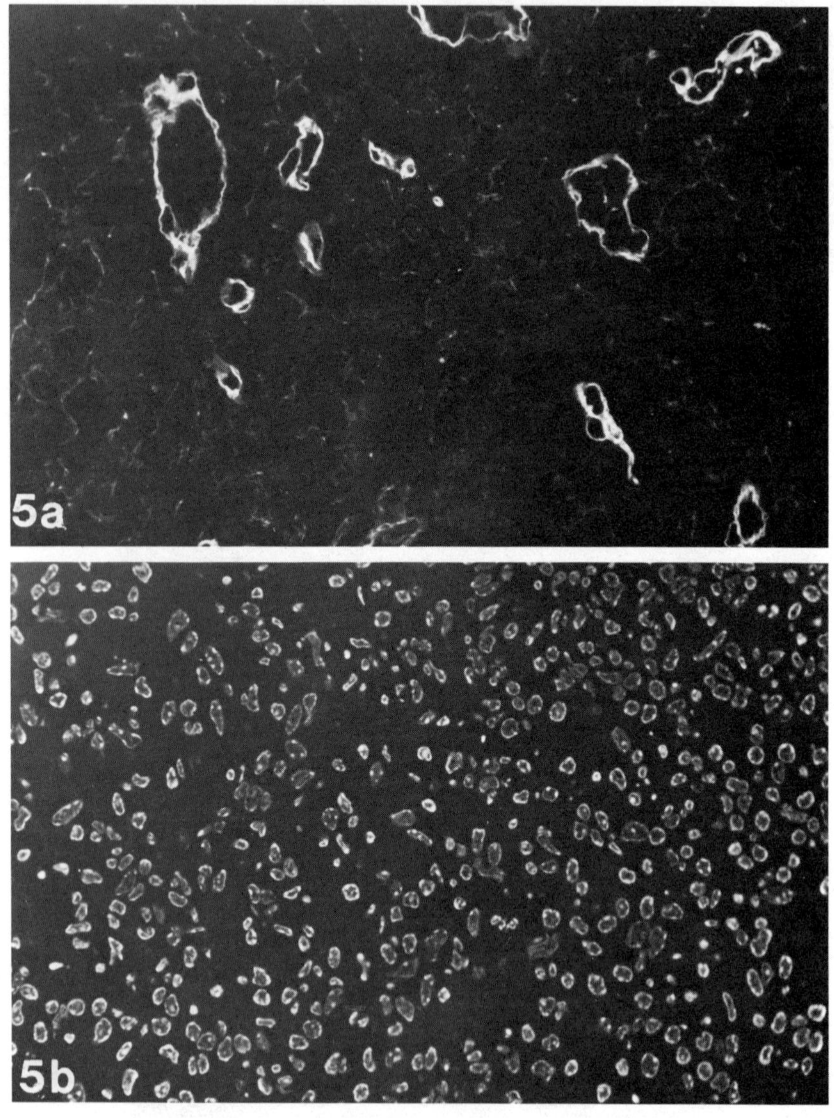

FIGURE 5. Mouse (BALB/c) Peyer's patch, 1 μm section, magnification 448×. (a) Anti-fibronectin (FITC) and (b) Hoechst 33258.

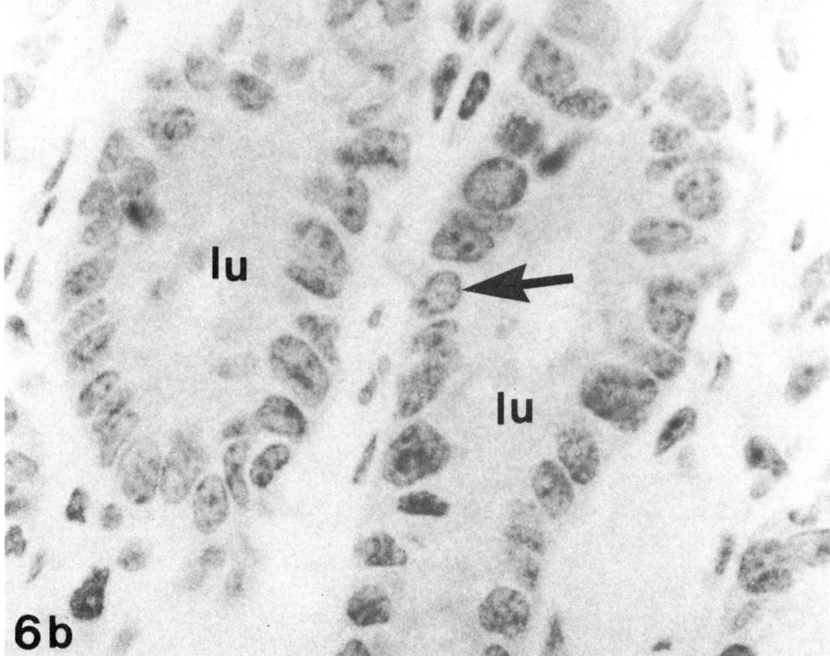

FIGURE 6. Mouse (BALB/c), 1 μm section, magnification 806×. (a) Anti-somatostatin (FITC) and (b) Gill's haematoxylin.

muscles) are separated from each other by a basal lamina, a PAS-positive structure made up of cylinders of basal membrane proteins (R. M. Franklin and R. Timpl, work in progress).[5] The submucosa is not labeled and is therefore deficient in fibronectin. The mucosa, however, has a rich network of fibronectin fibers that also form one component of the basal membrane separating the lamina propria from the columnar epithelial lining of each villus. The cells of the columnar epithelium can be readily distinguished in the pattern of Hoechst-stained nuclei. The crypts of Lieberkühn are also readily distinguished and delimited by a fibronectin-positive basal membrane. The central lacteal, the smaller lacteals or lymphatic vessels, and the blood capillaries of the villus lying close to the basal membrane are all delimited by the fibronectin network. In FIGURE 3 another feature of the lamina propria fibronectin network is shown, a central fiber system (cf) that is only seen at certain section levels. This central fiber system provides a support for the extension of the muscularis mucosae into the

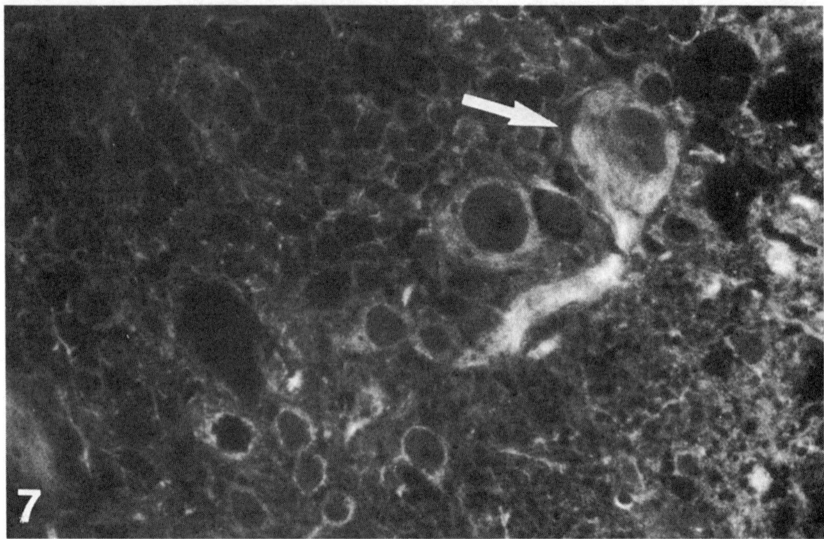

FIGURE 7. Mouse (CBA) cerebellum, 1 μm section, stained with antitubulin (FITC), magnification 705×.

lamina propria. This system of cells, rich in actin,[5] may be responsible for the motion of the villi. A third section of the duodenum, in this case a longitudinal section, is shown in FIGURE 4. The fibronectin-rich basal membrane delimiting the crypts of Lieberkühn is stained with a TRITC-labeled antibody and the secretory granules of the Paneth cells are stained with FITC wheat germ agglutinin. The secretory granules contain PAS-positive mucopolysaccharide, which may also be the substance to which the wheat germ agglutinin binds. A different type of fibronectin network is seen in Peyer's patches (FIGURE 5). There is a rich but disordered network of fibronectin here plus the very dense fibronectin network surrounding the lymphatic vessels and capillaries. Returning to the villi of the duodenum, an endocrine cell[2] producing somatostatin (SOM cell) is shown in FIGURE 6. The nucleus of this cell (arrow) is readily identified in the hematoxylin-stained section. This staining, as is the case for Hoechst 33258 staining, aids us in localizing the structure(s) of interest. In this case we see that the SOM cell is in the epithelial layer of the villus, projecting into the gut lumen (lu). The final illustration shows staining of the mouse cerebellum with anti-tubulin. A prominent Purkinje cell (arrow) reveals fine details of the microtubulin network, almost at the level of resolution seen in cultured cells.

The examples shown here should give a good idea of the potential of pre-embedding immunofluorescent staining. The drawbacks are the low penetrability of the immune reagents into the tissue slices, the necessity of using a separate staining for each antibody (except for double-label experiments), and the long incubation times. The major advantage is the excellent resolution achieved. This method may be used alone or in conjunction with conventional post-embedding staining methods to survey a tissue.

ACKNOWLEDGMENTS

Mr. R. Longato, Miss U. Schaub, and Mrs. M. T. Martin provided excellent assistance; Mrs. M. C. Gauler was responsible for the secretarial work.

REFERENCES

1. BENNETT, H. S., A. D. WYRICK, S. W. LEE & J. H. McNEIL. 1976. Science and art in preparing tissues embedded in plastic for light microscopy, with a special reference to glycol methacrylate, glass knives and simple stains. Stain Technol. **51:** 71–97.

2. BUCHAN, A. M. J. & J. M. POLAK. 1980. The classification of the human gasteroenteropancreatic endocrine cells. Invest. Cell. Pathol. **3:** 51–71.

3. FRANKLIN, R. M. 1982. Pre-embedding staining for immunohistochemistry at the light microscopic level. *In* Techniques in Immunocytochemistry. G. R. Bullock & P. Petrusz, Eds. **1:** 251–268. Academic Press. London.

4. FRANKLIN, R. M. & M. T. MARTIN. 1980. Staining and histochemistry of undecalcified bone embedded in a water-miscible plastic. Stain Technol. **55:** 313–321.

5. FRANKLIN, R. M. & M. T. MARTIN. 1981. Pre-embedding immunohistochemistry as an approach to high resolution antigen localization at the light microscopic level. Histochemistry **72:** 173–190.

6. FRANKLIN, R. M., M. T. MARTIN & R. LONGATO. 1981. A simple procedure for extracting methacrylic acid from water-miscible methacrylates. Stain Technol. **56:** 283–285.

7. HERMANNS, W., K. LIEBIG & L. C. SCHULZ. 1981. Post-embedding immunohistochemical demonstration of antigen in experimental polyarthritis using plastic embedded whole joints. Histochemistry **73:** 439–446.

8. KUHLMANN, W. D. & R. KRISCHAN. 1981. Resin embedments of organs and post-embedment localization of antigens by immunoperoxidase methods. Histochemistry **72:** 377–389.

9. POLAK, J. M., A. G. E. PEARSE & C. M. HEATH. 1975. Complete identification of endocrine cells in the gastrointestinal tract using semithin-thin sections to identify motilin cells in human and animal intestine. Gut **16:** 225–229.

10. RANDERATH, K. 1965. Dünnschicht Chromatographie. Verlag-Chemie. Weinheim.

Immunohistological Demonstration of Serum Proteins and Structural and Viral Antigens in Paraffin Sections of Nervous Tissues

HERBERT BUDKA

Division of Special Neuropathology
Neurological Institute
University of Vienna
A-1090 Vienna, Austria

INTRODUCTION

Numerous studies have indicated that many antigens remain detectable in tissue sections by immunohistological techniques even after routine formalin fixation and paraffin embedding. Using paraffin sections, a great volume of processed and stored material can be worked up in retrospective immunohistological studies. This is most important in conditions in which fresh material is only rarely available. Immunohistological examinations of nervous tissues now constitute a major part of the scientific and routine diagnostic work of neuropathological and neuroanatomical centers and a large number of publications is currently devoted to applications of immunohistological techniques to the study of the nervous system. This popularity, however, sometimes regrettably overrides the necessity that such studies must be performed in a carefully controlled manner. It is the aim of this paper to present our experience with immunostaining of routinely processed neuropathologic specimens. Emphasis will be given to studies of the influence of length of fixation, to the usefulness of enzyme digestion, and to the sensitivity of the methods employed.

MATERIALS AND METHODS

All immunostained sections have been stored as paraffin blocks in the neuropathological collection of this institute for periods up to 32 years. Fixation of tissues was performed by immersion into 10% neutrally buffered formalin for periods ranging from a few hours (neurosurgical biopsies) to one week or up to several weeks (autopsy material, mainly brains); occasionally material was studied after several years in formalin. Our immunostaining procedures have been previously described for direct and indirect immunofluorescence (IF)[7] and peroxidase-antiperoxidase (PAP) techniques.[8] To detect rabies virus antigen by PAP, we developed a four-layered method (first hamster anti-rabies serum, then rabbit anti-hamster serum, and the last two steps of the PAP procedure using a rabbit PAP complex). For digestion studies, protease

Abbreviations: ALD = adrenoleukodystrophy, EM = electron microscopy, GFAP = glial fibrillary acidic protein, H = hematoxylin, HSV = Herpes simplex virus, IF = immunofluorescence, MBP = myelin basic protein, MS = multiple sclerosis, PAP = peroxidase-antiperoxidase complex, PML = progressive multifocal leukoencephalopathy, and SSPE = subacute sclerosing panencephalitis.

type VII (*Bacillus subtilis*, Sigma Chemical Co.) was used[5,7,8] and counterstaining was performed according to standard histological stains. Antisera of previously reported characteristics were directed against various serum proteins,[1,5] secretory component,[4] rabies and Herpes simplex virus (HSV) type 1 antigens[7,8] and measles virus antigens.[5] Several anti-papovavirus sera with differing specificity also were used.[9] An anti-myelin basic protein (MBP) serum was kindly provided by Dr. H. Bernheimer, Vienna, and an anti-glial fibrillary acidic protein (GFAP) serum by Dr. D. McCormick, Belfast. Control materials usually included known positive specimens and normal tissue treated as the material to be investigated. Staining specificity was usually controlled by replacing the specific antiserum by pre- or non-immune sera at identical dilutions, by use of antiserum, which was adsorbed with the appropriate antigen, and in some studies also by two-step blocking tests with specific antisera from different species.

RESULTS

Influence of Protease Treatment

Enzymatic digestion studies were performed in a direct IF system for detection of rabies virus antigen. This system is especially suited for such studies since here immunostaining is mainly dependent upon digestion. When protease is omitted, no staining is seen (FIGURE 1 A); after protease for 15 min at 37°C, Negri and lyssa bodies are brilliantly stained in large hippocampal neurons (FIGURE 1 B). When digestion was prolonged to 30 minutes, tissue morphology becomes slightly damaged whereas it is well preserved after 15 minutes. Longer digestion periods at 37°C lead to significant damage of tissue morphology and eventual loss of the section. A protease concentration below 0.05% has little effectiveness whereas higher concentrations such as 0.5% have a similar or slightly weaker result as compared with 0.1%. When incubation was performed at room temperature, a 0.5% protease solution had to be used for 60 minutes to get a staining intensity comparable with that of 0.1% protease at 37°C; the section had a worse morphological appearance and showed a strong tendency to float away. Floating of tissue sections during and after protease treatment may constitute a major problem in some cases. It appears to us that this undesired effect is mainly seen when the material had been cut and mounted by a less experienced technician.

In a PAP technique for glial fibrillary acidic protein (GFAP), protease also had a beneficial effect; when the antiserum is used at identical dilutions, astroglial processes are much more prominent with protease (FIGURE 1 C) than without protease treatment (FIGURE 1 D). To obtain a better result without protease treatment, the antiserum usually had to be used at higher concentrations. In most systems, with protease a two- to fourfold higher dilution of antiserum can be used to achieve an intensity of immunostaining similar to that without protease.

Sensitivity Study

The sensitivity of immunostaining of formalin-fixed paraffin material is known to be much lower than with fresh material.[18] To obtain brilliant fluorescence of Negri bodies in fresh imprints of rabies brains, ten- to hundredfold higher dilutions of antiserum could be used than those for use on paraffin sections. When the sensitivity and reliability of direct IF staining for rabies antigen in paraffin sections were

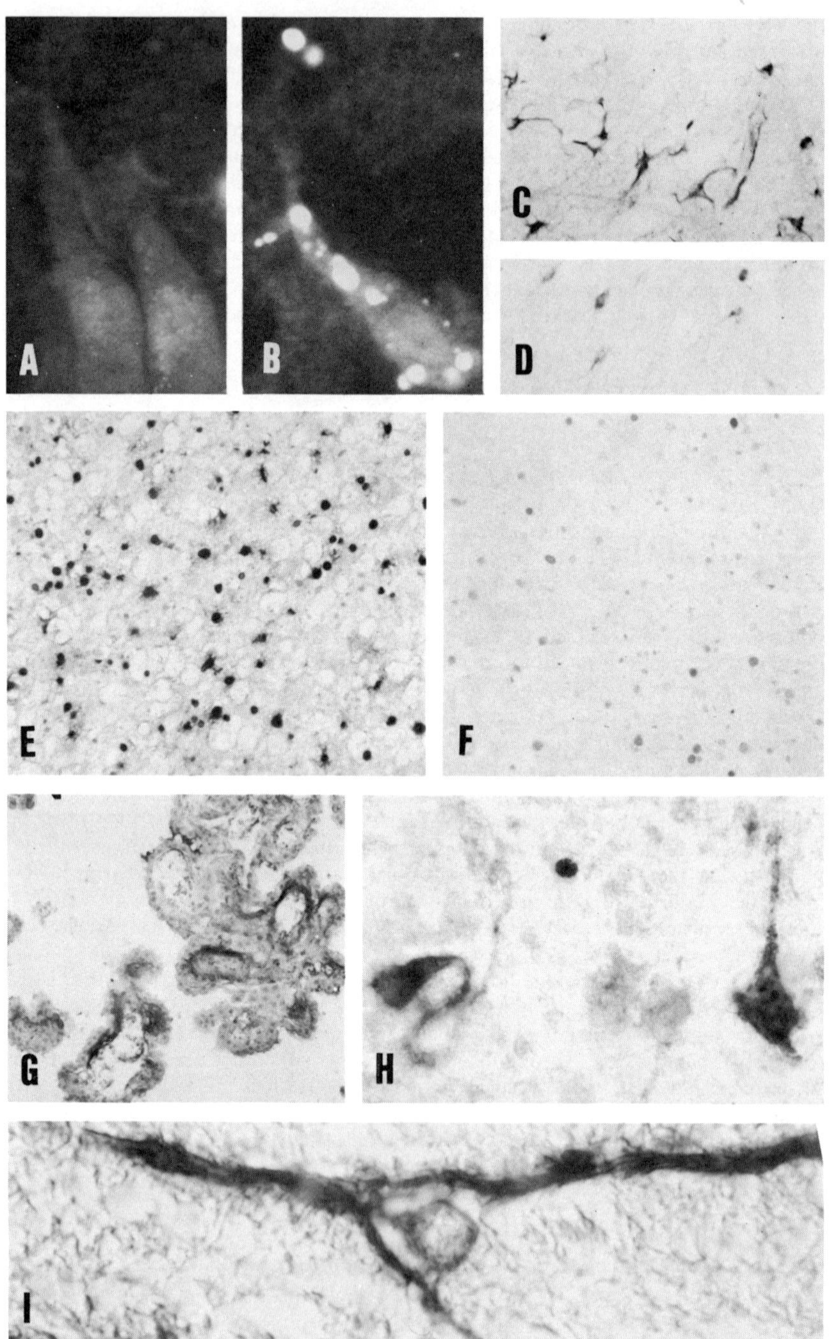

compared to conventional virological techniques for a rabies diagnosis in fresh tissues (direct IF of imprint preparations and inoculation of mice; study performed in collaboration with E. Scharfen, Austrian Institute for Control of Animal Diseases, Mödling, Lower Austria) no significant differences were found in a series of 120 animals with natural rabies. There was only one "false negative" paraffin block from a cow; since only a very small part of the hippocampus was embedded in paraffin, this case could well represent a sampling error.

Sensitivity of Various Immunohistological Techniques

It is well established that sensitivity is enhanced when several layers are used in an immunomorphological system; in the PAP technique, 100- to 1,000-fold higher antiserum dilutions may be used than in IF methods.[18] On formalin-fixed and protease-treated sections, we usually apply a 10- to 20-fold higher dilution for PAP than for indirect immunofluorescence with the same antiserum. Considering positive/negative results in immunostaining for rabies, Herpes simplex virus (HSV), and measles virus antigens we found no differences between PAP and IF techniques in 12 cases of rabies, 10 HSV encephalitides, and 28 cases of subacute sclerosing panencephalitis (SSPE). Some details, however, like prominent involvement of dendrites by virus, were more distinctly outlined by the PAP method.

Sensitivity of Virus Antigen Detection by Immunohistology Compared to Detection of Virus Particles with the Electron Microscope

Among 28 measles virus-antigen positive SSPE cases, one case with negative electron microscope (EM) search for virions clearly demonstrated virus-infected cells in low density when investigated by immunohistology. A similar result was found with papovavirus antigens in progressive multifocal leukoencephalopathy (PML);[9] in comparison with an EM search for virions (found in 16/18 cases), immunocytochemistry was able to detect viral investment in two more cases in which density of infected cells was low.

Sensitivity of Immunostaining for Myelin Basic Protein Compared to that of Conventional Myelin Stains

The distribution of myelin basic protein (MBP) corresponds closely to that of myelin.[14] In conventional myelin stains (Luxol fast blue or Heidenhain's method),

---

FIGURE 1. (A–D) Enhancement of immunoreactivity by protease treatment. (A and B): Rabies, roe deer, large hippocampal neurons. Without protease, no staining is found (A), whereas after protease, Negri bodies are brilliantly stained (B). Direct IF for rabies antigen, ×100. (C and D): White matter gliosis in adrenoleukodystrophy. With protease, astroglial processes are much more visible (C) than without protease pretreatment (D); PAP for GFAP, ×160. (E and F): Reduction of immunoreactivity by prolonged tissue storage in formalin. PML, white matter lesion. Numerous cells filled with papovavirus antigens easily detectable after two weeks of fixation (E); distinct reduction of immunoreactivity after one year in formalin (F). PAP with cross-reacting anti-SV40 serum, hematoxylin counterstain, ×100. (G) Plexus chorioideus in SLE. Immune complexes between vessels and epithelium. PAP for IgG, hematoxylin counterstain, ×100. (H) HSV encephalitis, imprint preparation from autopsy brain. Pyramidal neurons strongly labeled for virus antigen. PAP for HSV antigen, ×630. (I) Preterm infant, pontine tegmentum. An oligodendrocyte myelinates fibers running in different directions and shows MBP in cytoplasm and processes; labeled myelin sheaths are still thin. PAP for MBP, ×1,000.

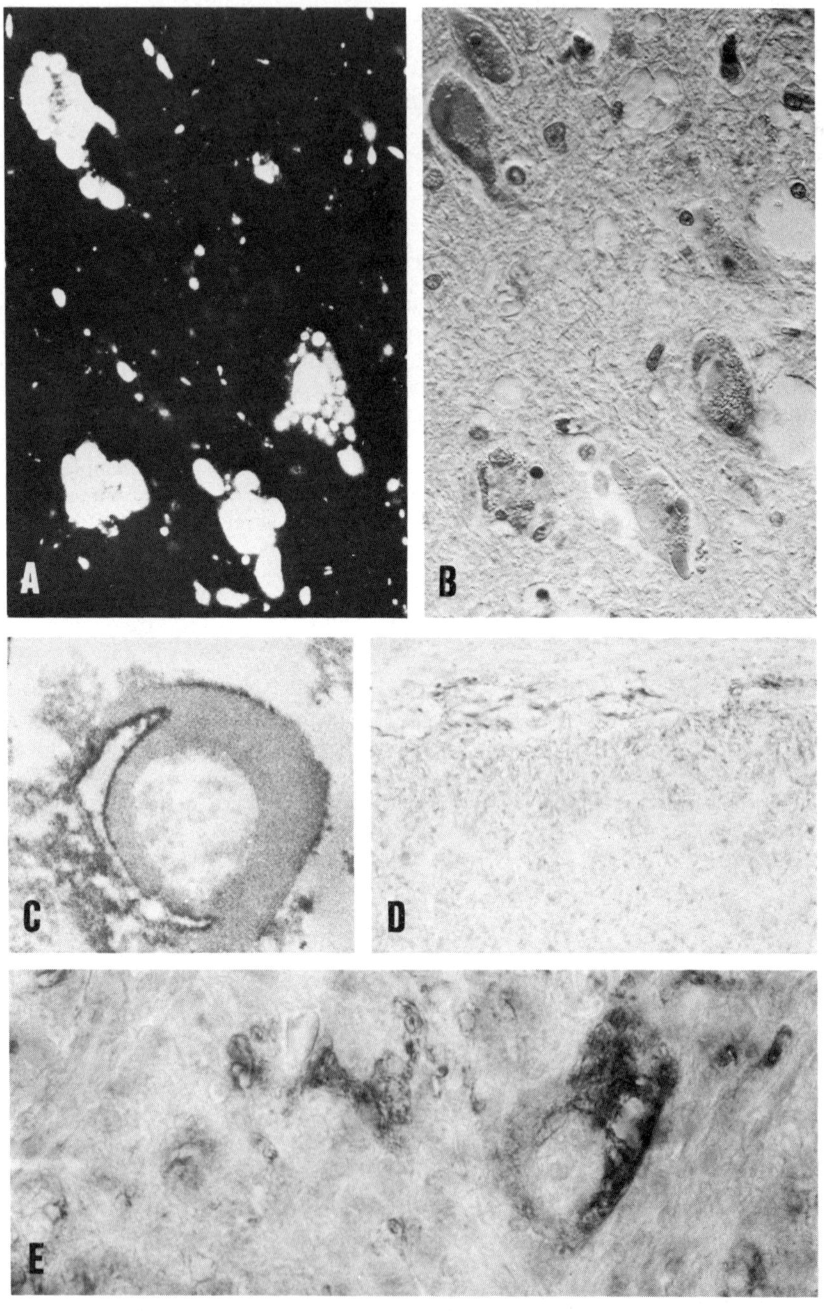

intensity of staining usually is proportional to the thickness of the myelin sheath; very thinly myelinated fibers, e.g. in upper layers of cerebral cortex, in cerebellar cortex, or fetal and myelinating fibers, are beautifully outlined by immunocytochemistry for MBP (FIGURE 1 I) but appear indistinct or are not detectable by conventional myelin stains.

Influence of Length of Fixation Time

In the papovavirus study in cases of progressive multifocal leukoencephalopathy (PML),[9] among other cases different regions of five brains were fixed with formalin for different periods of time. There was a marked loss of immunoreactivity after long fixation (one to five years) (FIGURE 1 F) compared to the result achieved after shorter fixation (one to four weeks) (FIGURE 1 E). When less potent antisera were used, even negative results in immunostaining could result after prolonged fixation. In some cases that were fixed in non-buffered formalin, immunoreactivity was relatively low even after short fixation periods.

Antibody Penetration of Tissue

When immunostaining paraffin sections for MBP, we observed that intensity of labeling of very thick myelin sheaths, especially when sectioned longitudinally, was somewhat less than expected. We thought that MBP immunolabeling might be accentuated in the outer myelin layers. To substantiate this consideration, a paraffin section immunostained by the PAP technique for MBP was removed from the glass slide by ultrasound and processed for transmission EM (Dr. H. Lassmann, Vienna). When the uncontrasted material was investigated by EM, the PAP reaction product was mainly confined to the outer surface and artefactual disruptions of the myelin sheath (FIGURE 2 C), suggesting that the antibodies had no access to the compacted myelin layers.

Examples of Immunostaining for Various Antigens

Some of our immunohistological findings were reported elsewhere[1,3–9] and will not be repeated here. Some previously unpublished data and examples of the application of immunohistology to the nervous system are briefly mentioned here.

In collagen vascular diseases, immune complex deposits were detected in the

---

FIGURE 2. (A and B) SSPE, midbrain tegmentum. Masses of measles virus antigen in nuclei, cytoplasm, and processes of neurons (A, indirect IF for measles virus antigen); when the same section is counterstained with H & E (B), tissue morphology seems well preserved (after protease treatment); although some nuclear and cytoplasmic inclusions are seen, viral investment appears much less intensive than shown by IF; ×400. (C) Normal adult cerebral cortex; paraffin section was stained by PAP for MBP and then processed for EM, uncontrasted. Dark reaction product mainly at outside and at artefactual disruptions of a cross-sectioned myelin sheath, ×30,000. (D) Cerebral astrocytoma. Tumor cells moderately rich in GFAP produce more GFAP when growing over the cortical surface into the leptomeninges (above); PAP for GFAP, ×100. (E) Brain metastasis of solid pleomorphic bronchial carcinoma. GFAP-like immunoreactivity in cytoplasm of a large tumor cell (right of center) and smaller cells. PAP for GFAP, hematoxylin counterstain, ×630.

choroid plexus (FIGURE 1 G) and in the walls of cerebral vessels.[3] Immunoglobulins, mainly IgG, were frequently labeled in the cytoplasm and processes of astrocytes, and inconstantly on axons, myelin sheaths, and nerve cells; this is interpreted as postmortal diffusion of serum from vessels into surrounding tissues or premortal barrier impairment.

Tissues from 83 patients were immunostained for MBP, including normal brain tissue and various disorders of the white matter. Myelin of central and peripheral nervous systems was strongly labeled; in the developing brain, cytoplasm and processes of myelinating oligodendroglia also were stained (FIGURE 1 I). In some demyelinating diseases, e.g. perivenous encephalitis and adrenoleukodystrophy (ALD), a few oligodendrocytes were labeled in a similar manner, suggesting their remyelinating function. This finding was corroborated by occasional very thin myelin sheaths in demyelinated lesions of multiple sclerosis (MS) (especially recent lesions) and ALD.[6] In 20 cases of white matter necroses (infarcts), MBP immunostaining outlined structural myelin up to two to three weeks after the insult. Phagocytized myelin debris showed similar staining properties for MBP and Luxol fast blue, being stainable up to a lesion age of 45 days.

In a study in collaboration with E. Scharfen, rabies virus antigen was found in 120 cases of natural rabies in wild and domestic animals and was restricted to nerve cells and their processes. When studied topographically, rabies virus antigen was found widespread, including the cerebral cortex in varying intensity. In three cases of necrotizing encephalitis, imprint preparations from biopsy or autopsy brain showed strong immunolabeling for HSV antigen (FIGURE 1 H).

Apart from 28 SSPE cases with masses of virus antigen (FIGURE 2 A,B), measles virus antigen was searched for in eight cases of perivenous postmeasles encephalitis, seven cases of postmeasles encephalopathy, and two cases of acute measles without neurological symptoms by means of the PAP technique. In these cases, measles virus antigen could not be detected.

Glial Fibrillary Acidic Protein in Tumors

In 40 gliomas of varying types and differentiation, GFAP was found in high amounts in cytoplasm and processes of astrocytoma cells but also in some glioblastomas whereas few labeled cells in oligodendrogliomas and ependymomas were interpreted as reactive gliosis. A previously unnoticed feature was the increase of GFAP immunoreactivity in tumor cells that had invaded either the leptomeninges (FIGURE 2 D) or the perivascular spaces. As a chance finding, we found in the brain metastasis of a bronchial carcinoma a distinct GFAP-like immunoreactivity in the cytoplasm of some tumor cells (FIGURE 2 E), deeply within the tumor and remote from peritumorous gliotic tissue or invading reactive gliosis.

DISCUSSION

It is evident from this brief outline that immunohistological studies add a new dimension to studies of structure and function of the nervous system. As an example, immunostaining for MBP, which currently appears as the most sensitive method to stain myelin in paraffin sections, might prove to be of value in studies of myelogenesis and cortical myeloarchitecture. Delineation of type and distribution of Ig-containing cells in inflammatory[1,5,12] and metabolic[2] diseases and demonstration of immune complexes in choroid plexus and brain vessels in collagen vascular diseases[3] contribute to the clarification of the immunopathology of these diseases. Most evident is the diagnostic value of virus antigen detection; a reliable diagnosis seems possible without the hazard of handling highly infectious fresh material, such as rabies brain. To obtain reliable and consistent immunohistological results from paraffin sections, some factors of substrate processing should be evaluated. Long fixation time in formalin was found to diminish or, in some cases, even abolish immunoreactivity.

The use of proteolytic enzymes on paraffin sections before immunostaining has been a matter of debate. Both trypsin[13,15,19] and pronase (protease type VII)[11,17] were recommended for IF and PAP techniques to enhance reliability and sensitivity. Although we did not test different enzyme solutions, we agree with Denk *et al.*[11] that the use of 0.1% protease for 15 minutes at 37°C is well suited in many systems; in our experience only MBP immunostaining was not enhanced by previous digestion. When enzyme treatment is performed in a controlled manner, tissue morphology may be well preserved.

Immunohistological determination of GFAP is currently considered an important tool to prove an astrocytic origin of tumors.[10] However, our finding of GFAP-like immunoreactivity in a cancer metastasis casts doubt on the specificity of GFAP for the nervous system. Most recently, GFAP-like immunoreactivity was found in pleomorphic adenoma of salivary gland as well as the S-100 protein (also previously considered specific for nervous tissue) in adenohypophysis, interdigitating reticular cells of lymph nodes, and myoepithelial cells of mammary and salivary glands.[16] It appears that with an increase of immunostaining experience in various organs, "cell markers" are likely to become markers for specific cellular functions (which theoretically could be performed by all somatic cells) and the significance of "histogenetic markers" may decrease. Further immunohistological studies should add a wealth of information in this respect in the near future.

SUMMARY

A brief outline is given of applications of immunohistological techniques to the study of normal and diseased nervous tissue. Protease treatment of paraffin sections usually enhances sensitivity and reliability both of IF and PAP techniques. Sensitivity of immunohistological examination of paraffin sections is comparable to that of virus detection by normal virological techniques in animal rabies and slightly superior to EM search for virions in SSPE and PML. Immunostaining for MBP appears to be the most sensitive method for myelin, especially for demonstration of very thin myelin sheaths, which are important in studies of myelogenesis and cortical myeloarchitecture. Prolonged fixation in formalin clearly diminishes or abolishes immunoreactivity. Compacted myelin stains less well for MBP than preparative myelin artefacts and the surface of myelinated fibers. GFAP production is enhanced when glioma cells invade surrounding mesenchymal structures. The chance finding of GFAP-like immunoreactivity in a cancer metastasis casts doubt on the astroglial specificity of GFAP.

REFERENCES

1. AUFF, E. & H. BUDKA. 1980. Immunohistologische Methoden in der Neuropathologie. Curr. Topics Neuropathol. **6:** 21–29.
2. BERNHEIMER, H., H. BUDKA & P. MÜLLER. 1983. Brain tissue immunoglobulins in

adrenoleukodystrophy: A comparison with multiple sclerosis and systemic lupus erythematosus. Acta Neuropathol. (Berl.) **59:** 95–102.

3. BUDKA, H. 1981. Brain pathology in the collagen vascular diseases. Angiology **32:** 365–372.

4. BUDKA, H. 1982. Hyaline inclusions (pseudopsammoma bodies) in meningiomas: Immunocytochemical demonstration of epithel-like secretion of secretory component and immunoglobulins A and M. Acta Neuropathol. (Berl.) **56:** 294–298.

5. BUDKA, H., H. LASSMANN & T. POPOW-KRAUPP. 1982. Measles virus antigen in panencephalitis. An immunomorphological study stressing dendritic involvement in SSPE. Acta Neuropathol. (Berl.) **56:** 52–62.

6. BUDKA, H., B. MOLZER, H. BERNHEIMER, H. LASSMANN, P. PILZ & K. TOIFL. 1983. Clinical, morphological and neurochemical findings in adrenoleukodystrophy and its variants. *In* Proc. Intern. Symp. on the Leucodystrophy and Allied Diseases. (Kyoto, 1981) T. Yonezawa, Ed. Neuropathology (Tokyo) Suppl. I: 209–224.

7. BUDKA, H. & T. POPOW-KRAUPP. 1981. Immunohistological studies in viral encephalitis. Acta Neuropathol. (Berl.) Suppl. VII: 142–144.

8. BUDKA, H. & T. POPOW-KRAUPP. 1981. Rabies and herpes simplex virus encephalitis. An immunohistological study on site and distribution of viral antigens. Virchows Arch. A (Pathol. Anat.) **390:** 353–364.

9. BUDKA, H. & K. V. SHAH. 1983. Papovavirus antigens in paraffin sections of PML brains. *In* Polyomaviruses and Human Neurological Disease. J. L. Sever and D. L. Madden, Eds.: 299–309. Alan R. Liss. New York.

10. DE ARMOND, S. J., L. F. ENG & L. J. RUBINSTEIN. 1980. The application of glial fibrillary acidic (GFA) protein immunohistochemistry in neurooncology. Path. Res. Pract. **168:** 374–394.

11. DENK, H., T. RADASZIEWICZ & E. WEIRICH. 1977. Pronase pretreatment of tissue sections enhances sensitivity of the unlabelled antibody-enzyme (PAP) technique. J. Immunol. Meth. **15:** 163–167.

12. ESIRI, M. M. 1980. Multiple sclerosis: A quantitative and qualitative study of immunoglobulin-containing cells in the central nervous system. Neuropathol. Appl. Neurobiol. **6:** 9–21.

13. HUANG, S., H. MINASSIAN & J. D. MORE. 1976. Application of immunofluorescent staining on paraffin sections improved by trypsin digestion. Lab. Invest. **35:** 383–390.

14. ITOYAMA, Y., N. H. STERNBERGER, M. W. KIES, S. R. COHEN, E. P. RICHARDSON, JR. & H. DEF. WEBSTER. 1980. Immunocytochemical method to identify myelin basic protein in oligodendroglia and myelin sheaths of human nervous system. Ann. Neurol. **7:** 157–166.

15. MEPHAM, B. L., W. FRATER & B. S. MITCHELL. 1979. The use of proteolytic enzymes to improve immunoglobulin staining by the PAP technique. Histochem. J. **11:** 345–357.

16. NAKAZATO, Y., H. YAMAGUCHI, K. TAKAHASHI & T. MORI. 1983. Immunohistochemical localization of nervous tissue proteins, S-100, GFAP and astroprotein in extraneural tissues. *In* Proc. XXXth Ann. Colloquium: Protides of the Biological Fluids (Brussels). H. Peeters, Ed.: 47–50. Pergamon Press. Oxford.

17. RADASKIEWICZ, T., B. DRAGOSICS, M. ABDELFATTAHGAD & H. DENK. 1979. Effect of protease pretreatment on immunomorphologic demonstration of hepatitis-B-surface antigen in conventional paraffin-embedded liver biopsy material: Qualitative evaluation. J. Immunol. Meth. **29:** 27–33.

18. STERNBERGER, L. A. 1979. Immunocytochemistry. 2nd edit. John Wiley & Sons, Inc. New York.

19. SWOVELAND, P. T. & K. P. JOHNSON. 1979. Enhancement of fluorescent antibody staining of viral antigens in formalin-fixed tissues by trypsin digestion. J. Infect. Dis. **140:** 758–764.

Identification of Rabies Antigen in Human and Animal Tissues[a]

PEGGY T. SWOVELAND AND KENNETH P. JOHNSON

Department of Neurology
University of Maryland School of Medicine
Baltimore, Maryland 21201

INTRODUCTION

Throughout the recorded history of Western civilization, rabies virus infection of humans and animals has been described and attributed to the bite of a "mad dog" or other animal.[12] At the present time, rabies is endemic in many parts of the world, including the United States. Among the wild animal populations that harbor rabies, skunks and bats are the most commonly infected.[11] Other significant populations include raccoons and foxes. Domestic animals most frequently reported to be rabid are cattle, dogs, cats, and horses.

Because of the continued presence of rabies and the threat it poses not only to the infected individual but also to others in contact with the individual, accurate diagnosis is imperative. Two sensitive diagnostic procedures are currently in use: mouse inoculation of the tissue in question and fluorescent antibody labeling. In the mouse inoculation test, weanling mice are inoculated intracerebrally with brain, salivary gland, or other tissue homogenates.[8] Infected mice become ill in 6 to 10 days. Fluorescent antibody labeling can be performed on corneal impressions, impression smears of biopsy tissue, or cryostat sections of biopsy or autopsy brain tissue.[10] The tissue must be either fresh frozen or glycerol impregnated. The mouse inoculation and fluorescent antibody methods, both of which require fresh or fresh-frozen tissue, have been shown to give identical results in detecting rabies virus or viral antigens in over 99% of specimens.[7]

The presence of Negri bodies, visible as eosinophilic cytoplasmic inclusion bodies in hematoxylin and eosin stained sections or as red inclusions with blue internal granules in Seller's stained material, has been considered to be diagnostic for rabies virus infection. However, 10 to 30% of rabies-positive material lacks inclusions visible by light microscopy.

Clinically, human rabies may assume two different courses.[4] The most characteristic is "furious" rabies in which the signs of nervous system involvement include hyperactivity, disorientation, hallucinations, or seizures. Paralytic or "dumb" rabies characterizes the clinical course in 5–20% of patients. The different clinical courses may be related to the virus strains carried by different animal species or to differing host factors. The reported incubation period for rabies ranges from 9 days to 19 years. In about 1% of well-documented cases, the incubation period is greater than one year,[3] and there are cases where no exposure to rabies is known.

Because there are cases of rabies in which the diagnosis is not suspected because of the unusual clinical course or potentially extremely long incubation period, it is possible that no fresh and fresh-frozen material will be available. Therefore, an alternate method is required to confirm the diagnosis. Immunofluorescence of viral

[a]Supported by funds from the Research Service of the Veterans Administration.

antigens in formalin-fixed tissue, embedded in paraffin for routine histology has been made possible by partial digestion of the tissue sections by trypsin prior to application of the antiserum and fluorescein conjugate.[5,13] The sensitivity and specificity of this method has been compared with mouse inoculations and immunofluorescence in cryostat sections and a large number of rabies-positive and rabies-negative human brains have been screened.

MATERIALS AND METHODS

Mouse Tissue

Mouse brains were acquired as coded specimens in formalin from the Centers for Disease Control (CDC) in Atlanta, Georgia, and from the Virus and Rickettsial Diseases Laboratory Section (VRDLS) of the California State Public Health Service in Berkeley, California. From the CDC, a total of 12 infected and uninfected brains were received.

In a collaborative experiment with the VRDLS, 28 mice were inoculated with the Coyote TO-577 strain of rabies virus. At daily intervals through seven days, four animals were sacrificed. Brains from two animals were fixed in 10% formalin and embedded in paraffin for histological examination and immunofluorescence on trypsin-treated sections. Brains from two animals were split for mouse inoculation and immunofluorescence on frozen sections. Three uninoculated mice were processed in the same manner. All formalin-fixed brains were received as coded specimens.

Other Animal Tissue

Other animal species whose brains were examined histopathologically and by immunofluorescence included skunk, rhesus monkey, and cat.

Human Tissues

Human brain tissues in formalin or in paraffin blocks from confirmed cases of rabies were provided by numerous sources. Dr. Kenneth M. Earle of the Armed Forces Institute of Pathology (AFIP) in Washington, D.C. provided paraffin blocks of eight cases of rabies dating from 1934 to 1951. Dr. R. W. Emmons of the VRDLS provided tissue from four cases of confirmed rabies from California dating from 1971 to 1979. Dr. Rangel, Dr. A. Escobar, and Dr. H. Alcala provided multiple tissue samples from rabies cases in Mexico. Dr. Karl T. Kleeman of the Virology Branch of the North Carolina Department of Human Resources, Division of Health Services, provided tissue from an undiagnosed case in which cytoplasmic inclusions characteristic of Negri bodies were visible in the hippocampus. Dr. V. E. Sangalang of Halifax, Nova Scotia supplied paraffin sections from a case in which rabies was not suspected until neuropathological examination of the brain revealed Negri bodies in the Purkinje cells of the cerebellum. Electron microscopy revealed rhabdovirus virions, but no further diagnostic confirmation of rabies was possible.

Control human tissue included sections of brain from subacute sclerosing panencephalitis (SSPE), which contains cytoplasmic and nuclear inclusions, herpes simplex encephalitis, and brain tissue from individuals with non-neurological diseases.

Antisera and Conjugates

Rabies hyperimmune serum prepared in rabbits was provided by Dr. R. W. Emmons of the VRDLS and diluted 1:100 for use. Measles hyperimmune serum and herpes simplex hyperimmune serum prepared in rabbits and normal rabbit serum were used as controls. Fluorescein isothiocyanate-conjugated goat anti-rabbit IgG (Meloy Labs, Springfield, VA) was used at a dilution of 1:40.

Tissue Preparation and Trypsin Digestion

Brain tissue was embedded in paraffin for routine histology; 7 μm sections were stained with hematoxylin and eosin (H and E). Adjacent unstained sections mounted on microscope slides were processed for immunofluorescence. Sections were deparaffinized in xylene, hydrated through alcohol, and rinsed in Ca^{2+}-Mg^{2+}–free phosphate-buffered saline (PBS). Slides were incubated in 0.25% trypsin (Gibco, Grand Island, NY) at 37°C for 1 hour. After rinsing in PBS, the sections were incubated for 30 minutes in rabbit anti-rabies serum or control serum, rinsed three times in PBS, incubated for 30 minutes in fluorescein-conjugated goat anti-rabbit IgG, rinsed three times in PBS, and coverslipped with 50% glycerol in PBS. Sections were examined by epi-illumination with a Leitz Dialux 20 fluorescence microscope equipped with a 100 watt mercury lamp.

RESULTS

Mouse Studies

In an initial study, infected and control mouse brains in 10% formalin were received as coded specimens from the CDC. The tissue was embedded in paraffin and adjacent sections were stained with H and E for light microscopy and with immunofluorescence reagents. Six of the twelve brains were positive by immunofluorescence with hyperimmune rabies serum. All brains were negative when reacted with anti-measles, anti-herpes simplex, or normal rabbit sera. Upon decoding, it was found that the 6 FA-positive brains were from infected animals, while the 6 FA-negative brains were from control animals. Histopathologically, the six FA-positive brains all contained inflammatory infiltrates. Negri bodies were not prominent in the H and E stained slides, although fluorescence clearly demonstrated large cytoplasmic inclusions. Viral antigen appeared to be restricted to neurons, and always in a cytoplasmic distribution. Viral inclusions were of varying sizes and viral antigen frequently was present far out into neuronal processes.

In a collaborative study with the VRDLS, coded specimens taken at daily intervals after inoculation were processed in a similar manner to the CDC specimens. Results of the various detection methods are shown in TABLE 1. Mouse inoculation of brain tissue, immunofluorescence on cryostat section, and immunofluorescence on formalin-fixed, trypsin-treated tissue were all positive three days after inoculation. The grading system of 1 to 4 used in the immunofluorescence studies with frozen sections and trypsin-treated sections represents relative numbers of positive cells in the brain sections. In all positive brains, the fluorescence intensity of the individual infected cells was equally bright.

Viral antigen was first identifiable in small foci of cortical neurons on day 3. No histological abnormalities were evident at that time. On successive days, the infected

TABLE 1. Rabies Virus Identification in Infected Mouse Brains

Days Post-inoculation	Infectivity Mouse LD_{50}	Immunofluorescence Frozen Sections	Immunofluorescence Trypsin-treated
1	$<10^{-1.0}$	0	0
2	$<10^{-1.0}$	0	0
3	$10^{-1.1}$	1+	1+
4	$10^{-2.3}$	1–2+	2+
5	$10^{-2.8}$	2–3+	3+
6	$10^{-5.1}$	3–4+	4+
7	$10^{-5.5}$	3–4+	4+
pre-inoc.	ND	ND	0

foci increased in size and antigen spread to the hippocampus, throughout the cerebral cortex and to the cerebellum. Inflammatory infiltrates became more pronounced each day.

Skunk, rhesus monkey, and cat brains have also been examined with equivalent results. The specimens tested have all been infected with various laboratory strains of rabies virus.

Human Studies

Multiple human autopsy brain tissue specimens from cases of confirmed or presumptive diagnosis of rabies were examined. TABLE 2 summarizes the results of this study. A total of 111 cases were examined, of which 93 were positive by immunofluorescence. Control brain tissue and control serum produced consistently negative results (data not shown).

Specimens received from the AFIP represented tissues that had been stored in paraffin blocks for 28 to 45 years. Two of the specimens were fixed in Zenkers fixative, and both were positive by immunofluorescence. One of the negative tissues was fixed in an unknown fixative. Specimens more than 30 years old had the weakest fluorescence, but were unequivocally positive.

Of the 94 specimens received from Mexico, 79 were positive. Within this group of specimens, tissue fixation and degree of tissue autolysis were variable, which could account for some of the negative results.

TABLE 2. Human Rabies Cases Examined by Immunofluorescence on Trypsin-treated Paraffin Sections

Source	Date of Autopsy	Number of Cases	Number of FA Positive
AFIP	1934–1951	8[a]	6
VRDLS	1971–1979	4	4
Mexico	1952–1980	94	79
Sangalang-Nova Scotia	1978	1	1
Hanrahan-Pittsburgh, PA	1979	1	1
Burton-Boise, Idaho	1979	2	2
Kleeman-Raleigh, NC	1979	1	0
Total		111	93

[a]Two cases fixed in Zenker's fixative, one in unknown fixative

The specimens of recent cases of rabies received from the VRDLS and various individuals were all intensely fluorescent. The case from North Carolina was one in which rabies was considered after histopathological examination of autopsy tissue. Eosinophilic inclusions were observed in the cytoplasm of hippocampal neurons, but immunofluorescence on trypsin-treated paraffin sections was negative, and serologic studies revealed no antibodies to rabies virus. No fresh tissue was available for mouse inoculation or immunofluorescence on frozen sections.

Two of the other cases represent circumstances in which the diagnosis of rabies was made retrospectively, based on the presence of Negri bodies, rhabdovirus virions visible by electron microscopy, and antibodies to rabies virus in one case.[6] Both contained rabies antigen demonstrable by immunofluorescence on paraffin sections.

Among the specimens examined, the histologic evidence of acute encephalitis varied from minimal to severe, although perivascular and parenchymal inflammatory cell infiltration was evident in most cases. The visibility of Negri bodies in H and E stained sections was also highly variable. Immunofluorescence staining of trypsin-treated paraffin sections demonstrated viral inclusions of varying size and number of the cytoplasm of neurons, and frequently in neuronal processes.

DISCUSSION

Rabies continues to be a public health problem in the United States and Europe, as well as in developing countries. The increased incidence in recent years of rabies in animals that live in close association with humans makes the need for rapid and accurate diagnosis a continuing issue. The report of 83 confirmed cases of cat rabies in Iowa in 1981 with 42 humans receiving post-exposure rabies vaccine after penetrating bites by these rabid cats exemplifies the continuing public health issue.[1] In addition, the incidence of rabies in raccoon populations in the South is increasing annually, and canine rabies along the U.S.-Mexican border remains a problem. In spite of these increasing numbers of animal rabies, it must be noted that the annual incidence of human rabies remains very low, ranging from 0 to 5 cases per year during the past 20 years.[11] This is attributable in large part to the highly effective post-exposure vaccination program.

Alternative methods of rabies diagnosis that can be used when fresh, fresh-frozen, or glycerol-impregnated tissues are not available must be comparable in sensitivity and specificity to established diagnostic techniques. The mouse study comparing mouse inoculation, immunofluorescence of frozen sections, and immunofluorescence in formalin-fixed, trypsin-treated brain tissue demonstrated equivalent results in terms of the earliest detection of virus or viral antigens. In addition, the virus titers from infected brains correlated with the subjective ratings of numbers of antigen-positive cells.

Of the recent cases of human rabies, two were not diagnosed until autopsy; therefore, no fresh tissue was available. Both contained Negri bodies by light microscopy and rhabdovirus by electron microscopy, and both were positive by immunofluorescence on trypsin-treated tissue.

A third case showed large cytoplasmic inclusions in the pyramidal cells of the hippocampus. The serum contained no detectable antibodies to rabies virus and the immunofluorescence was negative. The Negri body has been considered to be pathognomonic for rabies, but there are reports of similar inclusions in non-rabies diseases of unknown etiology.[2]

It has been shown that rabies antigen is stable for prolonged periods after fixation and embedding. While those materials that had been stored for more than 30 years were less consistent in the intensity of the staining reaction, there was no information available on the kind of fixative used, the length of time after death when the tissue was fixed, or the storage conditions.

Some of the older tissues received from the AFIP were fixed in Zenkers fixative. Treatment with iodine to remove the fixative had no adverse effect on the immuno-fluorescence. None of the tissues were fixed in Bouin's fixative. While it has been shown that Bouin's fixative is not a suitable fixative for trypsin immunofluorescence for measles virus,[13] there have been reports of good results with peroxidase-antiperoxidase (PAP) labeling for Herpes simplex virus.[9]

Using paraffin-embedded tissue blocks, large areas of tissue can be screened. When antigen is sparse, this allows selection of areas of brain in which histopathology is most pronounced or Negri bodies are evident. In the absence of pronounced histopathology, multiple large areas can be surveyed by immunofluorescence. In addition, the ability to survey multiple, large tissue sections should provide information on areas of the brain involved, types of neurons infected, and, in experimental situations, information on the progression of the infectious process into and through the central nervous system.

As a diagnostic tool, it must be emphasized that the methods of choice continue to be sampling of fresh, fresh-frozen, or glycerol-impregnated tissues. However, in circumstances where fresh tissue is not available, and in regions of the world where facilities are limited, immunofluorescence on formalin-fixed tissue provides a means of confirming a diagnosis.

ACKNOWLEDGMENTS

Expert technical assistance of Ms. Paula Fukushima and Mr. J. Woodie and advice and reagents from Dr. R. W. Emmons are gratefully acknowledged.

REFERENCES

1. Cat Rabies Exposures in Iowa—1981. 1982. Morbidity and Mortality Weekly Report **31:** 67–73.
2. DERAKHSHAN, I. 1978. Light microscopical diagnosis of rabies: A reappraisal. Lancet **1:** 302–303.
3. DUPONT, J. R. & K. M. EARLE. 1965. Human rabies encephalitis. Neurology **15:** 1023–1034.
4. HATTWICK, M. A. W. & M. B. GREGG. 1975. The disease in man. *In* The Natural History of Rabies. G. M. Baer, Ed. **2:** 281–304. Academic Press. New York.
5. HUANG, S., H. MINASSIAN & J. D. MOORE. 1976. Application of immunofluorescent staining on paraffin sections improved by trypsin digestion. Lab Invest. **35:** 383–390.
6. Human Rabies—Pennsylvania. 1979. Morbidity and Mortality Weekly Report **28:** 75–81.
7. KISSLING, R. E. 1975. The fluorescent antibody test in rabies. *In* The Natural History of Rabies. G. M. Baer, Ed. **1:** 401–416. Academic Press. New York.
8. KOPROWSKI, H. 1973. The mouse inoculation test. *In* Laboratory Techniques in Rabies. 3rd edit. M. M. Kaplan & H. Koprowski, Eds.: 85–93. World Health Organization. Geneva.
9. LASCANO, E. F. & M. I. BERRIA. 1980. Histological study of the progression of Herpes simplex virus in mice. Arch. Virol. **64:** 67–79.
10. LENNETTE, E. H., J. D. WOODIE, K. NAKAMURA & R. L. MAGOFFIN. 1965. The diagnosis of rabies by fluorescent antibody method (FRA) employing immune hamster serum. Health Lab. Sci. **2:** 24–34.

11. Rabies Surveillance Annual Summary—1979. 1981. U.S. Dept. of Health, Education, and Welfare. Centers for Disease Control.
12. STEELE, J. H. 1975. History of rabies. *In* The Natural History of Rabies. G. M. Baer, Ed. **1:** 1–29. Academic Press. New York.
13. SWOVELAND, P. T. & K. P. JOHNSON. 1979. Enhancement of fluorescent antibody staining of viral antigens in formalin-fixed tissues by trypsin digestion. J. Infect. Dis. **140:** 758–764.

Detection of Viral Antigens
in Formalin-fixed Specimens
by Enzyme Treatment

TAKESHI KURATA, RYO HONDO, SHOICHIRO SATO,
AKIRA ODA, AND YUZO AOYAMA

Department of Pathology
Institute of Medical Science
University of Tokyo
Shirokanedai 4-6-1, Minatoku
Tokyo 108, Japan

JOSEPH B. McCORMICK

Special Pathogens Branch
Division of Viral Diseases
Centers for Disease Control
Atlanta, Georgia 30333

INTRODUCTION

In 1962, Sainte-Marie[6] succeeded in demonstrating influenza A antigen by immunofluorescence (IF) in paraffin tissues embedded after cold alcohol fixation. Since the first application of enzyme treatment for IF by Huang[2] to detect hepatitis B antigens, several viral antigens[3,5,7] have been demonstrated in formalin-fixed paraffin materials. In a previous paper,[1] we examined the basic conditions for enzyme treatment of paraffin sections from cases infected with herpes simplex (HS), varicella-zoster (VZ), and cytomegalovirus (CM).

In this communication we expand these findings and report the detection of viral antigens by this method in paraffin materials from cases of HS, VZ, CM, subacute sclerosing panencephalitis (SSPE), Japanese B encephalitis (JE), progressive multifocal leukoencephalopathy (PML), measles, acute hemorrhagic conjunctivitis (AHC), and Lassa and Korean hemorrhagic fever (KHF). The effectiveness of enzyme treatment (protease or trypsin) for antigen recovery was examined in several virus-infected, formalin-fixed cells. HS virus-infected mouse brain was used to test the relationship between the duration of formalin fixation and antigen recovery.

MATERIALS AND METHODS

Biopsy or Autopsy Materials

Formalin-fixed and paraffin-embedded human materials were collected from 49 patients from several hospitals and institutions in Japan and the United States. In most cases, the diagnoses were established by IF in frozen materials, by virus isolation, serological tests, and/or electron microscopy. Virus was isolated in 21 cases. Virus infection was suspected in 28 cases from clinical and histopathological findings.

HS Virus-infected Mouse Brain

Ten-week-old female DDD mice were inoculated intracerebrally with type 1 HS virus (Tomioka strain, 10^4 plaque-forming units/0.01 ml/mouse). When mice were moribund, the brain was dissected and divided coronally into two parts for fixation in 10% buffered formalin and into *n*-hexane at $-70°C$ for frozen section.

Virus-infected Cells

E6 cell cultures (cloned line from Vero cells) were infected with type 1 (Tomioka) and 2 (Ogiwara) HS, measles (Edmonston), Lassa (Josiah), Ebola (Maleo), Marburg (Musoke), Congo (Matin), Rift Valley (Fort Detrick 80641), and KHF (Hantaan) viruses. The cells were harvested at optimal conditions of infection, mixed 1:4 or 1:3 with noninfected E6 cells, and were prepared on spot slides. The slides were fixed in acetone for 5 min, and then fixed in 10% buffered formalin solution for the indicated period. Except for HS and measles cells, all other infected cells were gamma-irradiated for 30 min after acetone fixation.

Preparation of Paraffin Sections

The method was previously described in detail elsewhere. Briefly, slides were coated with 0.1% neoprene in toluene (Polychlorprene, oil-resistant synthetic rubber, Ohken-Shoji Co., Ltd., Japan) to facilitate firm attachment of the sections. Thin paraffin sections were picked up on the neoprene-coated slides from a water bath and dried at 37°C for 2–3 days.

Enzyme Treatment

Deparaffinized sections were treated with 0.25% trypsin (trypsin 1:250, DIFCO) with 0.02% $CaCl_2$ in PBS (pH 7.62) or 0.1% protease (protease, SIGMA type VII) with 0.02% $CaCl_2$ in PBS (pH 7.5–7.6) at 37°C or room temperature and rinsed in PBS for 15 min.

Immunofluorescence

Direct- and indirect-IF were used in frozen and paraffin-embedded materials, respectively. Specificities of the antibodies were determined by complement fixation, IF, hemagglutination inhibition, and neutralization tests. Normal sera from rabbits, guinea pigs, rats, mice, and humans were used at the same dilution as normal controls. All sera were inactivated at 56°C for 30 min prior to use.

Antisera were applied to antigen slides or sections and incubated for 1 hr at 37°C or overnight at room temperature or at 4°C, respectively. Anti-IgG conjugated with fluorescein isothiocyanate (Miles Lab. Inc. or Burroughs Wellcome, Ltd.) were applied for 1 hr at 37°C. The specimens were observed under a Olympus fluorescence microscope. Conjugates were absorbed with tissue powder, if necessary.

Serial sections were stained with hematoxylin and eosin (HE) for histopathological study.

RESULTS

Formalin Fixation Period and Sensitivity of Antigen Recovery

Sensitivity of antigen detection was examined in virus-infected E6 cells that were fixed in formalin for the indicated period (TABLE 1). All tests were done by indirect IF. Specific antibodies were made by hyperimmunizing rabbits to HS type 1 (Tomioka strain) and type 2 (Ogiwara strain) viruses. Anti-measles (Edmonston strain) and rinderpest virus antibodies were obtained in rabbits immunized with highly purified virus particles (kindly supplied by Dr. K. Yamanouchi of our Institute). Anti-Lassa (Josiah strain) and Ebola (Boniface strain) antibodies were prepared in guinea pigs by infection with live virus in the Maximum Containment Laboratory at Centers for Disease Control. Monoclonal antibodies were prepared in mice immunized with Lassa and Ebola viruses inactivated by gamma-irradiation. Convalescent antibodies from human patients with hemorrhagic diseases were also used for antigen detection. In staining formalin-fixed cells, antibodies were used at a concentration two to four times higher than the dilution that optimally stained unfixed cells. Antigenicity decreased rapidly for up to 30 min during formalin fixation. With increasing periods of formalin fixation, antigenicity could slowly be recovered reaching a maximum response at 3 hr to 1 day of fixation when trypsin was used and 3 to 14 days of fixation with protease digestion. Further fixation resulted in a decline in antigenicity. There was no difference in recovery of HS virus 1 and 2 antigens. FIGURE 1 shows the specific fluorescence of measles antigen. Lassa is shown in FIGURE 2. Monoclonal antibodies to Lassa virus give two different staining patterns, smooth (FIGURE 3) and speckled in unfixed cells. Only the speckled pattern remains after fixation (FIGURE 4). With respect to cells infected with HS, measles, Lassa, and Ebola viruses, protease digestion seems more effective than trypsin in restoring antigen. When high concentrations of convalescent sera were used, no difference between protease and trypsin was observed in recovery of antigen in cells infected with four other hemorrhagic fever viruses (FIGURE 5 Ebola, FIGURE 6 Marburg, FIGURE 7 Congo, and FIGURE 8 KHF, stained by human antibodies). Antigen could be recovered by enzyme treatment after periods of formalin fixation of four weeks and storage for three months at 4°C or one month at room temperature.

Detection of HS Antigen in Formalin-fixed Mouse Brain

Stability of the viral antigen in tissues in formalin was tested in mouse brain infected with type 1 HS virus. Frozen sections from all brains were examined for viral antigen by direct IF. Seven brains with strong specific fluorescence were chosen, fixed in formalin for 3, 10, 20, 30, 45, 60, and 90 days, and embedded in paraffin. The sections were digested with protease or trypsin for 1 to 3 hr and used for indirect IF. The brightness of the specific fluorescence was compared in frozen and paraffin sections. HS antigens were stable in formalin-fixed tissues after three months. FIGURE 9 (protease) and FIGURE 10 (trypsin) show the detected antigens in nerve cells of the hippocampus and cerebral cortex, respectively, after 60- and 90-day fixation in formalin.

Detection of Viral Antigens in Human Tissues

Viral antigens were demonstrated in several organs by IF after enzyme digestion in formalin-fixed paraffin sections from autopsy or biopsy specimens. Results of IF from

TABLE 1. Formalin Fixation Time and Sensitivity of Antigen Detection in E6 Cells Infected with Different Viruses

	Hours				Days					
	0	0.5	1	3	1	3	7	14	21	28
Herpes simplex rabbit (1:64)[a]										
P[b]	(+ + +)[c]	+[d]	+ +	+ +	+ +	+ + +	+ + +	+ +	+ +	+ +
T[b]		+	+ +	+ + +	+ +	+	+	+	±	−
Measles rabbit (1:40)										
P	(+ + +)	−	−	±	+	+	+ +	+ +	+	+
T		−	±	+ +	+	+	±	−	−	−
Lassa human (1:32)										
P	(+ + +)	−	±	+ +	+ +	+ +	+ +	+ +	+	±
T		−	±	+	+ +	+	+	±	−	−
guinea pig (1:100)										
P	(+ + +)	−	±	+	+	+ +	+ +	+ +	+ +	+
T		±	+	+ +	+	−	−	−	−	−
mouse monoclonal (1:40)										
P	(+ + +)	−	±	+ +	+ +	+ + +	+ +	+ +	+ +	+
T		±	±	+	+ +	+	+	+	−	−
Ebola human (1:32)										
P	(+ + +)	±	+	+ +	+ +	+ +	+ +	+	+	+
T		−	−	+	+	±	−	−	−	−
guinea pig (1:16)										
P	(+ + +)	±	+	+ +	+ +	+ +	+ +	+ + +	+ +	+ +
T		±	+	+ +	+ +	+ +	+	+	+	+
mouse monoclonal (1:16)										
P	(+ + +)	±	+	+ +	+ +	+ +	+ +	+ +	+ +	+
T		±	+	+ +	+ +	+	+	+	+	±
Marburg human (1:16) P or T	(+ + +)		±	+	+	+	+ +	+ +	+ +	+
Congo human (1:16) P or T	(+ + +)		+ +	+ + +	+ + +	+ + +	+ + +	+ + +	+ +	+ +
Rift Valley human (1:16) P or T	(+ + +)		+	+ +	+ +	+ +	+ +	+ +	+ +	+ +
Korean hemorrhagic fever human (1:16) P or T	(+ + +)		+	+ +	+ +	+ +	+ +	+ +	+ +	+

[a]Source of the antibody and dilution used for the indirect immunofluorescence. [b]P; 0.1% protease, T; 0.25% trypsin. The cells were digested for 30 min at 37°C. [c](+ + +); brilliance of the fluorescence without any treatment. [d]−, +; The degree of the fluorescence is given on a scale from − to + + +.

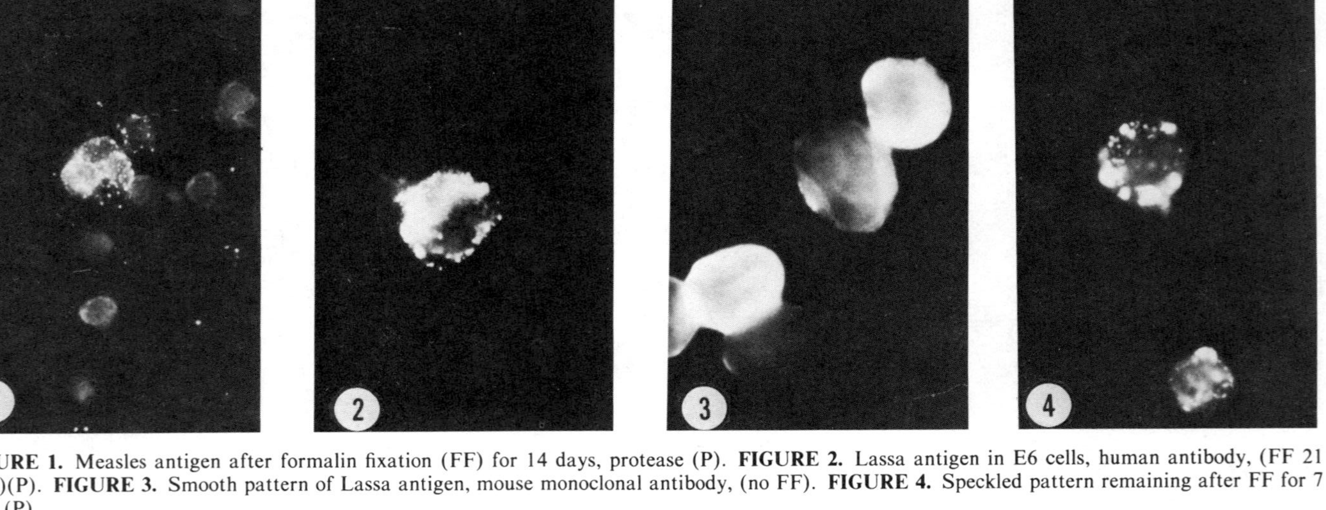

FIGURE 1. Measles antigen after formalin fixation (FF) for 14 days, protease (P). FIGURE 2. Lassa antigen in E6 cells, human antibody, (FF 21 days)(P). FIGURE 3. Smooth pattern of Lassa antigen, mouse monoclonal antibody, (no FF). FIGURE 4. Speckled pattern remaining after FF for 7 days (P).

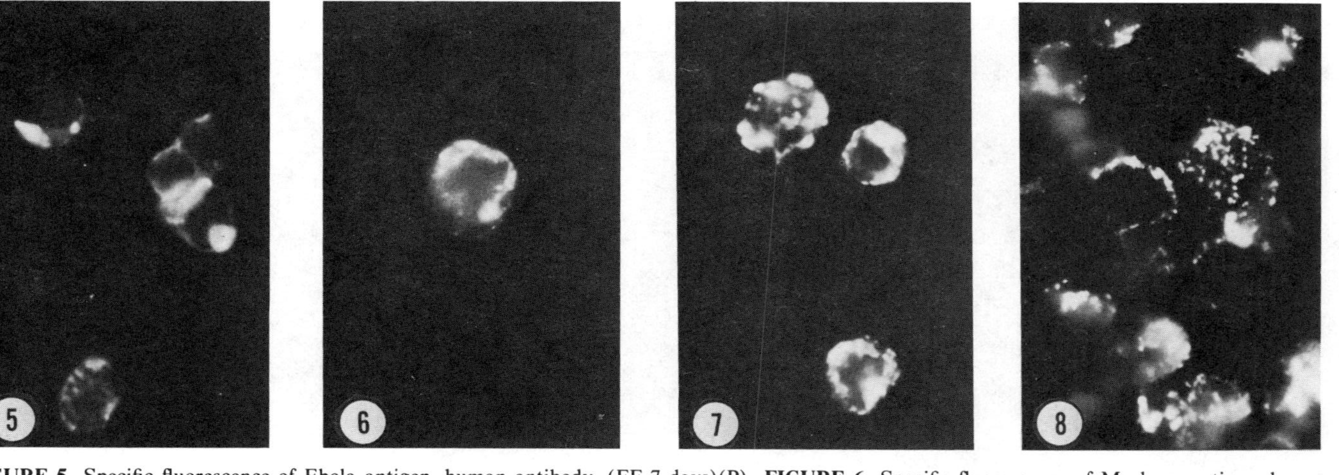

FIGURE 5. Specific fluorescence of Ebola antigen, human antibody, (FF 7 days)(P). **FIGURE 6.** Specific fluorescence of Marburg antigen, human antibody, (FF 7 days)(P). **FIGURE 7.** Specific fluorescence of Congo antigen, human antibody, (FF 7 days)(P). **FIGURE 8.** Specific fluorescence of Korean hemorrhagic fever antigen, human antibody, (FF 28 days)(P).

All tissues (**FIGURES 9–21**) were formalin-fixed, paraffin-embedded, and digested with enzyme before application for IF.

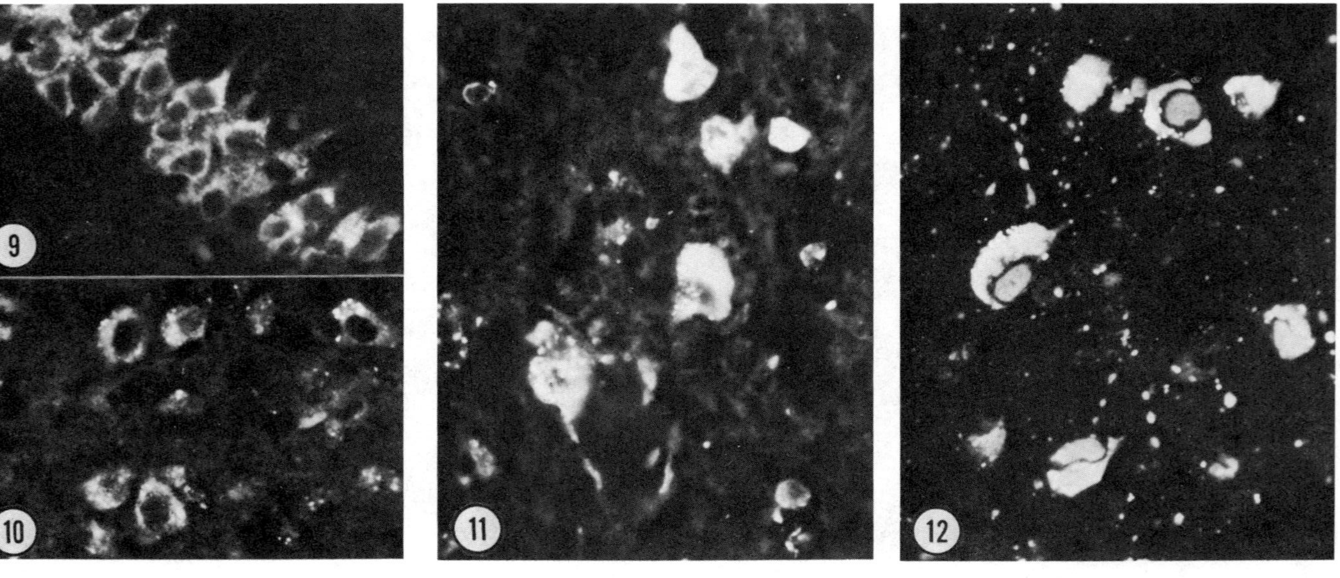

FIGURE 9. Herpes simplex virus (type 1) antigens in mouse brain, (FF 60 days) (P). FIGURE 10. Herpes simplex virus antigens in mouse brain, (FF 90 days) trypsin (T). FIGURE 11. Measles antigen in brain of SSPE case (Case 2, TABLE 2)(P). FIGURE 12. Specific antigen of herpes simplex encephalitis in neurons and glia in Case 4 (TABLE 2)(P).

encephalitis cases are summarized in TABLE 2. In HS encephalitis (HSE), type 1 virus was isolated from six cases. In Cases 7 and 8, HS virus infection was diagnosed on the basis of serology and a typical clinical course. Immunofluorescence with protease or trypsin revealed the specific antigen in these two brain specimens. Specific fluorescence was observed in the intranuclear and/or cytoplasmic areas of glial cells and neurons (FIGURE 11, Case 4), and less frequently in endothelia of capillaries and small vessels, but antigens were not detected in the dendrites of neurons or in the mononuclear cells in perivascular cuffs. In HSE cases, Cowdry type A inclusion bodies in neurons or glial cells, which are characteristic of herpes group-virus and measles-

TABLE 2. Demonstration of Viral Antigens in Formalin-fixed Brain Tissues from Encephalitis Patients

Case	Sex	Age	Course	Virus Isolation	Immunofluorescence	Remarks
Herpes simplex encephalitis (HSE)						
1.	M	22y	20d	HSV type 1	$+$[a]	Frozen[b], CF[c]
2.	F	57y	7w	HSV type 1	$+$	Frozen, CF
3.	F	28y	15d	HSV type 1	$++$	Frozen, ELISA[c]
4.	F	12y	15d	HSV type 1	$++$	EM[d]
5.	F	42y	2d	HSV type 1	$++$	EM
6.	M	32y	19d	HSV type 1	$++$	
7.	F	26y	10d	N.D.[e]	$++$	ELISA
8.	F	26y	24d	N.D.	$++$	ELISA
Subacute sclerosing panencephalitis (SSPE)						
1.	M	7y	3m	SSPE (measles)	$++$	Frozen, EM, CF
2.	M	28y	2y	N.D.	$+++$	CF
3.	M	6y	2.5m	N.D.	$++$	CF
4.	M	17y	2y4m	–	$+$	Frozen, CF, EM
Japanese B encephalitis (JE)						
1.	M	4m	4d	JE	$++$	
Progressive multifocal leukoencephalopathy (PML)						
1.	M	70y	(hepato-ma)	Papova (JC)	$++$	Frozen, EM

[a] $+,++$; a scale of brightness of the specific fluorescence in paraffin sections.
[b] Frozen; viral antigens were detected in frozen materials by IF.
[c] CF and ELISA; significant rise of antibody by complement fixation test and/or ELISA test in serum and/or cerebrospinal fluid.
[d] EM; viral particles were confirmed by electron microscopy.
[e] N.D.; not done.

virus infection are not always observed in HE-stained specimens. Therefore application of IF in paraffin sections might be useful in the diagnosis of suspected cases in which virus is not successfully isolated.

In four cases of SSPE with typical clinical courses, diagnoses were established by IF in frozen tissues and by serology and/or virus isolation. Virus antigens were demonstrated in the nucleus and cytoplasm of nerve and glial cells accompanied by fluorescence in dendrite-like dotted lines (FIGURE 12, Case 2). Positive fluorescence was obtained by using both anti-measles and rinderpest antibodies.

In the majority of JE cases, viral antigen has been observed in thalamus, substantia nigra, less frequently in the cortex or cerebellum in frozen materials (64 cases, 1963–1967).[4] In the Case of JE (TABLE 2), virus was isolated from the brain of a 4-year-old boy and antigen was detected in the basal nuclei (FIGURE 13). Anti-JE

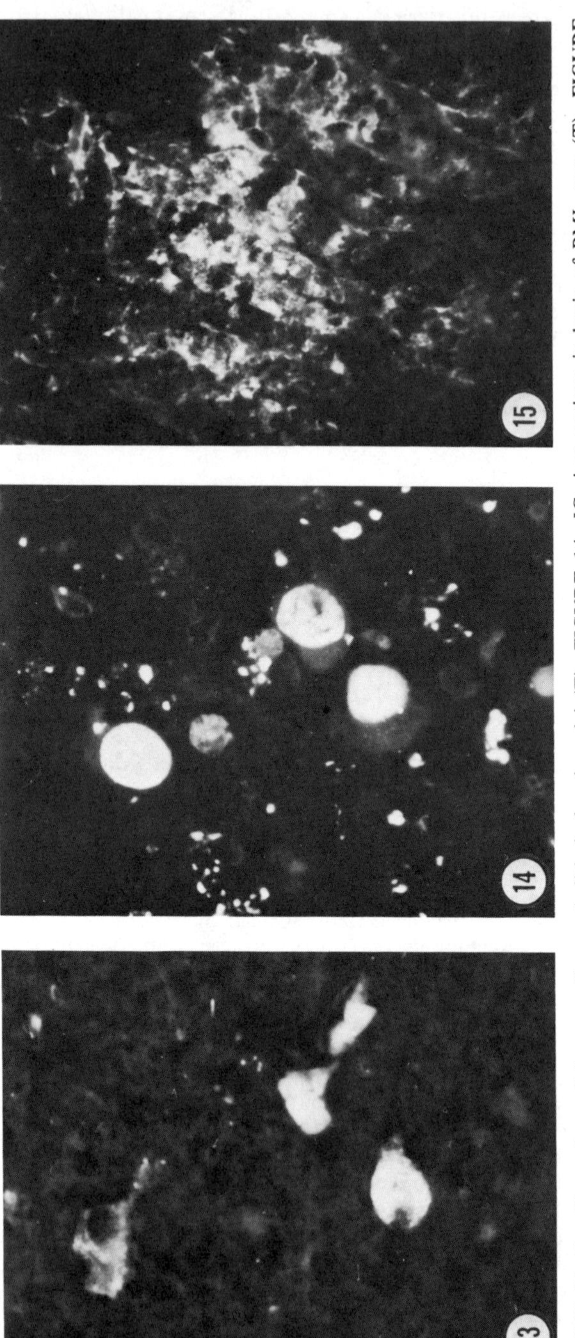

FIGURE 13. Specific antigen of Japanese B encephalitis in basal nuclei (T). FIGURE 14. JC virus antigen in brain of PML case (T). FIGURE 15. Herpes simplex virus antigen in adrenal gland of newborn (T).

antibody was prepared in rabbits by immunizing with the formalin-killed virus (Nakayama strain), which was purified from infected mouse brain.

In a 70-year-old man with hepatoma, JC-viral antigen was found in the nuclei of astrocytes and glial cells of cerebral white matter, which showed typical demyelinated lesions. The virus was isolated from the brain (Tokyo 1 strain (TABLE 2) (FIGURE 14). Anti-papova virus (JC, Tokyo 1) antibody was obtained from rabbits at 10 days after

TABLE 3. Demonstration of Herpes Simplex and Varicella-Zoster Virus Antigens in Autopsied Cases with Generalized Infection

| | | Virus | | Immunofluorescence | |
| | | | | Paraffin Section | Frozen Section |
Case	Sex	Age	Isolation		
Generalized herpes simplex infection in the newborn					
1.	F	10d	HSV 2	liver, adrenal gland	liver, adrenal gland[a]
2.	M	7d	HSV 2	lung	liver, adrenal gland, lung[a]
3.	F	9d	HSV 2	liver, adrenal gland, kidney, spleen	liver,[a] adrenal gland[b]
4.	F	12d	HSV 2	liver, adrenal gland, lung	liver,[a] adrenal gland[a]
5.	M	5d	HSV 1	liver, adrenal gland, lung, kidney	liver,[a] adrenal gland,[a] lung,[a] kidney[a]
6.	F	9d	N.D.[c]	liver	
7.	M	8d	N.D.	liver, adrenal gland	
8.	M	7d	N.D.	liver, adrenal gland, lung	
9.	F	11d	N.D.	liver, adrenal gland, lung, spleen	
Generalized varicella-zoster virus infection					
1.	M	5y	VZV acute lymphocytic leukemia[d]	liver, spleen, kidney	lung,[a] trigeminal ganglion,[a] liver pancreas, adrenal gland, testis
2.	F	11y (medulloblastoma)		spleen	liver, adrenal gland, lung trigeminal ganglion
3.	M	2y (Letterer-Siwe)		liver, skin, lung	liver, spleen, skin, bone marrow
4.	F	31y (malignant lymphoma)		skin, tongue, esophagus, stomach, lymph nodes, lung, ovary, uterus, liver, pancreas, spleen, kidney, rectum, adrenal gland	
5.	F	69y (malignant lymphoma)		skin, adrenal gland, kidney	
6.	M	7y (lymphoma)		skin, adrenal gland, liver, spleen, esophagus, thymus	

[a]Virus was isolated from this organ.
[b]The specimens were in paraffin for 14 years.
[c]N.D.; not done.
[d]Underlying disease.

immunization with purified virus (kindly supplied by Drs. H. Ogiwara and K. Yasui, Tokyo Metropolitan Institute of Neurological Science).

In TABLE 3, virus isolation and/or detection of antigens in autopsied cases with generalized infection of HS and VZ viruses are shown. Specific fluorescence in paraffin sections was observed in the hepatocytes of the liver, cortical layers of adrenal gland, alveolar or interstitial cells of the lung, and rarely in glomeruli of the kidney or

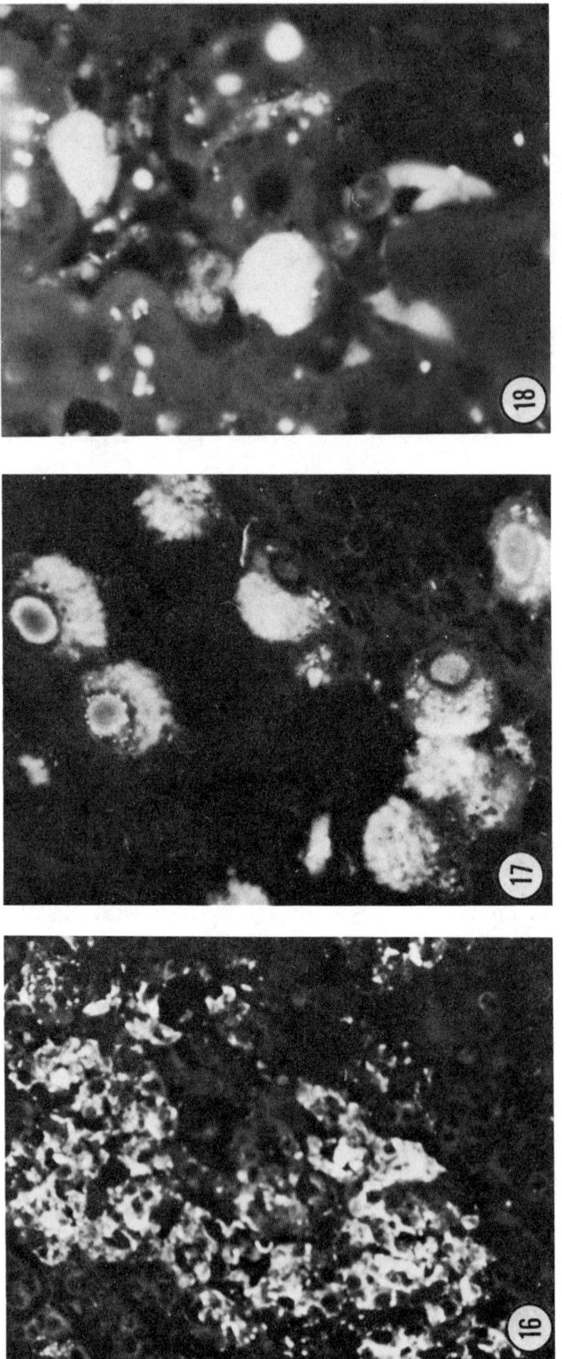

FIGURE 16. Varicella-zoster virus antigen in pancreas (Case 4, TABLE 3)(T). **FIGURE 17.** Cytomegalovirus antigen in renal tubular epithelia (Case 3, TABLE 3)(T). **FIGURE 18.** Lassa virus antigen in liver (P).

spleen from generalized HS infection of the newborn. The IF findings in formalin-fixed tissues and in frozen materials were similar. The antigens were revealed in the liver, adrenal gland, lungs, and spleen in four cases from which virus isolation was not done (FIGURE 15 Case 9). In Case 3, antigen was recovered from paraffin tissues after 14 years.

Antisera to VZ and CM viruses were made in rabbits by immunizing with alkali-eluted antigens from cells infected with varicella-derived H-S1 strain and AD 169 strain, respectively. All fatal cases of VZ virus infection had malignant tumor as

TABLE 4. Demonstration of Antigens in Autopsied Cases with Cytomegalovirus and Double Virus Infection

		Virus		Immunofluorescence	
Case	Sex	Age	Isolation	Paraffin Section	Frozen Section
Cytomegalovirus infection					
1.	F	48y	CMV	lung	kidney,[a] lung[a]
			(transplantation pneumonia)[d]		
2.	F	5m	CMV	lung, adrenal gland, spleen	liver,[a] lung,[a] spleen
			(hepato-spleno-megaly)	pituitary gland, pancreas lymph nodes	pancreas, adrenal gland, pituitary gland, lymph nodes
3.	M	6m	CMV	kidney, lung, adrenal gland	lung,[a] kidney,[a] adrenal gland
			(congenital cytomegalic inclusion disease)		
4.	F	41y	CMV	lung, stomach	lung, liver[a]
5.	F	2m	N.D.[c]	brain, lung	
			(bronchodysplasia)		
6.	M	81y		intestine	
			(disseminated intravascular coagulation)		
7.	F	25y	N.D.	placenta	
			(abortion)		
8.	F	48y	N.D.	retina, ovary, lung	
			(pneumocystis carinii, polyarthritis)		
9.	M	2y	N.D.	liver, lung	
			(acute lymphocytic leukemia)		
10.	F	62y	N.D.	lung, spleen, pancreas, rectum	
			(tuberculous meningitis)		
Dual infection with cytomegalovirus and herpes simplex virus					
				Detected Antigens by Immunofluorescence	
1.	M	85y	N.D.	HSV + CMV	adrenal gland
			(asthma bronchiale)	HSV	liver, urinary bladder
				CMV	lung
2.	M	66y	N.D.	HSV + CMV	esophagus
			(bronchitis, hypogamma-globulinemia)	CMV	adrenal gland
3.		46y	N.D.	HSV	esophagus, tongue
			(malignant lymphoma)	CMV	adrenal gland

See TABLE 3 for footnotes.

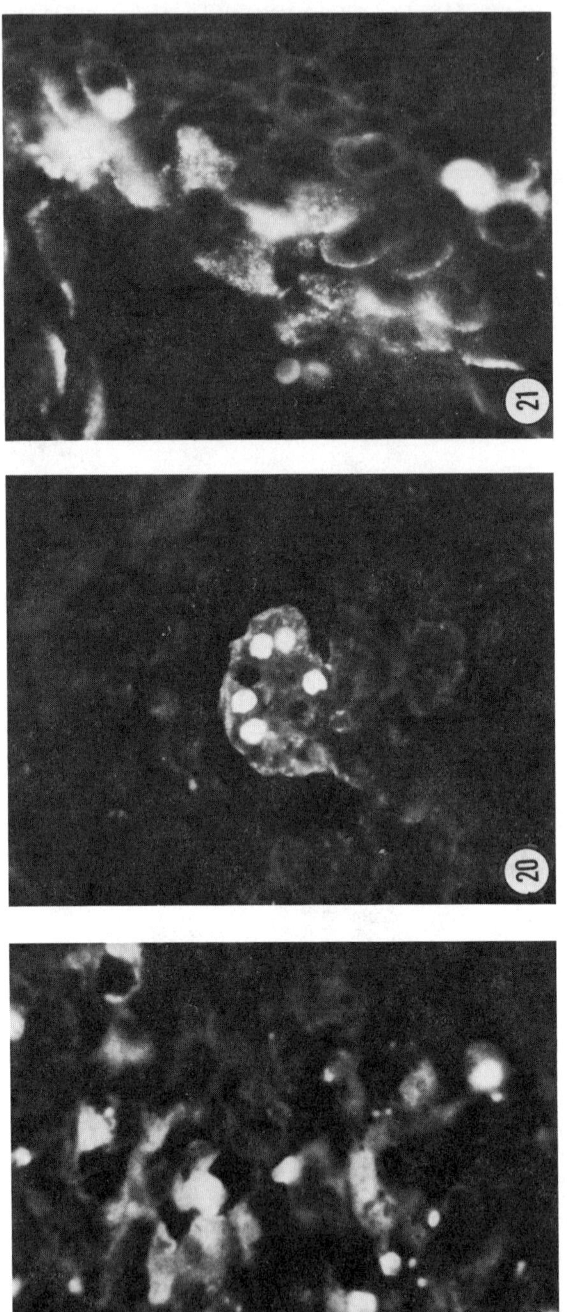

FIGURE 19. Korean hemorrhagic fever virus antigen in spleen (T). FIGURE 20. Measles antigen in lung (T). FIGURE 21. Enterovirus 70 antigen in conjunctiva of acute hemorrhagic conjunctivitis case (P).

an underlying disease. Use of large amounts of anti-tumor or immunosuppressant drugs results in generalized VZ-, CM-, or HS-infection in which histological features and distribution of antigen may vary greatly, especially in VZ and CM cases. Case 4 died 10 days after the onset of the rash of varicella and viral antigen was detected in all organs examined. FIGURE 16 shows the specific fluorescence in the glandular epithelia and interstitial cells of the pancreas. In VZ infection, viral antigens are observed mainly in interstitial areas and less frequently in parenchymal cells.

CM virus was isolated from four out of ten cases (TABLE 4). Cases 2 and 3 had congenital cytomegalic inclusion disease and the others had underlying diseases complicated by CM virus infection. The viral antigens were detected in alveolar epithelia or interstitial cells of the lung, adrenal medulla, pituitary gland (posterior lobe), β-cells or glandular cells of pancreas, tubular epithelia of the kidney (FIGURE 17, smooth ring-like staining in nucleus, and granular staining in cytoplasm), epithelial layer and/or capillary endothelia of intestines, stomach, rectum, and esophagus, etc. In Case 10, a patient with severe vasculitis, a rectal ulcer with cytomegalic inclusion lesions was found and fluorescence was observed in the swollen endothelia of submucosal capillaries.

In three cases (TABLE 4), dual virus infection was suspected from histological studies. In Case 1 and 2, both HS and CM viral antigens were simultaneously detected in the same organs, adrenal gland and esophagus, respectively. HS or CM antigen was also found in other organs. All cases had underlying disease. These cases may represent simultaneous generalized infection by two viruses.

Lassa viral antigen was demonstrated by guinea pig or mouse monoclonal antibodies in degenerating hepatocytes and epithelia of bile duct of a liver from which Lassa virus was isolated (FIGURE 18, 28-year-old woman). In the spleen and pituitary gland from a soldier who died of KHF during the Korean War in 1951, specific fluorescence was found (FIGURE 19). Naturally infected rat sera were used for IF (Tissues were kindly supplied by Dr. D. J. Wear, AFIP). A four-year-old girl who had glomerulonephritis as an underlying disease developed the rash of measles and died of giant cell pneumonia. Measles antigen was found in the nuclei of giant cells in the lung (FIGURE 20).

Detection of Viral Antigen in Biopsy Materials

The uterine cervix from three cases of genital herpes was examined for viral antigen. Specific fluorescence was observed in all cases after 20, 10, and 2.5 years in paraffin. In one case, type 2 virus had been isolated. HS viral antigen was also found in the epithelial layer of a lip biopsy from a 21-year-old woman. Type 1 HS virus was isolated from the lesion. In biopsied conjunctiva from a 40-year-old man who had a typical clinical course of AHC during an epidemic in Florida in 1981, viral antigen of enterovirus 70 was demonstrated (FIGURE 21). Antibody used for IF was prepared in rabbits (kindly supplied by Dr. J. Esposito, CDC). The virus was isolated in other patients (2/11) from the same outbreak.

DISCUSSION

Enzyme treatment (protease or trypsin) of paraffin-embedded sections made it possible to detect viral antigen by IF in several organs from cases of HS, VZ, CM,

SSPE, measles, JE, PML, Lassa, KHF, and AHC. Localization of antigens was very clear and without nonspecific staining facilitating comparison with histopathological observations. The relationship between the conditions of preservation of materials in formalin or paraffin and various periods of trypsin-digestion necessary to uncover viral antigen have been described in an earlier paper.[1] Viral antigens are stable in paraffin for long periods (10–20 years) if they have been fixed for only a short period in formalin. In our KHF case, antigen was recovered by enzyme treatment after 31 years of storage in paraffin. It is probable that the stability of antigenicity after formalin fixation differs among various viruses. Trypsin digestion was less successful than protease in recovering viral antigenicity when highly diluted antibody was used in IF, but no difference was observed when high antibody concentrations were used on cells infected with hemorrhagic fever viruses. In HS virus-infected mouse brains and human tissues, no clear difference between the two enzymes in uncovering antigen has been recognized to date. The observation that human antibodies could detect specific antigen in formalin-fixed cells infected with each of the hemorrhagic fever viruses suggests the possibility of application in diagnostic serology. Coating slides with neoprene provided a solution to the problem of attaching specimens during the lengthy periods of digestion and reaction. Enzymatic treatment of paraffin tissues for IF has proved helpful in the retrospective clinico-pathological diagnosis of virus infections that were suspected because of serological data or on the basis of histopathology. The technique is particularly valuable in the study of the pathogenesis of VZ and CM virus infections for which experimental systems are not available. With prior enzymatic treatment we have found it possible to examine antigen-bearing cells in IF-stained sections without loss of structure and have been able to make comparisons with histopathological changes in HE-stained sections. In cases of simultaneous infection with two viruses, usually only one virus, which replicates quickly is isolated, therefore antigen staining in paraffin tissues is very important to confirm the dual nature of infection. In conclusion, we are of the opinion that in tissues infected with either DNA or RNA viruses, it may be possible with proper enzyme treatment to detect viral antigens after paraffin embedding, providing optimal fixation in buffered formalin has occurred.

SUMMARY

Enzyme treatment (protease or trypsin) was applied to formalin-fixed paraffin-embedded materials and virus-infected cultured cells to detect viral antigens by immunofluorescence.

The viral antigens were demonstrated in several organs of autopsy or biopsy cases of which diagnoses had been established by immunofluorescence or virus isolation using frozen materials, or suspected on the basis of serology and/or histopathological findings. These included herpes simplex, varicella-zoster, cytomegalo, subacute sclerosing panencephalitis, progressive multifocal leukoencephalopathy, Japanese B encephalitis, measles, acute hemorrhagic conjunctivitis, Lassa and Korean hemorrhagic fever. Antigen could be recovered also in virus-infected cells (herpes simplex, measles, Lassa, Ebola, Marburg, Rift Valley, Congo and Korean Hemorrhagic fever) by enzyme treatment after periods of formalin fixation of four weeks and storage of three months. In herpes simplex virus–infected mouse brain, antigen was detected after fixation for three months in formalin.

ACKNOWLEDGMENTS

We thank Dr. K. Shimizu, National Children's Hospital; Dr. M. Hidano, Tokyo Womens Medical College; Drs. K. Yamaguchi and K. Nagashima, Tokyo University School of Medicine; Dr. Chen, Musashino Red Cross Hospital, Tokyo; Dr. M. Hamazaki, Shizuoka Children's Hospital, Shizuoka; Dr. H. Shigematsu, Shinshu University School of Medicine, Matsumoto; Dr. Mifune, Nagasaki University School of Medicine, Nagasaki; Dr. Y. Takei, Emory University School of Medicine, Atlanta; and Dr. D. J. Wear, Armed Forces Institute of Pathology, Washington, D.C. for their kind offers of important materials.

REFERENCES

1. HONDO, R., T. KURATA, S. SATO, A. ODA & Y. AOYAMA. 1982. Enzymatic treatment of formalin-fixed and paraffin-embedded specimens for detection of antigens of herpes simplex, varicella-zoster and human cytomegalo viruses. Japan. J. Exp. Med. **52:** 17–25.
2. HUANG, S-N., H. MINASSIAN & J. D. MORE. 1976. Application of immunofluorescent staining of paraffin sections improved by trypsin digestion. Lab. Invest. **35:** 383–390.
3. JOHNSON, K. P., P. T. SWOVELAND & R. W. EMMONS. 1980. Diagnosis of rabies by immunofluorescence in trypsin-treated histologic sections. J. Am. Med. Assoc. **242:** 42–43.
4. KAWAMURA, A., JR., Ed. 1977. Fluorescent Antibody Techniques and Their Applications. 2nd edit. University of Tokyo Press. Tokyo and University Park Press. Baltimore, MD
5. KUMANISHI, T., I. TAKESHITA, M. KOGA, F. IKUTA & T. TSUBAKI. 1977. Immunopathologic studies on herpes simplex virus type 1 encephalitis. Adv. Neurol. Sci. **21:** 512–518.
6. SAINTE-MARIE, G. 1962. A paraffin embedding technique for studies employing immunofluorescence. J. Histochem. Cytochem. **10:** 250–256.
7. SWOVELAND, P. T. & K. P. JOHNSON. 1979. Enhancement of fluorescent antibody staining of viral antigens in formalin-fixed tissues by trypsin digestion. J. Infect. Dis. **140:** 758–764.

Lymphocyte Activation Studies by Fluorescent Probes

R. C. NAIRN

Department of Pathology and Immunology
Monash University
Melbourne, Australia

Fluorescent cell probes provide new tools for studying early events in lymphocyte activation.[7,8] Depending on their chemical nature, probes bind to particular subcellular regions and reveal alterations in their local molecular environment by altered fluorescence emission. Within minutes of stimulation of lymphocytes by antigen or mitogen, there are molecular morphological changes in many parts of the cell. For example, membrane receptors are redistributed, membrane fluidity is increased because of greater fatty acid unsaturation in phospholipids, elements of the cytoskeleton are reorganized, and nuclear proteins are phosphorylated and acetylated causing altered conformation of DNA. By detection of these early events, fluorescent probes provide rapid assays of lymphocyte activation, compared with conventional methods based on lymphokine release after a few hours or on actual DNA synthesis after two to three days.

A variety of fluorescent probes is available for such studies, and their properties and use in detecting lymphocyte activation are reviewed elsewhere.[7] Our research group has concentrated on the use of four probes: N-phenyl-l-naphthylamine (NPN), tetramethylrhodamine isothiocyanate, acridine orange, and fluorescein.

USE OF NPN TO DETECT MITOGEN STIMULATION

NPN has provided one of the simplest and most useful practical probes for the study of early lymphocyte activation.[3,9,12] In aqueous solution, the dye is virtually non-fluorescent, but it becomes strongly fluorescent after binding to hydrophobic regions of cells, probably mainly in cell membranes. The NPN labeling causes no loss of cell viability as judged by dye exclusion or uptake of [³H]thymidine after stimulation with the mitogen concanavalin A (Con A).

Incubation of NPN-labeled lymphocytes with Con A for 30 minues at 37°C results in increased mean cell fluorescence intensity determined by microfluorimetry, compared with control labeled cells incubated in culture medium alone (FIGURE 1). This phenomenon has been investigated in detail to determine its biological significance.[12] Using mouse thymus cells, the increase in fluorescence on Con A stimulation has the same dose-response characteristics as the [³H]thymidine uptake assay, i.e., an optimal response at 5 μg/ml followed by decreasing activity at higher concentrations. The response was inhibited by pretreatment of the cells with α-methyl-mannoside, a competitive inhibitor of Con A and by cytochalasin B (FIGURE 1), an agent that disrupts microfilaments and inhibits membrane phospholipid metabolism. Fluorescence increase is also dependent on T cells, energy, and calcium.

Comparison with other lectins showed a similar fluorescence increase with phytohemagglutinin (PHA) and wheat germ lectin (WGL), but not with the divalent Con A derivative, succinyl Con A. Interestingly, pokeweed mitogen (PWM) caused decreased fluorescence, presumably because a different cellular event is detected. Our results might suggest that the fluorescence increase is associated with receptor aggregation into patches and caps, a feature of Con A, PHA, and WGL binding but not of succinyl Con A or PWM. However, using Con A labeled with fluorescein isothiocyanate to establish binding, capping was observed in the absence of calcium and in the presence of cytochalasin B, although NPN fluorescence was not increased in these circumstances. It therefore seems more likely that the early event detected in lymphocyte activation is increased phospholipid fatty acid turnover. Such fatty acid changes are known to be energy- and calcium-dependent, and are almost certainly inhibited by cytochalasin B. Increased cell-bound NPN fluorescence could be caused

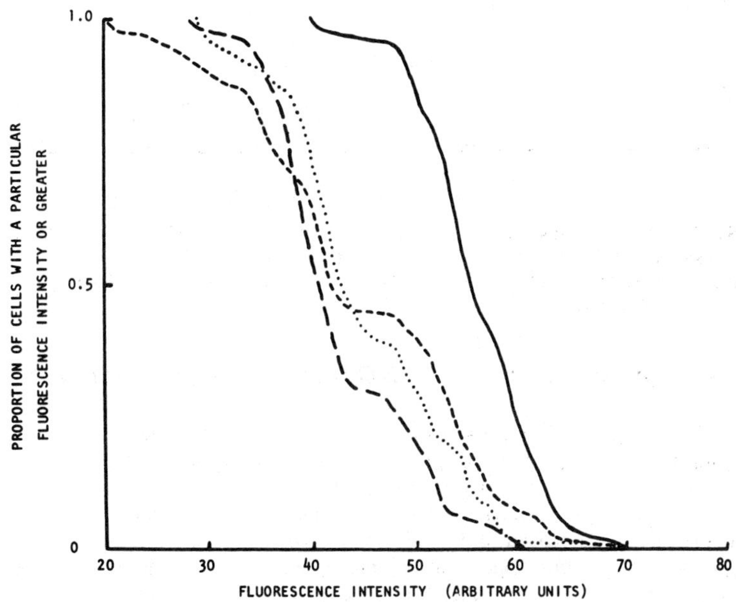

FIGURE 1. Cumulative distribution of fluorescence intensities of NPN-labeled mouse thymus cells incubated for 30 minutes at 37°C with Con A (——), culture medium (· · · ·), cytochalasin B + Con A (– –), or cytochalasin B (----). Con A is significantly different from culture medium ($p < 0.001$); no significant difference between cytochalasin B + Con A and cytochalasin B.

by either altered local fatty acid molecular polarity and rigidity, or by increased solution of NPN into the lipid from the aqueous environment.

Adaptation of the microfluorimetric analysis of early events detected by NPN to flow cytofluorimetry has been attempted using the Becton Dickinson Fluorescence Activated Cell Sorter. Flow analysis would permit rapid monitoring of large numbers of single cells and the possibility of separating and collecting responding cells for further study. However, the fluorescence increase with Con A stimulation could not be obtained reproducibly. In part, this was because of instability of NPN labeling in the flow system, but other effects such as distortion of the cells in the flow stream under pressure could well have masked the cellular fluorescence changes so reproducibly detected by microfluorimetry.

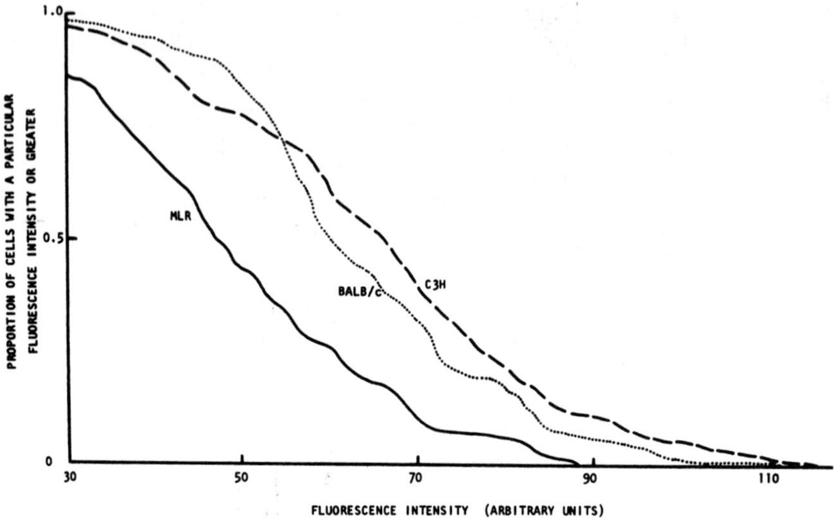

FIGURE 2. Cumulative distribution of fluorescence intensities of NPN-labeled BALB/c and C3H mouse spleen cells incubated for 30 minutes at 37°C together (MLR) or alone. Significant difference between MLR and controls (p < 0.001). Percentage fluorescence decrease = [(mean controls − mixed mean)/mean controls] × 100% = 25%.

USE OF NPN TO DETECT MIXED LYMPHOCYTE REACTION

Another early event in lymphocyte activation is detected by the NPN microfluorimetric method when lymphocytes are stimulated by allogeneic cells in a short-term mixed lymphocyte reaction (MLR).[9] When NPN-labeled spleen cell suspensions from inbred BALB/c and C3H mice are incubated together for 30 minutes at 37°C, mean cell fluorescence is decreased compared with control unmixed cultures (FIGURE 2).

Using several combinations of inbred mice, the extent of the fluorescence decrease was shown to depend on the genetic difference between the two strains of mice (TABLE 1). Results of the short term NPN-MLR test also correlated well with the survival of tail-skin grafts exchanged between the mice.[9] A large fluorescence change was caused by a genetic difference at the whole H-2 complex, or at the I region of this complex, correlating with rapid graft rejection. The smaller fluorescence change resulting from

TABLE 1. Correlation Between NPN-MLR Assay and Histocompatibility Difference for Different Combinations of Inbred Mice

Mouse Strain Combination	Genetic Difference	NPN-MLR: Mean % Fluorescence Decrease ± SD
BALB/c-C3H	M,H-2	28 ± 3 (8)[a]
A.TL-A.TH	I,S,T1a	19 ± 4 (6)
BALB/c-DBA/2	M	9 ± 3 (7)
A.TL-A.AL	H-2K	8 ± 4 (6)

[a] Number of experiments in brackets.

an H-2K or M locus difference correlated with more prolonged graft survival. Indeed, the NPN-MLR test correlated with allograft survival better than parallel conventional [3H]thymidine uptake assays, which detect differences at the M locus as well as at the H-2 complex.

We are currently testing the NPN-MLR assay clinically and have studied a series of human kidney transplants from living related donors. Post-transplant monitoring has shown a good correlation between graft rejection episodes and fluorescence decrease in the NPN test. Perhaps even more important has been the ability of the NPN-MLR test to provide an immediate cross-matching facility for marrow donations from non-twin siblings.

To investigate the cell types involved in the short-term MLR, C3H spleen cells were X-irradiated and incubated with untreated BALB/c spleen cells in a one-way MLR. With this system, it was shown that viable, mature T cells were required as responders, because reactivity was obtained using nylon wool-filtered (T cell-enriched) BALB/c spleen cells, but not if BALB/c and C3H thymus cells were used, or athymic BALB/c spleen cells. Depletion of monocytes and macrophages from stimulator and responder cells by adherence to plastic prevented the response, consistent with the important role of these cells in long-term cultures as stimulator cells because of their high surface density of relevant Ia antigens.

An additional role of adherent cells in the short-term MLR is the generation of soluble mediators of the response. If unlabeled BALB/c and C3H spleen cells were mixed for 30 minutes at 37°C, the cells removed by Millipore filtration, and the filtrate added to fresh NPN-labeled BALB/c cells, a fluorescence change equivalent to that found for the normal two-way response was observed. Using filtrates from one-way cultures, adherent cells in the responder but not the stimulator population were shown to be required for the generation of the filtrate. Thus, viable monocytes and macrophages must either produce the soluble factors themselves, or actively collaborate with secretory T cells.

USE OF TETRAMETHYLRHODAMINE ISOTHIOCYANATE

Tetramethylrhodamine isothiocyanate (TMRITC) is a hydrophilic probe of cellular proteins.[6] By labeling cells with TMRITC at pH 8.6, some covalent binding to proteins is achieved although a substantial proportion of the dye remains weakly attached. Labeling was not without effect on the function of the lymphocytes, as shown by a 24-hour delay in their [3H]thymidine uptake after Con A stimulation. It is possible that the labeling procedure at pH 7.4 described by Butcher and Weissman[1] would give better preservation of cell viability. Stimulation of TMRITC-labeled cells with PHA or PPD (purified protein derivative of tuberculin) caused increased mean cell fluorescence, and MLR resulted in decreased fluorescence.[6]

USE OF ACRIDINE ORANGE

Another probe used to study early lymphocyte activation is acridine orange (AO). It can detect alteration of lysosome membrane permeability[13] and altered DNA conformation.[4] Non-specific stimulation of cells with PHA and PWM, or specific stimulation of immune cells with ovalbumin resulted in an increased number of lysosomes stained red by AO in polymeric form. Stimulation with Con A of AO-stained cell suspensions revealed increased green nuclear fluorescence of AO in monomeric form, compared with unstimulated control cells. Thus AO can detect early changes in different regions of activated lymphocytes, although lysosome counting is tedious, and the nuclear staining method technically delicate.

THE FLUORESCEIN FLUORESCENCE POLARIZATION TEST

A probe technology of particular practical value detects altered fluorescence polarization of the cytoplasmic probe fluorescein. The test gives interesting information about early changes in PHA-stimulated lymphocytes, and discriminates between

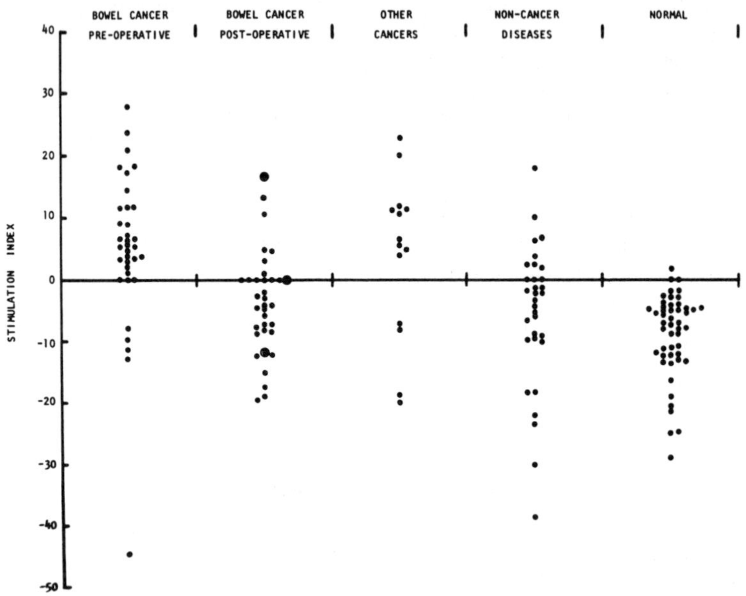

FIGURE 3. Fluorescence polarization changes in fluorescein-labeled cells after 40 minutes PHA incubation. Stimulation indices for 70 patients with colorectal carcinoma (35 pre-operative, 35 post-operative; ● recurrence confirmed), 14 patients with other cancers, 32 patients with other diseases, and 49 normal individuals.

cancer and non-cancer individuals.[5] The method used is that described by Pritchard *et al.,*[10] based on the procedure developed by Cercek and Cercek.[2] Briefly, blood lymphocytes are separated into two fractions on a Ficoll-Triosil gradient (density, 1.081 g/cm^3), those separating above the interphase (zone 1) and those separating below (zone 2). The cells are incubated with PHA for 40 minutes, fluorescein diacetate is added, and fluorescence polarization is measured using a polarizing spectrophotometer.

For non-cancer individuals, cells in zone 1 show a greater decrease in polarization after PHA stimulation than cells in zone 2, while cancer patients' cells from zone 2 show the best response. Thus, the stimulation index obtained by subtracting the zone 1 response from the zone 2 response is usually negative for normal donors and non-cancer patients but positive for cancer patients (FIGURE 3). The strong positive case in the non-malignant disease group is suspected of pre-clinical lymphoreticular malignancy, and four of the seven weak positive cases are ulcerative colitis, a well known pre-cancerous disease. Of the 49 cancer patients tested, 35 colorectal and 14 other types, there are only nine negative. In three cases retested post-operatively, the index showed a marked decrease as early as 24 hours after tumor removal.

The explanation for altered reactivity by cancer cells in this test has been sought by analysis of cell distribution in the two zones for normal and cancer bloods, and by incubation of normal cells in cancer plasma or cancer cell supernatant. The results of these studies are presented elsewhere in this volume.[11] It appears that altered cell activity rather than distribution is responsible for the subtle change in lymphocyte behavior detected by this fluorescent probe technique.

In conclusion, I believe that the fluorescent cell-probe technology is at just about the same stage as the fluorescent antibody technique was 20 years ago. It is providing new diagnostic assays and valuable techniques for identifying and quantifying the molecular events and regulation of early stages of the activation of lymphocytes. If it develops in the same way as the fluorescent antibody method has, we can expect the same expansion in its use and in the literature, and the same development of sophisticated instruments to work with.

REFERENCES

1. BUTCHER, E. C. & I. L. WEISSMAN. 1980. Direct fluorescent labeling of cells with fluorescein or rhodamine isothiocyanate. I. Technical aspects. J. Immunol. Methods **37:** 97–108.
2. CERCEK, L. & B. CERCEK. 1977. Application of the phenomenon of changes in the structuredness of cytoplasmic matrix (SCM) in the diagnosis of malignant disorders: A review. Eur. J. Cancer **13:** 903–915.
3. HALLIDAY, G. M., R. C. NAIRN, M. A. PALLETT, J. M. ROLLAND & H. A. WARD. 1979. Detection of early lymphocyte activation by the fluorescent cell membrane probe N-phenyl-l-naphthylamine. J. Immunol. Methods **28:** 381–390.
4. HALLIDAY, G. M., R. C. NAIRN & J. M. ROLLAND. 1981. Lymphocyte stimulation by concanavalin A studied by the fluorescent probe acridine orange. Cell Tissue Res. **217:** 117–126.
5. HOCKING, G. R., J. M. ROLLAND, R. C. NAIRN, E. PIHL, A. M. CUTHBERTSON, E. S. R. HUGHES & W. R. JOHNSON. 1982. Lymphocyte fluorescence polarization changes after photohaemagglutinin stimulation in the diagnosis of colorectal carcinoma. J. Natl. Cancer Inst. **68:** 579–583.
6. NAIRN, R. C., I. M. JABLONKA, J. M. ROLLAND, G. M. HALLIDAY & H. A. WARD. 1979. Rhodamine as a fluorescent probe of lymphocyte activation. Immunology **36:** 235–240.
7. NAIRN, R. C. & J. M. ROLLAND. 1980. Fluorescent probes to detect lymphocyte activation. Clin. Exp. Immunol. **39:** 1–13.
8. NAIRN, R. C. & J. M. ROLLAND. 1981. Fluorescent probes: a new way to study lymphocyte activation. Endeavour **5:** 167–171.
9. PALLETT, M. A., J. M. ROLLAND & R. C. NAIRN. 1983. Fluorescent probe assay of early mixed lymphocyte reaction to predict allograft survival in mice. Transplantation **35:** 243–248.
10. PRITCHARD, J. A., W. H. SUTHERLAND, J. E. SEAMAN, T. J. DEELEY, I. H. EVANS, I. J. KERBY, K. W. JAMES & I. C. M. PATERSON. 1978. Cancer-specific density changes in lymphocytes after stimulation with phytohaemagglutinin. Lancet **ii:** 1275–1277.

11. ROLLAND, J. M. 1983. Fluorescent lymphocyte probes for cancer detection. Ann. N.Y. Acad. Sci. (This volume.)
12. ROLLAND, J. M., R. L. BETTS, G. M. HALLIDAY, G. R. HOCKING & R. C. NAIRN. 1981. Early changes in concanavalin A-stimulated lymphocytes detected by the fluorescent probe N-phenyl-l-naphthylamine. Cell Tissue Res. **214:** 119–128.
13. ROLLAND, J. M., G. M. FERRIER, R. C. NAIRN & M. N. CAUCHI. 1976. Acridine orange fluorescence cytochemistry for detecting lymphocyte immunoreactivity. J. Immunol. Methods **12:** 347–354.

Fluorescent Lymphocyte Probes for Cancer Detection

J. M. ROLLAND, G. R. HOCKING, AND R. C. NAIRN

Department of Pathology and Immunology
Monash University
Melbourne, Australia

The principles of fluorescent cell probe assays for studying early events in lymphocyte activation are described elsewhere in this volume.[7] Briefly, fluorescent dyes localize to particular subcellular regions and can monitor very early changes in activated lymphocytes by alteration in their fluorescence emission.[8,9] We have found fluorescent probes to be useful tools in cancer diagnosis and investigation of host-cancer interaction.

FLUORESCENT PROBES TO DETECT CANCER

N-phenyl-l-naphthylamine (NPN) localizes to hydrophobic regions of cells, largely in cell membranes. Microfluorimetry of NPN-labeled spleen cells from a tumor-bearing rat revealed decreased mean cell fluorescence intensity 30 minutes after addition of a 3 M KCl extract of the tumor.[4] This fluorescence decrease was significant ($p < 0.001$) in 11 of 12 experiments. Incubation with an extract of a histogenetically different tumor caused no fluorescence change in three such experiments and when the experimental situation was reversed the same specific tumor correlation was obtained.

The NPN probe presumably detects the changes in membrane lipids reported to occur within minutes of lymphocyte activation. Automation of the technique and rapid data analysis by direct computer link-up with the microphotometer provide a potential screening test for cancer-reactive lymphocytes or for cancer antigens using known reactive lymphocytes.

Acridine orange (AO) is another fluorescent probe, which permits the detection of early events in lymphocyte activation. When viable lymphocytes in suspension are stained with AO, the dye is concentrated in lysosomes as the polymeric red fluorescent form and intercalated into DNA as the monomeric green fluorescent form. AO staining has also been used to detect cancer in rats. Lymphocytes from tumor-bearing rats incubated with the tumor cells for three hours showed an increased number of red lysosomes caused by increased lysosomal membrane permeability accompanied by concentration of AO within the lysosomes.[11] Conformational changes in DNA were revealed by increased nuclear fluorescence of AO-stained lymphocytes after only 30 minutes of incubation with tumor cells.[5]

Fluorescein provides another fluorescent probe technique of more practical value for clinical cancer diagnosis by the fluorescein fluorescence polarization (FFP) test, based on detection of altered response to phytohemagglutinin (PHA) by cancer patients' blood lymphocytes. The double-zone FFP test modified by Pritchard *et al.*[10] from the original method of Cercek and Cercek,[2] has been confirmed in our hands to discriminate between cancer and non-cancer lymphocyte reactivity.[6] Blood lympho-cytes are separated into two zones by density. Changes in fluorescence polarization of

TABLE 1. E-Rosette-forming Cells in Zones 1 and 2[a]

E-Rosette	Normal	Cancer
Zone 1		
E 4°C	57 ± 11	58 ± 15
E 22°C	40 ± 17	47 ± 19
E 37°C	4 ± 6	5 ± 5
Zone 2		
E 4°C	33 ± 24	38 ± 14
E 22°C	16 ± 22	24 ± 14
E 37°C	1 ± 2	2 ± 2

[a]Mean ± S.D. percentages for seven normal and six cancer cell donors.

intracellular fluorescein are measured after 45 minutes of incubation with PHA. For normal donors, zone 1 cells show a greater fluorescence polarization decrease than zone 2, but for cancer donors the reverse is true. Thus, the stimulation index expressed as zone 2 response minus zone 1 response is usually negative for normal and positive for cancer donors.[6]

INVESTIGATION OF THE BASIS FOR THE FLUORESCEIN FLUORESCENCE POLARIZATION TEST

To establish the reason for altered reactivity by cancer patients' lymphocytes in the FFP test, the distribution of cell types between the two zones was analyzed. No significant difference between the mean lymphocyte or granulocyte number in the two zones was observed in 49 normal and 49 cancer donors; more lymphocytes were obtained in zone 1 than in zone 2, and the granulocyte contamination was greater in zone 2 for both groups. These findings are in contrast to those of Balding *et al.,*[1] who suggested that altered reactivity by cancer donor lymphocytes in this test was attributable to a greater contamination with granulocytes.

Further analysis of lymphocyte subclasses also revealed no abnormal distribution of cancer lymphocytes between the two zones. Cells forming E rosettes at 4°C (total T), 22°C (active T), and 37°C (primitive T) (TABLE 1), cells binding OKT3 (total T), OKT4 (helper T), and OKT8 (suppressor/cytotoxic T) monoclonal antibodies, and surface immunoglobulin-positive B cells (TABLE 2) were counted.

Interestingly, [^{3}H]thymidine uptake after PHA stimulation was similar for zones 1

TABLE 2. T and B Cells in Zones 1 and 2[a]

Lymphocyte Subclass	Normal	Cancer
Zone 1 Total T	73 ± 6	71 ± 14
T Helper	45 ± 9	38 ± 22
T Supp/Cyt	21 ± 6	25 ± 8
Anti-Ig	20 ± 10	26 ± 17
Zone 2 Total T	68 ± 7	63 ± 17
T Helper	38 ± 9	33 ± 22
T Supp/Cyt	20 ± 5	21 ± 5
Anti-Ig	24 ± 9	29 ± 18

[a]Mean ± S.D. percentages for six normal and four cancer cell donors.

TABLE 3. PHA Stimulation of Cells from Zones 1 and 2[a]

	[³H]-Thymidine		Reduction of Fluorescence Polarization	
	Zone 1	Zone 2	Zone 1	Zone 2
Normal	103	109	29.7	11.3
	136	115	3.8	1.2
	401	288	14.7	7.7
Cancer	62	51	0	14.5
	326	329	0	9.1
	124	132	1.2	10.1

[a]Cells obtained from three normal and three cancer donors.

and 2 of normal and cancer cells, in contrast to the different reactivity detected by the short-term polarization test (TABLE 3).

One physical difference we have detected between normal and cancer donors' cells is shown by the unstimulated cells' polarization values (TABLE 4). For both normal and cancer cells, the mean polarization value for zone 2 is greater than for zone 1, but for zone 2 cells, the cancer values are greater than the normal values. Presumably, the higher polarization values of cancer zone 2 cells reflect altered orientation or concentration of particular intracellular components.

An alternative approach to understanding the basis of the FFP test for cancer was to test for factors in cancer patients' plasma that could convert the reactivity of normal lymphocytes to that of cancer lymphocytes. In two preliminary experiments, normal zone 1 and 2 cells were resuspended in cancer plasma for 30 minutes at 37°C and 30 minutes at room temperature. However, no increased response by zone 2 cells was observed in either experiment. The presence of such "cancer factors" in plasma has been reported by the Cerceks[3] using an experimental mouse tumor system.

In the same way, we attempted to detect "cancer factors" in supernatants from 3-day cultures of colorectal tumor cells. In one of four experiments, a concentration-dependent stimulation of normal zone 2 cell response was found, with little effect on zone 1 cells. Again, the Cerceks describe a factor released from tumor cells after only two to three hours of culture, which could modify normal cell response in their mouse system.[3] It is likely that a factor released by cancer cells or by host cells interacting with the cancer will induce a sequence of events which could include the triggering of an imbalance in normal lymphocyte regulatory factors. Our own preliminary studies suggest that changes in local cyclic nucleotide levels could alter the response of lymphocytes detected by the FFP test.

In conclusion, fluorescent cell probe techniques offer valuable, rapid tests useful in

TABLE 4. Fluorescence Polarization Values for Unstimulated Cells from Zones 1 and 2[a]

	Unstimulated Polarization Value	
Donor Group	Zone 1	Zone 2
Normal (49)	0.168 ± 0.015	0.172 ± 0.016
Non-malignant disease (32)	0.166 ± 0.016	0.177 ± 0.021
Colorectal cancer (35)	0.172 ± 0.021	0.189 ± 0.023
Other cancer (14)	0.170 ± 0.018	0.190 ± 0.021

[a]Mean ± S.D. with number of cases in parentheses.

cancer diagnosis. Refinement of some technical procedures and automation will make these tests applicable routinely. In particular, we are attempting to adapt the procedures to analysis by the Fluorescence Activated Cell Sorter. This would give greater accuracy and speed to the tests, and also permit analysis of response by individual cells in investigations of host-cancer relationships.

REFERENCES

1. BALDING, P., P. A. LIGHT & A. W. PREECE. 1980. Response of human lymphocytes to PHA and tumour-associated antigens as detected by fluorescence polarization. Br. J. Cancer **41:** 73–85.
2. CERCEK, L. & B. CERCEK. 1977. Application of the phenomenon of changes in the structuredness of cytoplasmic matrix (SCM) in the diagnosis of malignant disorders: A review. Eur. J. Cancer **13:** 903–915.
3. CERCEK, L. & B. CERCEK. 1981. Changes in SCM-responses of lymphocytes in mice after implantation with Ehrlich ascites cells. Eur. J. Cancer **17:** 167–171.
4. HALLIDAY, G. M., R. C. NAIRN, M. A. PALLETT, J. M. ROLLAND & H. A. WARD. 1979. Detection of early lymphocyte activation by the fluorescent cell membrane probe N-phenyl-l-naphthylamine. J. Immunol. Methods **28:** 381–390.
5. HALLIDAY, G. M., R. C. NAIRN & J. M. ROLLAND. 1981. Lymphocyte stimulation by concanavalin A studied by the fluorescent probe acridine orange. Cell Tissue Res. **217:** 117–126.
6. HOCKING, G. R., J. M. ROLLAND, R. C. NAIRN, E. PIHL, A. M. CUTHBERTSON, E. S. R. HUGHES & W. R. JOHNSON. 1982. Lymphocyte fluorescence polarization changes after phytohaemagglutinin stimulation in the diagnosis of colorectal carcinoma. J. Natl. Cancer Inst. (In press.)
7. NAIRN, R. C. 1982. Lymphocyte activation studies by fluorescent probes. Ann. N.Y. Acad. Sci. **368:** 579–583.
8. NAIRN, R. C. & J. M. ROLLAND. 1980. Fluorescent probes to detect lymphocyte activation. Clin. Exp. Immunol. **39:** 1–13.
9. NAIRN, R. C. & J. M. ROLLAND. 1981. Fluorescent probes. A new way to study lymphocyte activation. Endeavour **5:** 167–171.
10. PRITCHARD, J. A., W. H. SUTHERLAND, J. E. SEAMAN, T. J. DEELEY, I. H. EVANS, I. J. KERBY, K. W. JAMES & I. C. M. PATERSON. 1978. Cancer-specific density changes in lymphocytes after stimulation with phytohaemagglutinin. Lancet **ii:** 1275–1277.
11. ROLLAND, J. M., G. M. FERRIER, R. C. NAIRN & M. N. CAUCHI. 1976. Acridine orange fluorescence cytochemistry for detecting lymphocyte immunoreactivity. J. Immunol. Methods **12:** 347–354.

Fluorescent Probes for the Detection of Malignant Disease[a]

J. A. V. PRITCHARD, W. H. SUTHERLAND, J. E.
SIDDALL, A. J. BATER, I. J. KERBY, AND T. J. DEELEY

Immunology Department
Radiation Science Laboratories
South Wales Radiotherapy and Oncology Service
Velindre Hospital, Whitchurch
Cardiff, South Wales
United Kingdom

INTRODUCTION

Fluorescent probes can be used to detect a plethora of events associated with the early activation of lymphocytes by mitogens or antigens.[17] The cytoplasmic probe fluorescein diacetate (FDA) has been used in the development of an *in vitro* test for cancer, based on the differential response of lymphocytes to phytohemagglutinin (PHA) and to cancer basic protein (CaBP), as measured by a fluorescence polarization technique.[8] Various workers[11,12,15,18,20,21,26–28] have confirmed that the test discriminates between cancer and non-cancer, but two groups[2,16] reported failure and some possible reasons for this failure have been suggested.[4,7,9]

In these laboratories we have developed a double zone sampling technique in which cells are harvested from two regions of a Ficoll Triosil discontinuous gradient. We have shown that it is possible, on the basis of PHA response alone, to distinguish accurately between malignant and non-malignant disease.[23,24] Similar results can also be obtained with the membrane-bound probe N-phenyl-1-naphthylamine (NPN) by measuring fluorescence intensity changes in lymphocytes following stimulation by mitogen or antigen. The results reported here show that lymphocyte fluorophotometry (LFP) has an important role to play in the *in vitro* screening and monitoring of neoplastic cells.

METHODS

Lymphocyte Fluorophotometry Using Fluorescein Diacetate

The technique for separating lymphocytes responding to PHA from venous blood has been extensively documented.[23,24] Briefly, whole blood is rotated with 0.1 g carbonyl iron powder (Type SF, GAF Ltd.) and 5 ml is then layered on to 5 ml of a modified Ficoll Triosil density solution ($\rho = 1.081$ g/ml at 19°C and 320 mOsm/kg) and centrifuged at 1100g (interface) for 20 minutes. Zone 1 cells are removed by

[a]Supported by Tenovus, South Glamorgan Area Health Authority (T) and the South Wales Cancer Research Council.

Abbreviations: p = probability, I = intensity, P = fluorescence polarization index, LFP = lymphocyte fluorophotometry, I_{FP} = fluorophotometric index, FDA = fluorescein diacetate, PHA = phytohemagglutinin, CaBP = cancer basic protein, NPN = N-phenyl-1-naphthylamine, V_{coeff} = validity coefficient, HBSS = Hank's balanced salt solution, and n.s. = not significant.

holding a Pasteur pipette 1 mm above the visible interface and with a circular motion withdrawing approximately 1.5 ml of the liquid, which should include the upper part of the visible cell layer. Zone 2 cells are removed in an identical manner by placing the pipette at the interface and removing the remainder of the cell layer, together with about 1.5 ml of the liquid underneath the interface. Using aliquots containing 0.75×10^6 unstimulated or PHA-stimulated lymphocytes, fluorescence polarization indices ('P' values) are measured in a Perkin Elmer MPF4 spectrophotometer switched to fluorescence (AC mode–chopper ON) and equipped with an automatic polarization changer, as described elsewhere.[6,23]

Each test is interpreted according to the values obtained for two parameters. First, we define the fluorophotometric index (I_{FP}) as:

$$I_{FP} = Z_2 - Z_1,$$

where Z_1 is the percentage reduction in 'P' following stimulation of Zone 1 cells by PHA and Z_2 is the corresponding response from the Zone 2 cells. A positive value for this index occurs when the larger response is in Zone 2 and indicates the presence of malignant disease. In the absence of detectable malignancy the larger response occurs in Zone 1, giving a negative index. Up to the present time, cancer is the only condition that has been shown to give a positive index.

Low values for I_{FP} can result from a number of conditions, broadly described as 'defective lymphocyte function,' or from instrumental artefacts. In order to provide a single factor to evaluate the significance of any particular result, we define a second parameter, the 'validity coefficient,' as:

$$V_{coeff} = 100\,|Z_2 - Z_1|\,/K\,[1 + (8/K)^3],$$

$$\text{where } K = |Z_1| + |Z_2| + 0.001.$$

The validity coefficient is designed to range between the limits 0 and 100. The upper limit ("100 percent validity") is approached when a large percentage reduction occurs in one zone, accompanied by a zero or very small percentage reduction in the other zone. The lower limit ("zero validity") is reached when the cells in both zones yield equal percentage reductions, which may be anywhere from very small to very large. The term 0.001 is needed to prevent the coefficient from becoming indeterminate on the rare occasions when $Z_1 = Z_2 = 0$. The coefficient establishes an objective criterion for assessing the validity of the fluorescence polarization test in confirming the presence or absence of cancer. When a subject is in transition between these two states, there must be a period when the test can no longer confirm the absence of malignancy, but also cannot yet confirm its presence because the tumor burden is still below the threshold required for a fully positive index. Therefore, the validity coefficient should drop to zero during the period, however brief, when 'absence' or 'presence' of malignancy are both temporarily indeterminate. To confirm this aspect of the coefficient, we have applied it retrospectively to the results from an earlier experiment designed to define the 'detection threshold' of the LFP_{FDA} technique (LFP = lymphocyte fluorophotometry). The maximum and minimum dimensions of small basal cell skin carcinomas were measured and converted to diameters of circles of equivalent area for plotting against the measured fluorophotometric index in FIGURE 1, showing a detection threshold at about 2.5 mm equivalent diameter. These rather limited data confirm that the validity coefficient (upper broken curve) dips sharply to a minimum at the detection threshold, as expected, and rises very rapidly to full validity on either side of the minimum. The validity coefficient serves two main purposes: first, it defines a lower limit below which subjects must be referred for repeat tests. By applying the concept retrospectively to the 451 tests reported here we have arbitrarily placed the lower limit of acceptable validity at 15, with a further range of 'doubtful validity' from 15 to 30, in which repeat tests are not essential but are nevertheless recommended to confirm the result.

The second purpose of the validity coefficient is to act as an early warning against malfunction at any stage of the test protocol, which includes all the critical steps in the cell separation procedures and the total performance of the fluorescence spectropho-

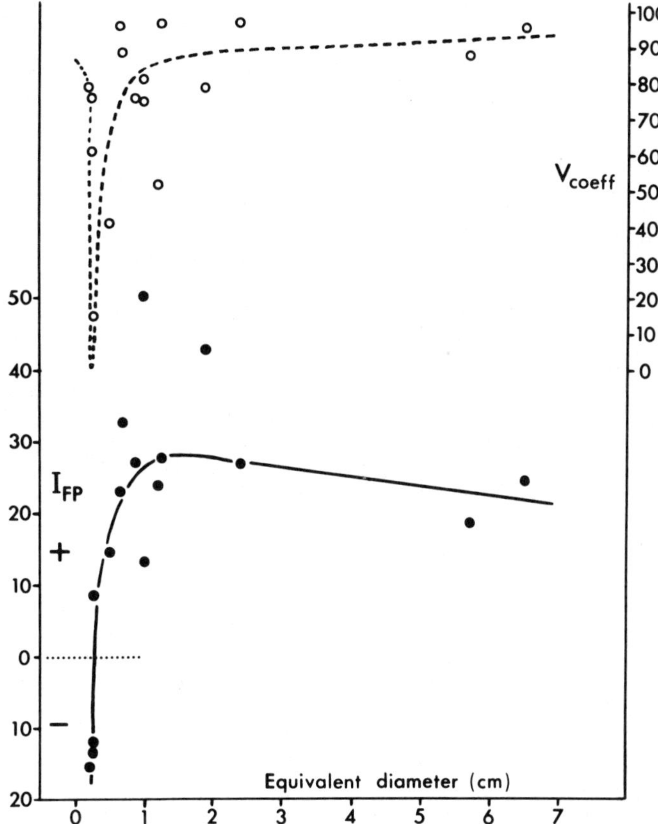

FIGURE 1. Fluorophotometric index (I_{FP}) and validity coefficient (V_{coeff}) plotted against the equivalent diameters of small basal cell skin carcinomas.

tometer. For this purpose, selected healthy donors are included on a regular schedule and watched for any significant fluctuations in the validity coefficients.

Lymphocyte Fluorophotometry Using NPN

Zone 1 lymphocytes are resuspended at a concentration of 10^7/ml in HEPES-buffered Hank's balanced salt solution (HBSS). Changes in the fluorescence intensity from the membrane-bound dye NPN in unstimulated and in PHA-stimulated lymphocytes is measured with a Leitz Dialux 20 microscope fitted with Ploemopak (filter block A) and Leitz MPV Compact system interfaced directly to a Commodore microcomputer for data analysis using a Wilcoxon Rank Sum test. Detailed procedures for lymphocyte handling and for the measurement of fluorescence intensity have been reported.[10,22] Briefly, 100 μl of a working solution of NPN (Eastman) prepared by adding 10 μl of a 200 mM stock solution in Analar Grade methanol to 10 ml of HBSS, was added to 300 μl of lymphocyte suspension. PHA (Gibco-Europe), at a protein concentration of 12.5 μg/μl, was used for stimulation and control and test samples were incubated at 37°C for 30 minutes before measuring fluorescence intensity. For the measurement, 40 μl of suspension was placed on a glass slide and overlaid with a coverslip. The fluorescence intensity of 100 single lymphocytes was measured. All samples were coded and measured blind to remove any possibility of operator bias.

RESULTS

Lymphocyte Fluorophotometry Using Fluorescein Diacetate

In this series of 51 controls and 400 cases of malignant and benign diseases, a total of 36 results (8%) were excluded from the final analysis because the validity coefficients were below the acceptance limit of 15. In normal clinical practice these would have been referred for repeat testing. These included a number of late-stage cancers that should not have been included in the evaluation of a technique designed for early detection of cancer, together with two cases of microinvasive carcinoma of cervix close to the validity limit (14 and 13) that were probably just below the detection threshold. A third case of microinvasive carcinoma had clearly crossed the threshold with an index of $+13.7$ and a validity coefficient of 49. Further, 15 subjects were excluded from analysis, and would have been referred for repeat testing in normal clinical practice, because the 'P' values were higher than the acceptance limit of 0.300. High 'P' values usually arise from excessive scattered light caused by red cell contamination, by bacteria, or by particulate debris from damaged cells. The results from the 400 subjects who produced values within the acceptance limits are shown in FIGURE 2. Nine patients with diagnosed malignant disease gave negative results that must be considered incorrect (2.3%). Because of the importance of negative results from subjects with clinically diagnosed cancer in evaluating a cancer test, these nine results are presented fully in TABLE 1. Five of the nine had late-stage cancers, which usually give weak test results and are not suitable subjects for evaluating a test aimed at early detection. For this reason, the true false-negative rate is probably closer to 1.0 percent when the LFP$_{FDA}$ technique is applied to the detection of early-stage malignant disease.

In 20 subjects, positive results were obtained where malignancy was not suspected (5.0 percent). Where possible, these will be followed up. A false negative rate of only 2.3 percent, possibly as low as 1.0 percent, under near-clinical conditions places the LFP$_{FDA}$ test amongst the most accurate techniques available for the detection of cancer. The reproducibility of the test is illustrated in TABLE 2, which shows the small range of variation in the 'P' values from three healthy individuals tested regularly over a period of three years. An example of the value of lymphocyte fluorophotometry for monitoring patients who have had conventional radiotherapy for cancer of the cervix is shown in FIGURE 3 (left). Tests were done before and after radiotherapy and at subsequent clinic attendances. In two instances (triangles) a change to a negative index

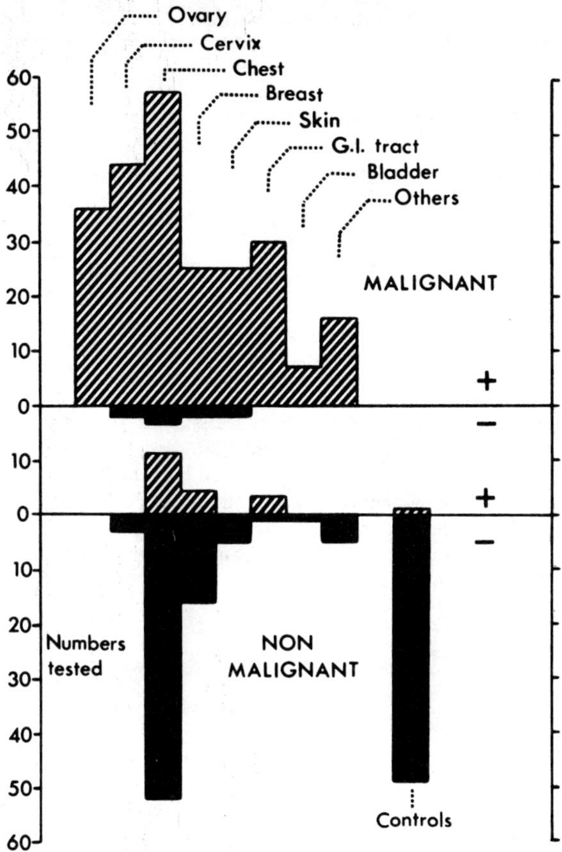

FIGURE 2. Overall results from 400 subjects tested with the double-zone LFP$_{FDA}$ technique.

TABLE 1. Negative Results from Subjects with Histologically Proven Malignant Disease

Site	Sex	Age	I_{FP}	V_{Coeff}	Diagnosis
Lung	M	64	−13.5	83	Squamous cell carcinoma, advanced
	F	57	−25.7	97	Oat cell carcinoma
	M	66	−36.3	99	Squamous cell carcinoma
Breast	F	42	− 3.5	16	Adenocarcinoma stage IV
	F	63	− 8.7	70	Adenocarcinoma stage IV
Cervix	F	47	− 6.7	34	Squamous cell carcinoma stage III
	F	34	−13.9	73	Adenocarcinoma and kidney failure
G.I. tract	M	75	− 8.9	56	Adenocarcinoma–stomach, advanced
	M	62	−23.2	96	Adenocarcinoma–colon

Five out of nine cases in advanced cancer category.

I_{FP} = fluorophotometric index and V_{coeff} = validity coefficient.

TABLE 2. Reproducibility of 'P' Value Measurement[a]

	Sex	Age	Zone 1		Zone 2	
			Unstimulated	Stimulated (PHA)	Unstimulated	Stimulated (PHA)
Control 1 N = 20	F	56	0.176 ± 0.013	0.134 ± 0.013	0.172 ± 0.011	0.174 ± 0.023
Control 2 N = 14	F	26	0.180 ± 0.022	0.127 ± 0.012	0.180 ± 0.040	0.174 ± 0.025
Control 3 N = 17	M	30	0.176 ± 0.009	0.133 ± 0.012	0.178 ± 0.011	0.173 ± 0.009

N = Number of tests done over 3-year period.
[a]Polarization values ± 1 standard deviation.

showed a response to therapy and tumor regression in keeping with the clinical observations. However, a later change in the index towards a positive value was shown to be correlated with recurrence of disease as proved by subsequent surgery. The third case (open circles) still shows a negative index and remains clinically free from disease. Similar results were obtained in the monitoring of two patients with ovarian cancer treated with the chemotherapeutic agent *cis*-platinum diamine dichloride, as shown in FIGURE 3 (right).

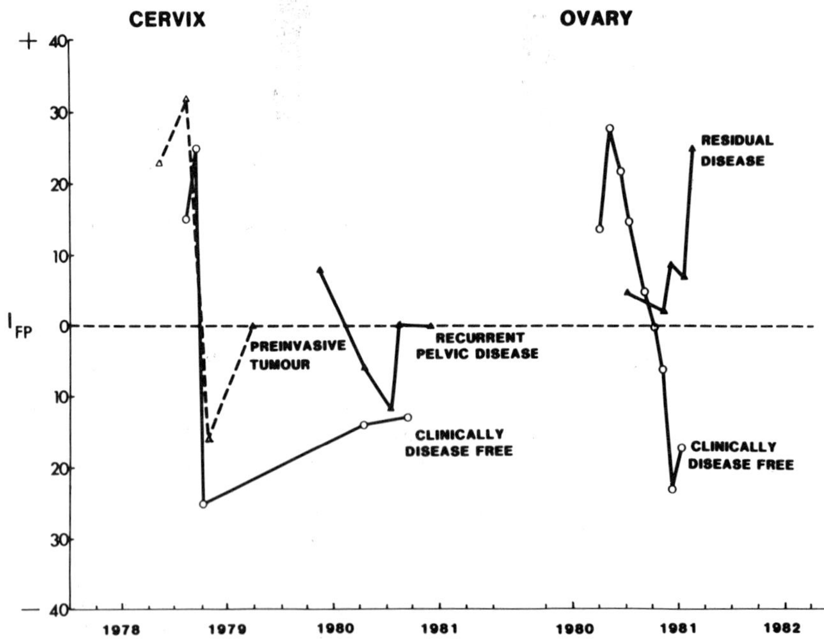

FIGURE 3. Examples of the application of lymphocyte fluorophotometry in post-treatment monitoring for recurrence of cancer of the cervix (left) and ovary (right).

Lymphocyte Fluorophotometry Using NPN

Typical computer-generated fluorescence intensity profiles for a healthy donor (upper) and a cancer patient (lower) are shown in FIGURE 4. A statistically significant increase ($p < 0.001$) in fluorescence intensity occurs after PHA stimulation of lymphocytes from non-malignant donors, whereas an opposite effect is seen with lymphocytes from patients with malignant disease. To establish whether this effect is reproducible, lymphocytes from 12 healthy donors and 11 patients with histologically proved carcinoma of the cervix were tested. The results are shown in TABLE 3. Significant increases in NPN fluorescence were found in all the PHA-stimulated cell suspensions from healthy donors, whereas PHA-stimulated lymphocytes from 8/11

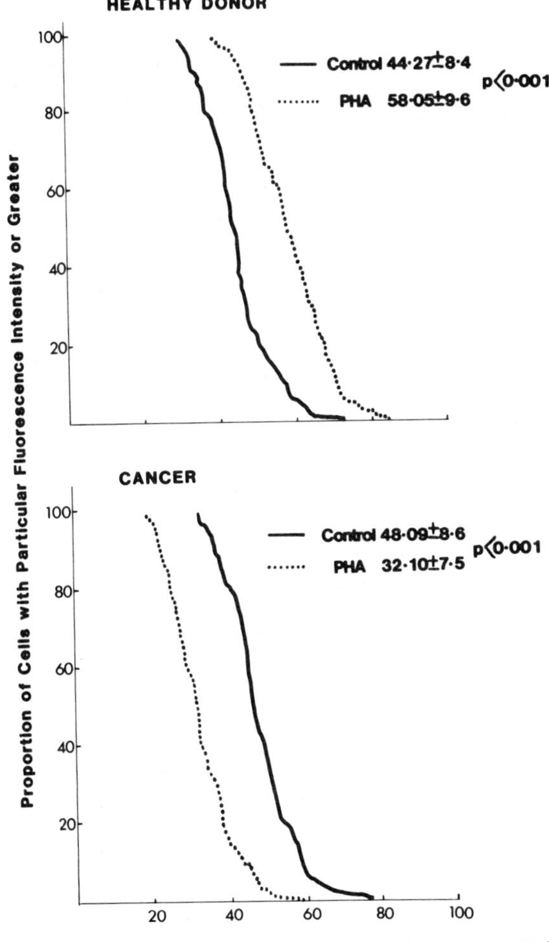

FIGURE 4. Fluorescence intensity profiles of lymphocytes before (solid) and after (dotted) stimulation by PHA, using the probe NPN. The overall intensity increased in cells from healthy donors (upper) but generally decreases in cells from cancer subjects.

cancer patients showed a significant decrease. The remaining three patients showed no response.

DISCUSSION

The biology of early activation events in lymphocytes is not yet fully understood but it seems that fluorescent probes can provide insight into the fundamental interactions of lymphocytes with antigens or mitogens. Whilst a great deal remains to be resolved, the potential and clinical application of techniques using fluorescent probes has escalated. The technique originally described eight years ago[8] was based on the differential response of Zone 1 cells to PHA and to an unstable tumor-derived extract.

TABLE 3. NPN Fluorescence Intensity Changes in PHA-stimulated Lymphocytes

	Age	I(control)	I(PHA)	Significance (p<)
Healthy Donor	42	94.8	110.9	0.001
Healthy Donor	25	91.3	97.0	0.001
Healthy Donor	30	87.5	107.5	0.001
Healthy Donor	57	67.5	90.5	0.001
Healthy Donor	41	47.6	60.1	0.001
Healthy Donor	24	35.1	39.7	0.001
Healthy Donor	36	41.0	46.1	0.01
Healthy Donor	42	34.4	57.2	0.001
Healthy Donor	59	38.9	53.4	0.001
Healthy Donor	32	40.6	64.9	0.001
Healthy Donor	44	45.0	79.1	0.001
Healthy Donor	44	44.2	58.0	0.001
Carcinoma cervix (Stage IB)	40	94.6	79.4	0.001
Carcinoma cervix *in situ*	34	78.9	65.9	0.001
Carcinoma cervix (Stage IIIB)	43	52.1	43.2	0.001
Carcinoma cervix (Stage IB)	47	52.4	55.7	n.s.
Carcinoma cervix (Stage IV)	48	64.3	54.4	0.001
Carcinoma cervix (Stage II)	35	72.7	52.7	0.001
Carcinoma cervix (Stage II)	59	56.0	50.1	0.001
Carcinoma cervix (Stage IB)	51	72.0	61.7	0.001
Carcinoma cervix (Stage IIB)	43	48.0	32.1	0.001
Carcinoma cervix (Stage II)	54	54.1	53.8	n.s.
Carcinoma cervix (Stage I)	46	54.4	51.2	n.s.

In contrast, the technique developed in these laboratories relies only on the differential response of two subpopulations of lymphocytes to PHA, a readily available and reasonably stable reagent. FIGURE 2 shows that fluorophotometry in PHA-stimulated lymphocytes, using the probe fluorescein diacetate, can give reproducible and accurate information on the presence or absence of cancer. The reasons for such discrimination are unclear at present. Preliminary work shows that the lymphocytes separated from both zones are 70 to 80 percent T cells.[13,23] Differences in PHA responsiveness have been attributed to density changes in the lymphocytes,[1,14] but no measurable changes can be found in the cell numbers in the two zones between cancer and non-cancer subjects, and therefore it must be assumed that the test depends on functional differences in the subpopulations of cells.[3] Some evidence exists to suggest that anatomical site-specificity may also be possible[5,21] but this still requires further investigation. Polarization measurements with lymphocytes have also been shown to be diagnostic for different types of leukemia.[29]

A number of reports have shown that lymphocyte fluorophotometry with membrane-bound NPN (LFP$_{NPN}$) can be used in animal models to measure early lymphocyte activation.[19,25] The results obtained with the LFP$_{NPN}$ technique complement the fluorescence polarization measurements in showing that the use of fluorescent probes to study the early stages of lymphocyte response to stimulation can have an important role in oncology. The value of mass screening for cancer remains controversial and it would certainly be premature to recommend its adoption based on the techniques described here, but it may well be that for certain high-risk populations benefit can be derived from limited screening programs. At this stage in the development of the techniques it is likely that the more prudent clinical application of fluorescent probes in a monitoring role will bring more immediate benefit to the cancer patient.

ACKNOWLEDGMENTS

The authors wish to thank Tenovus, South Glamorgan Area Health Authority (T) and the South Wales Cancer Research Council for their continued generous support.

REFERENCES

1. ADLER, W. H., T. TAKIGUCHI & R. T. SMITH. 1971. Phytohaemagglutinin unresponsiveness in mouse spleen cells induced by methylcholanthrene sarcomas. Cancer Res. **31:** 864–867.
2. BALDING, P., P. A. LIGHT & A. W. PREECE. 1980. Response of human lymphocytes to PHA and tumour-associated antigens as detected by fluorescence polarisation. Br. J. Cancer **41:** 73–85.
3. BATER, A. J. & J. A. V. PRITCHARD. 1980. Analysis of the lymphocyte separation techniques used in the double band SCM test. ICRS Med. Sci. **8:** 165.
4. CERCEK, B. 1980. Comments on:- Response of human lymphocytes to PHA and tumour associated antigens as detected by fluorescence polarisation. Br. J. Cancer **42:** 207–208.
5. CERCEK, L. & B. CERCEK. 1975. Apparent tumour specificity with the SCM Test. Br. J. Cancer **31:** 252–253.
6. CERCEK, L. & B. CERCEK. 1977. Application of the phenomenon of changes in the structuredness of cytoplasmic matrix (SCM) in the diagnosis of malignant disorders: A review. Eur. J. Cancer **13:** 167–171.
7. CERCEK, L. & B. CERCEK. 1980. The SCM test for cancer. An evaluation in terms of lymphocytes from healthy donors and cancer patients. Br. J. Cancer **42:** 947–948.
8. CERCEK, L., B. CERCEK & C. I. V. FRANKLIN. 1974. Biophysical differentiation between lymphocytes from healthy donors, patients with malignant diseases and other disorders. Br. J. Cancer **29:** 345–352.
9. CERCEK, L., J. A. V. PRITCHARD & W. H. SUTHERLAND. 1980. Response of human lymphocytes to PHA and tumour associated antigens as detected by fluorescence polarization. Br. J. Cancer **42:** 208–211.
10. HALLIDAY, G. M., R. C. NAIRN, M. A. PALLETT, J. M. ROLLAND & H. A. WARD. 1979. Detection of early lymphocyte activation by the fluorescent cell membrane probe *N*-phenyl-l-naphthylamine. J. Immunol. Methods **28:** 381–390.
11. HASHIMOTO, Y., F. TAKAKU & T. YAMANAKA. 1979. Changes in the structuredness of cytoplasmic matrix in single stimulated lymphocytes from healthy donors and patients with non-malignant and malignant diseases. Br. J. Cancer **40:** 156–160.
12. HASHIMOTO, Y., T. YAMANAKA & F. TAKAKU. 1978. Differentiation between patients with

malignant diseases and non-malignant diseases or healthy donors by changes of fluorescence polarisation in the cytoplasm of circulating lymphocytes. Gann. **69:** 145–149.

13. HOCKING, G. R., J. M. ROLLAND, R. C. NAIRN, E. PIHL, A. M. CUTHBERTSON, E. S. R. HUGHES & W. R. JOHNSON. 1982. Lymphocyte fluorescence polarization changes after phytohaemagglutinin stimulation in its diagnosis of colorectal cancers. J. Natl. Cancer Inst. **68**(4): 579–583.

14. HUBER, C., K. ZIER, G. MICHLMAYR, H. RODT, K. NILSSON, H. THEML & D. LUTZ. 1978. A comparative study of the buoyant density distribution of normal and malignant lymphocytes. Br. J. Haematol. **40:** 93–103.

15. KREUTZMANN, H., T. M. FLIEDNER, H. J. GALLA & E. SLACKMAN. 1978. Fluorescence polarisation changes in mononuclear blood leucocytes after PHA incubation: differences in cells from patients with and without neoplasia. Br. J. Cancer **37:** 797–805.

16. MITCHELL, H., P. WOOD, C. R. PENTYCROSS, E. ABEL & K. D. BAGSHAWE. 1980. The SCM test for cancer. An evaluation in terms of lymphocytes from healthy donors and cancer patients. Br. J. Cancer **41:** 772–777.

17. NAIRN, R. C. & J. M. ROLLAND. 1980. Fluorescent probes to detect lymphocyte activation. Clin. Exp. Immunol. **39:** 1–13.

18. NAIRN, R. C. & J. M. ROLLAND. 1981. Fluorescent probes: a new way to study lymphocyte activation. Endeavour **5**(4): 167–172.

19. NAIRN, R. C., J. M. ROLLAND, G. M. HALLIDAY, I. M. JABLONKA & H. A. WARD. 1978. Fluorescent probes to monitor early lymphocyte activation. *In* Immunofluorescence & Related Staining Techniques. W. Knapp, K. Holubar & G. Wick, Eds.: 57–66. Elsevier/North Holland.

20. ØRJASAETER, H., G. JORDFALD & I. SVENDSEN. 1979. Response of T lymphocytes to phytohaemagglutinin (PHA) and to cancer-tissue-associated antigens, measured by the intracellular fluorescence polarisation technique (SCM test). Br. J. Cancer **40:** 628–633.

21. PRITCHARD, J. A. V. & W. H. SUTHERLAND. 1978. Lymphocyte response to antigen stimulation as measured by fluorescence polarisation (SCM test). Br. J. Cancer **38:** 339–343.

22. PRITCHARD, J. A. V., J. E. SIDDAL, A. J. BATER, I. J. KERBY, W. H. SUTHERLAND & T. J. DEELEY. 1981. Cancer-dependent hydrophobicity changes in membranes of phytohaemagglutinin stimulated lymphocytes. IRCS Med. Sci. **9:** 711.

23. PRITCHARD, J. A. V., J. E. SEAMAN, I. H. EVANS, K. W. JAMES, W. H. SUTHERLAND, T. J. DEELEY, I. J. KERBY, I. C. M. PATERSON & B. H. DAVIES. 1978. Cancer-specific density changes in lymphocytes after stimulation with phytohaemagglutinin. Lancet **ii:** 1275–1277.

24. PRITCHARD, J. A. V., W. H. SUTHERLAND, J. E. SIDDALL, A. J. BATER, I. J. KERBY, T. J. DEELEY, G. GRIFFITH, R. SINCLAIR, B. H. DAVIES, A. RIMMER & D. J. T. WEBSTER. 1982. A clinical assessment of fluorescent polarisation changes in lymphocytes stimulated by phytohaemagglutinin (PHA) in malignant and benign disease. Eur. J. Cancer Clin. Oncol. **18:** 651–659.

25. ROLLAND, J. M., R. L. BETTS, G. M. HALLIDAY, G. R. HOCKING & R. C. NAIRN. 1981. Early changes in concanavalin A-stimulated lymphocytes deleted by the probe *N*-phenyl-l-naphthylamine. Cell Tissue Res. **214:** 119–130.

26. SCHNUDA, N. D. 1980. Evaluation of fluorescence polarisation of human blood lymphocytes (SCM) test in the diagnosis of cancer. Cancer **46:** 1164–1173.

27. STEWART, S., K. I. PRITCHARD, J. W. MEAKIN & G. B. PRICE. 1979. A flow system adaptation of the SCM test for detection of lymphocyte response in patients with recurrent breast cancer. Clin. Immunol. Immunopath. **13:** 171–181.

28. TAKAKU, F., T. YAMANAKA & Y. HASHIMOTO. 1977. Usefulness of the SCM test in the diagnosis of gastric cancer. Br. J. Cancer **36:** 810–813.

29. TSUDA, H., H. MAEDA & S. KISHIMOTO. 1981. Fluorescence polarization with FDA in leukaemic cells: A clear difference between myklogenous and lymphocytic origins. Br. J. Cancer **43:** 793–803.

Immunofluorescence Study of the Antigens
of the Basement Membrane
and the Peritumoral Stroma
in Human Colonic Adenocarcinomas

P. BURTIN AND G. CHAVANEL

Laboratoire d'Immunochimie
Institut de Recherches Scientifiques sur le Cancer
B.P. 8, 94802 Villejuif Cedex, France

J. M. FOIDART

Laboratoire de Dermatologie Experimentale
Tour de Pathologie
Université de Liège
4000 Liege, Belgium

INTRODUCTION

Epithelial basement membranes are complex structures that contain several antigens, recently characterized and isolated. These include type IV collagen,[10] laminin,[16] and a heparan sulfate–rich proteoglycan.[7] Fibronectin, well known as being associated with the plasma membrane of the fibroblasts[8,13] also seems to be associated with the basement membrane of epithelial cells, and is found in the connective tissue surrounding epithelia.[15] Other collagens are also present in this connective tissue, especially type I and type III.[10] Another component, known as hyaluronectin because of its specific affinity for hyaluronic acid,[3,4] is detected in fetal connective tissue and the tissue surrounding the tumors of all organs. Hyaluronectin is absent from all normal adult tissues, except for brain and kidney papilla. We report here a study of the antigens of the basement membrane together with those of the connective tissue in human colonic tumors.

MATERIALS AND METHODS

Tumors

Twenty-three colonic adenocarcinomas, all histologically verified and all large in size (Dukes B and C) were obtained at the time of surgery. Six lymph nodes were removed in the vicinity of colonic adenocarcinomas and contained metastatic foci.

Abbreviations: C III = collagen type III, C IV = collagen type IV, L = laminin, Pg = proteoglycan, HN = hyaluronectin, FN = fibronectin, PBS = phosphate-buffered saline, and NRS = normal rabbit serum.

Normal Colonic Mucosa

In five cases, a specimen of colonic mucosa was removed at the time of surgery at least 10 centimeters from the tumor margin. Three samples of colonic mucosa were also obtained during early autopsies from the victims of accidents maintained in a resuscitation unit until kidney removal for heterotransplantation.

Antisera

Type IV collagen (CIV), laminin (L), and a heparan sulfate–containing proteoglycan (Pg) were purified from the matrix of EHS sarcoma,[12] a murine tumor that produces a matrix of basement membrane.[5,7]

Antisera to type IV collagen, laminin, or a heparan sulfate–containing proteoglycan were raised in rabbits. The antibodies were purified by immunoabsorption and their specificity was tested by radioimmunoassay, immunofluorescence blocking, and

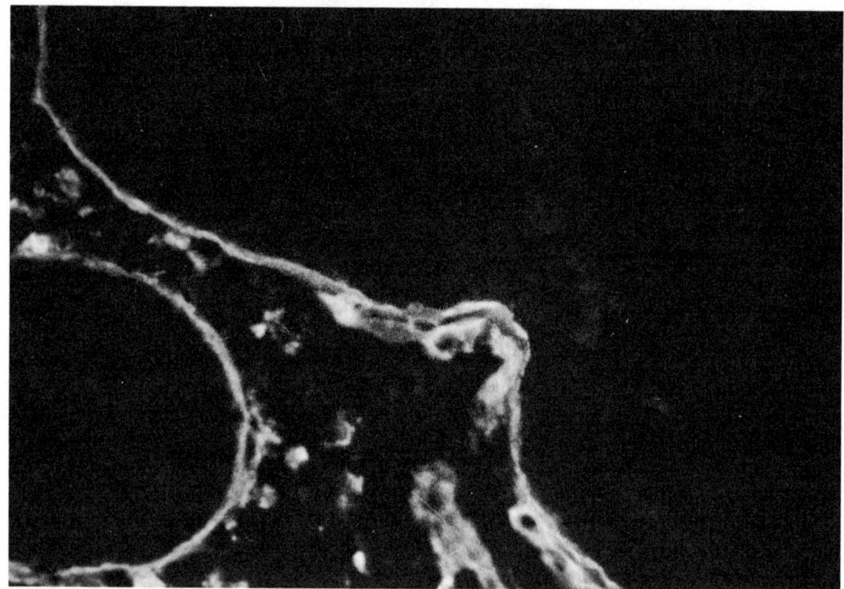

FIGURE 1. Section of normal (peritumoral) colonic mucosa stained by anti-laminin serum (×240).

absorption studies.[5,6] Guinea pig anti-laminin purified antibody was also prepared by immunoabsorption. Rabbit anti-hyaluronectin (HN) serum was a gift from Dr. B. Delpech (Rouen, France). Rabbit anti-human fibronectin (FN) was obtained through the courtesy of Behringwerke (Paris).

Immunofluorescence Method

A fragment of each tumor was cut with scissors, freed of fat and muscle, and immediately quick-frozen at −80°C. It was then cut in a cryostat. The same procedure was used for fragments of normal colonic mucosa. The sections were used either unfixed or fixed in cold 95% ethanol.

The sections were hydrated, covered with the first antiserum for 30 min at room temperature, washed several times with phosphate-buffered saline (PBS), and covered with a sheep anti-rabbit globulin conjugated with fluorescein isothiocyanate (Institut Pasteur, Paris) employed at a dilution of 1/50. As controls, other sections were treated with normal rabbit serum (NRS). All the sections were counterstained with 1% hematoxylin for 1 min. The antisera were used after dilution from 1/50 to 1/100, and compared to NRS diluted 1/50. Exceptions were anti-hyaluronectin and anti-proteoglycan, which were diluted, respectively, 1/5 and 1/10. Both were compared to NRS diluted 1/5.

In four cases, we attempted double-labeling experiments, using the anti-laminin guinea pig antibody, followed by a goat anti-guinea pig globulin serum conjugated with fluorescein (Nordic, Netherlands), then a rabbit antibody to another antigen (usually type IV collagen), followed by a sheep anti-rabbit globulin serum conjugated with tetramethylrhodamine isothiocyanate (Nordic). Preliminary experiments showed there was no cross-reactivity in this system, as the anti-guinea pig globulin serum did not react with rabbit globulin, nor did the anti-rabbit globulin serum react with guinea pig globulin.

The sections were examined with a Leitz Dialux microscope equipped with a Ploem illuminator. The same fields were photographed for fluorescence patterns under UV illumination, or for hematoxylin staining under white light, or both simultaneously.

In some cases the sections were treated with hyaluronidase (Sigma, type V) 600 U/ml in PBS for 1 hr at 37°C. They were then washed with PBS and stained as described before.

RESULTS

Normal Colonic Mucosa

Similar patterns were obtained with antisera against L, CIV, and Pg. Strong staining of the basement membrane, seen as a continuous and even line bordering the glands and the surface epithelium (FIGURE 1), was observed. In the chorion, only blood vessel walls were labeled, always very brightly. In contrast, anti-FN and CIII reacted strongly with the chorion, staining patterns were diffuse or more often discontinuous and net-like (FIGURE 2). Staining of the basement membrane by anti-FN was sometimes visible but was often barely distinguishable from the surrounding chorion. Anti-HN always gave negative results on normal colonic mucosa sections.

Colonic Tumors

All tumors of the colon were studied with anti-CIV and anti-L. Other antisera were used less frequently. Here again, antibodies against L and CIV gave similar results, producing staining of the basement membrane, which was strongly altered in all cases. It was weaker and more irregular than in normal mucosa, resulting in raveled or patchy patterns (FIGURE 3). At times, it disappeared almost entirely either on part of, or on the totality of, the periphery of the tumor glands (FIGURE 4).

Anti-Pg showed patterns similar to those previously described. However, of eight tumors that reacted with this antiserum, two exhibited staining of the basal membranes, which decreased less with anti-Pg than with anti-L and anti-CIV, and one showed an inverse reaction. For the five other tumors, staining was diminished in the same proportion, with all three antibodies.

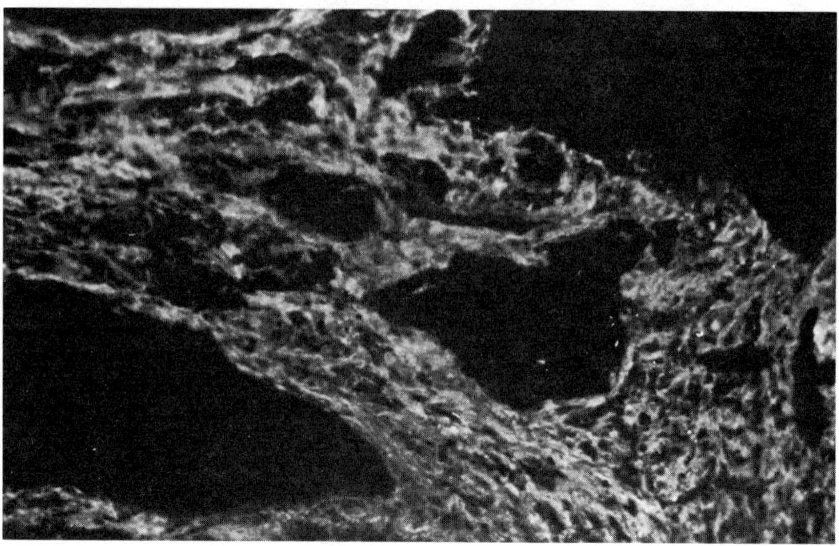

FIGURE 2. Section of a colonic tumor stained by anti-fibronectin serum. In the left lower corner, there is a normal gland ($\times 94$).

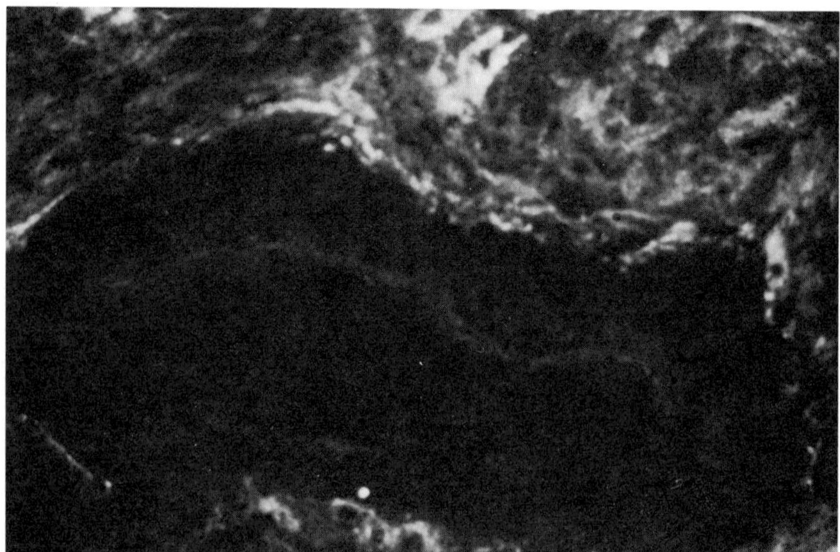

FIGURE 3. Section of a colonic tumor stained by anti-laminin serum. Note the remnants of basement membrane and the vessels ($\times 370$).

Antisera against CIV, L, and Pg strongly stained the wall of blood vessels present between tumor glands. The connective tissue surrounding these glands sometimes reacted with anti-CIV serum, but never with the other two antisera. Thus, brightly stained blood vessels were visible in an otherwise nonfluorescent area, which nevertheless contained several tumor glands (FIGURE 4).

Some correlation existed between the decreased staining of the basal membranes by these antisera and the dedifferentiation of the tumoral glands. The less differentiated glands often had the weakest staining of their basal membranes. Tumoral areas without a glandular structure were unstained by either antiserum. Treatment of sections by hyaluronidase did not change the results. Double-labeling studies were performed on four tumors, thus allowing direct comparison of the localization of L and CIV on the same glands. Their patterns of fluorescence were always superimposable.

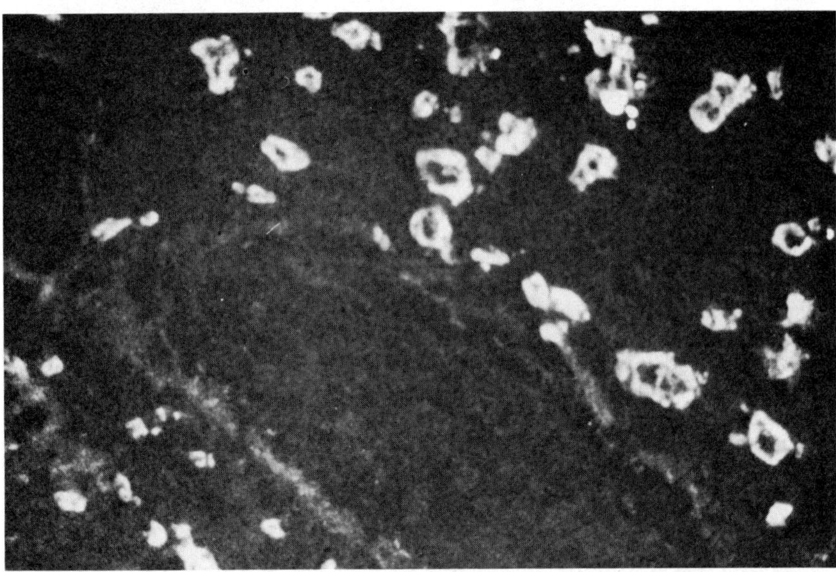

FIGURE 4. Section of a colonic tumor stained by anti-laminin serum. Note the strong fluorescence of vessel walls contrasting with the very weak staining (if any) of basement membranes ($\times 375$).

Anti-FN (FIGURE 2) and anti-CIII sera strongly stained the connective tissue surrounding the tumoral glands. The staining pattern was diffuse, often with a fibrillar structure. In general, it was at least as intense in the tumoral area as in normal mucosa for both antigens.

The labeling of the tumoral basement membranes by anti-FN serum was difficult to demonstrate, because of the strong staining of the connective tissue, except in one case, where weaker staining of the connective tissue made the staining of the basement membranes more easily visible. This was clear-cut, with a few interruptions. Anti-HN stained the conective tissue surrounding the tumor glands. In some areas, the reaction was as intense as with anti-FN serum, while in others it was weaker or negative. Moreover, 3 out of the 15 tumors studied with anti-HN gave negative results.

Lymph Nodes

The metastatic foci contained in these lymph nodes sometimes showed some remnants of basal membranes that were stained by anti-L and anti-CIV. In most cases no staining at all was seen surrounding tumoral foci. Anti-FN when it was used gave similar results. Each of these antisera strongly labeled the blood vessel walls, but never cancer cells themselves.

DISCUSSION

No study of the basement membrane antigens in colonic tumors has previously been reported. We are thus obliged to compare our data with those already published dealing with the same antigens, but in other organs, either tumoral or normal.

Stenman and Vaheri[15] reported that fibronectin was present in the stroma of various carcinomas. It seemed to be absent from the basement membrane, except for some differentiated carcinomas. Birembaut *et al.*[2] pointed out that staining with anti-fibronectin serum showed localized alterations of the basement membrane of breast carcinomas and also of severe dysplasias and adenomas. Chorion staining was not strong enough to impede visualization of the basement membranes.

Variable results have been published by several authors concerning laminin and type IV collagen. Foidart *et al.*[5] were the first to characterize L in normal human and mouse tissues by immunofluorescence. They obtained a strong labeling of the basement membranes of many glandular organs, but never observed staining of the cytoplasm. Ingber *et al.*[9] observed a bright staining of the basement membrane of normal rat pancreatic acini with anti-L and anti-CIV sera by immunofluorescence and a strong decrease in staining of the basement membrane in pancreatic cancers. No cytoplasmic staining was seen. In contrast, Albrechtsen *et al.*[1] and Siegal *et al.*[14] reported staining of both basement membranes and cytoplasm in human breast tumors. In Albrechtsen's experiments,[1] an anti-L serum was used by immunoperoxidase technique and this gave a strong cytoplasmic labeling, which disappeared after trypsin treatment, allowing a better visibility of the basement membranes. Alterations in the basement membranes at the periphery of tumor glands were clearly demonstrable. Siegal *et al.*[14] studied normal and tumoral breast stained with anti-L and anti-CIV sera using the immunoperoxidase technique. They found irregular and occasionally strong cytoplasmic labeling seen mainly in tumor cells. Furthermore, in metastatic lymph nodes, the cytoplasm of isolated tumor cells was stained by both antisera. In contrast, the staining of the basement membranes, easily seen in normal glands, was strongly reduced or absent in tumors. The discrepancies in these results could be related to the fact that different organs were examined in these studies.

The main point of our study and that of other groups was the immunological demonstration that antigens of the basement membranes are altered or apparently absent at the periphery of tumoral glands. These alterations are almost universal. Moreover, all the antigens that can be studied in basement membranes are modified at the same time, if not to the same extent. In fact, as previously described, the alterations in laminin and type IV collagen seemed to be superimposable. Those of the proteoglycan were at times less marked, but were always present. Fibronectin could not be studied as precisely as the other antigens because of the strong labeling of the connective tissue by anti-fibronectin serum.

In contrast, antigens of the connective tissues, fibronectin and type III collagen, were stained at least as well in tumoral as in normal tissues. Hence the phenomenon responsible for basement membrane alteration had a limited action.

We were able to confirm the data first reported by Delpech *et al.*[3] on the appearance of a new component, later to be called hyaluronectin in the peritumoral stroma. However, we had only 12 positive results out of 15 colonic tumors using anti-hyaluronectin serum. Among the three apparently negative tumors, two gave weak staining on Sainte-Marie's sections. The difference cannot be ascribed to the technique, since different tumor fragments were examined. For the last tumor, no sample was processed according to Sainte-Marie's method.

Several hypotheses might explain the mechanisms leading to the lack of basement membrane antigens: the absence of synthesis of the basement membrane antigens, a lack of binding of these antigens to an altered basement membrane, a local proteolytic degradation, or the masking of the basement membrane by an unidentified deposit (but treatment of sections by hyaluronidase did not change our results). In contradiction with the first hypothesis, results by Liotta *et al.*[11] and Foidart *et al.*[5] showed that metastatic cancer cells may still have laminin or type IV collagen in their cytoplasm and thus keep synthesizing it. The lack of binding of these antigens to basement membranes could lead to their deposition in the surrounding connective tisuses. We did not observe such patterns, especially for laminin which we never found outside the basement membranes and the vessel walls. As for proteolytic degradation, it should be viewed in the light of several studies that showed an increase in proteolytic activity in human tumors in comparison to the corresponding normal organ.

SUMMARY

Twenty-three colonic adenocarcinomas were studied by immunofluorescence with antisera against components of the basement membrane (type IV collagen, laminin, and heparan sulfate–rich proteoglycan) as well as antisera against antigens of the connective tissue (type III collagen, fibronectin, and hyaluronectin). Marked alterations of the basement membranes were consistently observed on staining with each one of the first three antisera. In contrast, staining of the normal components of connective tissue was in most cases as intense in tumors as in normal colonic mucosa. Hyaluronectin, a marker of peritumoral stroma, was found to be present in 13 out of 16 tumors studied.

In six metastatic lymph nodes, tumor foci were sometimes surrounded by antigens of the basement membrane. But these antigens were never found in the cytoplasm of cancer cells.

REFERENCES

1. ALBRECHTSEN, R., M. NIELSEN, U. WEWER, E. ENGVALL & E. RUOSLAHTI. 1981. Basement membrane changes in breast cancer detected by immunohistochemical staining for laminin. Cancer Res. **41:** 5076–5081.
2. BIREMBAUT, Ph., D. GAILLARD, J. J. ADNET, H. DOUSSET, B. KALIS, J. P. POYNARD, M. LEUTENEGGER, J. LABAT-ROBERT & L. ROBERT. 1981. La fibronectine. Etude immuno-morphologique de sa distribution en pathologie. Ann. Pathol. **1:** 140–146.
3. DELPECH, B., A. DELPECH & N. GIRARD. 1979. An antigen associated with mesenchyme in human tumors that cross-reacts with brain glycoprotein. Brit. J. Cancer **40:** 123–133.
4. DELPECH, B. & C. HALAVENT. 1981. Characterization and purification from human brain of a hyaluronic acid-binding glycoprotein, hyaluronectin. J. Neurochem. **36:** 855–859.
5. FOIDARD, J. M., E. W. BERE, M. YAAR, S. I. RENNARD, M. GULLINO, G. R. MARTIN &

S. I. KATZ. 1980. Distribution and immunoelectron microscopic localization of laminin. A non collagenous basement membrane glycoprotein. Lab. Invest. **42:** 336–342.

6. FOIDART, J. M. & A. H. REDDI. 1980. Immunofluorescent localization of type IV collagen and laminin during endochondral bone differentiation and regulation by pituitary growth hormone. Dev. Biol. **75:** 130–136.

7. HASSELL, J. R., P. GEHRON-ROBEY, H. J. BARRACH, J. WILCZEK, S. I. RENNARD & G. R. MARTIN. 1980. Isolation of a heparan sulfate-containing proteoglycan from basement membrane. Proc. Natl. Acad. Sci. USA **77:** 4494–4498.

8. HYNES, R. O. 1973. Alteration of cell-surface proteins by viral transformation and by proteolysis. Proc. Natl. Acad. Sci. USA **70:** 3170–3174.

9. INGBER, D. E., J. A. MADRI & J. D. JAMIESON. 1981. Role of basal lamina in neoplastic disorganization of tissue architecture. Proc. Natl. Acad. Sci. USA **78:** 3901–3905.

10. KEFALIDES, N. A. 1975. Basement membranes: structural and biosynthetic considerations. J. Invest. Derm. **65:** 85–92.

11. LIOTTA, L. A., J. M. FOIDART, P. GEHRON-ROBEY, G. R. MARTIN & P. M. GULLINO. 1979. Identification of micrometastasis of breast carcinomas by presence of basement membrane collagen. Lancet **2:** 146–147.

12. ORKIN, R. W., P. GEHRON, E. B. MCGOODWIN, G. R. MARTIN, T. VALENTINE & R. H. SWARM. 1977. A murine tumor producing a matrix of basement membrane. J. Exp. Med. **145:** 204–220.

13. RUOSLAHTI, E., A. VAHERI, P. KUUSELA & E. LINDER. 1973. Fibroblast surface antigen. A new serum protein. Biochem. Biophys. Acta **322:** 352–358.

14. SIEGAL, G. P., S. H. BARSKY, V. P. TERRANOVA & L. A. LIOTTA. 1981. Stages of neoplastic transformation of human breast tissue as monitored by dissolution of basement membrane components. An immunoperoxidase study. Invasion and Metastasis **1:** 54–70.

15. STENMAN, S. & A. VAHERI. 1981. Fibronectin in human solid tumors. Int. J. Cancer **27:** 427–435.

16. TIMPL, R., H. ROHDE, P. G. ROVEY, S. I. RENNARD, J. P. FOIDART & G. R. MARTIN. 1979. Laminin. A glycoprotein from basement membranes. J. Biol. Chem. **254:** 9933–9937.

Detection of Human Malignant Melanoma Antigens by Immunofluorescence and Autologous Postimmune Antimelanoma Sera[a]

STANLEY P. L. LEONG[b]

Division of Surgery
City of Hope National Medical Center
Duarte, California 91010
and
Department of Surgery
University of California Irvine Medical Center
Orange, California 92668

INTRODUCTION

The presence of human malignant melanoma-associated antigens and the extent of their polymorphism have been controversial. In general, alloantibodies in melanoma patients are low in titer and lack specificity.[3] Therefore, xenoantisera to human melanoma-associated antigens have been studied.[5,15,19,22,29] The problem with xenoantisera is that they contain antibodies to antihuman histocompatibility (HLA) antigens. Recently, mouse monoclonal antibodies to melanoma-associated antigens have been defined.[1,7,8,28,31] Several studies have shown that immunization of melanoma patients may result in the production of antimelanoma antibodies.[4,6,9,21,26,27] Our group has demonstrated that antisera from patients undergoing immunotherapy (using their irradiated tumor cells and, as adjuvant, Bacillus Calmette-Guerin (BCG)) react with the surface of a variety of cultured human melanoma cells lines as shown in immunofluorescence tests.[16] This report summarizes the work completed at Tulane University from 1975 to 1976, at Boston University from 1976 to 1978, and at City of Hope National Medical Center from 1979 to 1981. The objectives of this study are to demonstrate the specificity of autologous postimmune antimelanoma sera and to localize melanoma antigens on a cellular level.

[a]Supported in part by U.S. Public Health Service Research Grant NCI CA05108 at Tulane University, GRS 7-91500-545 at Boston University, and NIH BRSG RRO 5471 at City of Hope National Medical Center. The author's attendance at this conference was made possible by a travel grant from the Excerpta Medica Foundation, Amsterdam, the Netherlands.

[b]Present address: Surgery Branch, Bldg. 10, National Cancer Institute, National Institutes of Health, Bethesda, MD 20205.

Abbreviations: BCG = Bacillus Calmette Guerin, HLA = histocompatibility antigens, RPMI = Roswell Park Memorial Institute, EMEM = Earles minimal essential medium, FCS = fetal calf serum, PBS = phosphate-buffered saline, MA = melanoma antigen, and MIF = membrane immunofluorescence.

MATERIALS AND METHODS

Immunization of Melanoma Patients

Since HLA antibodies may be found in patients receiving allogeneic melanoma cells, these patients are excluded from this report. Only patients who were immunized with autologous melanoma cells are described here since their serological reactivity is most likely to be directed against possible tumor antigens rather than HLA antigens. Briefly, five patients with stage II or III melanoma that could be resected free of all gross tumor underwent autologous immunotherapy at the Department of Surgery, Tulane University. Approximately 1 to 4 $\times$ 10^6 irradiated (2 to 7 $\times$ 10^4 rads) melanoma cells in 0.5 ml of Roswell Park Memorial Institute (RPMI) 1640 medium plus BCG (1–3 $\times$ 10^5 viable organisms) were injected intradermally in about two-week

TABLE 1. Cell Lines Used in This Study

Cell Lines[a]	Established By
Melanoma (M)	
TU-M-BG	Tulane (TU)
TU-M-BW	Tulane
TU-M-FP	Tulane
TU-M-MR	Tulane
TU-M-PK	Tulane
TU-M-CY	Tulane
TU-M-PU	Tulane
CaCL 74-36	Dr. Liao
RPMI-8322	Dr. Moore
Skin fibroblast (SF)	
TU-SF-BG	Tulane
TU-SF-FP	Tulane
Others	
W1-38 (Human fetal lung fibroblast)	Commercial
HeLa	Commercial
Human kidney	Commercial
Monkey kidney	Commercial
Molt (leukemic T cells)	Commercial

[a]The cell lines were negative for mycoplasma.

intervals. The antibody responses to melanoma antigens as measured by immunofluorescence are summarized in FIGURE 2. TU-M-PK melanoma cell line was used as the target throughout the study. The technique used is described in *Immunofluorescence on Fixed Cells.*

Culture Cell Lines

The cell lines used in this study are summarized in TABLE 1. In general, the cells were grown in monolayers in plastic bottles (Falcon Plastics; 75-sq. cm 250-ml flasks) in RPMI medium 1640 or Earles Minimal Essential Medium (EMEM) containing 10% fetal calf serum (FCS) and antibiotics (penicillin, 100 units/ml and streptomycin, 100 μg/ml) at 37°C with 5% CO_2.

IMMUNOFLUORESCENCE STUDIES

Membrane Immunofluorescence

To demonstrate the presence of membrane antigens, a modification of Moller's technique[22] using viable cells was employed. When the cells in 75-sq. cm flask reached confluency, the medium was poured off and the cells were gently rinsed with phosphate-buffered saline (PBS, 8.8 g NaCl, 1.135 g Na_2HPO_4, and 0.276 g $NaH_2PO_4 \cdot H_2O$ in 1,000 ml distilled water, pH 7.4). The cells were then scraped off from the flask and washed in PBS. Approximately 1×10^5 cells were aspirated into a microcentrifuge tube (VWR, Scientific Inc., San Francisco, CA). Fifty μl of the serum was added to the tube and cells were shaken. The tube was incubated for 5 to 30 min with the cap closed. The longer the incubation time, the more likely capping would occur. Capping may be stopped by allowing the antigen-antibody reaction to occur at 4°C or in the presence of sodium azide (10 mM). The cells were then washed in PBS three times. Fifty μl of fluorescein isothiocyanate-conjugated antihuman IgG or IgM (Hyland Lab., Costa Mesa, CA) was added to the tube and the same procedure was repeated. The washed cells were suspended in one drop (50 μl) of buffered glycerol (95%, pH 8.4) and transferred onto a slide and covered by a 2-mm square cover slip. Fluorescence was read under a Leitz Wetzler fluorescence microscope Osram HBO 200 mercury lamp, UG-1 exciter filter and 430 barrier filter).

Membrane Fluorescence Inhibition

Similar steps as mentioned in the previous section were followed, except that the test sera had been previously incubated with the absorbents, which could be whole cells, cellular components, or purified proteins. A positive inhibition would result in a negative immunofluorescence reading.

Immunofluorescence on Fixed Cells

Frosted-end slides (Scientific Products, McGraw Park, IL) were cleaned and dried. Ten circles, each 5 mm in diameter, were drawn with a wax pencil or a correction fluid (Liquid Paper Corp., Dallas, TX). The latter was preferred since it did not wash off easily. The importance of circling is that when the serum is applied it will stay within the circle without diffusing throughout the slide, so that the cells will not become dried out during incubation. After the cells were washed as described above, one microdrop of cell suspension was pipetted into each of the circles. The cells were allowed to settle for 5 min and the slide was gently tilted to dab off the overlying PBS. After the cells were completely dried, different fixatives (TABLE 3) were pipetted over the cells for 10 min. If only one fixative was used, the whole slide could be immersed in the fixative. The slide was rinsed in PBS several times and again dried. Six μl of serum were applied onto each circle and incubation was allowed to take place in a moist chamber at room temperature or 37°C for 20–30 min. The serum was dabbed off and the slide was washed with PBS two to three times, each for 20–15 min. The slide was dried again and the same amount of fluorescein isothiocyanate-conjugated goat antihuman IgG or IgM was added to each circle. Similar incubation and washing steps were followed. The cells were then mounted on buffered glycerol (95%, pH 8.4) and read for fluorescence as described above. To study the cells in their attached form, they were grown in a slide chamber (Lab-Tek, Miles Lab, Inc., Naperville, IL). When

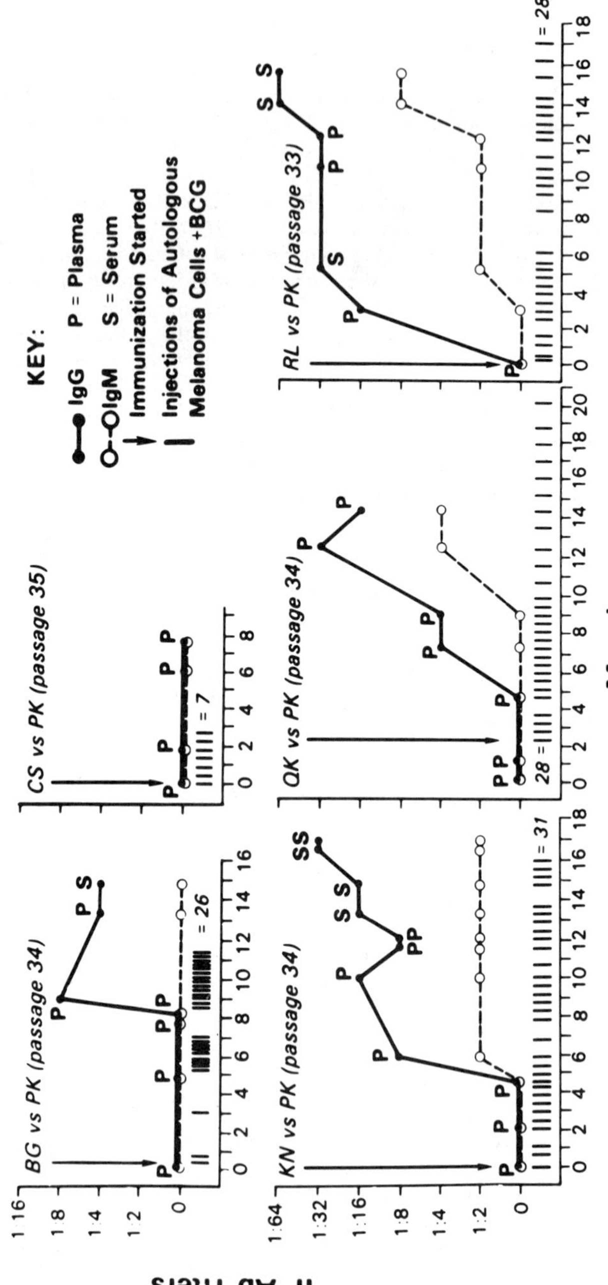

FIGURE 1. Antibody responses to autologous immunization. Both IgG and IgM were measured. Number above abscissa indicates total number of immunizations for each patient. (From Leong *et al.*[16] By permission of *Cancer Research*.)

confluency was reached, the cells were washed, circled, and stained as described above. It was found that a 5-mm diameter circle was quite adequate for microscopic examination of cells and yet enabled us to use a minute amount of serum, approximately 6 μl in contrast to 1 drop of serum (50 μl) if the circle was made larger. A positive reading was defined as unequivocal fluorescence in at least over 25% of the cells examined at an antiserum titer of 1:8 or greater. This criterion was applied to both viable and fixed cells.

Trypsinization of Melanoma Cells

The monolayers of melanoma cells were dispensed by trypsinization (3 to 5 ml of 0.25% trypsin in RPMI 1640 for 1 to 5 min). The cells were then washed with PBS and prepared for immunofluorescence study.

Demonstration of Complement-Fixation Ability of Antisera

Postimmune antisera were heated at 56°C for 30 min and used in complement-fixing experiments. After initial antigen-antibody reaction was completed as described under immunofluorescence on fixed cells, nonheated normal human sera were applied as source of complement for 30 min at room temperature. The slides were washed and fluorescein isothiocyanate-conjugated goat antihuman C_3 (Hyland Lab., Costa Mesa, CA) was applied for 30 min. The slides were washed and prepared for immunofluorescence microscopy.

Seroepidemiologic Study

Sera from 65 nonimmunized melanoma patients and 140 nonmelanoma patients were tested against positive melanoma cells.

RESULTS

Antibody Response to Autologous Immunization

This study has been previously reported.[16] Both IgG and IgM antibodies were followed during the course of immunotherapy in five melanoma patients. The results are shown in FIGURE 1. The predominant antibody was IgG. The highest postimmune IgG titer of 1/64 was that of RL in reference to TU-M-PK melanoma line. CS had no antibody response. All preimmune sera or plasmas were negative for fluorescence. In general, repeated immunizations were required to induce antibody production and its drop was usually associated with cessation of immunization or progression of disease. In another study,[30] preliminary evidence indicated that high antibody titers were related to a longer disease-free interval.

Membrane Immunofluorescence and Kinetics of Capping

When viable cells were incubated with postimmune serum at room temperature, sequential capping (FIGURE 2b), polarization, and extrusion (FIGURE 2c) of antigen-

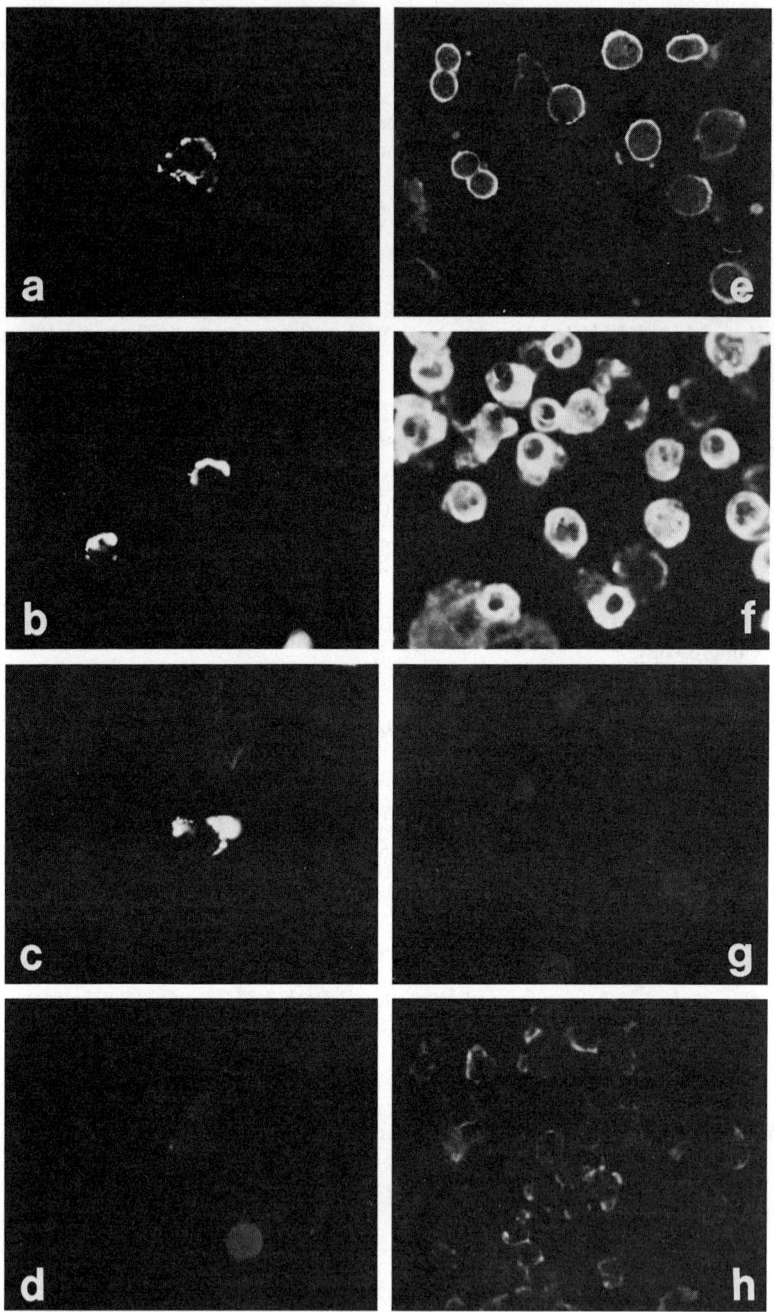

antibody complexes could be demonstrated. Capping took place within minutes (FIGURE 3). During the first 2 min, 29% of the cells showed full membrane fluorescence (FIGURE 2a), 8% of the cells showed early stages of capping, and 2% of the cells showed extrusion of antigens. Over a period of 30 min the number of full membrane fluorescence cells decreased with a concomitant increase in number of cells with capping. This was followed by sequential increase in cells that were either extruding the antigen-antibody complexes or had lost membrane fluorescence. Comparing between viable and formalin-fixed cells, it was found that about 35–40% of viable cells showed membrane fluorescence binding whereas about 70% of formalin-fixed cells were found to show ring-like membrane fluorescence (FIGURE 2e). These results have been reported elsewhere.[12]

Four of five "autologous" postimmune sera (BG, KN, QK, and RL) were reactive with surface membranes of five of seven melanoma cell lines. TU-M-CY and TY-M-PU were negative. These seven cell lines were established at Tulane University. KN and RL postimmune sera were tested against two other melanoma cell lines (CaCL 74-36[18] developed by Dr. S. K. Liao, McMaster University, Hamilton, Ont. and RPMI-8322[17] developed by Dr. G. Moore at University of Colorado, Denver, CO). Both KN and RL antisera were reactive against the two melanoma lines. Other negative controls consisted of two human skin fibroblast, W1-38, HeLa, human kidney, monkey kidney, and Molt cell lines.

Membrane Fluorescence Inhibition

Results are shown in TABLE 2. It should be noted that BG constitutes an autologous system with both the autologous melanoma and fibroblast cell lines.

Effect of Different Fixatives and Localization of Melanoma Antigens

Two postimmune sera, KN and RL, were tested on two melanoma cell lines using various fixatives.[11] Ethanol, methanol, formalin, trichloroacetic acid (TCA), and

---

FIGURE 2. Fluorescein-labeled goat anti-human IgG (1:8) was used for fluorescence staining throughout. All photographs were taken on Kodak Tri-X 135 black and white film at 30-sec exposures. (a) An unfixed viable melanoma cell showing full membrane speckled fluorescence after incubation with postimmune RL antimelanoma plasma and fluorescence staining. (From Leong et al.[12] By permission of Cancer Research.) (b) Two unfixed viable melanoma cells showing capping after incubation with postimmune RL antimelanoma plasma and fluorescence staining. (From Leong et al.[12] By permission of Cancer Research.) (c) An unfixed viable melanoma cell showing polarization and extrusion of antigen-antibody complexes after incubation with postimmune RL antimelanoma plasma and fluorescence staining. (From Leong et al.[12] By permission of Cancer Research.) (d) Unfixed viable melanoma cells showing no fluorescence after incubation with preimmune RL serum and fluorescence staining. (e) Formalin-fixed melanoma cells showing ring-like membrane fluorescence after incubation with postimmune RL antimelanoma plasma and fluorescence staining. (Leong et al.[14] By permission of Journal of Surgical Research.) (f) Melanoma cells being fixed with isooctane or isopentane showing cytoplasmic fluorescence after incubation with postimmune RL antimelanoma plasma and fluorescence staining. (From Leong et al.[11] By permission of Oncology.) (g) Formalin-fixed melanoma cells showing no fluorescence after incubation with preimmune RL serum and fluorescence staining. (h) Formalin-fixed melanoma cells grown in serum-free medium showing dotty to patchy membrane fluorescence after incubation with postimmune RL antimelanoma plasma and fluorescence staining.

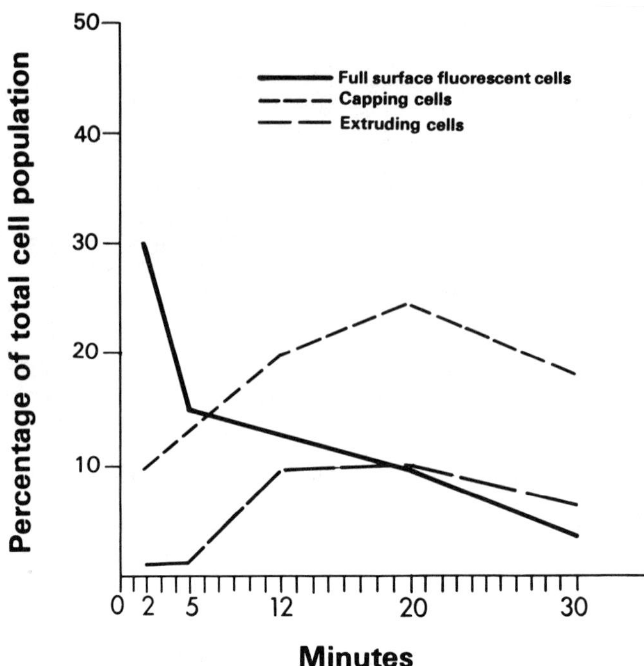

FIGURE 3. Percentage of full membrane fluorescence, capping and extruding cells as a result of immunofluorescence staining with fluorescein-labeled goat anti-human IgG of unfixed human melanoma cells after sequential incubation with postimmune antimelanoma plasma over a period of 30 minutes (From Leong *et al.*[12] By permission of *Cancer Research.*)

TABLE 2. Immunofluorescence Results of Absorption Experiments

	Target Melanoma Cells			
Postimmune plasmas or sera (1/4 dilution)	TU-M BG	TU-M FP	TU-M-MR	TU-M PK
BG plasma absorbed with				
BG skin fibroblasts	+	+	+	+
BG melanoma	−	NT	−	−
PK melanoma	−	NT	−	−
BCG	+	NT	NT	+
KN serum absorbed with				
FP melanoma	−	−	NT	NT
Culture medium RPMI 1640 + 20% FCS	+	+	NT	+
100% FCS	+	+	NT	+
RL serum absorbed with				
BG melanoma	−	NT	NT	NT
PK melanoma	NT	NT	NT	−
Molt T-cells	+ +	+ +	NT	+ +
BCG	NT	+ +	NT	+ +
Culture medium RPMI 1640	+ +	+ +	NT	+ +
100% FCS	+ +	+ +	NT	+ +

Number of cells: serum volume = 2×10^6: 1/30 ml, 1 hr at room temperature and 23 hr at 4°C. Volume of medium or BCG: serum volume = 1:1, same incubation time as above.

+, bright full membrane fluorescence; + +, very bright full membrane fluorescence; −, no fluorescence; NT, not tested.

(From Leong *et al.*[16])

TABLE 3. Fluorescence Patterns Observed with Different Fixatives

Fixatives	Membrane Fluorescence	Cytoplasmic Fluorescence
95% ethanol	+ +	−
95% methanol	+ +	−
10% formalin	+ +	−
5% TCA	+	−
95% acetone	+ +	+
10% glutaraldehyde	±	±
95% isopentane	−	+ +
95% iso-octane	−	+ +
2% paraformaldehyde	−	−

Cell line: CaCL 74-36, passage 25-27. Antiserum: Postautoimmune anti-melanoma serum (RL) at 1:8 in PBS (pH 7.4). + + = Bright fluorescence; + = positive fluorescence; ± = weak fluorescence; − = no fluorescence.
(From Leong *et al.*[11])

acetone yielded sharp ring-like membrane fluorescence. Cytoplasmic fluorescence was noted only after acetone fixation. Highest membrane fluorescence antibody titer of 1/512 was obtained with formalin. Bright cytoplasmic fluorescence was seen upon fixation in isopentane and isooctane (FIGURE 2f). Very weak, diffuse cytoplasmic fluorescence was seen with glutaraldehyde and no fluorescence was noted with paraformaldehyde fixation. The negative controls (FIGURE 2g) consisted of four from nonimmunized melanoma patients' sera and PBS. Different fluorescence patterns are summarized in TABLE 3. TABLE 4 shows the membrane antibody titers observed in preparations using various fixatives.

Effect of Serum Factors on the Expression of Human Melanoma Antigens

RL postimmune serum was used in this experiment. A preliminary report has been published elsewhere.[13] FIGURE 4 shows the experimental protocol as well as the results.

Effect of Trypsin on Melanoma Membrane Antigens

Trypsinized melanoma cells showed patchy or dotty membrane fluorescence after incubation with postimmune antimelanoma sera.

TABLE 4. Highest Membrane Fluorescence Antibody Titers Observed with Different Fixatives

Fixatives	Highest Antibody Titers versus CaCL 74-36 Cell Line
Formalin	1/512
Acetone	1/128
Ethanol	1/128
Methanol	1/64
TCA	1/8

Antisera used were the same as in TABLE 3.
(From Leong *et al.*[11])

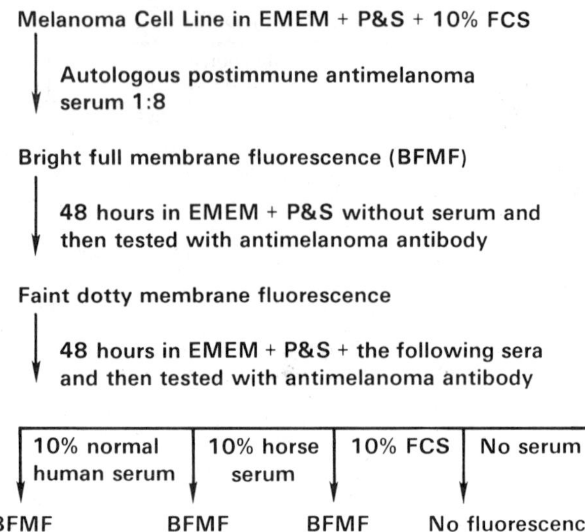

FIGURE 4. Experimental protocols to show shedding of membrane antigens in serum-free medium and resynthesis of antigens in the presence of human, horse, and fetal calf serum. EMEM = Earl's Minimal Essential Media and P&S = penicillin and streptomycin.

Complement Fixation

Four positive postimmune antisera were found to fix human complement component C_3. No cytotoxicity experiments were performed in this study.

Seroepidemiological Studies

Sera from 65 nonimmunized melanoma and 140 nonmelanoma patients showed no reactivity to the positive melanoma cell lines at a titer of 1:4, except for one non-melanoma serum, which was reactive at 1:4.

DISCUSSION

This report summarizes a series of studies that we have conducted to define melanoma-associated antigens using autologous postimmune antimelanoma sera. The humoral response in this group of patients was elicited only after multiple immunizations as shown in FIGURE 1, suggesting that melanoma antigens are weakly antigenic. One of five melanoma patients (CS) showed no response after several immunizations. This lack of response could be explained by fewer immunization doses for this patient than for other patients.

Theoretically, autologous antisera should contain antibodies specific to tumor-associated antigens, since in an autologous system the host will respond only to those tumor antigens that are not present on normal cells. In this study, the postimmune antisera reacted with melanoma-associated antigens present in most of the melanoma cell lines tested. In a separate study using mixed hemadsorption microassay and RL postimmune antiserum, we were able to demonstrate the presence of common melanoma antigens in 5 of 5 melanoma cell lines tested.[20] In another study by Livingston *et al.*,[21] 2 of 13 patients produced antibody to shared melanoma antigens after receiving melanoma autologous cell vaccines. In contrast, Ikonopisov *et al.*[9] could only produce antibody to individual specific melanoma antigen using irradiated, fresh melanoma cell suspension. Whether these "common" melanoma-associated antigens represent similar molecular structures is not known because the antibodies could be heterogeneous. Within the conditions set in this report, melanoma-associated antigens in tissue culture show the characteristics as listed in TABLE 5. Additional immunochemical methods, such as immunobiochemical purification, immunoprecipitation, and immunodepletion may further elucidate the nature of these antigens on a molecular level. Thus a meaningful comparison of these antigens may be made between different laboratories. To date, different immunization protocols have been adopted in the production of postimmune antimelanoma sera.[4,9,16,21,26,27] Furthermore, numerous immunologic techniques have been used in defining the serologic reactivity. Therefore, it is often difficult to compare these data in an organized fashion. The conclusive evidence for the existence of common or individual melanoma-associated antigens has yet to be established.

As the autologous melanoma cells have been grown in tissue culture, it is of concern that the tumor-associated antigens expressed by cultured melanoma cells are truly those antigens present under *in vivo* conditions. In addition, new antigenic determinants may be expressed either by absorbing fetal calf serum components[10] or being induced by culture condition. Fetal calf serum could be ruled out since it did not absorb out the antimelanoma activity (TABLE 2). The most crucial experiment yet to be performed is to test the reactivity of the postimmune sera against a variety of *in vivo* melanoma specimens and the presence of circulating antigens in serum or body fluid, since the establishment of melanoma antigens *in vivo* is of vital importance for immunodiagnosis and immunotherapy. The pursuit of these experiments has been hampered by the limited supply of antisera. Recent development of monoclonal antibodies against human melanoma antigens holds promises to unlimited and reproducible supply of specific antibodies. Although specific melanoma antigens have been detected by monoclonal antibodies in melanoma specimens removed from operations,[7] the exact nature of these antigens *in vivo* has not been determined. It is important that such monoclonal antibodies should be exchanged among different laboratories for comparative studies.

Although immunofluorescence is not as sensitive as some other techniques such as radioimmunoassay, it is an excellent method to demonstrate the localization of antigens on a cellular level. Membrane immunofluorescence is an important technique to demonstrate the ability of an antigen to cap. Understanding this phenomenon is

TABLE 5. Characteristics of Melanoma-Associated Antigens as Defined by the Postimmune Antimelanoma Sera

1. Antigenic in the autologous host. Repeated immunizations are required.
2. Common in 77% (7/9) of the melanoma cell lines tested.
3. Membrane antigens are induced to cap on the cell surface by the antibody.
4. Membrane antigens are shed into the extracellular milieu.
5. Antigenicity may be affected by fixatives.
6. Partially digested by trypsin.
7. Not related to fetal calf serum, BCG (TABLE 2) mycoplasma or HLA.
8. Fixing human complement C_3.

important as such antigenic modulation may result in tumor escape from the host's immune attack since the malignant cell is now denuded of its tumor antigen. Furthermore, the antigen-antibody complexes may paralyze the effective arm of the immune reaction, thus, leading to enhancement of tumor growth.

Membrane fluorescence inhibition may be used as an assay for isolation of antigens.[14] For large-scale screening of sera against target melanoma cell lines, it is much easier to fix the cells than to use viable cells. Therefore, a proper choice of fixatives that preserve the membrane antigen is important. We have found that among these fixatives, formalin gives the highest fluorescence antibody titer (TABLE 4). These data on formalin-fixed melanoma cells are in agreement with other studies using formalin to preserve human melanoma membrane antigens[23,24] and EL-4 murine lymphoma antigens.[2] Comparison between formalin-fixed and viable melanoma cells by indirect immunofluorescence shows a dramatic increase in antigenicity of fixed (70%) over viable (40%) cells. This difference may be explained by increase in density of surface antigen when the cells are fixed on to the surface of a slide and the apparent negative viable cells may contain antigens embedded in the membrane or antigens that are sterically obstructed from binding the antibody. Fixation may result in the exposure of these "cryptic" antigens. Additional experiments are needed to prove these hypotheses. Several advantages are obvious in the use of fixed cells: increase of sensitivity as detected by membrane immunofluorescence; feasibility of large-scale, routine screening of antisera; and feasibility of a safer form of vaccine since viable cells may potentially result in tumor growth in the host.

Without the addition of antibody, melanoma membrane antigens may be shed into the culture medium as shown in FIGURE 4. It is interesting that synthesis of the antigens is restored when sera are added back to the serum-free medium. Exploitation of this shedding phenomenon enables us to collect melanoma antigens from culture medium rather than extracting them from the cells.[14] The *in vivo* implication is that such membrane antigens may be shed into the circulation and potentially could be detected by specific antibody. These antigens may also block specific immune response against tumor.

SUMMARY

The objectives of this study are to demonstrate the specificity of sera from melanoma patients undergoing autologous immunization and to localize melanoma antigens on a cellular level. Five melanoma patients were immunized with autologous melanoma cells and Bacillus Calmette-Guerin. This immunization program was conducted at Tulane University. Indirect immunofluorescence using both viable and fixed melanoma cells was employed. Four of five postimmune sera were reactive to five of seven melanoma cell lines. Two of the four reactive antisera showed positive binding with two additional melanoma lines obtained from other laboratories. All these sera were negative against seven nonmelanoma lines. Negative controls consisted of sera from 65 nonimmunized melanoma patients and 140 nonmelanoma patients. Membrane immunofluorescence (MIF) demonstrated sequential full MIF, capping, polarization, and extrusion of antigen-antibody complexes on the cell surface. MIF inhibition showed shedding of melanoma antigens in the culture medium. Ethanol, methanol, formalin, trichloroacetic acid, and acetone yielded sharp MIF. Isopentane and isooctane gave bright cytoplasmic fluorescence. In conclusion, this study provides suggestive evidence for the existence of common melanoma antigens as defined by the postimmune antimelanoma sera. These antigens may be localized in the membrane or within the cytoplasm.

REFERENCES

1. BROWN, J. P., J. D. TAMERIUS & I. HELLSTROM. 1979. Indirect ^{125}I-labeled protein A assay for monoclonal antibodies to cell surface antigens. J. Immunol. Methods **31**: 201–209.

2. DRAKE, W. P., P. C. UNGARO & M. R. MARDINEY. 1972. Presentation of cellular antigenicity of tumor cells by the use of formalin fixation. Cancer Res. **32**: 1042–1044.

3. FERRONE, S. & M. A. PELLEGRINO. 1970. Serological detection of human melanoma-associated antigens. *In* Immunodiagnosis of Cancer. R.B. Heberman & K.R. McIntyre, Eds.: 588–632. Marcel Dekker. New York.

4. FRITZ D., D. H. KERN, C. R. DROGEMULLER & Y. H. PILCH. 1976. Production of antisera with specificity for malignant melanoma and human fetal skin. Cancer Res. **36**: 458–466.

5. GOODWIN, D. P., M. O. HORNUNG, S. P. L. LEONG & E. T. KREMENTZ. 1972. Immune responses induced by human malignant melanoma in the rabbit. Surgery **72**: 737–743.

6. GUPTA, R. K., R. F. IRIE, D. O. CHEE, D. H. KERN & D. L. MORTON. 1979. Demonstration of two distinct antigens in spent tissue culture medium of a human malignant cell line. J. Natl. Cancer Inst. **63**:347–356.

7. HERLYN, M., W. H. CLARK, JR., M. J. MASTRANGELO, D. GUERRY, IV., D. E. ELDAR, D. LA ROSSA, R. HAMILTON, E. BONDI, R. TUTHILL, Z. STEPLEWSKI & H. KOPROWSKI. 1980. Specific immunoreactivity of hybridoma-secreted monoclonal anti-melanoma antibodies to cultured cells and freshly derived human cells. Cancer Res. **40**: 3602–3609.

8. HERLYN, D., M. HERLYN, Z. STEPLEWSKI & H. KOPROWSKI. 1979. Monoclonal antibodies in cell-mediated cytotoxicity against human melanoma and colorectal carcinoma. Eur. J. Immunol. **9**: 657–659.

9. IKINOPISOV, R. L., M. G. LEWIS, I. D. HUNTER-CRAIG, D. C. BODENHAM, T. M. PHILLIPS, C. I. COOLING, J. PROCTOR, H. G. FAIRLEY & P. ALEXANDER. 1970. Autoimmunization with irradiated tumor cells in human malignant melanoma. Brit. Med. J. **2**: 752–754.

10. KERBEL, R. S. & D. BLAKESLEE. 1976. Rapid adsorption of a fetal calf serum component by mammalian cells in culture. A potential source of artifacts in studies of anti-sera to cell-specific antigens. Immunology **31**: 881–891.

11. LEONG, S. P. L., S. R. COOPERBAND, P. J. DECKERS, C. M. SUTHERLAND, J. F. CESARE & E. T. KREMENTZ. 1979. Effect of different fixatives on the localization of human melanoma antigens by immunofluorescence. Oncology **36**: 202–207.

12. LEONG, S. P. L., S. R. COOPERBAND, P. J. DECKERS, C. M. SUTHERLAND, J. P. CESARE & E. T. KREMENTZ. 1979. Antibody-induced movement of common melanoma membrane antigens on the surface of unfixed human melanoma cells. Cancer Res. **39**: 2125–2131.

13. LEONG, S. P. L., S. R. COOPERBAND, P. J. DECKERS, C. M. SUTHERLAND & E. T. KREMENTZ. 1980. Effect of serum factors from human, horse and fetal calf on the expression of human melanoma antigens. Proc. Am. Assoc. Cancer Res. **21**: 218.

14. LEONG, S. P. L., S. R. COOPERBAND, C. M. SUTHERLAND, E. T. KREMENTZ & P. J. DECKERS. 1978. Detection of human melanoma antigens in cell-free supernatants. J. Surg. Res. **24**: 245–252.

15. LEONG, S. P. L., M. O. HORNUNG & E. T. KREMENTZ. 1976. Immunofluorescent studies on chimpanzee humoral responses to human melanoma cells. Oncology **33**: 246–249.

16. LEONG, S. P. L., C. M. SUTHERLAND & E. T. KREMENTZ. 1977. Immunofluorescent detection of common melanoma membrane antigens by sera of melanoma patients immunized against autologous or allogeneic cultured melanoma cells. Cancer Res. **37**: 4035–4042.

17. LIAO, S. K., P. B. DENT & P. B. MCCULLOCH. 1975. Characterization of human malignant melanoma cell lines. I. Morphology and growth characteristics in culture. J. Natl. Cancer Inst. **54**: 1037–1044.

18. LIAO, S. K., P. B. DENT & P. B. MCCULLOCH. 1976. Cellular morphology of human malignant melanoma in primary culture. In Vitro **12**: 654–657.

19. LIAO, S. K., P. C. KWONG, J. C. THOMPSON & P. B. DENT 1979. Spectrum of melanoma antigens on cultured human malignant melanoma cells as detected by monkey antibodies. Cancer Res. **39**: 183–192.

20. LIAO, S. K., S. P. L. LEONG, C. M. SUTHERLAND, P. B. DENT, P. K. KWONG & E. T. KREMENTZ. 1978. Common human melanoma membrane antigens detected by mixed hemadsorption microassay with serum from a patient undergoing immunotherapy with autologous tumor cells. Cancer Res. **38:** 4395–4400.
21. LIVINGSTON, P. O., T. WATANABE, H. SHIKU, A. N. HOUGHTON, A. ALBINO, T. TAKAHASHI, L. A. RESNICK, R. MICHITSCH, C. M. PINSKY, H. F. OETTGEN & L. J. OLD. 1982. Serological response of melanoma patients receiving melanoma cell vaccines. I. Autologous cultured melanoma cells. Int. J. Cancer **30:** 413–422.
22. METZGAR, R. S., P. M. BEYOR, M. A. MORENO & H. F. SEIGLER. 1973. Melanoma specific antibodies produced in monkeys by immunization with human melanoma cell lives. J. Natl. Cancer Inst. **50:** 1065–1068.
23. MÖLLER, G. 1961. Demonstration of mouse isoantigens at the cellular level by the fluorescent antibody technique. J. Exp. Med. **114:** 415–434.
24. MÜLLER, C. & C. SORG. 1975. Use of formalin-fixed melanoma cells for detection of antibodies against surface antigens by a micro-immune adherence technique. Am. J. Immunol. **5:** 175–178.
25. ROSS, C. E., A. J. COCHRAN, D. E. HOYLE, R. M. GRANT & R. M. MACKIE. 1975. Formalinized tumor cells in the leukocyte migration inhibition test. Clin. Exp. Immunol. **22:** 126–132.
26. SEIGLER, H. F., W. W. SHINGLETON, R. S. METZGAR, C. E. BUKLEY, III. & P. M. BERGOC. 1973. Immunotherapy in patients with melanoma. Ann. Surg. **178:** 352–359.
27. SHIBATA, H. R., L. M. JERRY, G. M. LEWIS, P. W. A. MANSELL, A. CAPEK & G. MARQUIS. 1976. Immunotherapy of human malignant melanoma with irradiated tumor cells, oral Bacillus Calmette-Guerin, and levamisole. Ann. N.Y. Acad. Sci. **277:** 355–366.
28. STEPLEWSKI, Z., M. HERLYN, D. HERLYN, W. H. CLARK & H. KOPROWSKI. 1979. Reactivity of monoclonal antimelanoma antibodies with melanoma cells freshly isolated from primary and metastatic melanoma. Eur. J. Immunol. **9:** 94–96.
29. STUHLMILLER, G. M. & H. F. SIEGLER. 1975. Characterization of a chimpanzee antihuman melanoma antiserum. Cancer Res. **35:**2132–2137.
30. SUTHERLAND, C. M., M. O. HORNUNG, S. P. L. LEONG, J. W. C. HOLMES & E. T. KREMENTZ. 1983. Autoimmunization for patients with malignant melanoma. (Submitted for publication).
31. WOODBURY, R. G., J. P. BROWN, M. Y. YEH, I. HELLSTROM & K. E. HELLSTROM. 1980. Identification of a cell surface protein, p97, in human melanomas and certain other neoplasms. Proc. Natl. Acad. Sci. USA **77:** 2183–2187.

Typing of Leukemic Cells with Monoclonal Antibodies[a]

W. KNAPP, O. MAJDIC, P. BETTELHEIM, K. LISZKA,
W. ABERER, AND G. STINGL

Departments of Immunology, Internal Medicine I, and Dermatology I
University of Vienna
Vienna, Austria

INTRODUCTION

The immunofluorescence detection of surface-linked antigenic determinants by heteroantisera has greatly contributed to a better understanding of the heterogeneity of benign and malignant hemopoietic cells. The difficulties encountered in rendering these heteroantisera specific for a unique cell surface determinant have been a major drawback to a more rapid development of immunologic cell typing and, in particular, to a typing of non-lymphoid cells. The strategy ordinarily employed involved the extensive absorption of these heteroantisera with large numbers of cells of different cellular lineages. As a consequence, the resulting antibody titers were frequently low, thereby limiting the utility of these antisera for biological or clinical investigations. Finally, given the diversity of the immune response and the complexity of the absorption procedures, it has been difficult to reproduce antibodies with identical specificity, which frequently hindered the comparison of data obtained from different laboratories. With the introduction of the antibody-producing hybridoma technology by Köhler and Milstein,[7] most of these technical problems have been overcome. This technology provides for the first time uniform antibody reagents that are highly specific and of unlimited supply. The use of such antibodies should therefore considerably improve immunologic cell typing.

Using this technology, we were able to raise a number of monoclonal antibodies against cell type–specific differentiation antigens. These antibodies are of considerable help in the diagnosis and classification of human leukemia.

MATERIALS AND METHODS

Monoclonal Antibodies

The monoclonal antibodies described in this paper were obtained after fusion of X63-Ag8.653 Balb/c myeloma cells with spleen cells of previously immunized mice. The anti-myeloid monoclonal antibodies were obtained after immunization with KG1, THP-1, or K-562 cells, the antibodies VIL-A1 and VIB-C5 after immunization with Reh-6 cells, and VIE-G4 after immunization with a human thymus cell suspension. Detailed procedures are described elsewhere.[2,6,8,9]

[a]Supported by Fonds zur Förderung der wissenschaftlichen Forschung in Österreich (Austria).

Normal and Leukemic Hemopoietic Cells

Mononuclear cells, lymphocytes, granulocytes, monocytes, and thrombocytes were isolated as described before.[9] The diagnosis of leukemia was made using standard clinical, morphological, and cytochemical criteria. All neoplastic preparations selected for this study showed >75% abnormal cells.

Cell samples from patients with the following diseases were studied: acute myeloid leukemia (AML), acute myelo-monocytic leukemia (AMML), acute monocytic leukemia (AMoL), chronic myeloid leukemia in stable phase (Ph^{1+}CML), chronic myeloid leukemia in myeloid blast crisis (CML-BC "M"), chronic myeloid leukemia in lymphoid blast crisis (CML-BC "L"), acute lymphoblastic leukemia of T cell type (T-ALL), B cell type (B-ALL), and non-T-non-B cell type (Non T-Non B-ALL), erythroid leukemia (Erythr. L.), hairy cell leukemia, Non Hodgkin lymphoma, chronic lymphocytic leukemia of B cell type (B-CLL), multiple myeloma, and Sezary syndrome.

Immunofluorescence

The reactivity of our antibodies was tested in indirect immunofluorescence using, as secondary reagent, FITC-labeled rabbit $F(ab')_2$ antibodies against mouse IgG-$F(ab')_2$. It was essential for our studies that the antisera used to prepare these antibodies were previously extensively absorbed with human Ig to remove cross-reactive antibodies.

Before use, the reactivity of these FITC-labeled anti-mouse $F(ab')_2$ antibodies with mouse IgM, IgG1, IgG2a, IgG2b, and IgG3 was confirmed.

All readings were performed with a Leitz Ortholux fluorescence microscope equipped with a Ploem Pak II illuminator (Leitz, Wetzlar, GFR).

RESULTS

Anti-Myeloid Cell Monoclonals

So far we have established 11 clones of hybrid cells that secrete antibodies reactive with myeloid but not with lymphoid cells. Their reaction pattern with peripheral blood cells is shown in TABLE 1. Three reaction patterns can be distinguished. Group 1 antibodies (Clones VIM-D5, VIM-C6, VIM-4, VIM-6, VIM-7, VIM-9, VIM-1, VIM-11) react with granulocytes but not with monocytes, thrombocytes, and erythrocytes. Group 2 antibodies (VIM-2 and VIM-10) bind to granulocytes and to monocytes. Group 3 antibodies (VIM-D2 and VIM-5) bind only to monocytes sparing granulocytes as well as all other cells of the peripheral blood.

The most thoroughly studied among these anti-myeloid monoclonals is the VIM-D5 antibody from group 1.

Bone marrow stainings revealed that the VIM-D5 antibody reacts with the entire spectrum of the myeloid series ranging from the promyelocyte to the granulocyte. In addition, cell line studies indicated that this antibody, although not reactive with mature monocytic cells, binds to immature monoblast-like cells, such as the THP-1 and U-937 cell lines. Lymphoid cell lines of all differentiation stages were completely negative. K-562 cells were weakly positive.

From this reaction pattern, we concluded that the antigen defined by VIM-D5 is present on the precursors of all non-lymphoid hemopoietic cell lineages[9] and assumed

TABLE 1. Reaction Patterns of Anti-Myeloid Cell Monoclonal Antibodies with Peripheral Blood Cells

Clone Designations	Polymorpho-nuclear granulocyte	Monocyte	Reactivity Thrombocyte	Lymphocyte	Red Blood Cell
VIM-D5	+	−	−	−	−
VIM-C6	+	−	−	−	−
VIM-4	+	−	−	−	−
VIM-6	+	−	−	−	−
VIM-7	+	−	−	−	−
VIM-9	+	−	−	−	−
VIM-1	+	−	−	−	−
VIM-11	+	−	−	−	−
VIM-2	+	+	−	−	−
VIM-10	+	+	−	−	−
VIM-D2	−	+	−	−	−
VIM-5	−	+	−	−	−

that it is then lost during monocytic and erythroid differentiation, and only maintained by the granulocytic cell lineages.

However, in further studies we have been unable to detect this antigen on cells of the erythroid lineage with the exception of weakly reactive K-562 cells, which were originally used for immunization. We conjecture therefore that VIM-D5 is only aberrantly expressed on the K-562 cell.

Studies on leukemic cells seem to be in agreement with the observations on normal cells and cell lines. As shown in FIGURE 1, VIM-D5 reacts with myeloid leukemic cells at a rather advanced stage of differentiation, i.e. CML in stable phase. It also reacts with 80% of the cell samples of AML patients, although in most instances only a proportion of blast cells are VIM-D5 positive. The reactivity, both in terms of the percentage of reactive blast cells and the proportion of positively reacting patients, is higher in AMML than in AML. In agreement with results obtained with hemopoietic cell lines, we find also a considerable proportion of AMoL patients to be VIM-D5 positive.

Surprisingly, two out of 142 lymphatic leukemia patients tested were VIM-D5 positive. On the basis of morphological and cytochemical criteria, both these patients were classified as acute lymphatic leukemias. Both had TdT-positive blast cells, but still reacted positively with VIM-D5 antibody. In one of these patients, no other conclusive surface markers could be found. The other one was also VIL-A1 positive, which would indicate a cALL type of leukemia. At the moment, we have no interpretation for this finding. It might be that we are dealing with an aberrant antigen expression, but it could also be that these two leukemias represent a rare and so far unclassified subtype of AL.

Anti-Lymphoid Cell Monoclonals

The most frequent form of acute lymphatic leukemia is characterized by the expression of a 96,000–100,000 MW glycoprotein on the malignant blast cell surface.[3] The monoclonal antibody VIL-A1 recognizes this protein.[6] Our results with this antibody when screening various forms of leukemia are summarized in FIGURE 2.

Cells from more than 80% of all non B-non T ALL patients and all CML patients in lymphoid blast crisis tested displayed VIL-A1 reactivity. All other leukemias were negative. Positive reactions were also found in 4 of the 18 Non Hodgkin's B-lymphoma patients. All positively reacting patients were of Burkitt's type.

Another antibody, which is particularly helpful in the characterization of leukemic cells, is the VIB-C5 antibody. This antibody is, like all other antibodies described in this paper, of the IgM class. It is the only monoclonal antibody that we have obtained so far that lyses cells not only in the presence of rabbit, but also of human, complement.

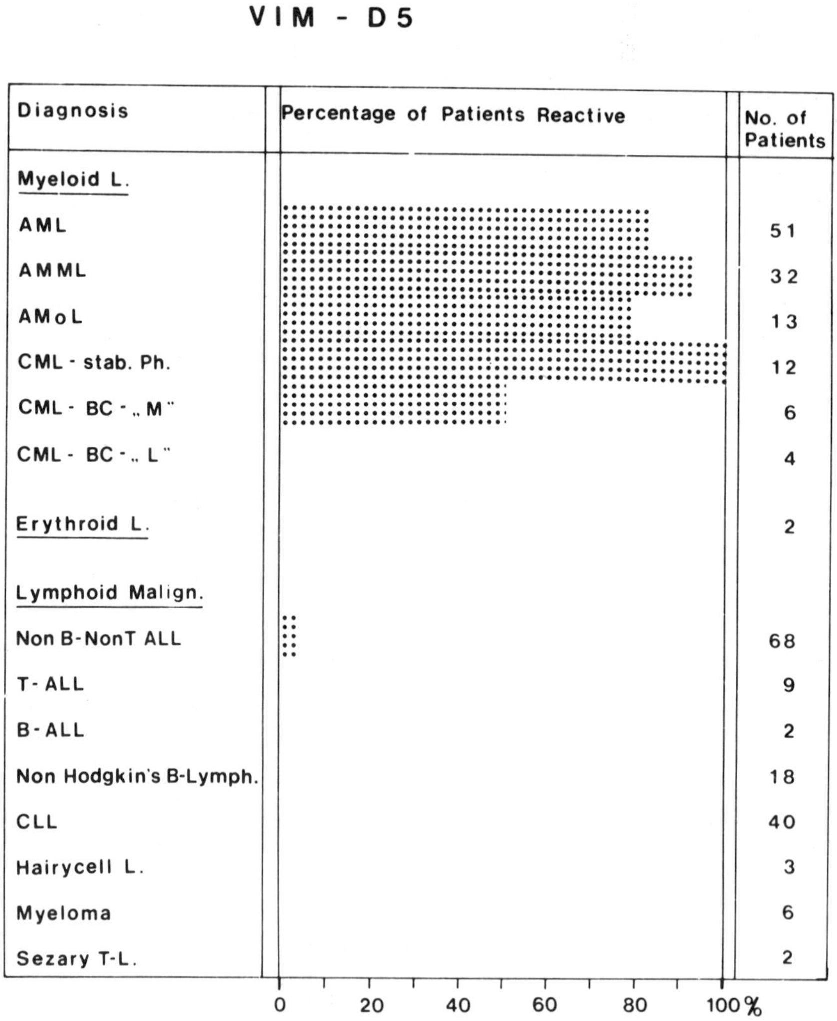

VIM − D 5

Diagnosis	Percentage of Patients Reactive	No. of Patients
Myeloid L.		
AML		51
AMML		32
AMoL		13
CML - stab. Ph.		12
CML - BC - „M"		6
CML - BC - „L"		4
Erythroid L.		2
Lymphoid Malign.		
Non B-NonT ALL		68
T- ALL		9
B- ALL		2
Non Hodgkin's B-Lymph.		18
CLL		40
Hairycell L.		3
Myeloma		6
Sezary T-L.		2

FIGURE 1. Reaction pattern of VIM-D5 antibody with leukemic cells

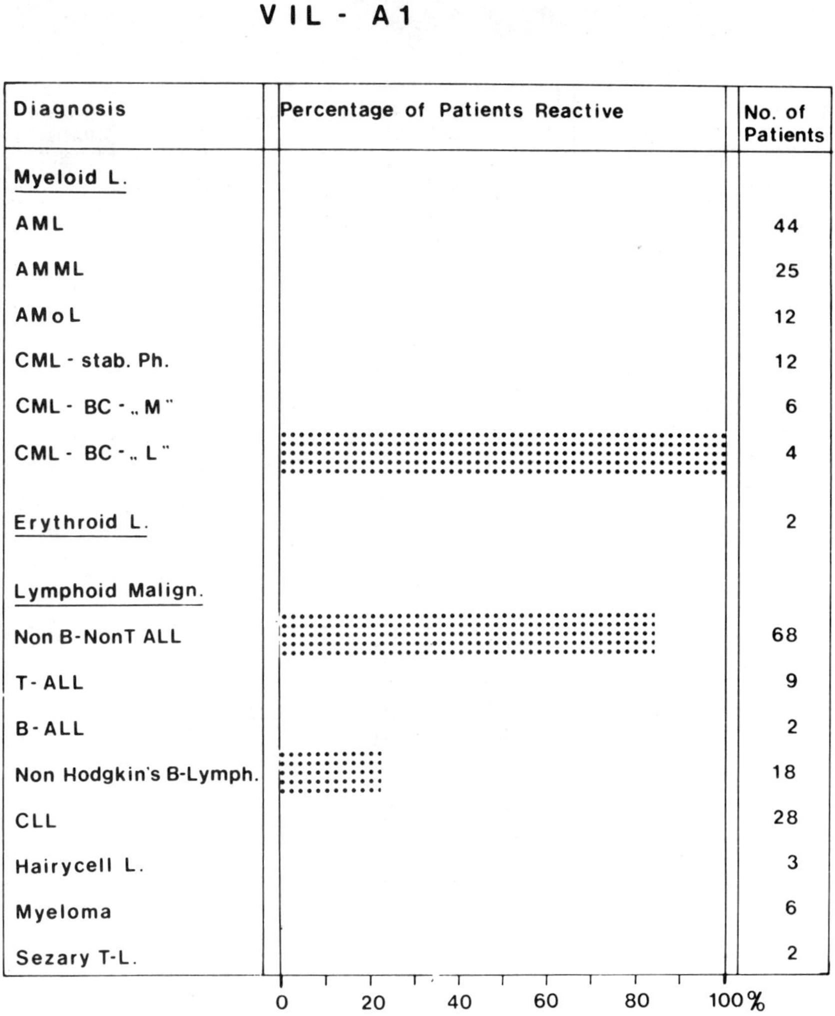

V IL - A 1

Diagnosis	Percentage of Patients Reactive	No. of Patients
Myeloid L.		
AML		44
AMML		25
AMoL		12
CML - stab. Ph.		12
CML - BC - „ M ″		6
CML - BC - „ L ″	▓▓▓▓▓▓▓ (0–100%)	4
Erythroid L.		2
Lymphoid Malign.		
Non B-NonT ALL	▓▓▓▓▓▓ (0–85%)	68
T- ALL		9
B- ALL		2
Non Hodgkin's B-Lymph.	▓▓ (0–25%)	18
CLL		28
Hairycell L.		3
Myeloma		6
Sezary T-L.		2

0   20   40   60   80   100 %

FIGURE 2. Reaction pattern of VIL-A1 antibody with leukemic cells

The VIB-C5 antibody strongly binds to peripheral blood B lymphocytes. It does not recognize T cells, monocytes, thrombocytes, or erythrocytes, nor does it react with the precursors of these cells. It does react, however, with mature granulocytic cells. Since these cells can be easily recognized by their morphology, this does not pose a diagnostic problem. Within the B cell series, VIB-C5 seems to detect a differentiation antigen that is expressed by B lymphocytes, pre-B cells, and most of the common ALL positive cells. It is lost during differentiation from the B cell to the plasma cell level. The reactivity of this antibody with leukemic cells is shown in FIGURE 3.

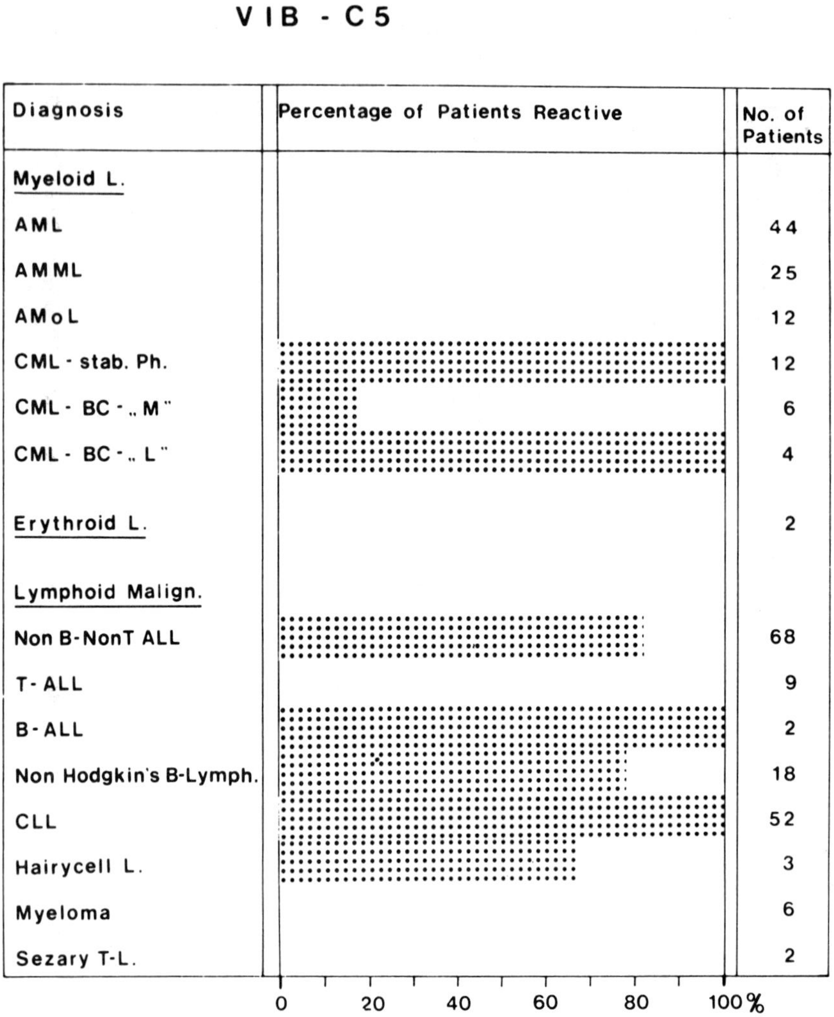

V I B - C 5

Diagnosis	Percentage of Patients Reactive	No. of Patients
Myeloid L.		
AML		44
AMML		25
AMoL		12
CML - stab. Ph.		12
CML - BC - „ M "		6
CML - BC - „ L "		4
Erythroid L.		2
Lymphoid Malign.		
Non B-NonT ALL		68
T- ALL		9
B- ALL		2
Non Hodgkin's B-Lymph.		18
CLL		52
Hairycell L.		3
Myeloma		6
Sezary T-L.		2

0   20   40   60   80   100 %

FIGURE 3. Reaction pattern of VIB-C5 antibody with leukemic cells

Anti-Erythroid Cell Monoclonal

Glycophorin A is the major sialoglycoprotein on human red blood cells. It is also expressed on erythroid precursor cells but not on non-erythroid cell lineages. This membrane protein seems thus to represent an ideal marker for erythroid leukemias. Our results with a monoclonal anti-glycophorin A antibody (VIE-G4) support this notion. As can be seen from FIGURE 4, VIE-G4 selectively reacts with erythroid leukemias, but not with any other form of leukemia.

DISCUSSION

Phenotypic characterization of leukemic cells by immunofluorescence is an attractive adjuvant, or even, an alternative to morphological and cytochemical methods currently employed.

Rapid progress in the field of hybridoma technology should lead to the development of reagents that selectively recognize particular malignant cell subsets.

The most urgent problem in leukemia diagnosis is the classification of acute leukemias. From a diagnostic and prognostic viewpoint, it is of great importance that a

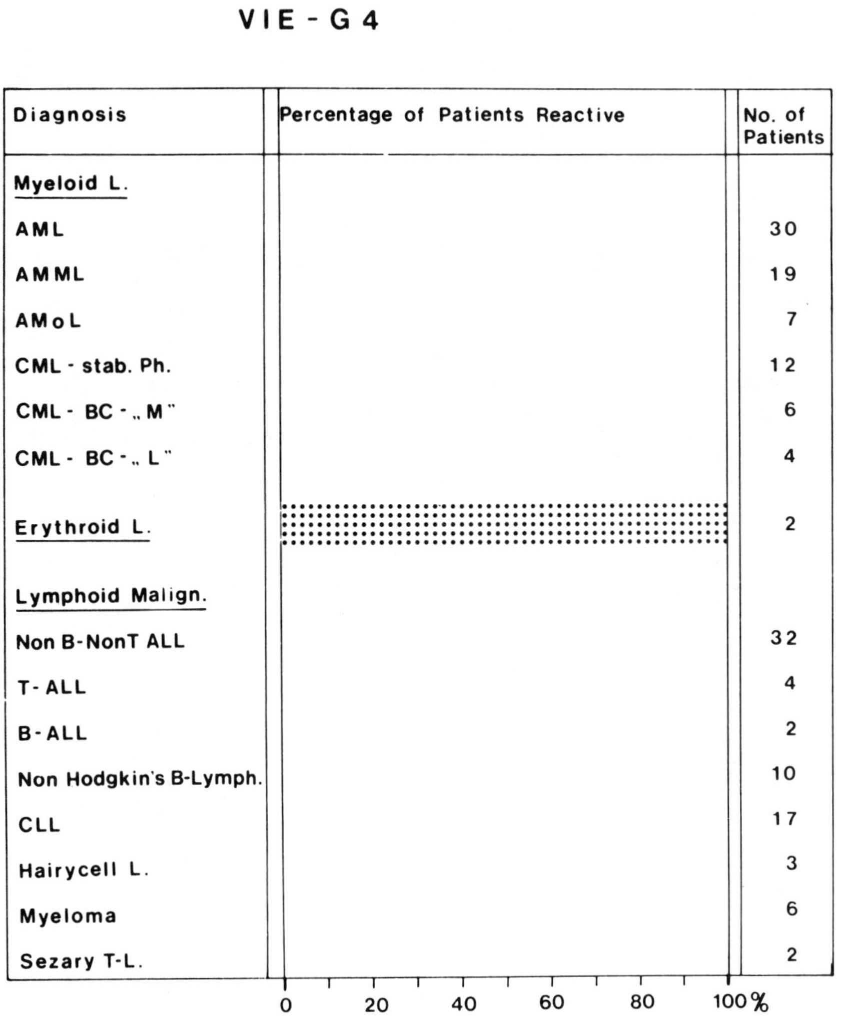

VIE - G 4

FIGURE 4. Reaction pattern of VIE-G4 antibody with leukemic cells

clear distinction between acute lymphoid and non-lymphoid leukemias is made, while the clinical implications of a subclassification of lymphatic leukemias have yet to be established.[5]

More than 50% of all cases of acute leukemia are of the myeloid type. Therefore, the monoclonal antibodies developed in our laboratory that react with myeloid, but not with lymphoid cells, promised to be powerful tools for leukemia diagnosis and classification. The most thoroughly studied anti-myeloid monoclonal is the VIM-D5 antibody. Polyacrylamide gel electrophoresis of the VIM-D5–reactive component on granulocytes shows two bands of 150,000 and 105,000 daltons, respectively (P.A.T. Tetteroo, Amsterdam, personal communication). Studies on human × mouse myeloid cell hybrids allowed the assignment of the gene(s) for the VIM-D5–defined antigen to human chromosome 11 (A.H.M. Geurts van Kessel, personal communication).

VIM-D5 IgM monoclonal antibody detects myeloid leukemias, particularly those in which the majority of malignant cells has reached the later developmental phases, i.e. CML in stable phase. It also reacts with more than 80% of the cell samples of AML patients, although in most instances only a proportion of blast cells are positive. The incidence of false positives (if they are false) is with 1% at an acceptable level.

VIM-D5, however, does not allow one to distinguish between myeloid and monocytic leukemias. Such a subclassification seems to be feasible with monocyte-specific antibodies like our VIM-D2 antibody (data not shown).

The great majority of the non-myeloid acute leukemias are lymphoid leukemias. Among these, common acute lymphoblastic leukemias (CALL) predominate. The malignant cells of this form of leukemia express neither surface immunoglobulins nor T cell markers. They are characterized by the expression of a 96,000–100,000 MW glycoprotein on their surface that is not expressed by normal B or T lymphocytes. This glycoprotein has been termed common ALL-associated antigen (CALLA).[3]

Monoclonal antibodies reactive with this membrane protein, e.g. like our VIL-A1 IgM antibody, are therefore very useful reagents for the diagnosis and classification of CALL. Let us caution, however, that CALLA does not seem to be a leukemia-specific antigen. Low percentages of weakly CALLA-positive cells have also been found in some non-leukemic bone marrow samples.[4] These cells most likely represent lymphoid precursor cells. This indicates that CALLA is expressed during early lymphoid differentiation and disappears with further maturation. An expression seems to occur again, however, at the site of peripheral B cell amplification, i.e. the germinal center in that both centroblasts and centrocytes have recently been found to be VIL-A1 positive.[10] Further, malignant cells constituting centroblastic-centrocytic lymphomas, the most common variant of non-Hodgkin's B-lymphomas, display VIL-A1 reactivity.

In contrast to VIL-A1, VIB-C5 antibody not only reacts with very early lymphoid differentiation stages, but also with more mature cells of the B cell series, including the peripheral blood B lymphocytes. The antigen recognized by VIB-C5 disappears with further maturation to plasma blasts/plasma cells.

Among leukemic cells, VIB-C5 correspondingly reacts with immature lymphoid B cells of Non B-Non T ALL patients and CML patients with lymphoid blast crisis. It also detects more mature malignant B cells found in patients with CLL, Hairy cell leukemia, Non Hodgkin's B lymphoma, and B cell ALL. T cell leukemias are completely negative. Since VIB-C5 also detects mature cells of the myeloid series, it is not surprising that cell samples from CML patients also exhibit VIB-C5 reactivity, whereas acute myeloid leukemias are consistently negative.

Morphologically recognizable erythroid leukemias are rare. Results obtained by L. Andersson *et al.* with a rabbit anti-glycophorin-A antiserum indicate, however, that erythroid leukemias might be more frequent than originally thought.[1] He found that about 20% of acute leukemias express glycophorin A and should thus be classified as erythroid.

We sought to determine whether similarly high percentages of glycophorin A-positive leukemias could also be detected with a monoclonal anti-glycophorin A antibody (VIE-G4), which we have recently raised in our laboratory. As seen in FIGURE 4, VIE-G4, in contrast to the antiserum used by Andersson *et al.,*[1] reacts only with morphologically recognizable erythroid leukemias.

We feel that the data presented demonstrate the present and potential utility of immunofluorescence procedures with monoclonal antibodies in leukemia cell typing.

While the visual readings used to obtain these results are quite time-consuming and demand a certain degree of experience, the development of flow-cytofluorometric technology will certainly overcome these barriers to efficient and accurate application of immunological leukocyte typing methods. Cell typing may thus become one of the most widely used applications of immunofluorescence.

With this perspective in mind a word of caution concerning potential sources of error is warranted. A common, but avoidable, source of error is nonspecific or unwanted binding of either monoclonal antibodies or fluorochrome-labeled anti-mouse immunoglobulin reagents to cells.

Nonspecific staining is frequently due to binding of antibody molecules to receptors for the Fc portion of IgG present on various cell types. No intact IgG antibodies should therefore be used for membrane fluorescence studies. Thus, monoclonal antibodies of the IgM class are preferable to those of the IgG class. Further, second step anti-immunoglobulin antibodies should be converted to $F(ab)'_2$ fragments by pepsin digestion.

The main reason for unwanted staining of cells is cross-reactivity of fluorochrome-labeled anti-mouse immunoglobulin antibodies with human immunoglobulins thus revealing not only surface Ig-positive B cells, but also all cells that carry human Ig passively adsorbed via their Fc receptors. It is therefore essential to use conjugates from which all cross-reactive antibodies have previously been removed. Addition of human serum to cell-conjugate incubation mixtures, as proposed by some commercial sources, is certainly not a solution and makes reliable interpretation even more difficult.

Additional criteria relevant to the use of anti-mouse Ig reagents, e.g. reactivity with all mouse Ig classes and subclasses, are covered elsewhere in this volume. In spite of the exquisite immunologic specificity of monoclonal antibodies, the antigenic determinants they recognize may not be exclusively confined to a particular cell type. It is therefore an absolute requirement that the cell reaction spectrum of individual monoclonals is thoroughly investigated before using them as diagnostic reagents. New monoclonal antibodies should be evaluated in parallel with existing discriminating markers both against normal and malignant hemopoietic cells and tissues. Further, since monoclonal antibodies are probes for individual and exposed antigenic determinants, the absence of binding does not necessarily indicate the absence of the structure that normally bears the determinant.

Thus, whenever possible, a battery of different, well-defined monoclonal antibodies for the definition of each particular cell type or differentiation stage should be employed. In many instances only the reaction pattern of various antibodies rather than the reactivity of a single antibody will provide reliable answers. For the study of individual cells double marker or even multimarker studies will become increasingly important.

The enthusiasm that greeted the discovery and development of hybridoma-derived monoclonal antibodies is certainly well warranted. Indeed, they are unique and elegant tools for the diagnosis and, perhaps, therapy of hemopoietic malignancies. It is, however, important that we realistically appraise the limitations currently encountered in their applicability, particularly those imposed by the lack of exclusive reactivity of monoclonal reagents developed to date with given malignant cell types. In view of the expanding research efforts in this field, our optimistic goals may eventually be realized.

ACKNOWLEDGMENTS

We wish to thank Mrs. Agathe Kaltenegger and Mrs. Susanne Beranek for their skillful technical assistance.

REFERENCES

1. ANDERSSON, L. C., C. G. GAHMBERG, L. TEERENHOVI & P. VUOPIO. 1979. Glycophorin A as a cell surface marker of early erythroid differentiation in acute leukemia. Int. J. Cancer **23:** 717–720.
2. BETTELHEIM, P., E. PAIETTA, O. MAJDIC, H. GADNER, J. SCHWARZMEIER & W. KNAPP. 1982. Expression of a myeloid marker on TdT-positive acute lymphocytic leukemia cells: Evidence by double-fluorescence-staining. Blood **60:** 1392–1396.
3. GREAVES, M. F., G. BROWN, N. T. RAPSON & T. A. LISTER. 1975. Antisera to acute lymphoblastic leukemia cells. Clin. Immunol. Immunopath. **4:** 67–84.
4. GREAVES, M. F., D. DELIA, G. JANOSSY, N. RAPSON, J. CHESSELLS, M. WOODS & G. PRENTIC. 1980. Acute lymphoblastic leukemia associated antigen. IV. Expression on non-leukaemic "lymphoid" cells. Leukemia Res. **4:** 15–32.
5. HUBER, H. 1981. Prognostic and clinical relevance of marker studies in leukemias and lymphomas (Round Table Discussion). *In* Leukemia Markers. W. Knapp, Ed.: 549–552. Academic Press. London.
6. KNAPP, W., O. MAJDIC, P. BETTELHEIM & K. LISZKA. 1982. VIL-A1, a monoclonal antibody reactive with acute lymphatic cells. Leukemia Res. **6:** 137–147.
7. KÖHLER, G. & C. MILSTEIN. 1975. Continuous cultures of fused cells secreting antibody of predefined specificity. Nature **256:** 495–497.
8. LISZKA, K., O. MAJDIC, P. BETTELHEIM & W. KNAPP. 1983. Glycophorin A expression in malignant haematopoiesis. Amer. J. Hematology. (In press.)
9. MAJDIC, O., K. LISZKA, D. LUTZ & W. KNAPP. 1981. Myeloid differentiation antigen defined by a monoclonal antibody. Blood **58:** 1127–1133.
10. STEIN, H., J. GERDES & D. Y. MASON. 1982. The normal and malignant germinal centre. Clinics in Haematology. **11:** 531–559.

S-100 Protein: A Marker for Melanocytic Tumors[a]

DUAN-REN WEN,[b,c] SUNITA BHUTA,[d]
HARVEY R. HERSCHMAN,[f] RICHARD B. GAYNOR,[e]
AND ALISTAIR J. COCHRAN[b,c,d,g]

[b]*Armand Hammer Laboratories, Division of Surgical Oncology*
[c]*Departments of Surgery,* [d]*Pathology,* [e]*Medicine, and*
[f]*Biological Chemistry*
[f]*Laboratory of Biomedical and Environmental Sciences*
Center for the Health Sciences
University of California
Los Angeles, California 90024

INTRODUCTION

Markers are available for a variety of human tumors and they have proved of value in diagnosis and in monitoring patients for recurrent disease and for the effects of therapy. Melanin could be regarded as a marker for melanocytic tumors, but unfortunately a proportion of melanomas, especially metastases, do not synthesize melanin. A specific marker for all melanomas has not yet been developed, but our recent studies, and those of others[4,6,7] suggest that S-100 protein, an interesting protein originally described in glial cells by Moore,[11] may be a candidate for this role.

S-100 protein, an acidic protein of unknown function, is named for its solubility in saturated ammonium sulfate at neutral pH.[11] It has been demonstrated in glial cells,[10–12] Schwann cells in the peripheral nervous system, and satellite cells in the peripheral ganglia.[5] It demonstrates major serological cross-reactivity among a variety of vertebrate species.[9] We have shown it to be present in melanoma cells[6,7] by complement fixation and immunofluorescent assays. To be of major practical use as a marker, S-100 protein should be detectable in routinely fixed and embedded melanoma cells. We now routinely employ an immunoperoxidase assay to examine tumors for S-100 protein content. This paper describes our initial experience with this approach.

MATERIALS AND METHODS

Sections were cut from formalin-fixed, paraffin-embedded tissues obtained from tumor patients seen at the Center for Health Sciences, University of California at Los Angeles. Tissue blocks had been stored for up to seven years. The histological diagnosis was not known to the individuals performing the assays and all slides were coded prior

[a]Supported by grants 1 R01 CA-29938-01 PTHB and CA 12582 from the Department of Health and Human Services and Contract DE-AM03-76-SF 00012 from the Department of Energy.

[g]Send all correspondence to A. J. C., University of California at Los Angeles, Factor Bldg., 9th fl., 10833 Le Conte Avenue, Los Angeles, CA 90024.

to being read, the code remaining unbroken until the readers had completed their assessment. Where the histogenesis of the tumor was debatable, results of electron microscopy and immunohistochemistry for "lymphoma markers" were withheld until the S-100 protein analysis was completed.

Immunoperoxidase Technique

Slides were deparaffinized by exposure to xylene for 10 minutes and washed for 3 minutes in 100% ethyl alcohol three times. The slides were then washed in each of the following solutions for 2 minutes: 95% ethyl alcohol, 75% ethyl alcohol, and distilled water. Endogenous peroxidases in the tissue were quenched by exposure to 3% hydrogen peroxide for 20 minutes. The slides were then washed in running tap water for 1 minute, dipped in distilled water and rinsed three times in a Tris-saline solution. The slides were exposed to 5% egg albumin for 20 minutes and then rinsed in Tris-saline. Rabbit anti-serum to bovine S-100 protein was applied for 1 hour at dilutions ranging from 1:50 to 1:200. As negative control, slides were treated with normal rabbit serum at dilutions from 1:20 to 1:100, or a first antibody was omitted. A positive control was provided by exposing slides containing a known reactive melanoma or cutaneous nerves to both antibodies and developer. Absorption of the anti–S-100 serum with S-100 protein removed all reactivity against melanoma and nerve. The slides were then rinsed three times in Tris-buffer. Swine anti-rabbit serum diluted 1:20 (DAKO, Inc., Santa Barbara, CA) was applied for 30 minutes and the slides then rinsed three times in Tris-buffer. A peroxidase–anti-peroxidase preparation diluted 1:100 (DAKO, Inc., Santa Barbara, CA) was applied for 30 minutes and the slides washed in Tris-buffer for 5 minutes. The slides were then immersed in substrate solution for 10 minutes. The substrate solution consists of 6 ml of aminoethyl carbazole (AEC) developer, 50 ml of 2.02 M sodium acetate buffer, and 0.4 ml of a 3% solution of hydrogen peroxide. The AEC developer is prepared by dissolving 167 mg of 3-amino, 9-ethyl carbazole in 100 ml of dimethyl sulfoxide. The slides were cleaned in running tap water, dipped in distilled water, counterstained by hematoxylin for 3 minutes, washed again in tap water, and finally dipped in a saturated aqueous solution of sodium bicarbonate. The slides were finally rinsed in distilled water, and a cover glass was applied.

The staining is remarkably durable, but for uniformity we try to read the preparations within 24 hours. Aminoethyl carbazole produces a red color that is readily discernible from the range of browns represented by the melanins.

Antiserum to bovine S-100 protein was prepared against antigen purified as described previously.[22] The serological specificity of our antiserum was shown by double diffusion against soluble protein extracts of a variety of tissues as antigen.[6,7]

RESULTS

We tested tissues from 43 human cutaneous malignant melanomas, 33 nevocytic nevi, and 37 human non-neural, non-melanocytic tumors (TABLE 1). All melanomas and all nevi tested contained S-100 protein, whereas only 1 of the 37 non-neural, non-melanocytic tumors tested reacted positively. Melanoma tissues were derived from 17 primary tumors, 15 lymph node metastases, and 11 visceral metastases. All stained positively regardless of the organ from which the tissue was derived.

We tested 33 melanocytic nevi, and all showed some degree of positivity regardless

TABLE 1. Proportions of Human Tumors of Various Types That Contained S-100 Protein

Tumor Types	Number Tested	Number Positive
Malignant melanoma	43	43
Primary	17	17
Metastatic	26	26
Melanotic nevi	33	33
Nevocytic	29	29
Blue	3	3
Epithelioid	1	1
Non-neural/Non-melanocytic	37	1
Ectodermal histogenesis[a]	14	1
Mesodermal histogenesis[b]	12	0
Endodermal histogenesis[c]	11	0
Granular cell myoblastoma	11	11

[a]12 breast cancer, 2 skin cancer.

[b]4 ovarian cancer, 2 prostatic cancer, 2 lymphoma, 2 sarcoma, 1 renal cancer, and 1 bladder cancer.

[c]4 gastric carcinoma, 2 lung carcinoma, 2 thyroid cancers (not C-cell cancers), 1 hepatocellular carcinoma, 1 laryngeal cancer, and 1 tongue cancer.

TABLE 2. Distribution of S-100 Protein in Various Normal Human and Murine Tissues

Tissues	Human	Mouse
Epidermal melanocytes	±	±
Schwann cells	+ +	+ +
Glial cells	+ +	+ +
Epidermal Langerhans cells	+	+
Dermal Langerhans cells	+	+
Lymph node interdigitating cells	+	+
Chondrocytes	+	+
Hepatocytes	0	0
Kupffer cells	0	0
Lymphocytes	0	0
Monocytes	0	0
Lipocytes	0	0
Endothelium	0	0
Keratinocytes	0	0
Sebaceous glands	0	0
Sweat glands	0	0
Myoepithelium	0	0
Muscle	0	0
Gut lining cells	0	0
Transitional epithelium	0	0
Lymph node sinus cells	0	0
Lymphocytes	0	0

Antiserum to S-100 tested at 1:100 dilution.

of whether they were junctional, compound, or intradermal. The tumor cells of three blue nevi showed weak staining and a single epithelioid cell nevus (Spitz tumor) was also positive.

Non-neural, non-melanocytic tumors reacted negatively with anti–S-100 serum, except for a single infiltrating ductal adenocarcinoma of breast, which stained maximally in its intraductal positions. The tumors studied were negative, regardless of histologic type, organ of growth, or germ layer of origin.

We also stained normal human and murine tissues (TABLE 2). With the exceptions of melanocytes, epidermal and dermal Langerhans cells, glial and Schwann cells, interdigitating cells of lymph nodes, and chondrocytes, all normal tissues were negative.

DISCUSSION

These data confirm the presence of S-100 protein in melanocytic tumors.[2,6,7,13] That S-100 protein may be detected in conventionally fixed and embedded cells greatly increases its potential usefulness as a marker for histogenesis. We have examined blocks stored for up to 7 years, and Nakajima *et al.*[13] reported on tissue up to 10 years old. S-100 protein expression is not directly related to the presence of melanin, as amelanotic tumor deposits may contain the protein, and there may be a negative correlation between melanin content and the S-100 protein expression. S-100 protein reactivity is certainly more difficult to detect in the presence of substantial amounts of melanin. S-100 protein is not a marker for malignancy since it may also be detected in some cells of melanocytic nevi. Expression of S-100 protein in nevocytes is not uniform, and in some nevi it is maximal in deeper, more neurotized areas. We examined three blue nevi and in contrast to others[2,20] found a proportion of the blue nevus cells to stain with anti–S-100 serum.

We did not detect S-100 protein in most non-neural, non-melanocytic tumors. While Schwannomas and neurofibromas contain S-100 protein,[8,15,19] other tumors that might be confused with amelanotic melanoma, such as renal adenocarcinoma, clear cell squamous carcinoma, and large-cell lymphomas, do not express S-100 protein. The histiocytosis X cells of Letterer-Siwe disease[16,20] and the neoplastic histiocytes of histiocytic medullary reticulosis[20] contain S-100 protein, but neither is likely to be confused clinically with melanoma. Other S-100 protein-positive tumors continue to be identified. For instance, it is now well established, and confirmed by our data, that granular cell myoblastoma contains S-100 protein. This supports a neural crest (Schwannian) histogenetic background for this previously enigmatic tumor.[1,14,18,19] Cytology, especially nuclear features, should allow separation of melanoma cells from those of granular cell myoblastomas. Although further S-100 protein containing tumors are likely to be identified, separation of usually striking melanoma cell reactivity from the total non-reaction of the cells of most common malignancies seems likely to remain clear.

Our data on normal tissues confirm the previous localization of S-100 protein in glial cells,[10] Schwann cells,[5] melanocytes, Langerhans cells,[3] chondrocytes,[17] and the interdigitating reticulum cells of lymph nodes.[21] The expression of S-100 protein in normal melanocytes is less than in nevus cells or melanoma cells, requiring a higher concentration of antiserum to detect it (1/50 relative to 1/200–1/400 for nevocytes and melanoma cells).

In practice, the technique is valuable in three separate areas. The most common is the separation of metastatic amelanotic melanoma from carcinoma and lymphoma.

Combined with electron microscopy and a battery of immunohistochemical markers for lymphoma, we have been able to confirm or refute the diagnosis of melanoma in all tissue samples now tested. A second application is to demonstrate the melanocytic nature of cutaneous tumors where the normal hallmarks of malignant melanoma are absent or where epidermis is not available to assess junctional origin. We are also attempting to identify tumor cells in tissues where small numbers of suspicious cells are seen on light microscopy. Identification of small foci of melanoma cells in lymph nodes may increase the accuracy of staging and prognostication. Perhaps the most useful example of this approach to date was a liver biopsy in which single amelanotic melanoma cells were present that were not easily discernible by conventional staining. Microscopy after staining with anti–S-100 serum made the diagnosis certain.

The detection of S-100 protein has, thus, clear applications in the investigation of melanoma patients and in basic research in the embryology and biology of anomalies of the melanocyte and other neural crest–derived cells.

REFERENCES

1. ARMIN, A. R., E. M. CONNELLY & G. ROWDEN. 1982. An immunoperoxidase investigation of S-100 protein in granular cell myoblastomas: Evidence for Schwann cell derivation. Am. J. Clin. Pathol. (In press.)
2. CLARK, H. B., D. SANTA CRUZ, B. K. HARTMAN & B. W. MOORE. 1982. S-100 protein, an immunohistochemical marker for malignant melanoma and other melanocytic lesions. Lab. Invest. **46:** 13A (abstract).
3. COCCHIA, D., F. MICHETTI & R. DONATO. 1981. Immunohistochemical and immunocytochemical localization of S-100 antigen in normal human skin. Nature **294:** 85–87.
4. COCHRAN, A. J., D-R. WEN, H. R. HERSCHMAN & R. B. GAYNOR. 1982. Detection of S-100 protein as an aid to the identification of melanomatous tumors. Int. J. Cancer. (In press.)
5. ENG, L. F., J. C. KOSEK, L. FORNO, J. DECK & J. BIGBEE. 1976. Immunohistochemistry of brain protein in fixed paraffin embedded tissue. Trans. Am. Soc. Neurochem. **7:** 211.
6. GAYNOR, R., R. F. IRIE, D. L. MORTON & H. R. HERSCHMAN. 1980. S-100 protein in cultured human malignant melanomas. Nature **286:** 400–401.
7. GAYNOR, R., H. R. HERSCHMAN, R. F. IRIE, P. JONES, D. L. MORTON & A. J. COCHRAN. 1981. S-100 protein: A marker for malignant melanomas? Lancet **1:** 869–871.
8. JACQUE, C. M., M. KUJAS, A. PROEAU, M. RAOUL, P. COLLIER, J. RACADOT & N. J. BAUMANN. 1979. GFA and S-100 protein levels as an index for malignancy in human gliomas and neurinomas. J. Natl. Cancer Inst. **62:** 479–483.
9. KESSLER, D., L. LEVINE & G. FASSMAN. 1968. Some conformation and immunological properties of bovine brain acidic protein. Biochemistry **7:** 758–764.
10. LUDWIN, S. K., J. C. KOSEK & L. F. ENG. 1976. The topographical distribution of S-100 and GFA proteins in adult rat brain. J. Comp. Neurol. **165:** 197–208.
11. MOORE, B. W. 1965. A soluble protein characteristic of the nervous system. Res. Commun. **19:** 739–747.
12. MOORE, B. W. & D. MACGREGOR. 1965. Chromatographic and electrophoretic fractionation of soluble proteins of brain and liver. J. Biol. Chem. **240:** 1647–1653.
13. NAKAJIMA, T., S. WATANABE, Y. SATO, T. KAMEYA & Y. SHIMOSATO. 1981. Immunohistochemical demonstrations of S-100 protein in human malignant melanoma and pigmented nevi. Gann **72:** 335–336.
14. NAKAZOTA, Y., J. SIHIZEKI, K. TAKAHASHI & H. YAMAGUCHI. 1982. Immunohistochemical localization of S-100 protein in granular cell myoblastoma. Cancer **49:** 1624.
15. PFEIFFER, S. E., P. L. KORNBLITH, H. L. CARES, J. SEALES & L. LEVINE. 1972. S-100 protein in human acoustic neuromas. Brain Res. **41:** 187–193.
16. ROWDEN, G., E. M. CONNELLY & R. K. WINKELMANN. 1982. Cutaneous histiocytosis X: The presence of S-100 protein and its use in diagnosis. (Submitted for publication.)

17. STEFANSSON, K., R. L. WOLLMAN & B. W. MOORE. 1982. S-100 protein in human chondrocytes. Nature **295:** 63–64.
18. STEFANNSON, K. & R. L. WOLLMAN. 1982. S-100 protein in granular cell tumors (granular cell myoblastomas). Cancer **49:** 1834–1838.
19. STEFANSSON, K., R. WOLLMAN & M. JERKOVIC. 1982. S-100 protein in soft tissue tumors derived from Schwann cells and melanocytes. Am. J. Pathol. **106:** 261–268.
20. WATANABE, S., T. NAKAJIMA, Y. SHIMOSATO & T. ISE. 1981. A case report of histiocytic medullary reticulosis defined as a neoplasm of T-zone histiocytes. Jpn. J. Clin. Oncol. **11:** 411–418.
21. WATANABE, S., T. NAKAJIMA, Y. SHIMOSATO, K. SHIMAMURA & H. SAKUMA. 1982. T-zone histiocytes with S-100 protein: development and distribution in human fetuses. Acta Pathol. Jpn. (In press.)
22. ZUCKERMAN, J., H. R. HERSCHMAN & L. LEVINE. 1970. Appearance of a brain specific antigen, S-100 during fetal development. J. Neurochem. **17:** 247–251.

Basement Membrane-Producing Tumors as Antigenic Substrate for the Demonstration of Anti–Basement Membrane Antibodies[a]

G. WICK AND V. MUNRO[b]

Institute for General and Experimental Pathology
University of Innsbruck, Medical School
Innsbruck, Austria

W. GEBHART

Second Clinic for Dermatology
University of Vienna Medical School
Vienna, Austria

R. TIMPL

Max-Planck-Institut für Biochemie
Martinsried
Munich, Federal Republic of Germany

INTRODUCTION

Autoantibodies (AAb), which react with basement membranes (BM), occur in the sera of human patients with different autoimmune diseases. Such BM-AAb are characteristic for Goodpasture's disease, affecting lung and kidney, and for bullous pemphigoid, a condition leading to blister formation in skin and oral mucosa. In both instances immunohistochemical techniques are of prime importance for the demonstration of these AAb and thus for diagnosis of the respective diseases. Interestingly, BM-AAb in the sera of patients with Goodpasture's disease react only with BM of renal glomeruli and lung alveoli, and not with those of squamous epithelia, while the reverse staining pattern is observed with sera from patients with bullous pemphigoid.[1] This type of reactivity indicates that constituents common to all BM, such as type IV collagen[7] and laminin—another large well-defined non-collagenous BM-specific glycoprotein[15]—are unlikely to be the candidates for antigens homologous for the BM-AAb in these cases. It rather seems likely that noncollagenous and non-laminin glycoprotein fractions, which are specific for BM at certain sites in the body (such as kidney or skin, respectively), will contain the relevant antigen(s). The exact nature of the BM-antigens involved in Goodpasture's syndrome and bullous pemphigoid is still unknown, or at least subject to controversy, although several authors claim to have identified the autoantigen playing a role in bullous pemphigoid as a non-collagenous

[a]Part of this work was supported by grant from the Austrian Research Council, Project Nr.P 4879.
[b]Present address: Department of Pathology, St. Vincent's Hospital, Darlinghurst, Sidney, Australia.

protein that is present in the zona lucida of BM epithelia.[3,12] The "Goodpasture antigen" has also not yet been identified. Mahieu *et al.*[8] presented data that speak for a collagenous nature of this antigenic material while Marquardt *et al.*[9] rather favor the possibility that the antigen(s) consist(s) of one or several glycoproteins.

These unsolved questions reflect the general problems encountered in trying to isolate the chemical components of BM derived from authentic tissues, such as kidney, lung, skin, etc. The present communication concerns our attempts to find other sources for the demonstration, identification, and characterization of BM constituents.

GOODPASTURE'S SYNDROME

Orkin *et al.*[11] have described a transplantable, BM-producing mouse tumor, the EHS sarcoma, from which collagen type IV and another large non-collagenous glycoprotein, laminin (chains of 220 and 440 Kd), can be isolated without proteolytic treatment (FIGURE 1). Furthermore, a biochemically still not extensively characterized pool of other BM-specific non-collagenous, non-laminin glycoproteins has been extracted from this tumor but has not yet been processed for more detailed analyses. Specific rabbit antibodies against type IV collagen[14] and laminin[15] purified from the EHS tumor have been characterized by a variety of methods including indirect immunofluorescence (IIF).[16,18] Such antibodies react with all BM when tested on frozen sections of mouse tissues, since collagen type IV and laminin are ubiquitous constituents of BM throughout the body (FIGURE 2a). Furthermore, these antibodies show complete cross-reaction with human BM in all tissues tested sofar (FIGURE 2b).

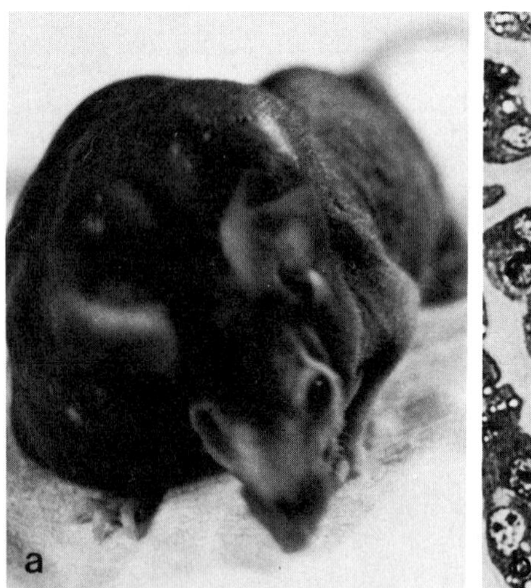
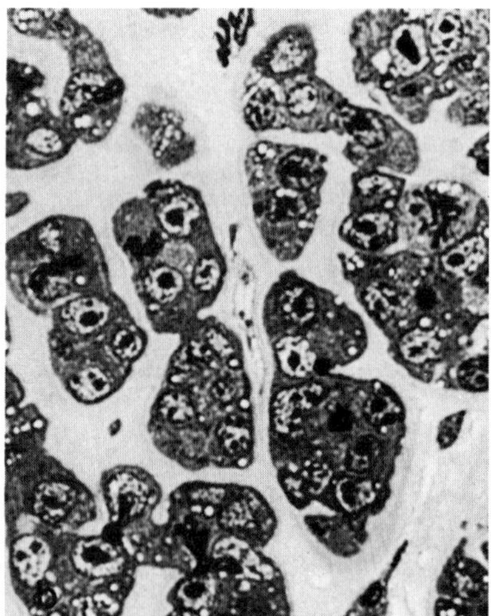

FIGURE 1. (a) Macroscopic appearance of an EHS tumor growing for three weeks in a nude mouse. Tumors also grow in conventional mice with different genetic backgrounds; a nude mouse was selected as a recipient in this case for better demonstration of tumor size. (b) Hematoxylin-eosin–stained histologic section of EHS tumor showing abundant strands of hyaline, basement membrane–like matrix (bright areas). Photograph courtesy of Dr. G. R. Martin. Paraffin section, ×400.

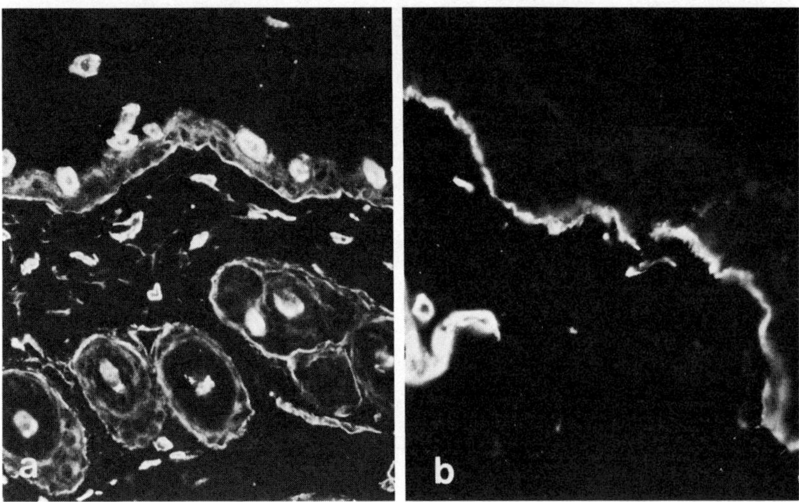

FIGURE 2. Indirect immunofluorescence tests on frozen, unfixed tissue sections using specific rabbit antibodies to basement membrane components isolated from the EHS tumor. (a) Anti-laminin on mouse skin, ×200. (b) Anti-type IV collagen on human skin, ×400. Note staining of all (epithelial, vascular, skin appendages, etc.) basement membranes and strong interspecies cross-reaction.

In addition to basic investigations on the distribution of BM components at various sites, we also use these antibodies—together with antibodies that specifically recognize interstitial types of collagen (types I, II, and III) (FIGURE 3a) and procollagen—for diagnostic purposes.[17] Thus, it was shown that BM in the skin of mice and humans with protoporphyria contain increased amounts of type IV collagen and laminin that express normal antigenic determinants.[19] In all normal and pathological instances type IV collagen and laminin show complete co-distribution. The only exception was observed in fibrotic and cirrhotic human livers, where strong perisinusoidal staining with anti-type IV collagen preceded that with anti-laminin during the course of the disease. The most impressive finding in the context of the present contribution, however, was the observation that sera from patients with Goodpasture's syndrome react with the BM matrix of the EHS tumor (FIGURE 3b). This material therefore seems to contain the BM antigen(s) that bind to BM-AAb in these sera. Since type IV collagen and laminin also occur in BM that do not stain with sera from Goodpasture's syndrome patients (skin, oral mucosa, etc.), it was not unexpected that radioimmunological analyses led to the exclusion of these components as the antigen(s) in question.[20] We are, therefore, now in the process of trying to identify one (or several) of the above-mentioned non-collagenous, non-laminin glycoproteins as the antigen(s) homologous for BM-antibodies in the sera of Goodpasture's syndrome patients. Cross-absorption experiments with mouse or human kidney and EHS tumor homogenates support the notion that the tumor contains the same BM antigens as the authentic tissues: both murine and human kidney and the tumor material were able to absorb out BM antibodies from the sera patients with Goodpasture's syndrome. As a matter of fact, we now use frozen, unfixed sections of this tumor regularly as an antigenic substrate, in addition to kidney and lung, for the diagnostical evaluation of Goodpasture's syndrome sera. Our final goals are to develop a simple and rapid immunoassay, such as radio-, and enzyme-, or immunofluoroassay, for routine diagnosis of the Goodpasture's syndrome; and to possibly combine the current therapy of plasmapheresis with an affinity chromatography step to remove the pathogenic BM-AAb only.

BULLOUS PEMPHIGOID

Similar problems to those that have just been discussed for the identification of "Goodpasture antigens" were encountered during attempts to define BM components that react with sera of patients with another human autoimmune disease, viz. bullous pemphigoid.[1] Diaz *et al.*[3,4] reported the isolation of bullous pemphigoid antigens from extracts of human skin and also from the urine of a patient with this disease. The

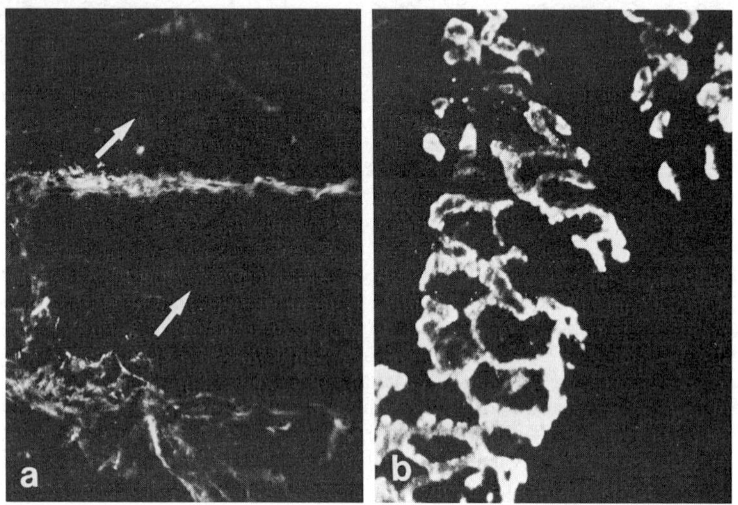

FIGURE 3. Indirect immunofluorescence tests on frozen, unfixed sections of EHS tumor. (a) Rabbit anti-type III collagen antibodies staining the interstitium only, but leaving the tumor mass (arrows) unstained. (b) Reaction of the serum of a patient with Goodpasture's syndrome with the tumor cell matrix. ×400.

molecular weight of this material was estimated to be approximately 20,000 with some larger aggregates. In contrast, Stanley *et al.*[12] purified an antigen that is synthesized by human epidermal cells and consists of two polypeptide chains linked by disulfide bonds and has a molecular weight of approximately 220,000. For the same reasons discussed above for Goodpasture's syndrome, it is not surprising that these authors did not find any reaction of antibodies to laminin or fibronectin with these antigens. Since fibronectin also has a molecular weight of 220,000 and laminin consists of polypeptide chains of about 220,000 and 440,000, the latter findings can be considered as very crucial controls. Nevertheless, the question of the exact nature of the bullous pemphigoid antigen(s) is still not settled, one reason being the use of different, very heterogeneous sources of tissues for their preparation and the rather low yield of final purified material thereof. Similar to the situation with Goodpasture's syndrome, a tumor cell line producing large amounts of bullous pemphigoid antigen(s) would be an ideal source for the definitive clarification of this tissue. We analyzed human cylindroma in this respect and found evidence that it may fulfill these requirements.[5] The cylindroma is a rather rare condition in which multiple skin tumors of considerable size may be formed. Histologically the tumor cells are surrounded by a BM matrix that has long been identified as PAS-positive "hyalinous" bands.[2,10] Furthermore, this material is also found as hyalinous granules within the tumor cells (FIGURE 4).

Electron microscopic evaluation of cylindromas shows characteristic features of BM in the matrix surrounding the tumor cells and the hyalinous granules within the cytoplasm. Directly adjacent to the surface of the tumor cells a lamina rara can be demonstrated followed by a lamina densa showing an extensive increase, protrusions, branching, and duplications. This material accounts for a significant portion of the total tumor mass. Between the various layers of the lamina densa, numerous fibrils can be found that fulfill the ultrastructural characteristics of anchoring fibrils. Towards the surrounding interstitial connective tissue the hyalinous material is sharply demarcated by collagen fibers and the cytoplasmic extension of fibroblasts.

IIF tests on frozen unfixed sections of cylindroma revealed a strong positive reaction with all sera from patients with bullous pemphigoid while no staining was observed in sera from patients with pemphigues vulgaris or Goodpasture's syndrome. Further analyses on the nature of the hyalinous intra- and extra-tumorous material were performed with guinea pig antibodies to type IV collagen, and rabbit antibodies to laminin, as well as antibodies to the interstitial collagen types I and III. FIGURE 5a shows the characteristic staining of interstitial collagen type I fibers sparing the BM structures around the tumor cells and the intracellular hyalinous globules. FIGURE 5b shows that guinea pig anti-type IV collagen antibodies bind to the central portion of both the pericellular and intracellular hyalinous material. Double-staining experiments using FITC anti-guinea pig and TRITC anti-rabbit IgG conjugates made it possible to visualize the expected co-distribution of collagen type IV and laminin at these locations. Sera from patients with bullous pemphigoid display a unique staining pattern, viz. a sharp and continuous line at the junction between the hyalinous BM-like extracellular and intracellular structures and the surface of the tumor cells (FIGURE 5c).

Attempts to block the binding of antibodies from bullous pemphigoid sera to cylindroma sections by pretreatment with anti-type IV collagen or anti-laminin, and vice versa, failed. Thus, the human cylindroma does contain the "bullous pemphigoid antigen(s)" in large quantities and may be a useful source for its isolation and subsequent biochemical characterization.

No attempts have been made so far to propagate cylindroma cells as a continuous line or to transplant them into nude mice, but further steps in this direction may overcome the limitations set by the low incidence of this disease.

Finally, we have also excluded the possibility that binding of antibodies to BM-like structures of the murine EHS tumor or the human cylindroma may have occurred via an Fc receptor. This was recently done by means of the mixed hemagglutination technique using rabbit IgG-coated sheep red blood cells as indicator cells. Adherence of these indicator cells to BM structures was never observed in both of the tumors, but rather to the interstitial areas only.

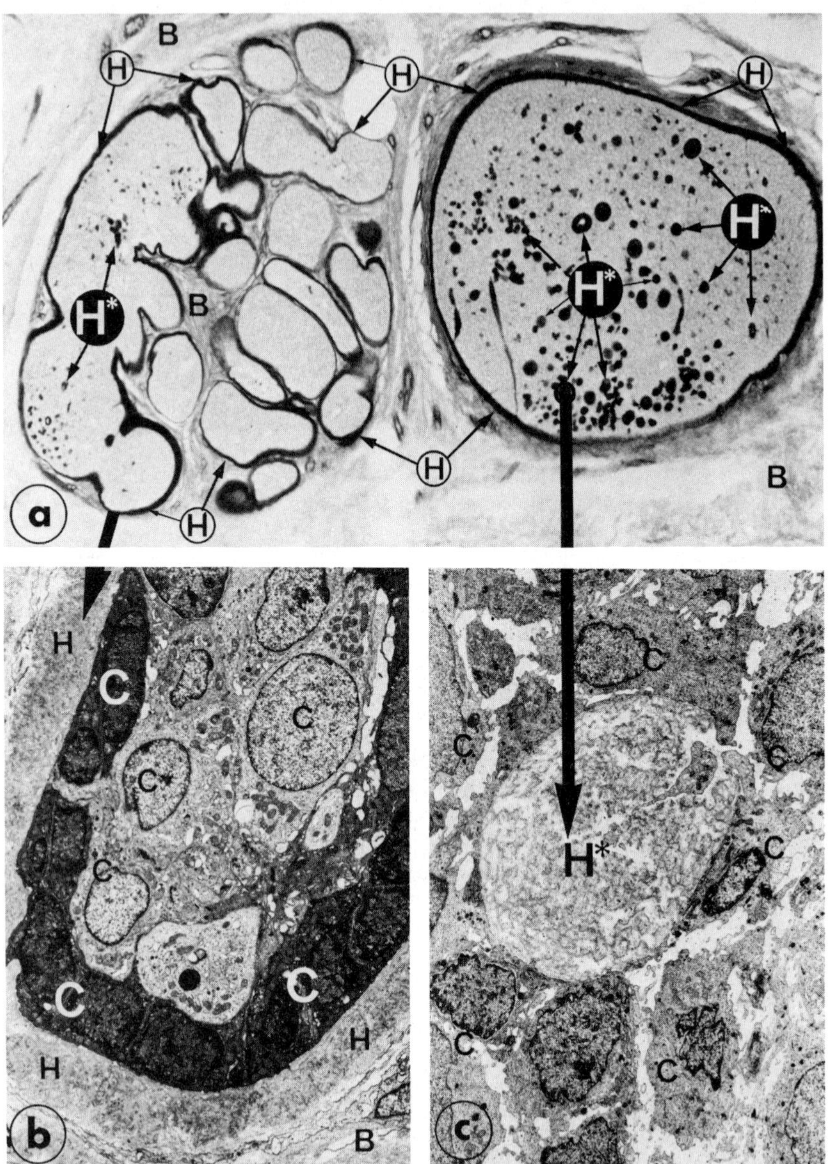

FIGURE 4. (a) Histologically, cylindroma is characterized by a pronounced PAS-positive "hyaline" band (H) surrounding the individual dermal tumor islands. A variable amount of PAS-positive "droplets" (H*) is also present within the tumor cell nests. Paraffin section, PAS-stain, ×180. (b) Electron microscopically, the undifferentiated basaloid tumor cells of cylindroma (C) exhibit a 10–15 μm broad marginal zone corresponding to the "hyaline" band (H). Epon section, ×2,000. (c) The "hyaline" droplets (H*) within the tumor are also surrounded by basaloid cylindroma cells (C) and consist of inhomogeneous electron-dense and electron-translucent material. Epon section, ×2,250.

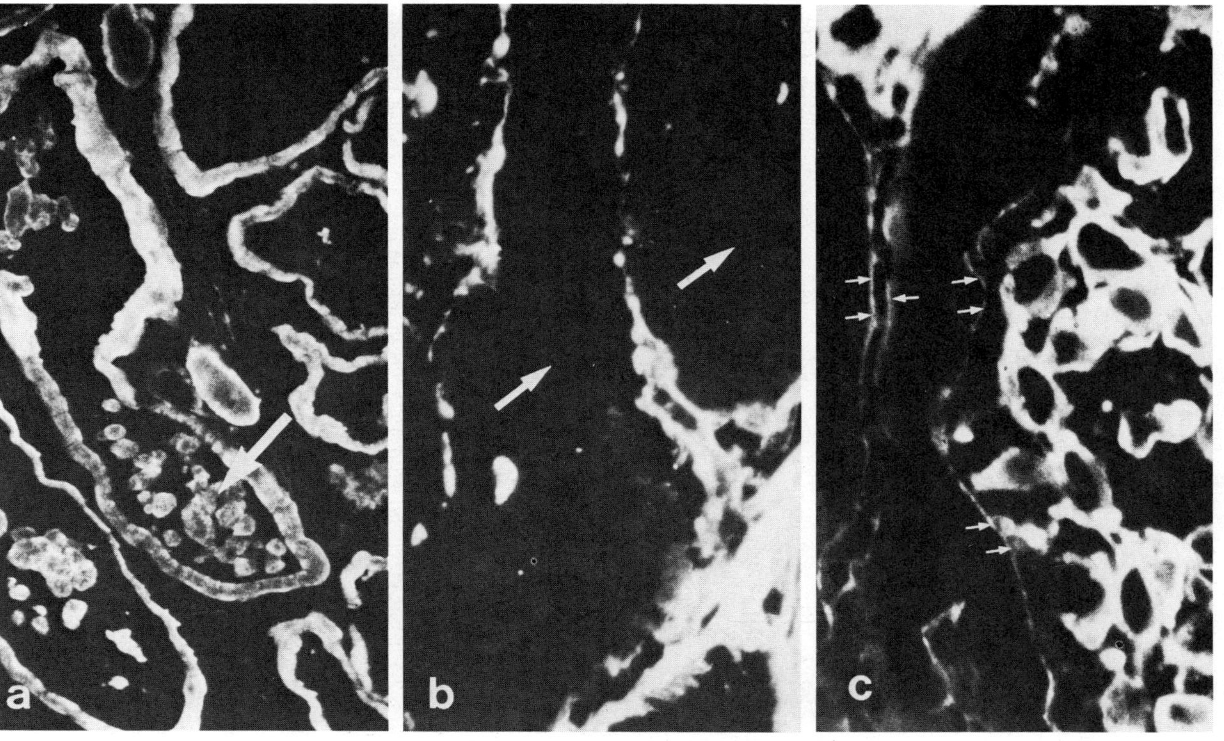

FIGURE 5. Indirect immunofluorescence tests on frozen, unfixed sections of cylindroma. (a) Rabbit anti-type IV collagen antibodies binding to the pericellular basement membrane and the globular structures within the tumor cells (large arrow). (b) Rabbit anti-type I collagen antibodies staining the interstitial areas only, but not the tumor cells (arrows) or basement membranes. (c) Serum of a patient with bullous pemphigoid staining a thin layer of the basement membrane near the surface of the tumor cell aggregates (small arrows; zona lucida?), the individual tumor cells and the intracellular hyaline material. ×500.

[NOTE ADDED IN PROOF: Recent, still preliminary, data suggest that the globular, non-collagenous portion of the Type IV collagen molecule is the "Goodpasture antigen."]

REFERENCES

1. BEUTNER, E. H., T. P. CHORZELSKI & S. F. BEAN. 1979. Immunopathology of the Skin. 2nd edit. John Wiley and Sons. New York.
2. BRAUN-FALCO, O. 1955. Histochemische Untersuchungen zur Charakterisierung des Hyalin von Spiegler'schen Tumoren. Arch. Klin. Exp. Dermatol. **202:** 56–68.
3. DIAZ, L. A., N. J. CALVANICO, T. B. TOMASI & R. E. JORDON. 1977. Bullous pemphigoid antigen: Isolation from normal human skin. J. Immunol. **118:** 455–460.
4. DIAZ, L. A., H. PATEL & N. J. CALVANICO. 1979. Bullous pemphigoid antigen. II. Isolation from the urine of a patient. J. Immunol. **122:** 605–608.
5. GEBHART, W., E. M. KOKOSCHKA & G. WICK. 1973. The cylindroma: A model for human epithelial basement membranes. J. Invest Dermatol. **64:** 386.
6. HAHN, E., G. WICK, D. PENCEV & R. TIMPL. 1980. Distribution of basement membrane proteins in normal and fibrotic human liver: Collagen type IV, laminin and fibronectin. Gut **21:** 63–71.
7. KEFALIDES, N. A. 1973. Structure and biosynthesis of basement membranes. Int. Rev. Conn. Tiss. Res. **6:** 63–104.
8. MAHIEU, P. M., P. H. LAMBERT & G. R. MAGHUIN-ROGISTER. 1973. Primary structure of a small glycopeptide isolated from human glomerular basement membrane and carrying a major antigenic site. Eur. J Biochem. **40:** 399–606.
9. MARQUARDT, H., C. B. WILSON & F. J. DIXON. 1973. Human glomerular basement membrane. Selective solubilization with chaotropes and chemical and immunologic characterization of its components. Biochemistry **12:** 3260–3264.
10. MUNGER, B. L., J. H. GRAHAM & E. B. HELWIG. 1962. Ultrastructure and histochemical characteristics of dermal eccrine cylindroma (turban tumor). J. Invest. Dermatol **39:** 577–585.
11. ORKIN, R. W., P. GEHRON, B. McGOODWIN, G. R. MARTIN, T. VALENTINE & R. SWARM. 1977. A murine tumor producing a matrix or basement membrane. J. Exp. Med. **145:** 204–220.
12. STANLEY, J. R., P. HAWLEY-NELSON, S. H. YUSPA, E. M. SHEVACH & S. T. KATZ. 1981. Characterization of bullous pemphigoid antigen—a unique basement membrane protein of stratified squamous epithelia. Cell **24:**897–904.
13. THUNOLD, S., O. TÖNDER & J. N. WIIG. 1973. Localization in mouse lymphoid tissue of receptors for immunoglobulin. Scand. J Immunol **3:** 135–145.
14. TIMPL, R., G. R. MARTIN, P. BRUCKNER, G. WICK & H. WIEDEMANN. 1978. Nature of the collagenous protein in a tumor basement membrane. Eur. J. Biochem. **84:** 43–52.
15. TIMPL, R., H. RHODE, P. GEHRON-ROBEY, S. I. RENNARD, J. M. FOIDART & G. R. MARTIN. 1979. Laminin-aglycoprotein from basement membranes. J. Biol. Chem. **254:** 9933–9937.
16. TIMPL, R., H. RHODE, G. WICK, P. GEHRON-ROBEY, S. I. RENNARD, J. M. FOIDART & G. R. MARTIN. 1979. Characterization of laminin, a major glycoprotein of basement membranes. Coll. Internat. CNRS; Biochimie des Tissues Conjonctifs Normaux et Pathologiques. **287:** 225–228.
17. TIMPL, R., G. WICK & S. GAY. 1977. Antibodies to distinct types of collagens and procollagens and their application in immunohistology. J. Immunol. Methods **18:** 165–182.
18. WICK, G., R. W. GLANVILLE & R. TIMPL. 1979. Characterization of antibodies to basement membrane (type IV) collagen in immunohistological studies. Immunobiology **156:** 372–381.
19. WICK, G., H. HÖNIGSMANN & R. TIMPL. 1979. Immunofluorescence demonstration of type IV collagen and a non-collagenous glycoprotein in thickened vascular basal membranes in protoporphyria. J. Invest. Dermatol. **73:** 335–338.
20. WICK, G. & R. TIMPL. 1980. Study on the nature of the Goodpasture antigen using a basement membrane-producing mouse tumour. Clin. Exp. Immunol. **39:** 733–738.

Recent Advances in the Study of Autoimmune Endocrine Diseases by the Use of Immunofluorescence

G. F. BOTTAZZO,[a] W. A. SCHERBAUM,
AND T. HANAFUSA

Department of Immunology
The Middlesex Hospital Medical School
London WIP 9PG, England

INTRODUCTION

The indirect immunofluorescence test (IFT) is widely used for the diagnosis of autoimmune diseases and offers a powerful tool for the study of new antibody specificities in disorders of unknown origin. Endocrine autoimmunity is not uncommon in the general population and the spectrum of organ-specific autoimmune endocrinopathies is given in TABLE 1. Women are more often affected than men and the prevalence increases with age in both sexes.

Clinical expression of an autoimmune disease may occur years after the appearance of the relevant autoantibodies in the serum and patients with symptoms usually constitute a small proportion of subjects in whom these serological markers can be detected. Prospective studies are now under way to follow the natural history of autoimmune diseases and the various immunological phenomena occurring in their preclinical stages. Screening for autoantibodies in cases of suspected endocrine disorders and in genetically predisposed members of affected families will help to make early diagnoses and to institute more rational treatment. The four-layer doublefluorochrome test is now in use. Briefly, the IFT sandwich with the patient's serum is combined with the animal antihormone sandwich as used in immunocytochemistry but omitting the fixation. The two sandwiches are applied successively to the same section and the anti-Ig conjugates are, respectively, green (FITC) and red (rhodamine), allowing direct visual assessment of whether the patient's serum and the rabbit anti-hormone stain the same or different cells. This extension of the conventional IFT permits characterization of antibodies reacting with single cell types in complex endocrine organs such as the pancreas, gut, pituitary, and hypothalamus.

In another approach, immunofluorescence studies on cultured endocrine cells have highlighted the importance of surface-reactive immunoglobulins present in the sera of patients with endocrine autoimmune diseases. The complex anatomical structure of some endocrine organs has required more sophisticated culture procedures that now allow a precise localization of the autoantibodies reacting with the corresponding antigens.

---

[a]Address correspondence to: G.F.B., Department of Immunology, The Middlesex Hospital Medical School, Arthur Stanley House, 40–50 Tottenham Street, London WIP 9PG, U.K.

Abbreviations: IFL = immunofluorescence, FCS = fetal calf serum, ICA = islet cell (pancreatic) antibodies, CF-ICA = complement-fixing ICA, PRL = prolactin, GH = growth hormone, DI = diabetes insipidus, SON = supraoptic nuclei, and PVN = paraventricular nuclei.

TABLE 1. Organ-specific Autoimmune Endocrinopathies

Thyroid	Graves' thyrotoxicosis, endocrine exophthalmos
	Hashimoto's thyroiditis
	Primary myxoedema
Stomach	Pernicious anemia
	Some forms of fundal and antral gastritis
Pancreas	Type IA and IB diabetes mellitus
Adrenal	Addison's disease
Gonads	Hypergonadotrophic hypogonadism
Parathyroid	Some cases of first degree hypoparathyroidism
Pituitary	Possibly some cases of GH deficiency and hyper- or hypoprolacti-nemia
Hypothalamus	Central diabetes insipidus

Endocrine autoimmunity has been extensively reviewed[10] and this paper will focus on recent developments in the field.

THYROID

The first demonstration of thyroid autoantibodies in Hashimoto's thyroiditis[19] opened a new venue in the investigation of autoimmune phenomena in human disease and thyroid autoimmunity is still the most extensively investigated subject. Thyroglobulin and thyroid microsomal antibodies are important serological markers for the diagnosis of thyroid autoimmune disorders and the hemagglutination technique using commercial kits has now replaced the IFT for diagnostic purposes.

Antibodies reacting with the surface of thyroid cells in suspension were first described by Fagraeus and Jonsson.[11] In an extension of this work, Khoury et al.[14] were able to demonstrate complete correspondence between surface IFT staining of thyroid monolayers and 'microsomal' hemagglutination titers, thus indicating that the intracytoplasmic-membrane antigen is also expressed on the cell surface. It could be further shown that these sera gave a positive surface reaction confined to the 'microvillar' portion of the apical pole of the cells and that immunoglobulins were sometimes already fixed *in vivo* on the microvillar border in some acini.

These results have been confirmed using human thyrotoxic glands removed from patients with Graves' disease. The specimens were gently digested in order to obtain semi-disrupted follicles and stained by direct IFT. Immunoglobulins were attached *in vivo* only on the internal surface of the follicles. The external wall was not stained and the addition of a microsomal antibody-positive serum increased the immunofluorescence on the microvilli-rich pole of the follicles.

Cells in monolayer cultures may react differently from those normally organized *in vivo* and it is for this reason that a more physiological approach is required, i.e. the use of intact human thyroid follicles. Following experiments carried out with rat thyroid cells,[17] it was also possible to show that human thyroid cells cultured in a low concentration (0.5%) of fetal calf serum (FCS) reacquire their 'physiological' tridimensional follicular structure. If the same culture system is implemented with high concentrations (10%) of FCS, a gradual change in the polarity of the cells is observed. This reversed polarity was confirmed by electron microscopy, which showed the microvillar redistribution on the outer surface of reconstituted follicles.[13] With this culture system, it is now possible to examine the effect of autoreactive lymphocyte subsets and of microsomal and other thyroid antibodies upon various parameters of thyroid cell metabolism, which are only maintained when thyroid cells are grown in their normal organized position.

PANCREAS

Specific antibodies to pancreatic islet cells (ICA) were first detected by IFT in diabetic polyendocrine patients.[4] They react with all cell types of the islet. As the islet antigens have only been partially characterized,[1] cytoplasmic ICA are still detected by standard IFT and anti-human IgG is applied as a conjugate (ICA-IgG). Group O human pancreas is the substrate of choice and unfixed tissue should be used because fixatives such as Bouin's may destroy the putative autoantigens or reveal new antigenic sites and produce doubtful reactions on the islets.

ICA are present in over 80% of patients with Type I diabetes at the time of diagnosis. The incidence decreases to 50% within two months and to 15–20% after the first five years. Patients in whom these antibodies tend to persist often suffer from other endocrine autoimmune disease.[9]

The application of an immunofluorescent complement-fixation test (ICFT) has shown that 50–55% of ICA-positive sera fix complement (CF-ICA).[3] Some CF-ICA selectively stain beta cells and these specificities seem to be of more predictive value for the future development of overt disease in predisposed individuals.[6]

In a prospective family study comprising 700 individuals, 20 (7%) out of 313 parents and 34 (12%) out of 288 siblings of Type I diabetics had ICA-IgG in their sera. The striking observation was that all but one of seven relatives who developed clinical diabetes during a four-year follow-up period also had persistent CF-ICA (TABLE 2). These antibodies were detected during the entire follow-up period.[12]

PITUITARY

Antibodies to pituitary prolactin (PRL)[7] and growth hormone (GH) cells[5] are detected by the IFT where they give the typical granular cytoplasmic staining on pituitary sections. The endocrine nature of these cells is revealed by applying specific anti-hormone sera in the four-layer doublefluorochrome IFT.

PRL-cell antibodies occur in 8% of patients with endocrine autoimmune disorders, with or without partial defects of pituitary function, but they are not found in cases of

TABLE 2. Development of Diabetes Mellitus on Follow-up in Seven Relatives of DM Children: Genetic and Immunological Features

Relation to Proband	HLA Haplotype Concordance	Age at Diagnosis	Islet Cell Abs.[a]		Minimum Duration Latency (months)
			IgG	CF	
Sister	2	18	+	+	9
Brother	1	5	+	+	28
Brother	1	20	+	−	23
Sister	2	21	+	+	30
Brother	2	9	+	+	36
Father		47	+	+	4
Mother		46	+	+	22

[a]ICA was tested with undiluted serum by standard indirect immunofluorescence on unfixed group O human pancreas. CF-ICA was performed on parallel cryostat sections with addition of normal fresh human serum as a source of complement, using FITC anti-C3 conjugate.

severe panhypopituitarism. These markers might indicate the presence of an active 'hypophysitis' process. As already mentioned, prospective studies in diabetic families have provided strong evidence for a long prediabetic interval between the initiation of beta cell destruction, the development of biochemical decompensation, and the onset of overt diabetes.[12] This long, symptom-free period could be a time when abnormalities in other endocrine glands might also occur.

A number of reports have described children with unusually tall stature presenting with Type I diabetes, indicating that pituitary and possibly other endocrine factors may play a fundamental role in the pathogenesis of the disease. Therefore pituitary antibodies were tested in patients with recently diagnosed Type I diabetes and in their first-degree relatives (TABLE 3); 17% of the ICA-positive, newly diagnosed diabetics had pituitary antibodies and in contrast to the polyendocrine cases, they reacted with

TABLE 3. Prevalence of Pituitary Cell Antibodies in Type I Diabetes and Selected Relatives

			Characterization by Double Immunofluorescence Technique				
Groups Tested	Number Tested	Total Positive[d] (%)	Single Cells PRL	GH	LH	Multiple Cells PRL + GH ?Others	Unidentified Cells ?FSH ?TSH ?Others
Type I juvenile IDDM							
Duration up to 1 yr.							
ICA[a] positive	51	10 (16.6)	1	0	0	5	4[b]
ICA negative	10	0	0	0	0	0	0
Duration 3–23 yrs[c]							
ICA positive	24	1 (2.2)	0	0	0	1	0
ICA negative	24	0	0	0	0	0	0
First-degree relatives[c]							
ICA positive	33	12 (36.4)	1	0	0	4	7[b]
ICA negative	30	1 (3.3)	1	0	0	0	0
Type Ib							
polyendocrine IDM	117	9 (7.7)	8	0	0	1	0
Normal controls	48	0	0	0	0	0	0

[a]ICA = Islet-cell antibodies. [b]One serum also showed PRL cells. [c]Probands and relatives from Barts-Windsor IDDM family study. [d]Pituitary antibodies were detected on unfixed sections of human pituitary glands obtained at hypophysectomy with undiluted serum using specific anti-Ig G, A, and M in the standard indirect immunofluorescence test.

more than one cell type. Surprisingly, these antibodies were also positive in 36% of genetically predisposed ICA-positive relatives, half of whom became diabetic during a four-year follow-up period.[15] The sera of longstanding insulin-independent diabetics were positive in only 2% of the cases and pituitary antibodies were not found in normal controls. These results suggested that in some cases temporary autoimmune processes may involve simultaneously the pituitary gland and the endocrine pancreas, a phenomenon also observed in virally induced diabetes in animals.[18] Pituitary antibodies of the multiple cell IFT pattern could indicate the presence of an underlying stimulating immune process resembling that of Graves' thyrotoxicosis. If this hypothesis could be confirmed, the long latency period in Type I diabetes could be explained by a concomitant presence of destructive and reparatory factors. If stimulating antibodies did exist, would they act directly on the pituitary cells or through the hypothalamus? It could be envisaged that such antibodies directed to hypothalamic cell receptors might enhance the output of the peptide that was recently shown to stimulate insulin secretion.[2]

HYPOTHALAMUS

Central diabetes insipidus (DI) is a disorder characterized by a shortage of vasopressin hormone, which is mainly produced in the supraoptic (SON) and paraventricular nuclei (PVN) of the hypothalamus. About 30–50% of the cases remain of unknown etiology and, by analogy with other endocrine disorders, autoimmunity may play a role in certain conditions that had previously been considered idiopathic in origin. The possibility that an autoimmune variety of DI might exist is supported by some case reports where idiopathic central DI was found to be associated with other autoimmune endocrine disorders.

The indirect IFT was applied on unfixed cryostat sections of human fetal and fresh post-mortem hypothalamus. Oxytocin cells were differentiated from vasopressin cells by specific antihormone sera raised in rabbits. About 30% of patients with idiopathic central DI reacted with secretory cells of the SON and PVN. The four-layer doublefluorochrome IFT with the patient's serum and fluoresceinated anti-human immunoglobulin in the first layers and anti-vasopressin serum and rhodaminated anti-rabbit immunoglobulin in the next layers revealed that the autoantibody reaction was directed against vasopressin cells. The identity of the reacting cells could be proven by double-exposure photography, which showed a yellow color when the same section was photographed sequentially with a red and a green filter.[20] The same procedure with anti-oxytocin serum showed that the sera of some patients also reacted with oxytocin cells. These new antibodies were never detected in cases of 'symptomatic' DI with a clearly established cause or in the 143 control patients tested so far.

FUTURE PROSPECTS

Probably the most intriguing challenge of the advances in the study of endocrine autoimmunity is the rapidly developing area of autoimmune phenomena to pituitary and hypothalamus and their possible implication in new disease states. A blood-brain barrier does not exist in most parts of the hypothalamus[8] and antibodies can reach this area of the central nervous system. There are close interactions between central and peripheral endocrine organs as well as between closely located endocrine areas, e.g. the entero-insular axis. It may well be that the hypothalamus also regulates the function of peripheral endocrine organs. It has recently been shown that hypothalamic factors exert a direct effect on the physiological regulation of insulin secretion.[2] In this context it is of interest that autoantibodies to gastric inhibitory peptide (GIP) cells and secretin cells of the duodenum are present in sera of Type I and II diabetics and these specificities may underlie a more profound immunological defect in the pathogenic events leading to beta cell damage.[16]

Pituitary autoantibodies may recognize receptor antigens also present in the hypothalamus. Preliminary experiments showed that 'multiple cell' positive sera react with the cytoplasm of small cells scattered in the ventromedial and dorsolateral hypothalamus, distinct from vasopressin and oxytocin cells. These cells have not yet been characterized, but if the results are confirmed a new experimental approach becomes available and its future implications will certainly shed new light on the fine interactions between different endocrine systems.

REFERENCES

1. BAEKKESKOV, S., J. H. NIELSEN, B. MARNER, T. BILDE, J. LUDVIGSSON & A. LERNMARK. 1982. Autoantibodies in newly diagnosed diabetic children immunoprecipitate human pancreatic islet cell proteins. Nature **298:** 167–169.
2. BOBBIONI, E. & B. JEANRENAUD. 1982. Effect of rat hypothalamic extract administration on insulin secretion *in vivo*. Endocrinology **110:** 631–636.
3. BOTTAZZO, G. F., B. M. DEAN, A. N. GORSUCH, A. G. CUDWORTH & D. DONIACH. 1980. Complement-fixing islet-cell antibodies in Type I diabetes: possible monitors of active beta cell damage. Lancet **i:** 668–672.
4. BOTTAZZO, G. F., A. FLORIN-CHRISTENSEN & D. DONIACH. 1974. Islet-cell antibodies in diabetes mellitus with autoimmune polyendocrine deficiencies. Lancet **ii:** 1279–1283.
5. BOTTAZZO, G. F., C. MCINTOSH, W. STANFORD & M. PREECE. 1980. Growth-hormone-cell antibodies and partial growth-hormone deficiency in a girl with Turner's syndrome. Clin. Endocrinol. **12:** 1–9.
6. BOTTAZZO, G. F., R. MIRAKIAN, B. M. DEAN, J. M. MCNALLY & D. DONIACH. 1982. How immunology helps to define heterogeneity in diabetes mellitus. *In* Genetics of Diabetes Mellitus. 2nd edit. P. B. Tattersall & J. K. Koberling, Eds.: 79–90. Academic Press. London.
7. BOTTAZZO, G. F., A. POUPLARD, A. FLORIN-CHRISTENSEN & D. DONIACH. 1975. Autoantibodies to prolactin-secreting cells of human pituitary. Lancet **ii:** 97–101.
8. BROADWELL, R. D. & M. W. BRIGHTMAN. 1976. Entry of peroxidase into neurons of the central and peripheral nervous system from extracerebral and cerebral blood. J. Comp. Neurol. **166:** 257–284.
9. DONIACH, D. & G. F. BOTTAZZO. 1981. Polyendocrine autoimmunity. *In* Clinical Immunology Update. E. C. Franklin, Ed.: 95–121. Elsevier. New York.
10. DONIACH, D., G. F. BOTTAZZO & H. A. DREXHAGE. 1982. The autoimmune endocrinopathies. *In* Clinical Aspects of Immunology. P. A. Lachmann & D. K. Peters, Eds. **2:** 903—937. Blackwell Scientific Publications. Oxford.
11. FAGRAEUS, A. & J. JONSSON. 1970. Distribution of organ antigens over the surface of thyroid cells as examined by the immunofluorescence test. Immunology **18:** 413–416.
12. GORSUCH, A. N., K. M. SPENCER, J. LISTER, J. M. MCNALLY, B. M. DEAN, G. F. BOTTAZZO & A. G. CUDWORTH. 1981. The natural history of Type I (insulin-dependent) diabetes mellitus: evidence for a long pre-diabetic period. Lancet **ii:** 1363–1365.
13. HANAFUSA, T., R. PUJOL-BORRELL, L. CHIOVATO, L. J. HAMMOND & G. F. BOTTAZZO. 1982. Culture of human thyroid follicles with normal and reversed polarity. Ann. Endocrinol. Paris **43:**25A.
14. KHOURY, E. L., L. J. HAMMOND, G. F. BOTTAZZO & D. DONIACH. 1981. Presence of the organ-specific 'microsomal' autoantigen on the surface of human thyroid cells in culture: Its involvement in complement-mediated cytotoxicity. Clin. Exp. Immunol. **45:** 316–328.
15. MIRAKIAN, R., A. G. CUDWORTH, G. F. BOTTAZZO, C. A. RICHARDSON & D. DONIACH. 1982. Autoimmunity to anterior pituitary cells and the pathogenesis of Type I (insulin-dependent) diabetes mellitus. Lancet **i:** 755–759.
16. MIRAKIAN, R., C. A. RICHARDSON, G. F. BOTTAZZO & D. DONIACH. 1981. Humoral autoimmunity to gut-related endocrine cells. Clin. Immunol. Newsletter **2:** 161–167.
17. NITSCH, L. & S. H. WOLLMAN. 1980. Suspension culture of separated follicles consisting of differentiated thyroid epithelial cells. Proc. Natl. Acad. Sci. USA **77:** 472–476.
18. ONODERA, T., U. R. RAY, K. A. MELEZ, H. SUZUKI, A. TONIOLO & A. L. NOTKINS. 1982. Virus-induced diabetes mellitus: autoimmunity and polyendocrine disease prevented by immunosuppression. Nature **297:** 66–68.
19. ROITT, I. M., D. DONIACH, P. N. CAMPBELL & R. V. HUDSON. 1956. Autoantibodies in Hashimoto's disease (lymphadenoid goitre). Lancet **ii:** 820–822.
20. SCHERBAUM, W. A., G. F. BOTTAZZO. 1983. Autoantibodies to vasopressin cells in idiopathic diabetes insipidus: evidence for an autoimmune variant. Lancet **i:** 897–901.

Immune Complex–Mediated Disease and Immunofluorescence

B. ALBINI, E. PENNER, A. FAGUNDUS, D. KATZ,
AND C. NEULAND

Department of Microbiology
School of Medicine
State University of New York at Buffalo
Buffalo, New York 14214

This presentation includes a brief account of the early history of the concept of immune complex (IC)–mediated pathology and a report on the use of immunofluorescence techniques in the study of IC-mediated diseases as illustrated by the results of two recent projects performed in our laboratories. More background details are given in reviews from this[11–17] and other laboratories.[1–10]

Development of the Concept of Immune Complex–Mediated Pathology

The immunopathological potential of IC was first suggested in studies on human serum sickness (SS).[18] Behring and Kitasato[19] had described the effect of antitoxic sera in animals in 1890. One year later, Kitasato[20] and Wernicke[21] introduced immune sera of animal origin into the therapy of infectious diseases. Only a few years later, the same approach was taken in France.[22] Serum therapy was widely used in the treatment of diphtheria and tetanus, as well as in some other infectious diseases.

Early in the "serum therapy era," the beneficial effects of immune sera were overwhelming. Behring[20] in 1892 claimed that immune sera "in the amounts and modes of application used are as unharming as a sterile sodium chloride solution for humans." Unfortunately, this statement soon proved to be wrong. Mild reactions (urticaria-like exanthems) had been observed already in the first patients receiving serum therapy.[23] Fatal side effects first were reported in 1895[24] and dramatically came to the attention of the general public by the tragic death of the son of the noted pathologist Langerhans immediately after prophylactic inoculation with heterologous serum in 1896.[25] Fatal reactions, however, were rather rare events. Already in 1895, the most salient features of serum sickness had been reported.[26–31] These may be summarized as follows:[32] erythema and itching at the injection site, lymphadenopathy, skin rashes, fever, headache, and, less frequently, facial edema, joint pain, petechiae or purpura, neurological symptoms, cardiac arrhythmias, pericarditis, and splenomegaly, as well as gastrointestinal complications. Albuminuria and hemorrhagic nephritis were reported early,[33,39] but it is difficult to establish a direct link between SS and the kidney manifestations described in the early reports. Nevertheless, glomerulonephritis with proteinuria presently is a well-established, albeit rare, entity in SS-associated pathology.[35] The incidence of side effects of serum therapy was 23.3% in 1896[36] and 15% in 1920.[37] In the 1890s, the side effects of serum therapy were thought to be associated with either antibody specificities of the sera;[38] or other serum properties.[39] The latter notion was substantiated when Johannesson[40] induced the symptoms seen in patients treated with immune sera by injection of normal horse serum.

In the beginning of this century, Arthus[41] described the effect of multiple foreign serum injections into animals. He realized the relation of this effect with anaphylaxis, a phenomenon described by Portier and Richet.[42] Participation of the immune system in the pathogenesis of SS had already been suggested, although indirectly, when Tschistovitsch[43] reported on the *in vitro* precipitation of horse serum by antiserum obtained from a rabbit previously immunized to horse serum. These findings subsequently were confirmed and extended.[44–46] In addition, it was observed that clinical symptoms, upon multiple injection of hyperimmune serum produced in the same animal species, appeared faster and were more severe than after the first injection.[47,48] After the term "serum exanthem" had been used in 1895 by Colla,[49] Pirquet and Schick[50] coined the term "serum sickness" for the syndrome observed in some patients injected with foreign sera. Interestingly, Pirquet and Schick[51] did reject the possibility that "precipitins" are directly responsible for the clinical manifestations in serum sickness. Precipitins were found only in some of the patients and they were demonstrable in the serum only after clinical symptoms became manifest. These two authors wrote: "It is not the formation of a precipitate . . . which creates the serum sickness but another kind of clinical interaction between horse serum and antibodies. . . ."[51] Nevertheless, an immune mechanism for the pathogenesis of serum sickness seemed

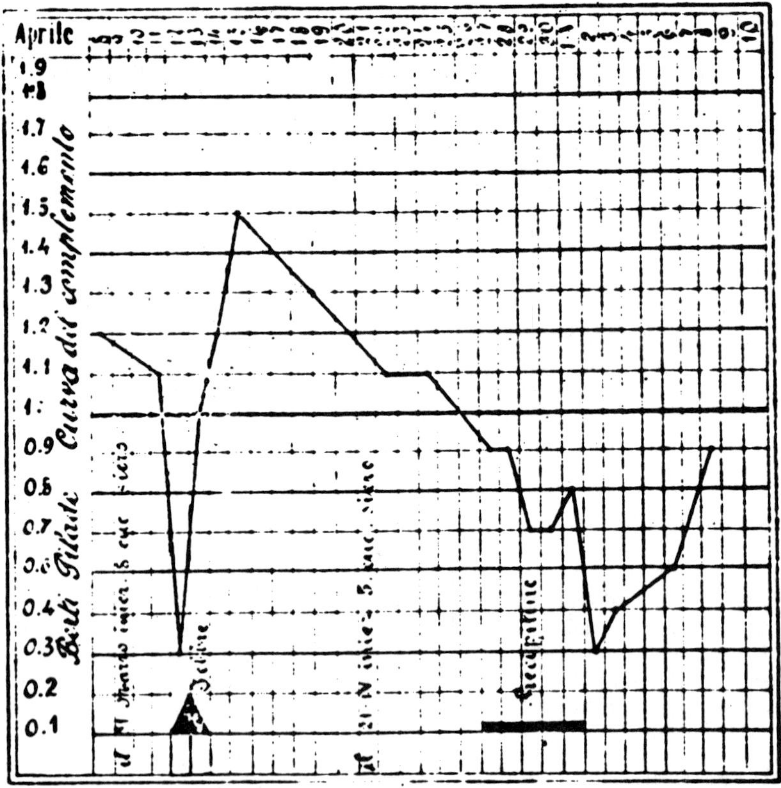

FIGURE 1. Decrease of hemolytic activity of patients serum during serum sickness (From Francioni.[54])

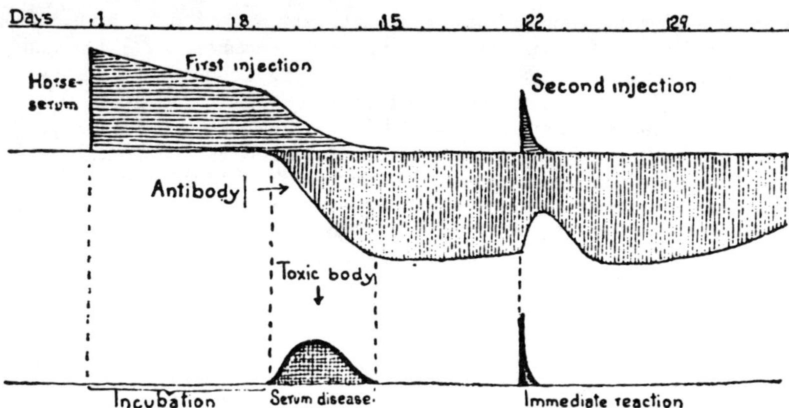

FIGURE 2. Antigen, antibody, and clinical manifestations in the course of serum sickness. (From Pirquet.[18] With permission from *Arch. Internal Med.*)

obvious and was proposed in both classical monographs on serum sickness published by Francioni[52] and Pirquet and Schick.[51]

In 1906, Pfeiffer and Moreschi[53] reported decreased complement concentrations in sera of rabbits injected with foreign serum. In 1908, Francioni published a paper[54] documenting decreased complement activity in the serum of patients with serum sickness (FIGURE 1). Chronological measurements of antigen and antibody in sera of patients with serum sickness were reported by Pirquet in 1911.[18] The graphs published (FIGURE 2) show manifestation of serum sickness at the time when antigen just ceases, and antibody just becomes detectable in circulation. Pirquet[18] summarized: "the reaction (i.e., serum sickness) . . . resulted from a combination of the antibody with the horse serum, forming a *toxic compound. . . .*"

Changes in the phagocytic cell compartment were also suggested early in this century. Bienenfeld[55] reported on a sharp decrease of circulating granulocytes in patients with serum sickness. This decrease followed immediately after antigen injection (FIGURE 3). She speculated on the possibility that these phagocytic cells, in this condition, "may migrate . . . to visceral organs." Almost simultaneously, Menabuoni[54] described impaired phagocytosis by cells from patients with serum sickness.

In the following half century, experimental studies in animals led to major breakthroughs in the understanding of immunological and immunopathological phenomena of serum sickness.[1,11,56,57] Ovalbumin[58] and serum[59] had been injected into animals many decades before serum therapy had been implemented. Albuminuria and sometimes death of the animals ensued immediately upon injection. These early experiments were discussed extensively by Creite[60] and Weiss.[61] The albuminuria seen may have resulted from protein overload, and descriptions of some experiments suggest anaphylaxis as cause of death; in some cases, heterophile antibodies also may have contributed to the animal's distress, making reliable interpretation difficult or impossible. The first animal experiments leading to a disease comparable in some aspects to human serum sickness were reported by Beclere, Chambon, and Menard,[62] who saw urticarial, measles-like rashes, and arthralgias develop in cattle injected with horse serum. It is surprising, however, that these manifestations had already developed four days after injection of foreign serum. In 1906, Lemair[63] reported weight loss during the second week after injection of horse serum into rabbits (i.e., at the same time as reported earlier for a decrease of serum complement concentrations[53]). The study of experimental serum sickness was then continued by Longcope,[64] Longcope and Rackemann,[65] Mackenzie and Leake,[66] Rich and Gregory,[56,67–70] Hopps and Wissler,[71] Ehrich *et al.*,[72] and Hawn and Janeway.[73] The latter researchers introduced purified serum components as antigens and thus made interpretation of results much easier.

Some three decades ago, Germuth[5,74–77] and Dixon[1,2,7,78–83] and their collaborators established convincingly the sequence of serological events in experimental serum sickness as well as the correlation of tissue pathology and antibody and antigen concentrations in sera. In one-shot acute serum sickness, concentration of antigen was shown both by measurement of antigen nitrogen in blood (FIGURE 4) and of

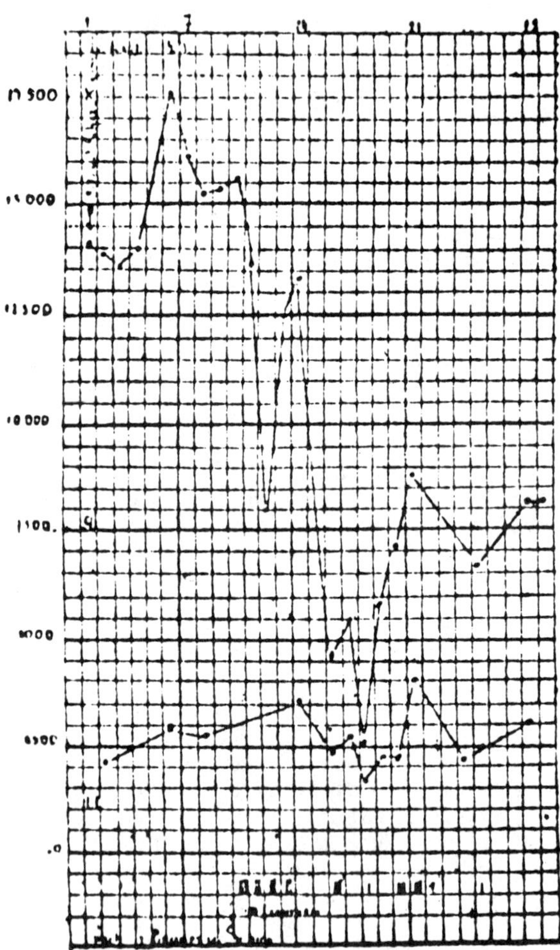

FIGURE 3. Initial increase and subsequent decrease of the number of circulating granulocytes in a patient with serum sickness. (From Bienenfeld.[55] With permission from *Jahrb. Kinderheilkunde.*)

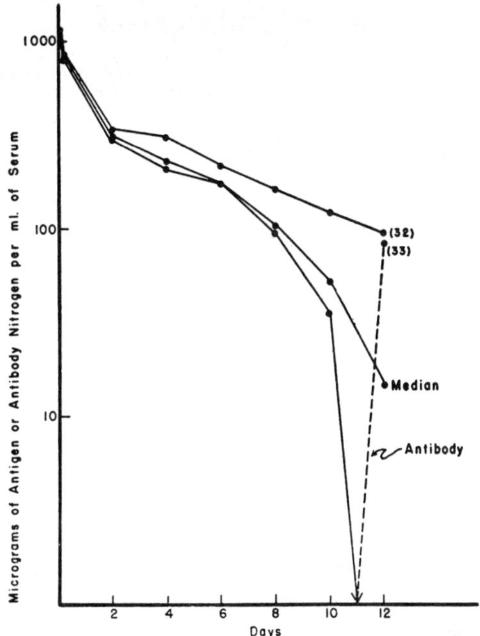

FIGURE 4. Antigen and antibody in one-shot serum sickness. (From Germuth. 1953. *J. Exp. Med.* **97:** 257, with permission.)

radiolabeled antigen (FIGURE 5) to decrease rapidly prior to development of serum sickness (the stage of "immune elimination of antigen from serum"). The tissue injury was shown to occur at the later phases of the immune elimination stage and during the very early period, when circulating antibody to the specific foreign species antigen became demonstrable (FIGURE 6). More recently, availability of assays for the detection of circulating immune complexes made it possible to define the period in which immune complexes were present in circulation (FIGURE 7). Thus, meticulous experimental research of the 1950s to 1970s confirmed the sequence of events postulated by Pirquet in 1911 (FIGURE 2).[18] The mode of the formation of immune complex deposits in tissues, namely the deposition of circulating immune complexes versus formation of immune complexes *in situ,* has been discussed since the formulation of the notion of immune complex–mediated pathology. The possible alternatives have been defined in the classical paper by Dixon *et al.*[80] The preference of investigators has vacillated from one to the other hypothesis, with little direct and much indirect, difficult-to-interpret evidence. The interesting recent work on Heymann nephritis[87,88] did reopen this discussion, but it remains rather doubtful that the Heymann nephritis is indeed a representative model for immune complex–mediated disease.

Immunohistology and Immune Complex–Mediated Diseases

Thorough studies of the immunopathology of immune complex–mediated diseases were only possible after immunohistological techniques became available. Immuno-

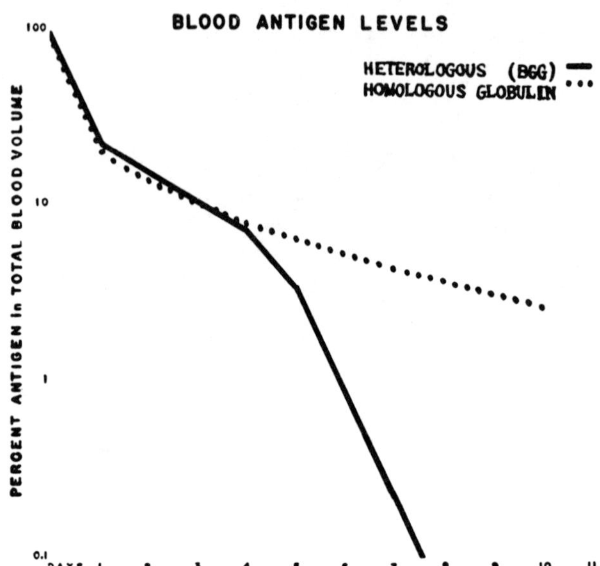

FIGURE 5. Radiolabeled antigen in serum of one-shot serum sickness. (From Talmage *et al.* 1951. *J. Immunol.* **67:** 243, with permission.)

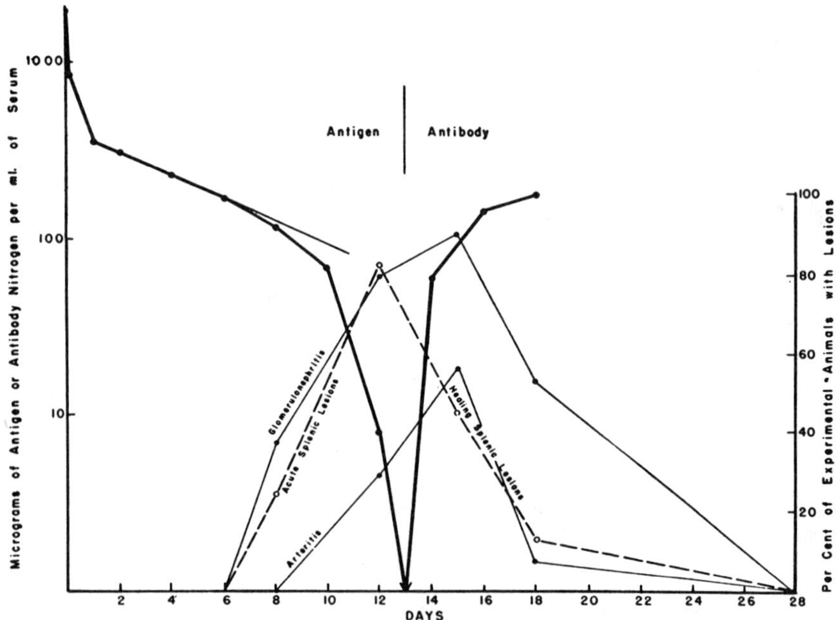

FIGURE 6. Chronology of circulating antigen and antibody and appearance of tissue pathology in experimental serum sickness. (From Germuth. 1953. *J. Exp. Med.* **97:** 257.)

fluorescence was the first such technique to allow for reliable identification of antigen and antibody in tissue structures,[15,84,85] and made it possible to localize the site of activation of inflammatory mediators, such as complement and components of the blood coagulation system. "Classical" immunopathology postulates that immune complex deposits are associated with a granular staining pattern in immunofluorescence.[1,5,11,15,80] This pattern is seen in the majority of lesions of chronic serum sickness of rabbits. Other patterns of immune deposits, however, are not infrequent; these include the ribbon-like staining and a patchy distribution of antigen and antibody.[15] These patterns of staining were contrasted to the linear immunofluorescence seen in experimental kidney disease mediated by nephrotoxic sera and in human Goodpasture's disease, and have been used widely to classify immunologically mediated tissue injury as antibody (antibasement membrane) or immune complex–mediated injury.[86] Despite the great appeal of this simple principle of differentiation and the great

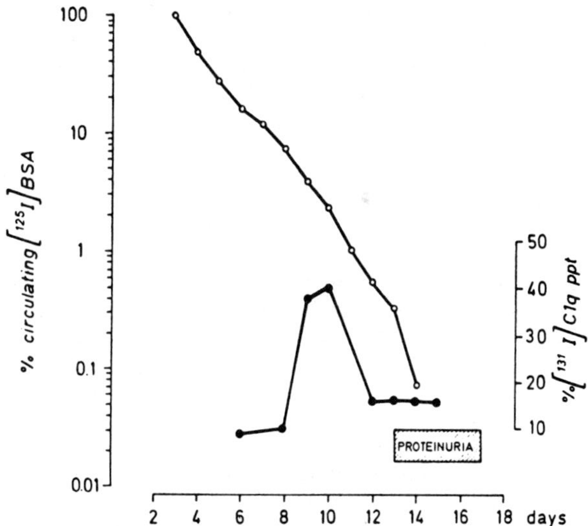

FIGURE 7. Chronology of appearance of circulating immune complexes in acute serum sickness. (From Nydegger.[4] With permission from *Review of Physiology, Biochemistry, and Pharmacology*.)

contributions made applying this "rule of thumb" to immunopathology and diagnostics, it soon became obvious that there were many exceptions to the granular versus linear staining theorem. First, there may be technical difficulties: dense, finely granular deposits in basement membranes may not be readily distinguishable from linear staining and heavy ribbon-like deposits also may be confounded with "broad" linear staining. Second, it has been documented some time ago that one component of the immune deposit may appear linear whereas another component may be granular in the same lesion; Germuth and associates[5] showed that linear deposits of immunoglobulin may be accompanied by granular staining for complement. Finally, the recent work on Heymann nephritis[87,88] seems to indicate that antibody interacting directly with antigens of the basement membrane or podocytes may be distributed in a strictly granular fashion; antibody reacting with discontinuously and granularly distributed antigen obviously may give rise to granular immunofluorescence with conjugated antisera to immunoglobulins. Thus, immunofluorescence patterns are of help in the evaluation of immunopathology but have to be interpreted with the appropriate critical considerations.

In addition to the documentation of "characteristic" patterns associated with immune complex deposition in tissues, double staining in immunofluorescence allows for directly testing the presence of two or three immune reactants at the same site. Thus, antibody and antigen, and very often complement, may be demonstrated in the same immune deposit. Since immunofluorescence can be performed with a minimal amount of expenses, it still is the favored and most popular technique used in immunohistology and immunopathology. It seems to be, in many immunopathological analyses, a better diagnostic indicator than other more elaborate techniques, e.g., electron microscopy.[89] Immunofluorescence has also been essential in formulating classifications of diseases on the basis of immunopathological parameters rather than on pure morphological criteria, the most prominent of such classifications being to our knowledge that proposed for kidney diseases by McCluskey and Andres.[86] However, immunoelectron microscopy has been essential for confirming the presence of antigen in the electron-dense deposits seen in basement membranes thus linking the presence of immune reactants to electron-dense deposits.[90]

The kidney has been the main organ used in studies on immune complex–mediated pathology, but more and more other organs and organ systems are being subjected to search for immune complex–mediated pathology. The association of many human diseases with immune complexes has indeed been suggested first by immunofluorescence studies. The animal equivalents of such pathology became available with the description of a number of experimental and spontaneous animal models with "systemic" or multi-organ deposition of immune complexes.[14,91–96] It may be of special interest that an experimental systemic immune complex disease can be induced in rabbits and rats by administration of low doses of a rather common environmental pollutant, namely mercuric chloride.[97–99] More recently, immunofluorescence has also been used to study the changes induced by pathological processes, e.g., deposition of immune complexes, in the structure of tissues. The elegant work on changes of antigenic components in basement membranes of a number of immunologically mediated kidney diseases and prior work on changes in collagen distribution in liver diseases open a new perspective in immunopathology.[100,101]

With the use of immunofluorescence, immune complex–mediated pathology has been suggested in a number of different organs and the predominant tissue structures in which immune complexes are deposited can now be enumerated with a degree of certainty (TABLE 1). As an illustration of the use of immunofluorescence in research on immune complex–mediated pathology, brief reports on two studies from our laboratories are presented in the following sections.

Use of Immunofluorescence in Studies on Early Deposition of Immune Complexes in Kidneys of Rabbits with Chronic Serum Sickness

Little is known about the mode of early antigen or immune complex deposition in kidneys in experimental serum sickness. The classic papers by Wilson and Dixon[81,82] documented the amounts of antigen deposited in the kidneys of rabbits with acute and chronic serum sickness at 24 hours after administration of the radiolabeled antigen. More recently, similar studies of antigen uptake in the systemic immune complex–mediated disease described in rabbits injected with high doses of bovine serum albumin (BSA) over a prolonged period of time established the amount of antigen present in a number of organs of such animals and confirmed the data published previously.[102] Still, studies on the dynamics of antigen deposition and correlation with staining patterns were limited, until recently, to injection of antigen or aggregated immunoglobulin into otherwise normal animals (see Haakenstad and Mannik's review[3]). A new methodology involving double-radiolabel studies with the use of a computer-coupled gamma camera and evaluation of plasma values over a period of 24 to 48 hours was applied to lightly anesthetized rabbits at various stages of chronic serum sickness induced by BSA.[103] The uptake of BSA by the tissues varied in its kinetics according to tissue

TABLE 1. Tissue Structures Preferentially Involved in Immune Complex Deposition

Eye	vessels of iris
	uvea
	ciliary body
Heart	vessels
Intestine	vessels
	basal lamina
Kidney	vessels
	mesangium
	glomerular basement membrane
	tubular basement membrane
	Bowman's capsule
Liver	vessels
	sinusoids (Disse's space)
Lung	vessels
	alveolar basement membrane
Ovary	vessels
	zona pellucida
	basement membrane between granulosa and theca
Plexus chorioideus	vessels
Serosae	vessels
Skin	vessels
	dermo-epidermal junction
Spleen	vessels
	basement membrane of red pulp
Synovia	vessels
Testis	vessels
	basal lamina of tubules
Thyroidea	vessels
	basal lamina of follicles
Uterus	vessels
	basal lamina of glands

structure in which the antigen was incorporated. Thus uptake in cells and presence as intravascular aggregates followed a rapid uptake with peak values of radiolabel localized to a tissue immediately following injection and up to three hours. Immunofluorescence staining on tissues obtained during this period clearly showed staining of the cytoplasm of phagocytic cells. On the other hand, deposition in basement membranes (e.g., glomerular basement membrane, GBM) followed protracted kinetics with a latency period of up to 24 hours. Again, immunofluorescence showed deposition in basement membranes (BM). To avoid any confusion caused by immune deposits already in place before the injection of radiolabeled BSA, autoradiographies

TABLE 2. Deposition of BSA in Kidneys of Rabbits With Chronic Serum Sickness: Immunofluorescent Staining

	Group I			Group II			Group III		
Proteinuria	0 to ±			+ to + +			+ to + +		
Circ. Ag	0			0 or ±			0 or +		
Circ. Ab	+ +			±			0 or +		
	IVP	C	BM	IVP	C	BM	IVP	C	BM
Hours									
1.5	+ + +	+	0	+ +	+	+	0	0	+ + +
6.0	+ + +	+ +	±	+ +	±	+	0	0	+ + +
24.0	±	0	0	0	0	+ +	0	0	+ + +

IVP = intravascular precipitate, C = intracellular staining, and BM = basement membrane staining.

were performed and confirmed the results seen in immunofluorescence in animals with early disease but not in rabbits with longstanding chronic phase proteinuria, which showed strong BSA deposits in immunofluorescence but only minimal deposition of radiolabeled BSA in autoradiography.[103] In a more detailed study on the early phase deposition of immune complexes and antigen in the kidney of rabbits with chronic serum sickness, Fagundus[104] obtained the following pattern of staining in immunofluorescence using kidney biopsies and autopsies performed on two rabbits in the stage preceding chronic phase proteinuria, two rabbits in early chronic proteinuric phase, and two rabbits in late stages of the disease (TABLE 2): rabbits tested in the earliest stage of the disease showed first strong uptake of BSA in cells and in vessels as immunoprecipitates, which disappeared completely after six hours post injection; only minimal granular staining for BSA in BM was seen six hours after injection in two of

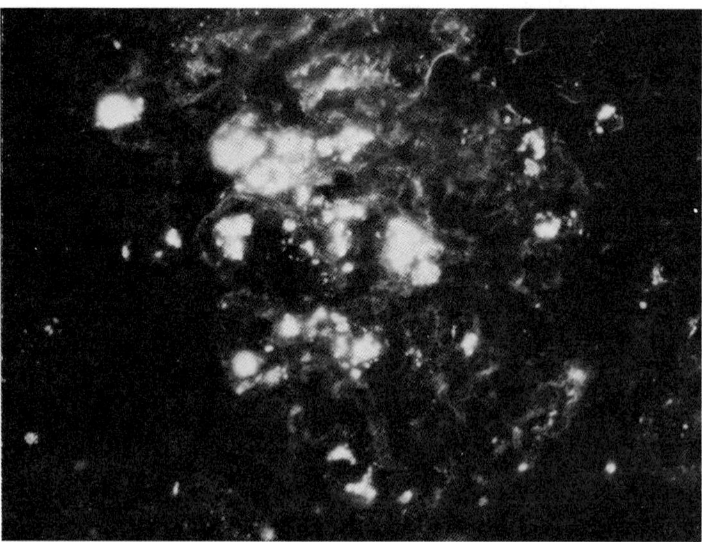

FIGURE 8. Kidney of rabbit 865. Biopsy taken 1.5 hours after injection of BSA. Immunofluorescent staining for BSA. The antigen is localized in cells and forms intravascular precipitates. Only BM is stained equivocally. 400×.

the rabbits. In rabbits during the early proteinuric stage, (i.e., with low antibody or antibody and antigen detectable in circulation) cellular uptake also was pronounced during the first hours to disappear later, but the deposition in basement membranes increased markedly over the observation period (FIGURES 8 and 9). Rabbits in the late stage of the chronic proteinuric phase did not show any changes in the immunofluorescence intensity or pattern over the observation period. Thus the sequence of patterns of uptake of antigen in tissues seems to follow the order: vascular precipitates and cellular deposition in early phases of the disease, combined later with "late" uptake in basement membranes; these early basement membrane deposits are detectable only over a short period of time; in rabbits with proteinuria, antigen deposits persist in immune deposits in the basement membranes late after injection of antigen. Late in the

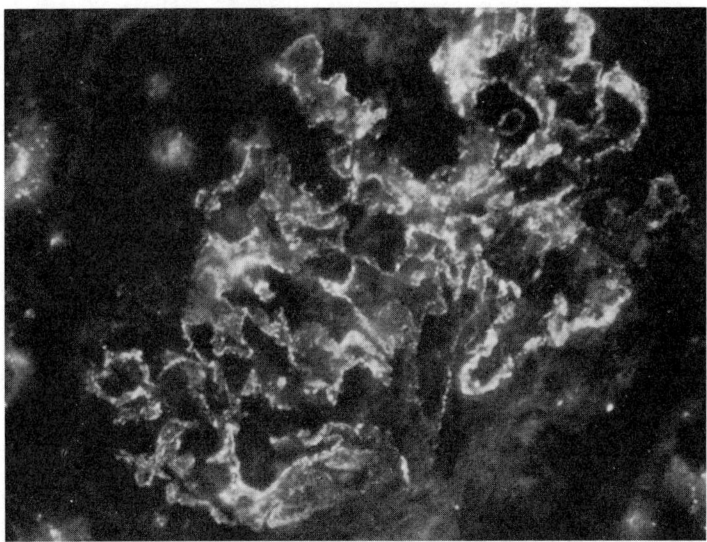

FIGURE 9. Kidney of rabbit 865. Biopsy taken 24 hours after injection of BSA. Immunofluorescent staining for BSA. The antigen is localized in BM only. 400×.

disease, the old basement membrane deposits seem to persist without any appreciable changes and no cellular uptake is observed.

Use of Immunofluorescence in Studies on the Identity of Antigens Involved in Tissue-Deposited Immune Complexes

Methods of using antigen excess to solubilize immune complexes have been suggested by the first description of the precipitation curve by Heidelberger and have been used for studies with circulating immune complexes.[105,106] This principle has been applied to tissue-bound immune complexes using kidney tissue sections from patients with membranous glomerulopathy, lupus nephritis, and post-streptococcal glomerulonephritis and specimens obtained from rabbits with acute and chronic serum sickness.[107] The immunofluorescence results of comparison of tissue sections incubated

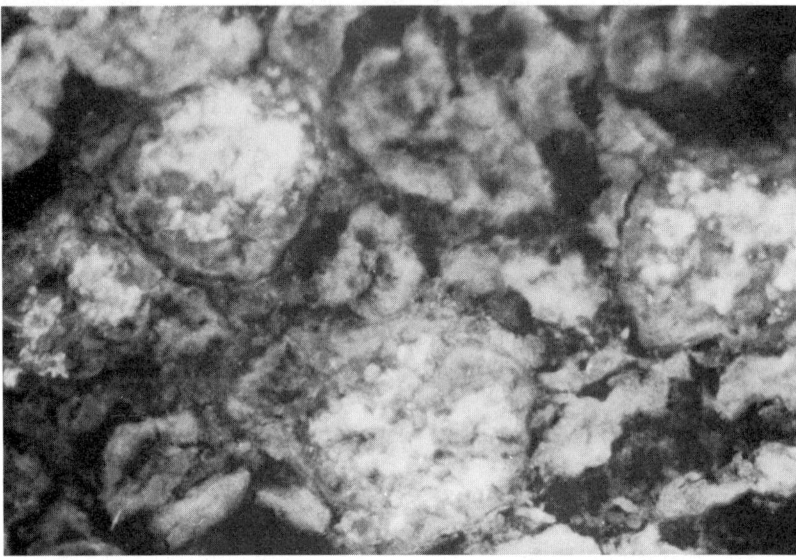

FIGURE 10. Kidney of NZB/NZW F1 mouse. Immunofluorescent staining of mesangial IgG deposits. 250×.

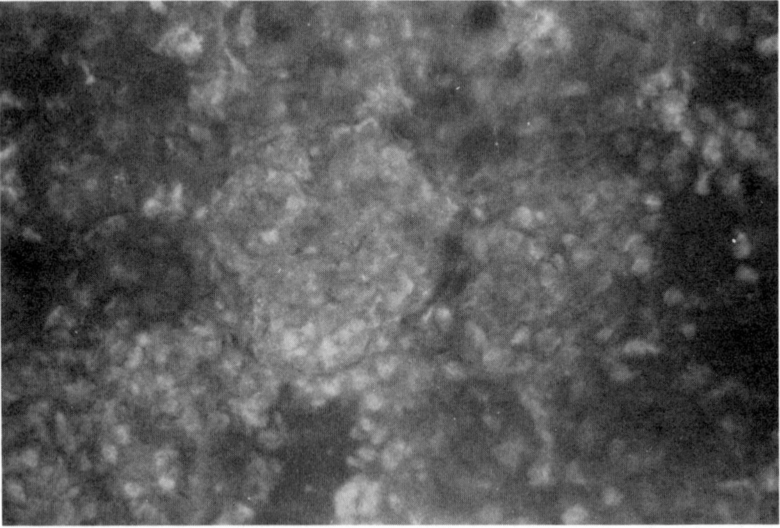

FIGURE 11. The same kidney of NZB/NZW F1 mouse as shown in FIGURE 10 after incubation for six hours with 10 mg of aggregated mouse gammaglobulin/ml. No staining for IgG is detectable. 250×.

with high doses of antigen for prolonged periods of time showed solubilization of immune complexes in specimens from chronic serum sickness rabbits. Similarly, mesangial IgG-IgM deposits seen in NZB/WF1 mice could be resolved by repeated incubation with 10 mg/ml aggregated mouse gammaglobulin (FIGURES 10 and 11). Together with other reports,[15,107] these results strongly suggest the participation of rheumatoid-like factors in the tissue-bound immune complex deposits of a number of human and animal kidney diseases. It is of interest that monomeric IgG did not have the same effect as aggregated immunoglobulin, differentiating our findings from the phenomenon of immune complex solubilization with monomeric C3b and other monomeric reagents.[108]

Conclusion

Immunofluorescence has been crucial to identify and classify immune complex–mediated pathology and to correlate serological with immunopathological derangements in spontaneous and experimental diseases. It can be applied to many open questions of immune complex–mediated pathogenesis, such as the kinetics of antigen deposition and the nature of antigen involved in immune complex formation. Together with other immunohistological techniques, immunofluorescence is bound to remain a mainstay in research and clinical laboratory applications to immune complex–mediated diseases.

ACKNOWLEDGMENTS

We thank Drs. Andres, Beutner, and Milgrom for many helpful discussions.

REFERENCES

1. DIXON, F. J. 1963. Harvey Lect. **58:** 21.
2. COCHRANE, C. G. & D. KOFFLER. 1973. Adv. Immunol. **16:** 185.
3. HAAKENSTAD, A. O. & M. MANNIK. 1977. *In* Autoimmunity. N. Talal, Ed.: 278. Academic Press. New York.
4. NYDEGGER, U. E. 1979. Rev. Physiol. Biochem. Pharmacol. **85:** 63.
5. GERMUTH, F. G., JR. & E. RODRIGUEZ. 1973. Immunopathology of the Renal Glomerulus. Little, Brown & Co. Boston.
6. MCCLUSKEY, R. T. & G. A. ANDRES. 1978. Immunologically Mediated Renal Diseases. Marcel Dekker. New York.
7. THEOFILOPOULOS, A. N. & F. J. DIXON. 1979. Adv. Immunol. **228:** 89.
8. KANO, K. & F. MILGROM. 1980. Vox Sang. **38:** 121.
9. ZUBLER, R. B. & P-H. LAMBERT. Orogr. Allergy **24:** 1.
10. WILLIAMS, R. C., JR. 1984. Immune Complexes in Clinical and Experimental Medicine. Harvard University Press. Cambridge, Mass.
11. ALBINI, B., J. R. BRENTJENS & G. A. ANDRES. 1979. Immunopathology of the Kidney. Arnold. London.
12. ALBINI, B., D. KERJASCHKI & B. NOBLE. 1982. *In* Immunofluorescence Technology. G. Wick, K. Trail & K. Schauenstein, Eds.: 267. Elsevier. Amsterdam.
13. ANDRES, G. A., B. ALBINI, J. R. BRENTJENS, B. K. NOBLE & F. MILGROM. 1981. *In* Proceedings Immunology of the Eye. Workshop II. R. J. Helmsen, A. Suran, I. Gery & R. B. Nussenblatt, Eds.: 145. Suppl. Immunology Abstracts.
14. ALBINI, B., A. FAGUNDUS, I. GLURICH, D. KERJASCHKI, C. NEULAND B. NOBLE, K.

OLSON, E. OSSI & E. PENNER. 1981. *In* Clinical Immunology and Allergology. C. Steffen & H. Ludwig, Eds.: 177. Elsevier. Amsterdam.

15. PENNER, E. & B. ALBINI. 1982. *In* Immunofluorescence Technology. G. Wick, K. Trail & K. Schauenstein, Eds.: 349. Elsevier. Amsterdam.

16. ALBINI, B. & E. PENNER. 1982. *In* Methods of Nuclear Medicine in Rheumatology. G. Kolarz & N. Thumb, Eds.: 11. Schottauer. Stuttgart.

17. ALBINI, B., A. FAGUNDUS & A. VLADUTIU. 1983. *In* Molecular Immunology. Z. Attassi, D. Absolom & C. J. van Oss, Eds. Marcel Dekker. New York. (In press.)

18. PIRQUET, C. V. 1911. Arch. Internal Med. **7:** 259.

19. BEHRING, E. V. & S. KITASATO. 1890. Dtsche. Med. Wschr. **1113.**

20. BEHRING, E. V., O. BOER & H. KOSSEL. 1893. Dtsch. Med. Woschr. 339.

21. ENGELHARDT, A. V. 1940. Behringwerke Mtlg. 10.

22. ROUX, E., L. MARTIN & I. CHARLTON. 1894. Ann. Inst. Pasteur **8:** 640.

23. ROTTER, J. 1892. *In* Die Blutserumtherapie. E. V. Behring, Ed. **2:** 84. Deuticke. Leipzig.

24. GUINON, L. & N. ROUFILANGE. 1896. Le Semain Med. p. 300.

25. LANGERHANS, R. 1896. Berl. Klin. Woschr. **33:** 602.

26. BAGINSKY, A. 1894. Berl. Klin. Woschr. **31:** 1173.

27. CANON, P. 1894. Dtsch. Med. Woschr. p. 500.

28. KATZ, O. 1894. Berl. Klin. Woschr. p. 667.

29. VOSWINCKEL, E. 1894. Dtsch. Med. Woschr. p. 479.

30. LUBLINSKI, W. 1894. Dtsch. Med. Woschr. p. 857.

31. MOIZARD, P. 1895. La Semain Med. p. 298.

32. SMITH, J. M. & P. G. H. GELL. 1975. *In* Clinical Aspects of Immunology. 3rd edit. P. G. H. Gell, R. R. A. Coombs & P. J. Lachmann, Eds.: 903. Blackwell. Oxford.

33. HANSEMANN, D. 1894. Berl. Klin. Woschr. **31:** 1127.

34. TIBIERGE, G. 1895. Lasemain Med. p. 300.

35. NORMAN, P. S. 1976. *In* The Principles and Practice of Medicine. A. M. Harvey, R. J. Johns, A. H. Owens, Jr. & R. S. Ross, Eds.: 1349. Appleton-Century-Crofts. New York.

36. GOTTSTEIN, A. 1896. Therapeut. Mohft. p. 269.

37. BORDET, J. 1920. Traite de l'immunite dans les maladies infectieuses. Masson et Cie. Paris.

38. KOSSEL, H. 1894. Zschr. Hyg. **17:** 489.

39. HUEBNER, E. 1894. Dtsch. Med. Woschr. p. 701.

40. JOHANNESSON, A. 1895. Dtsch. Med. Woschr. p. 855.

41. ARTHUS, M. C. R. 1903. Seances Soc. Biol. **55:** 817.

42. PORTIER, P. & C. RICHET. 1902. Comp. Rend. Spc. Biol. **54:** 170.

43. TSCHISTOWITSCH, N. G. 1895. St. Petersburg Med. Woschr. **12:** 324 & 330.

44. HAMBURGER, F. & E. MORO. 1903. Wien. Klin. Woschr. **16:** 445.

45. MARFAN, A. 1905. Rev. Mens. Med. Del'enfant **23:** 337.

46. ROVERE, G. 1906. Arch. Gen. Med. p. 6.

47. LEHNDORFF, N. 1905/1906. Moschr. Kinderheilkinde **4:** 545.

48. GOODALL, E. W. J. 1907. Hyg. **7:** 607.

49. COLLA, V. 1895. Dtsch. Med. Woschr. p. 51.

50. PIRQUET, C. V. & B. SCHICK. 1903. Wien Klin. Woschr. p. 758.

51. PIRQUET, C. V. & B. SCHICK. 1905. Die Serumkrankheit. Deuticke, Leipzig and Vienna.

52. FRANCIONI, C. 1904. Sperimentale **58:** 767.

53. PFEIFFER, R. & C. MORESCHI. 1906. Berl. Klin. Woschr. **43:** 33.

54. FRANCIONI, C. 1908. Rev. Clin. Ped. p. 321.

55. BIENENFELD, B. 1907. Jahrb. Kinderheilkunde **65:** 174.

56. RICH, A. R. 1956. Bull. Johns Hopkins Hosp. **98:** 120.

57. BLITTERSDORF, F. 1952. Sudhoffs Arch. **36:** 149.

58. BARNARD, C. Lecons sur les proprietes physiologiques et les alterations pathologiques des liquides de l'organism. **2:** 459. Paris.

59. BERZELIUS. (cit. Stokvis, A.) 1864. Centralblatt Med. Wschft. p. 597.

60. CREITE, A. 1869. Ztschr. Human Rat. Med. **36:** 90.

61. WEISS, O. 1896. Pflugers Arch. **65:** 215.

62. BECLERE, A., F. CHAMBON & V. MENARD. 1896. Ann. Inst. Pasteur 10: 567.
63. LEMAIR, H. 1906. Recherches cliniques et experimentales sur les accidants serotoxiques. Thesis. Paris.
64. LONGCOPE, W. T. 1913. J. Exp. Med. 18: 678.
65. LONGCOPE, W. T. & F. M. RACKMANN. 1918. J. Exp. Med. 27: 341.
66. MACKENZIE, G. M. & W. H. LEAKE. 1921. J. Exp. Med. 33: 601.
67. RICH, A. R. & J. E. GREGORY. 1943. Bull. Johns Hopkins Hosp. 72: 65.
68. RICH, A. R. & J. E. GREGORY. 1943. Bull. Johns Hopkins Hosp. 73: 239.
69. RICH, A. R. & J. E. GREGORY. 1947. Bull. Johns Hopkins Hosp. 81: 312.
70. RICH, A. R. 1947. Harvey Lect. 42: 106.
71. HOPPS, H. C. & R. W. WISSLER. 1946. J. Lab. Clin. Med. 31: 939.
72. EHRICH, W. E., J. SEIFTER & C. FORMAN. 1949. J. Exp. Med. 89: 23.
73. HAWN, C. V. & C. A. JANEWAY. 1947. J. Exp. Med. 85: 571.
74. GERMUTH, F. G., JR. 1953. J. Exp. Med. 97: 257.
75. GERMUTH, F. G., JR. & G. E. McKINNON. 1953. Johns Hopkins Med. J. 101: 13.
76. GERMUTH, F. G., JR., M. G. PACE & M. J. C. TIPETT. 1955. J. Exp. Med. 101: 35.
77. GERMUTH, F. G., JR., C. FLANAGAN & M. R. MONTENEGRO. 1957. Bull. Johns Hopkins Hosp. 101: 149.
78. DIXON, F. J., S. C. BUKANTZ, G. J. DAMMIN & D. W. TALMAGE. 1951. Proc. Soc. Exp. Biol. Med. 78: 123.
79. DIXON, F. J., J. J. VAZQUEZ, W. O. WEIGLE & C. G. COCHRANE. 1958. Arch. Pathol. 65: 18.
80. DIXON, F. J., J. D. FELDMAN & J. J. VAZQUEZ. 1961. J. Exp. Med. 113: 899.
81. WILSON, C. B. & F. J. DIXON. 1971. J. Exp. Med. 134: 7s.
82. WILSON, C. B. & F. J. DIXON. 1970. J. Immunol. 105: 279.
83. COCHRANE, C. G., E. R. UNANUE & F. J. DIXON. 1965. J. Exp. Med. 122: 99.
84. SEEGAL, B. C., K. S. HSU, E. FIASCHI & G. A. ANDRES. 1959. Rass. Fis. Clin. Ter. 38: 1.
85. McCLUSKEY, R. T. 1971. J. Exp. Med. 134: 242s.
86. McCLUSKEY, R. T. & G. A. ANDRES. 1978. Immunologically Mediated Renal Diseases. Marcel Dekker. New York.
87. VAN DAMME, B. J. C., G. J. FLEUREN, W. W. BAKKER. R. L. VERNIER & P. J. HOEDEMAEKER. 1978. Lab. Invest. 38: 502.
88. COUSER, W. G., D. R. STEINMULLER M. M. STILMANT, D. J. SALANT & L. M. LOEWENSTEIN. 1978. J. Clin. Invest. 62: 1275.
89. ZOLLINGER, H. U. & M. J. MIHATSCH. 1978. Renal Pathology in Biopsy. Springer. Berlin.
90. ANDRES, G. A., B. C. SEEGAL, K. C. HSU, M. S. ROTHENBERG & M. L. CHAPEAU. 1963. J. Exp. Med. 117: 691.
91. BRENTJENS, J. R., D. W. O'CONNELL, I. B. PAWLOWSKI, K. C. HSU & G. A. ANDRES. 1977. J. Exp. Med. 140: 105.
92. BRENTJENS, J. R., B. ALBINI, D. MENDRICK, C. NEULAND, D. O'CONNELL, E. OSSI, L. ACCINNI & G. A. ANDRES. 1979. *In* Immunopathology. F. Milgrom & B. Albini, Eds.: 202. Karger. Basel.
93. ARISZ, L., B. NOBLE, M. MILGROM, J. R. BRENTJENS & G. A. ANDRES. 1979. Int. Arch. Allergy 60: 80.
94. NOBLE, B., K. A. OLSON, M. MILGROM & B. ALBINI. 1980. Clin. Exp. Immunol. 42: 255.
95. ACCINNI, L. & F. J. DIXON. 1979. Am. J. Pathol. 96: 477.
96. ACCINNI, L., B. ALBINI, G. ANDRES & F. J. DIXON. 1980. Am. J. Pathol. 99: 589.
97. ROMAN-FRANCO, A. A., M. TURIELLO, B. ALBINI, E. OSSI, F. MILGROM & G. A. ANDRES. 1978. Clin. Immunol. Immunopathol. 9: 464.
98. DRUET, P., E. DRUET, F. POTDEVIN & C. SAPIN. 1978. Ann. Immunol. (Inst. Pasteur) 129C: 777.
99. ALBINI, B., I. GLURICH & G. A. ANDRES. 1982. *In* Nephrotoxic Mechanisms of Drugs and Environmental Toxins. G. A. Porter, Ed.: 413. Plenum Med. Book Comp. New York.
100. WICK, G., H. FURTHMAYER & R. TIMPL. 1975. Int. Arch. Allergy 48: 664.
101. SCHEINMAN, J. I., J. M. FOIDART & A. F. MICHAEL. 1980. Lab. Invest. 43: 373.

102. NEULAND, C., B. ALBINI, J. R. BRENTJENS, A. I. GROSSBERG & G. A. ANDRES. 1981. Int. Arch. Allergy **64:** 385.
103. NEULAND, C. 1978. Kinetic studies on antigen deposition in rabbits with systemic serum sickness. Thesis. Buffalo, N.Y.
104. FAGUNDUS, A. 1981. Physicochemical and immunochemical studies on immune complexes in systemic chronic serum sickness of the rabbit. Thesis. Buffalo, N.Y.
105. DELIRE, M., C. L. CAMBIASO & P. L. MASSON. 1978. Nature **272:** 632.
106. NISHIMAKI, T., K. KANO & F. MILGROM. 1978. Arthritis and Rheum. **21:** 639.
107. PENNER, E., B. ALBINI, I. GLURICH, G. A. ANDRES & F. MILGROM. 1982. Int. Arch. Allergy **67:** 245.
108. TAKAHASHI, M., S. TAKAHASHI & M. HIROSE. 1980. Progr. Allergy **27:** 134.

Auto-antibodies to Actin:
Recent Findings[a]

A. FAGRAEUS, R. NORBERG, R. THORSTENSSON,
G. UTTER, AND C. ÖRVELL

National Bacteriological Laboratory
S-105 21 Stockholm, Sweden

Actin forms an important part of the contractile proteins in eukaryotic cells. Spontaneously occurring high-titered human anti-actin antibodies have been used to study the localization of actin.[4] We have mainly studied lymphoid cells and fibroblasts. The preferred target cells have been Epstein-Barr virus–transformed lymphoblastoid cells. Early on in these studies we found it necessary to use chelating agents when preparing cell smears and heat-inactivated sera when staining the actin structures using indirect immunofluorescence (IFL) technique.[2] Recent research has clarified why this was necessary.[8,9]

Actin in cells is present as bundles and networks of polymerized actin, F-actin, and a large portion in an unpolymerized state, G-actin. In resting lymphocytes the monomeric form represents 70–80% of the total actin whereas activation of the cells leads to an increased polymerization[7] and bundles of filaments can be visualized by IFL staining. The indirect IFL technique makes use of acetone-fixed Ca^{2+}-chelated cell smears and seems mainly to detect filamentous actin organized in bundles and in networks.[5]

The proteins involved in cell motility in the lymphoblastoid cell have been localized as follows by IFL staining:[9] human anti-actin antibodies reveal the long microvilli; goat antibodies to actin binding protein (ABP) react with the basal part of the microvillus, where ABP is supposed to cross-link actin filaments; goat anti-gelsolin antibodies stain the microvilli and the cytoplasm of the cells; and goat anti-myosin antibodies give a dotted IFL staining of the peripheral cytoplasm.

Staining for actin and actin-associated proteins in the microvilli was no longer evident in the presence of Ca^{2+}, whereas the cytoplasmic staining remained unaffected. All sera from human and other species contain a heat-labile protein, a 93,000 dalton polypeptide. It is called actin depolymerizing factor (ADF)[6] or brevin[3] on account of its ability to shorten filamentous actin. ADF is structurally and functionally related to gelsolin and is supposed to affect humoral actin, e.g. after cell destruction.[8] Gelsolin and ADF fragment actin filaments in the presence of Ca^{2+} and these fragments are eluted from cell preparations during the staining procedures.[8,9] This fragmentation explains why Ca^{2+} must be chelated and why sera must be heat-inactivated when staining the cell smears for actin or actin-associated proteins present in the microvilli.

Actin is a highly conserved protein changing very little from species to species. This might explain why it is difficult to produce potent antisera and necessary to use denatured actin for immunization. In view of the amount of actin existing in all cells and the release of actin when cells break down, tolerance should be expected and it is surprising that auto-antibodies to actin are produced. Virus infections have been discussed as a mechanism for breaking tolerance perhaps by formation of virus-actin complexes.

[a]Supported by grants from the Swedish Medical Reseach Council B82-16X-04976-06.

TABLE 1. Anti-actin Antibody Stains[a]

| Anti-actin Antibodies | Lymphoblastoid Cells | Fibroblasts | | Tissue Section Smooth Muscle |
		Sparse	Semiconfluent	
M372 ascitic fluid	microvilli + + +	nuclei + + + filament bundles +	nuclei + filament bundles + +	±
AS eluate	microvilli + + +	nuclei − filament bundles + +	nuclei − filament bundles + + +	+ + +

[a]Purified polyclonal antibodies (AS Eluate) give strong staining with bundles of cytoplasmic microfilaments and of smooth muscle in tissue sections. The staining by M 372 was weaker. Note the strong nuclear staining by M 372 of growing fibroblasts. This was not given by any of the polyclonal antisera.

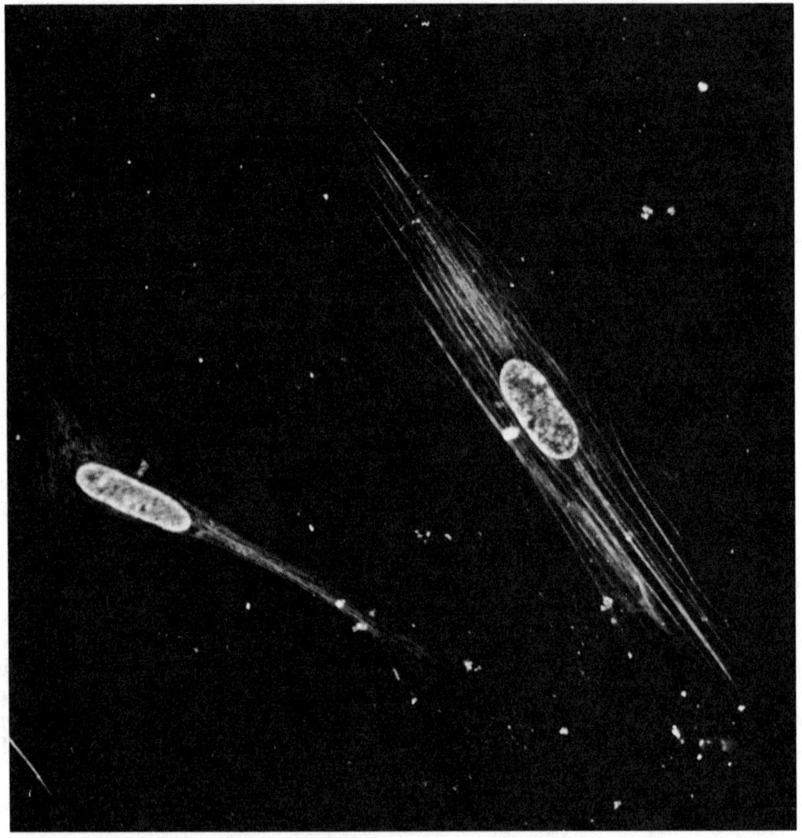

FIGURE 1. M 372 gives a granular nuclear staining and a faint staining of microfilament bundles of fibroblasts taken on the second day after subcultivation.

Actin was first found in paramyxoviruses within the virions of Sendai,[11] and then in measles virions.[10] IgM antibodies to actin are transiently observed after measles infections. These findings gave the impetus to test hybridomas established from mice immunized with purified paramyxoviruses. One donor, M 372, producing anti-actin antibodies of IgM class, was found after immunization with parainfluenza type 3 virions. It was tested by ELISA, (titer 1/2000) and by IFL technique on lymphoblas-

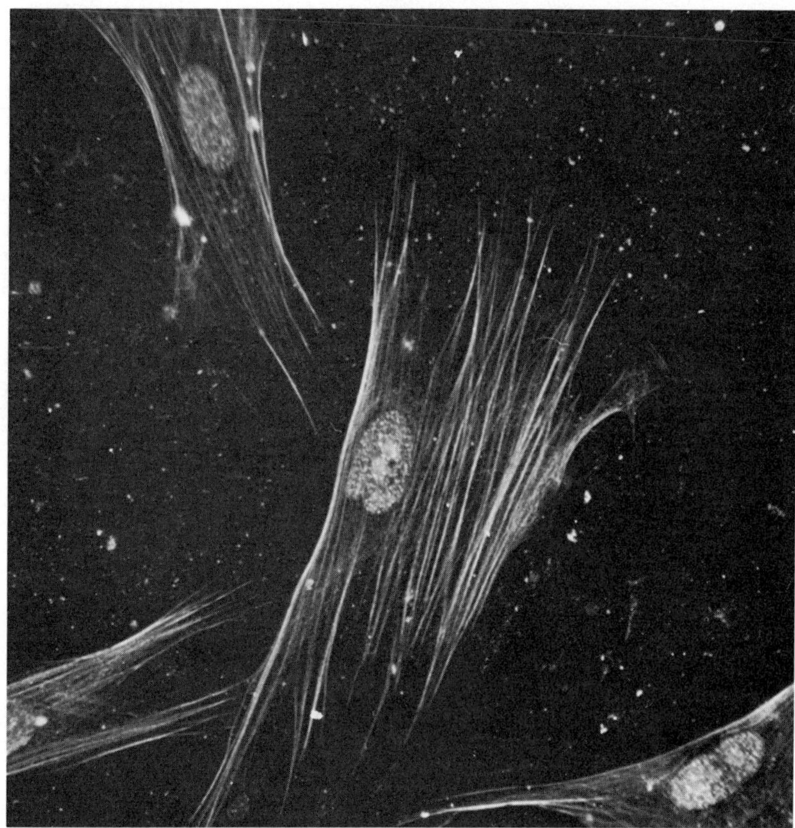

FIGURE 2. A fainter nuclear staining and a more prominent staining of filament bundles given by M 372 on fibroblasts, grown for three days.

toid cells (titer 1/1000), sparse or confluent cultures of lung fibroblasts, and tissue sections (rat stomach and kidney and human[1] thyroid) and the staining pattern was compared with that given by one rabbit anti-actin serum and 20 high-titered human anti-actin sera (TABLE 1, FIGURE 1).

As can be seen from TABLE 1 and FIGURES 1–3, there are similarities but also differences between the staining patterns given by polyclonal human antibodies and the monoclonal antibodies.

The M 372 reacted more faintly with smooth muscle in tissue sections. Secondly the monoclonal antibodies gave a distinct staining of the nucleus but a rather weak staining of actin filaments in growing fibroblasts. The polyclonal anti-actin sera investigated did not stain the nuclei, but strongly stained the microfilament bundles. One explanation could be that M 372 reacts with an epitope, to which human anti-actin sera have few antibodies. Or it may be a difference in structural arrange-

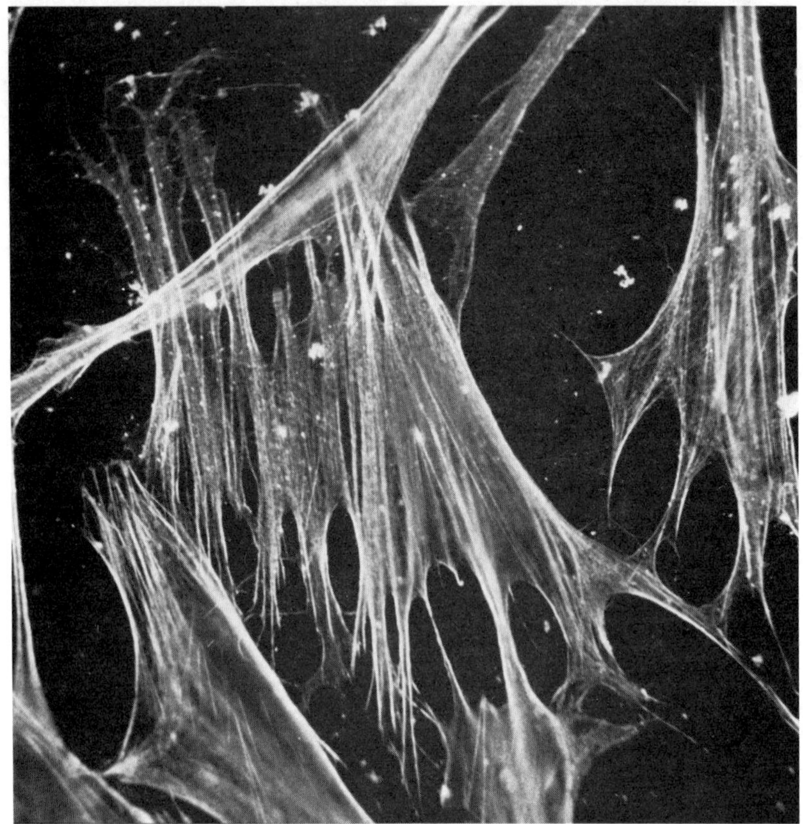

FIGURE 3. Fibroblasts grown for three days as in **FIGURE 2** but stained by AS eluate. The microfilament bundles are strongly stained but no nuclear reaction can be demonstrated.

ment and accessability of actin. Purified actin absorbed all anti-actin reactivity given by M 372, since the epitope reacting with the monoclonal antibody is available in muscle actin. The clone M 372 gives us the possibility to study actin within the nucleus. The research to find additional clones continues. Two have been established after immunization with distemper virions. These clones do not produce antibodies giving the same staining pattern as M 372.

ACKNOWLEDGMENTS

The advanced technial assistance of Miss Anita Östborn is gratefully acknowledged.

REFERENCES

1. BIBERFELD, G., A. FAGRAEUS & R. LENKEI. 1974. Reaction of human smooth muscle antibodies with thyroid cells. Clin. Exp. Immunol. **18:** 371–377.
2. FAGRAEUS, A., K. LIDMAN & R. NORBERG. 1975. Indirect immunofluorescent staining of contractile proteins in smeared cells by smooth muscle antibodies. Clin. Exp. Immunol. **20:** 469–477.
3. HARRIS, D. A. & J. H. SCHWARTZ. 1981. Characterization of brevin, a serum protein that shortens actin filaments. Proc. Natl. Acad. Sci. USA **78:** 6798–6802.
4. LIDMAN, K., G. BIBERFELD, A. FAGRAEUS, R. NORBERG, R. THORSTENSSON, G. UTTER, L. CARLSSON, J. LUCA & U. LINDBERG. 1976. Anti-actin specificity of human smooth muscle antibodies in chronic active hepatitis. Clin. Exp. Immunol. **24:** 266–272.
5. NORBERG, R., G. BIBERFELD, A. FAGRAEUS, K. LIDMAN, R. THORSTENSSON & G. UTTER. 1977. The reaction of cells with anti-actin sera in relation to the amount of cellular actin. Clin. Exp. Immunol. **28:** 512–516.
6. NORBERG, R., R. THORSTENSSON, G. UTTER & A. FAGRAEUS. 1979. F-actin depolymerizing activity of human serum. Eur. J. Biochem. **100:** 575–583.
7. THORSTENSSON, R. 1982. Thesis. Karolinska Institute. Stockholm.
8. THORSTENSSON, R., G. UTTER & R. NORBERG. 1982. Eur. J. Biochem. **126:** 11–16.
9. THORSTENSSON, R., G. UTTER, R. NORBERG, A. FAGRAEUS, J. H. HARTWIG, H. L. YIN & T. P. STOSSEL. 1982. Exp. Cell Res. **140:** 395–400.
10. TYRELL, D. & E. J. NORRBY. 1978. Structural polypeptides of measles virus. J. Gen. Virol. **39:** 219–229.
11. WANG, E., B. A. WOLF, R. A. LAMB, P. W. CHOPPIN & A. R. GOLDBERG. 1976. *In* Cell Motility. R. Goldberg, R. Pollard & J. Rosenbaum, Eds.: 671. Cold Spring Harbor Laboratory. Cold Spring Harbor, N.Y.

The Value of Immunofluorescence in the Study of Renal Disease[a]

ROBERT T. McCLUSKEY AND A. BERNARD COLLINS

Department of Pathology
Harvard Medical School
Massachusetts General Hospital
Boston, Massachusetts 02114

Based largely on studies of experimental models and on immunofluorescence findings in human renal disease, it was widely accepted by the end of the 1960s that there were two major primary immunological mechanisms capable of producing renal injury (excluding that seen in allografts): the trapping of circulating immune complexes in glomeruli or, less commonly, arteries, and the combination of circulating autoantibodies with components of the glomerular basement membrane (GBM). Both forms are characterized by abundant extracellular deposits of immunoglobulins and can therefore readily be identified by immunofluorescence. Immunologic mechanisms were also thought to be responsible for certain forms of drug-induced acute interstitial nephritis; however, the immunofluorescence findings were generally not distinctive and the pathogenetic mechanisms were not understood.

During the 1970s, experimental studies and observations in humans provided evidence for additional primary immunopathogenetic mechanisms (TABLE 1). Immunofluorescence led to the recognition of two of these: immune complex tubulointerstitial nephritis and anti-tubular basement membrane (TBM) nephritis.[3] Clearly, however, immunofluorescence has not helped in the evaluation of the possible role of cell-mediated mechanisms or of antibodies directed against cell surface antigens. It is hoped that the use of monoclonal antibodies with immunoperoxidase techniques, which permit the identification in frozen sections of surface antigens of T and B cell subsets, as well as other surface antigens, will provide relevant information.[4,21]

Immunofluorescence studies are of value not only in identifying immunologic mechanisms, but in classification and diagnosis, especially in the heterogeneous category of immune complex glomerulonephritis. Most of the currently accepted criteria for diagnosis were established by about 1970[19] and will not be reviewed here. We will instead discuss selected problems in the interpretation of immunofluorescence findings and then evaluate the use of immunofluorescence in the identification of antigens in glomeruli.

IgA NEPHROPATHY

Evaluation of the predominant immunoglobulin class in glomerular deposits led to the identification of the condition called IgA nephropathy (Berger's disease). This condition is diagnosed in a patient with primary glomerular disease with conspicuous IgA mesangial deposits (IgG and IgM are also often present). Nevertheless, it is clear

[a]Supported by the National Institutes of Health Grant No. AM18729.

Abbreviations: GBM = glomerular basement membrane, TBM = tubular basement membrane, HSP = Henoch-Schonlein purpura, HBV = hepatitis B virus, and HBsAg, HBeAg, and HBcAg = designated hepatitis B antigens Ag.

that an immune complex disease cannot be defined on the basis of the predominant immunoglobulin class. For one thing, more than one class is usually found in a given specimen and the relative amounts are difficult to measure. Furthermore, IgA is found in glomeruli in several diseases, notably Henoch-Schonlein purpura (HSP), cirrhosis, and lupus nephritis (where IgG predominates). A recent study from Japan[14] also provides evidence that IgA nephropathy is not a single disease: patients with severe progressive disease were found to have an association with HLA DR4, whereas this was not found in patients with milder disease. There is also uncertainty concerning the relationship between IgA nephropathy and HSP. Progress awaits the discovery of more specific hallmarks, such as antigens in the glomerular deposits or distinctive primary host defects. To date attempts to identify characteristic immunoregulatory defects in patients with IgA nephropathy have not given consistent results.[10]

The conclusion that a given immunoglobulin class predominates is based on the subjective grading of the relative intensity of staining. It is doubtful that improvements in quantification would be helpful in classification because even within a given disease wide variations in intensity of staining for different immunoglobulin classes are seen.

TABLE 1. Primary Immunopathogenetic Mechanisms[a] in Renal Disease

I. Glomerular diseases
    A. Immune complex glomerulonephritis
        1) Due to trapping of circulating immune complexes
        2) Due to *in situ* formation
    B. Anti GBM nephritis
    C. T cell-mediated glomerular injury
II. Tubulointerstitial nephritis (TIN)
    A. Immune complex TIN
    B. Anti TBM nephritis
    C. T cell mediated TIN
    D. Antibody-mediated cytotoxicity
        1) Antibody-dependent, cell-mediated cytotoxicity (ADCC)
        2) Complement-mediated cell lysis
III. Arteritis
    A. Immune complex
    B. Cell mediated (not well documented, except in allografts)

[a]Primary immunopathogenetic mechanisms refer to those that initiate the injury within the kidney; excluded from consideration are secondary pathogenetic mechanisms, such as complement activation or neutrophil accumulation, as well as causative factor (etiologic agents or underlying predisposing host factors).

Another approach to IgA nephropathy has been to study the IgA subclasses found in glomeruli. It was hoped that the findings might be of diagnostic value and also have implications concerning pathogenesis. However, there have been conflicting reports concerning the prevalence of IgA_1 and IgA_2 in glomeruli. Based on a study from France using rabbit antisera, Andre *et al.*[2] reported that IgA_2 was the major constituent in patients with IgA nephropathy, HSP, and cirrhosis, whereas IgA_1 predominated in lupus nephritis. In contrast, Conley *et al.*,[9] through the use of monoclonal antibodies, found IgA_1 but not IgA_2 in glomeruli in cases of IgA nephropathy, HSP, and lupus nephritis, in patients from the United States. Recently we have studied, by previously described immunofluorescence techniques,[6,27] specimens from 13 patients classified as IgA nephropathy and one case of HSP nephritis, using monoclonal antibodies against IgA subclasses and fluorescein-labeled goat anti-mouse IgG (Cappel Laboratories, Inc., Cochranville, PA); in all cases IgA_1 was found in the deposits, whereas IgA_2 was detected in only two cases, and in these the staining was faint. The specificity of the monoclonal antibodies we used was assessed by the manufacturer (Bethesda Research Laboratories, Inc., Gaithersburg, MD) by solid phase enzyme-linked immunoabsorbent assay (ELISA) with purified IgA_1, IgA_2, and secretory component of IgA. The manufacturers state that the antibodies do not cross-react with other human immunoglobulin heavy chains or with light chains and that the antibodies have a stability of three months (use after this time, in our hands, has failed to give positive staining). We obtained additional evidence of specificity by showing that only the anti-IgA_1 antibody reacted with splenic plasma cells, whereas both stained plasma cells in the small intestine. In the study of Andre et al.,[1] in which heteroantisera were used, specificity was also examined by several criteria. It should be noted that in their study both IgA_1 and IgA_2 were detected in the glomerular deposits in most cases and that the conclusion that one subclass predominated was based on estimation of the intensity of staining. However, five of the cases of Berger's nephropathy showed no staining for IgA_1.

These discordant findings raise the question as to whether results obtained with monoclonal antibodies should automatically be given priority over these obtained with heteroantisera. Although monoclonal antibodies generally have far greater specificity, unexpected cross-reactivities have been described.[17] (However, it seems unlikely that a monoclonal antibody that has been characterized as being reactive with only one IgA subclass would, on occasion, unexpectedly react with the other subclass). Conversely, because they fail to detect certain determinants, monoclonal antibodies may give false-negative results. With respect to the latter point, it can be noted that in our studies, as well as in those of Conley et al.,[9] the anti-IgA_2 antibodies reacted with tubular casts. Although the results obtained with monoclonal antibodies favor the conclusion that the predominant and often the only IgA subclass present in glomerular deposits in IgA nephropathy is IgA_1, further studies of such cases from various geographic areas, using additional monoclonal antibodies, are needed to resolve this issue. Why one subclass might be selectively deposited in glomeruli, presumably in immune complexes, is not clear; possible explanations have been suggested by Conley et al.[9]

Aside from the studies just cited, monoclonal antibodies have not been widely used to detect immunoglobulin classes in tissue sections. For most purposes conventional heteroantisera appear to be satisfactory. However, there is reason to believe that at least some results of staining for IgE are not reliable. An early report describing IgE glomerular deposits in lipoid nephrosis has not been confirmed.[11] Nagai et al.[23] described IgE in glomerular deposits in a patient with asthma and glomerulonephritis. We have observed staining for IgE in deposits in 16 of 86 cases of membranous glomerulonephritis using heteroantisera to IgE (Cappel Laboratories, Cochranville, PA). However, with a Fc-specific monoclonal antibody (Hybrid-tech, Inc., La Jolla, CA) affinity constant of 10^9 L/M, no staining was observed. It seems likely that the results obtained with the heteroantiserum were due to undesired cross-reactivity; in support of this interpretation we found that the heteroantiserum failed to produce cytoplasmic staining of plasma cells in a frozen section of nasal polyp, whereas the monoclonal antibody resulted in the staining of an appreciable number of plasma cells.

ANTI–BASEMENT MEMBRANE NEPHRITIS

Although the immunofluorescence features in most cases of anti-GBM nephritis are typical, in some cases diagnosis is difficult. The diagnosis obviously requires the demonstration of anti-GBM antibodies. The typical linear accumulation of Ig (almost always IgG) along the GBM provides evidence for *in vivo* bound antibodies. However, non-immunologically bound IgG can produce a similar picture, as is seen frequently in autopsy tissue or in biopsy specimens from diabetics, and not infrequently in other conditions. Thus, when linear GBM staining for IgG is found, tests should be performed to detect antibodies in the circulation or, if possible, in eluates of renal tissue. Anti-GBM antibodies have generally been detected by indirect immunofluorescence, but recently described radioimmunoassays tests[32] appear to be more sensitive and reliable. Unfortunately, these assays are not widely available.

In the typical case of anti-GBM disease with crescentic nephritis and bright linear staining for IgG along the GBM (and often the TBM), the diagnosis usually presents no problem; such cases, however, are rare. McPhaul and Mullins[22] have suggested that anti-GBM nephritis accounts for many cases of glomerulonephritis, including some with minor glomerular damage; this view has not been substantiated, nor completely excluded, largely because of the difficulties in detecting anti-GBM antibodies.

Is there reason to hope that modification of the usual immunofluorescence techniques will permit the distinction between anti-GBM antibodies and non-immunologically bound IgG? Greater sensitivity in the detection of bound IgG would not help and identification of IgG subclasses would not appear to be promising. Perhaps different washing techniques might selectively remove nonspecifically bound IgG. Recently, Sisson *et al.*[29] have reported on ultrastructural studies employing peroxidase-labeled antibodies, which have revealed that anti-GBM antibodies are reactive only with the lamina rara interna. Thus, high resolution studies of the distribution of IgG may be useful in diagnosis.

IDENTIFICATION OF ANTIGENS IN IMMUNE COMPLEX GLOMERULAR DISEASES

In all forms of immune complex glomerulonephritis immunofluorescence shows irregular or granular deposits of immunoglobulins and complement components within glomeruli. Although these findings provide evidence for immune complexes, proof requires demonstration that at least some of the immunoglobulins represent specific antibodies combined with antigen. This is best accomplished by demonstrating and measuring specific antibodies eluted from renal tissue. Evidence of this sort has been obtained in only a few human diseases, notably in severe lupus nephritis, where DNA–anti-DNA complexes have been shown to be present.[16] Less direct evidence for immune complexes can be provided by demonstrating specific antigens within deposits. However, because false-positive results are frequently obtained by immunofluorescence, claims that antigens have been demonstrated should be scrutinized carefully. In any case, antigens have been reported to be present in only a small percentage of cases (excluding lupus nephritis). Moreover, some of the findings, such as the presence of brush border antigens in membranous nephritis,[26] have not been confirmed in larger series.[6] Autologous, possibly tumor-associated, antigens have been described in some patients with neoplasia and presumed immune complex glomerulonephritis. However, the evidence for the specificity of these findings is open to serious question.

Why has the search for antigens in glomerular deposits been so unsuccessful? One reason is that in most cases there is no clue to the identity of the antigens and therefore no way of screening for them. However, even when a certain antigen is suspected, documentation of its presence in glomeruli has often been unconvincing. For example, there is conflicting evidence concerning the presence of streptococcal antigens in deposits in post-streptococcal glomerulonephritis.[20] It is possible that masking by excess antibodies makes detection of microbial antigens difficult in some cases.

Another explanation for the failure to identify antigens in glomerular deposits is that some are not recognized as antigens. In particular, immunoglobulins may serve as antigens and combine with rheumatoid factor[13] or antiidiotypes.[28] At the moment, prospects for rapid progress in identification of specific antibodies or antigens in glomerular deposits do not appear promising. The development of high titered monoclonal antibodies against microbial agents should permit more effective screening. Development of sensitive techniques for the detection of antibodies in eluates may make it possible to search for specific antibodies obtained from small biopsy specimens. Although the search for specific antibodies and antigen is indicated, it is possible that the basic problem in some forms of immune complex glomerulonephritis is an inability of the host to dispose of immune complexes of a variety of specificities, in which case no one antigen antibody system may be of paramount importance.

To end this discussion in an area where positive results have been obtained, we will review evidence concerning the role of hepatitis B antigens in glomerulonephritis. In view of the established role of viruses in several forms of immune complex glomerulonephritis in animals, viruses are considered likely causes of some cases of chronic immune complex glomerulonephritis in humans. However, to date only hepatitis B (HBV) infection appears to be reasonably well established as a causative agent. There are geographic differences in the importance of HBV infection in this respect. Thus, the incidence of HBV infection in patients with glomerulonephritis (generally membranous or membranoproliferative) has been reported to be high in areas where the carrier rate is high, such as Poland, Hungary, and Japan.[5,25,30,33] In contrast, the frequency in the United States or Great Britain appears to be quite low.[15]

Although hepatitis antigens have been detected in glomerular deposits by immunofluorescence in some cases, there is disagreement about the identity of the antigen: the designated hepatitis B antigens Ag (HBsAg, HBeAg, and HBcAg) have all been described.[5,8,24,30,31] Slusarczyk et al.[30] studied 24 HBsAg seropositive children with glomerulonephritis; about equal numbers were found to have HBcAg alone, HBsAg alone, or both antigens in the deposits. Staining for HBeAg was not attempted. In contrast, Ito et al.,[12] in a study of six HBsAg seropositive children with membranous glomerulonephritis, found HBeAg in glomeruli in four cases and HBsAg in only one; HBcAg was not demonstrable. These authors presented reasons for concluding that the antisera employed by Slusarczyk et al.[30] probably contained anti-HBe as well as anti-HBc antibodies.

We have recently studied a 29-year-old homosexual male with chronic hepatitis B infection and immune complex glomerulonephritis.[7] Glomerular immunoglobulin-containing deposits were found in both peripheral loops (predominantly epimembranous) and mesangial sites. Immunofluorescence studies, which included the use of a monoclonal antibody, showed that HBsAg was present in mesangial deposits but not in peripheral loop deposits. Immunoperoxidase studies employing HRP-conjugated anti-HBc and anti-HBe showed that the epimembranous deposits contained HBe but not HBc. We presented evidence that the staining was not due to binding of the probe antibodies to IgM antibodies with rheumatoid factor activity in the deposits, which has been reported as a cause of false-positive staining for HBsAg.[18] Our finding of HBeAg, but not HBcAg in the peripheral loop deposits, is consistent with the observations of Ito et al.[12] and with their interpretation that HBeAg is the most important hepatitis antigen involved in the formation of epimembranous deposits. Further studies are needed, employing well-defined antibodies, especially monoclonal antibodies, to assess the importance and nature of hepatitis antigens in glomerulonephritis.

REFERENCES

1. ANDRE, C., F. ANDRE & M. C. FARGIER. 1978. Distribution of IgA_1 and IgA_2 plasma cells in various normal human tissues and in the jejunium of plasma IgA-deficient patients. Clin. Exp. Immunol. **33:** 327–331.
2. ANDRE, C., F. C. BERTHOUX, F. ANDRE, J. GILLON, C. GENIN & J. C. SABATIER. 1980. Prevalence of IgA_2 deposits in IgA nephropathies. N. Engl. J. Med **303:** 1343–1346.
3. ANDRES, G. A. & R. T. McCLUSKEY. 1975. Tubular and interstitial renal disease due to immunologic mechanisms. Kid. Int. **7:** 271–289.
4. BHAN, A. K., L. M. NADLER, P. STASHENKO, R. T. McCLUSKEY & S. F. SCHLOSSMAN. 1981. Stages of B cell differentiation in human lymphoid tissue. J. Exp. Med. **154:** 737–749.
5. BRZOSKO, W. J., K. KRAWCZYNSKI, T. NAZAREWICZ, M. MORZYCKA & A. NOWOSLAWSKI. 1974. Glomerulonephritis associated with hepatitis B surface antigen immune complexes in children. Lancet **2:** 477–481.
6. COLLINS, A. B., G. ANDRES & R. T. McCLUSKEY. 1981. Lack of evidence for a role of renal tubular antigen in human membranous glomerulonephritis. Nephron **27:** 297–301.
7. COLLINS, A. B., A. K. BHAN, J. L. DIENSTAG, R. B. COLVIN, G. T. HAUPERT, JR., I. K. MUSHAHWAR & R. T. McCLUSKEY. 1983. Hepatitis B immune complex glomerulonephritis: simultaneous glomerular deposition of hepatitis B surface and E antigens. Clin. Immunol. and Immunopathol. **26:** 137–153.
8. COMBES, B., P. STASTNY, J. SHOREY, E. H. EIGENBRODT, A. BARRERA, A. R. HULL & N. W. CARTER. 1971. Glomerulonephritis with deposition of Australia antigen—antibody complexes in glomerular basement membrane. Lancet **2:** 234–237.
9. CONLEY, M. E., M. D. COOPER & A. F. MICHAEL. 1980. Selective deposition of immunoglobulin A_1 in immunoglobulin A nephropathy, anaphylactoid purpura nephritis, and systemic lupus erythematosus. J. Clin. Invest. **66:** 1432–1435.
10. COSIO, F. G., S. LAM, A. O. FOLAMI, M. E. CONLEY & A. F. MICHAEL. 1982. Immune regulation of immunoglobulin production in IgA nephropathy. Clin. Immunol. Immunopathol. **23:** 430–436.
11. GERBER, M. A. & F. PARONETTO. 1971. IgE in glomeruli of patients with nephrotic syndrome. Lancet **1:** 1097–1099.
12. ITO, H., S. HATTORI, I, MATUSDA, S. AMAMIYA, H. HAJIKANO, H. YOSHIZAWA, Y. MIYAKAWA & M. MAYUMI. 1981. Hepatitis B e antigen-mediated membranous glomerulonephritis. Lab. Invest. **44:** 214–220.
13. KANO, K. & F. MILGROM. 1980. Immune complex disease. Vox Sang. **38:** 121–137.
14. KASAHARA, M., K. HAMADA, T. OKUYAMA, N. ISHIKAWA, G. OGASAWARA, H. IKEDA, T. TAKENOUCHI, A. WAKISAKA, M. AIZAWA, Y. KATAOKA, R. MIYAMOTO, M. KOHARA, S. NAITO, N. KASHIWAGI & Y. HIKI. 1982. Role of HLA in IgA nephropathy. Clin. Immunol. Immunopathol. **125:** 189–195.
15. KOHLER, P. F., R. E. CRONIN, W. S. HAMMOND, D. OLIN & R. I. CARR. 1974. Chronic membranous glomerulonephritis caused by hepatitis B antigen—antibody immune complexes. Ann. Intern. Med. **81:** 448–451.
16. KRISHNAN, C. & M. H. KAPLAN. 1967. Immunopathologic studies of systemic lupus erythematosus. II. Antinuclear reaction of α globulin eluted from homogenates and isolated glomeruli of kidneys from patients with lupus nephritis. J. Clin. Invest. **46:** 569–579.
17. LAFER, E. M., J. RAUCH, C. ANDRZEJEWSKI, JR., D. MUDD, B. FURIE, B. FURIE, R. S. SCHWARTZ & D. STOLLAR. 1981. Polyspecific monoclonal lupus autoantibodies reactive with both polynucleotides and phospholipids. J. Exp. Med. **153:** 897–909.
18. MAGGIORE, Q., F. BARTOLOMEO, A. L'ABBATE & V. MISEFARI. 1981. HBsAg glomerular deposits in glomerulonephritis factor artifact? Kidney Int. **19:** 579–586.
19. McCLUSKEY, R. T. 1971. The value of immunofluorescence in study of human renal disease. J. Exp. Med. **134:** 242s–255s.
20. McCLUSKEY, R. T. 1978. An evaluation of current knowledge of glomerular diseases. *In* Prevention of Kidney and Urinary Tract Diseases. C. Coggins & N. B. Cummings, Eds.:49. National Institute of Health. Maryland.

21. McCLUSKEY, R. T. & A. K. BHAN. 1982. Cell mediated mechanisms in renal disease. Kidney Int. **21:** S6–S12.
22. McPHAUL, J. J. JR. & J. D. MULLINS. 1976. Glomerulonephritis mediated by antibody to glomerular basement membrane. Immunological, clinical and histopathologic characteristics. J. Clin. Invest. **57:** 351–361.
23. NAGAI, T., T. TAMURA, K. OGINO & T. KITA. 1973. IgE deposits in glomeruli with membranous nephropathy and marked asthmatic predisposition in humans. Jap. Circula. J. **37:** 1227–1232.
24. NAGY, J., A. PAR, G. BAJTAI, M. AMBRUS & G. DEAK. 1976. Membranous glomerulonephritis induced by HBs (Australia) antigen—antibody complexes. Act. Morph. Acad. Sci. Hung. **24:** 129.
25. NAGY, J., G. BAJTAI, H. BRASCH, T. SULE, M. AMBRUS, G. DEAK & A. HAMORI. 1979. The role of hepatitis B surface antigen in the pathogenesis of glomerulopathies. Clin. Nephrol. **12:** 109–116.
26. NARUSE, T., Y. MIYAKAWA, K. KITAMURA & S. SHIBATA. 1974. Membranous glomerulonephritis mediated by renal tubular epithelial antigen antibody complex. J. Allergy Clin. Immunol. **54:** 311–318.
27. PETTERSSON, E. E., A. K. BHAN, E. E. SCHNEEBERGER, A. B. COLLINS, R. B. COLVIN & R. T. McCLUSKEY. 1978. Glomerular C_3 receptors in human renal disease. Kidney Int. **13:** 245–252.
28. ROSE, L. M. & P. H. LAMBERT. 1980. The natural occurrence of circulating idiotype—antiidiotype complexes during a secondary immune response to phosphorylcholine. Clin. Immunol. Immunopathol. **15:** 481–492.
29. SISSON, S., N. K. DYSART, JR., A. J. FISH & R. L. VERNIER. 1982. Localization of the Goodpasture antigen by immunoelectron microscopy. Clin. Immunol. Immunopathol. **23:** 414–429.
30. SLUSARCZYK, J., T. MICHALAK, T. NAZAREWICZ-DE MEZER, K. KRAWCZYNSKI & A. NOWOSLAWSKI. 1980. Membranous glomerulopathy associated with hepatitis B core antigen immune complexes in children. Am. J. Path. **98:** 29–39.
31. TAKEKOSHI, Y., M. TANAKA, Y. MIYAKAWA, H. YOSHIZAWA, K. TAKAHASHI & M. MAYUMI. 1979. Free "small" and IgG-associated "large" hepatitis B e antigen in the serum and glomerular capillary walls of two patients with membranous glomerulonephritis. N. Engl. J. Med **300:** 814–819.
32. WILSON, C. B., H. MARGUADT & F. J. DIXON. 1973. Radioimmunoassay for circulating antiglomerular basement membrane (GBM) antibodies. Kidney Int. **6:** 114a.
33. YAMASHITA, F., H. MATSUU, N. FUJISAWA, Y. ITO, S. SHINDO, T. FUJIMOTO, K. YOSHIMOTO, K. NAGAYAMA, H. ARAKI & M. NAKANO. 1979. HBs antigen associated nephropathy in children. First Asian Pacific Congress of Nephrology. (Tokyo, 1).

Rectification of Immunological Abnormalities and Lupus Nephritis by the Transfer of Bone Marrow Cells[a]

MASAKI SHIRAKI, MICHIO FUJIWARA,
AND KYOICHI KANO[b,c]

Department of Immunology
Institute of Medical Science
University of Tokyo
Tokyo 108 Japan
and
[b]Department of Microbiology
School of Medicine
State University of New York at Buffalo
Buffalo, New York 14214

Various attempts have been made to treat mice suffering from symptoms similar to those of SLE. In the present study, MRL/Mp-*lpr/lpr* (MRL/1, a strain of mice homozygous for *lpr* gene; severe lupus nephritis) mice[11] were employed as recipients of bone marrow grafts from a congenic strain MRL/Mp +/+ (MRL/n, a strain of mice lacking *lpr* gene; mild lupus nephritis). The MRL/1 strain is homozygous for the *lpr* gene that accelerates development of lymphoproliferation and autoimmunity. They form anti-DNA antibodies and immune complexes and suffer from a rather severe lupus nephritis with 50% mortality at 5–6 months of age, whereas the congenic MRL/n mice lacking the *lpr* gene suffer from a mild form of nephritis in the second year of their lives.[1,11,14–16]

MATERIALS AND METHODS

MRL/1 mice at two months of age were irradiated at 850 rads in the ^{137}Cs irradiator (γ-cell 40, Atomic Energy of Canada, Ottawa). Subsequently, they received 5×10^6 bone marrow cells from either MRL/n and MRL/1 mice. They were designated as (n → 1) and (1 → 1), respectively. After the treatments, they were maintained in sterilized cages with filter caps on an Isorack® (Sanki Kagoku Co.,

[a]Supported by a research grant from the Ministry of Health and Welfare, the Government of Japan, and the U.S.-Japan Cooperative Cancer Research Program, sponsored by the National Cancer Institute and the Japan Society for the Promotion of Science.

[c]Address correspondence to Dr. Kyoichi Kano, Department of Immunology, Institute of Medical Science, University of Tokyo, 4-6-1 Shiroganedai, Minatoku, Tokyo 108, Japan.

Abbreviations: BMC = bone marrow cells, MRL/1 = strain of mice homozygous for lpr gene; severe lupus nephritis, MRL/n = strain of mice lacking lpr gene; mild lupus nephritis, IgSC = immunoglobulin secreting cells, Thy-1 = antigen on T cell membrane, HE = hematoxylin and eosin staining, PAS = periodic acid Schiff staining, ds-DNA = double-stranded DNA, and ss-DNA = single-stranded DNA.

Tokyo), which supplied filtered sterile air. During the first month of experimentation, they were given drinking water containing 1 mg/ml of oxytetracycline. Urinary protein was examined by Albustix tip (Miles-Sankyo Co., Tokyo) and over 300 mg/dl was considered as positive.

Immunoglobulin-secreting cells (IgSC) were detected by plaque assay[6] using protein A–coated sheep erythrocytes as target cells. Plaque-forming cells corresponding to IgSC were enumerated in a Cunningham's chamber.[4] Cells positive for Thy-1 antigen were enumerated by indirect immunofluorescence technique as described previously.[8]

Serum levels of anti-DNA antibodies were measured by enzyme immunoassay. Commercially obtained DNA (Miles Laboratory, Elkhart, IN) was reextracted with chloroform-isoamylalcohol. Single-stranded DNA (ss-DNA) was prepared by heating ds-DNA at 100°C for 15 min and then immediately cooling in an ice bath. The test was performed essentially following the procedures described in Aotsuka *et al.*[2] and Mathiesen *et al.*[9]

Sections of formalin-fixed kidney were stained with hematoxylin and eosin (HE) or with periodic acid Schiff (PAS). Immunofluorescence studies were performed following the procedure described in Kawamura[7] using fluorescein-conjugated rabbit antisera to mouse IgG and C3.

RESULTS

Recipient mice were sacrificed five months after bone marrow cell (BMC) transfer and their mesenteric lymph nodes were weighed and the number of splenic IgSCs was determined. As shown in FIGURE 1a, the weight of mesenteric lymph nodes of the (n → 1) mice was far less than that of (1 → 1) mice; mean ± SD was 43.2 ± 29.8 mg for the (n → 1) mice and 293 ± 122.1 mg for the (1 → 1) mice (p < 0.001). As seen in FIGURE 1b, the number of IgSC of (n → 1) mice was significantly smaller than (1 → 1) mice; mean ± SD was 1.19 ± 0.68 and 3.89 ± 1.44 (× 10^6/spleen), respectively

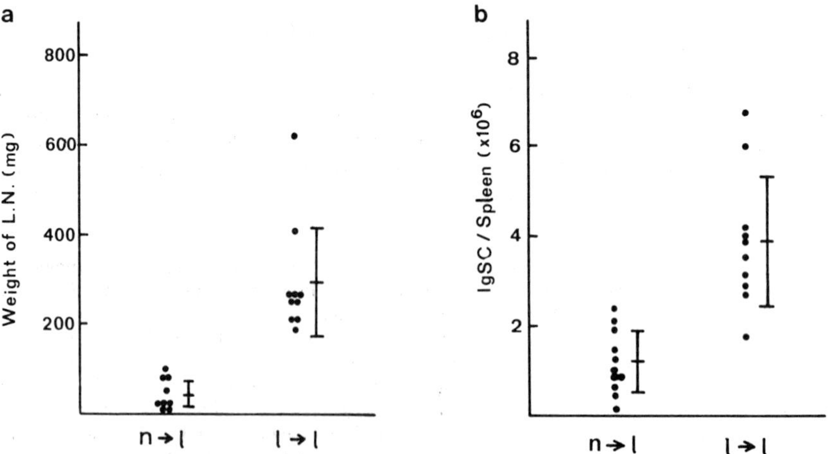

FIGURE 1. Weight of mesenteric lymph nodes of (n → 1) and (1 → 1) mice excised at sacrifice (a) and number of splenic IgSC in the mice (b).

TABLE 1. Proteinuria and Intensity of Immunofluorescence (IF) Staining Within Glomeruli

		#1	#2	#3	#4	#5	#6	#7	#8	#9	#10	#11
(1→1)	Proteinuria	+[a]	+	+	−	−	−	−	−	−	−	
	IgG	3	3	4	3	4	4	3	3	3	1	
	IF[b]											
	C3	3	4	4	3	4	4	3	3	4	1	
(n→1)	Proteinuria	+	−	−	−	−	−	−	−	−	−	−
	IgG	4	4	3	1	1	1	1	1	1	1	1
	IF											
	C3	4	3	2	2	2	1	1	1	1	1	1

[a]Urinary protein more than 300 mg/dl.
[b]Grades of intensity of IF staining: (1) focal scanty fluorescence, (2) general diffuse fluorescence, (3) intermediate between 2 and 4, and (4) marked diffuse fluorescence.

($p < 0.001$). Thy-1–positive cells were also determined in spleens of some of these mice. The mean values were 51.7% for (n → 1) and 67.2% for (1 → 1) mice.

Serum levels of anti-ss DNA antibodies were then studied. At the time of sacrifice, the levels of (1 → 1) mice were quite high in all mice showing the mean O.D. value of 0.7, whereas those of (n → 1) mice varied from <0.1 to 1.0.

Subsequently, development of lupus nephritis was studied in 11 (n → 1) and 10 (1 → 1) mice. The results are summarized in TABLE 1. Severe proteinuria was observed in three (1 → 1) mice, whereas only one of the (n → 1) mice developed such a degree of proteinuria. Intensity of fluorescence for both anti-IgG and anti-C3 reagents was grade 3 or grade 4 for the kidneys of the majority of (1 → 1) mice, while kidneys of most of the (n → 1) mice showed only focal, scanty fluorescence (grade 1).

As seen in FIGURE 2a, intense granular deposits of IgG were observed along capillary walls and in the mesangium of glomeruli of the kidney from (1 → 1) mouse. On the other hand, only weak and focal mesangial staining was noted in the glomeruli of a (n → 1) mouse FIGURE 2b. Similar staining patterns were observed with anti-C3 conjugate. Histologic examination of the kidneys of (1 → 1) mice showed diffuse endocapillary proliferation, marked increase of PAS-positive materials in the glomerular mesangium and capillaries, adhesion of the glomerular tuft to the parietal epithelium of Bowman's capsule, and prominent infiltration of the interstitium with mononuclear cells (FIGURE 2c). Although mild endocapillary proliferation was noted, deposition of PAS-positive materials and mononuclear cell infiltration were significantly less pronounced in the kidney of (n → 1) mice (FIGURE 2d).

DISCUSSION

The results of this study showed that the early onset of abnormalities, such as massive lymphoproliferative syndrome and severe lupus nephritis in MRL/1 mice, was restrained by transfer of BMC from the congenic strain MRL/n mice. The establishment of true bone marrow chimerism in the (n → 1) mice could not be assessed at the present time because of lack of reagents to detect *lpr* gene products. However, unexpected observation that some (n → 1) mice, especially females, tended to die earlier than the control (1 → 1) due to what appeared to be chronic graft versus host reactions indicated a chimeric state of the (n → 1) mice.

Since MRL/n mice do develop chronic immune complex nephritis and vascular

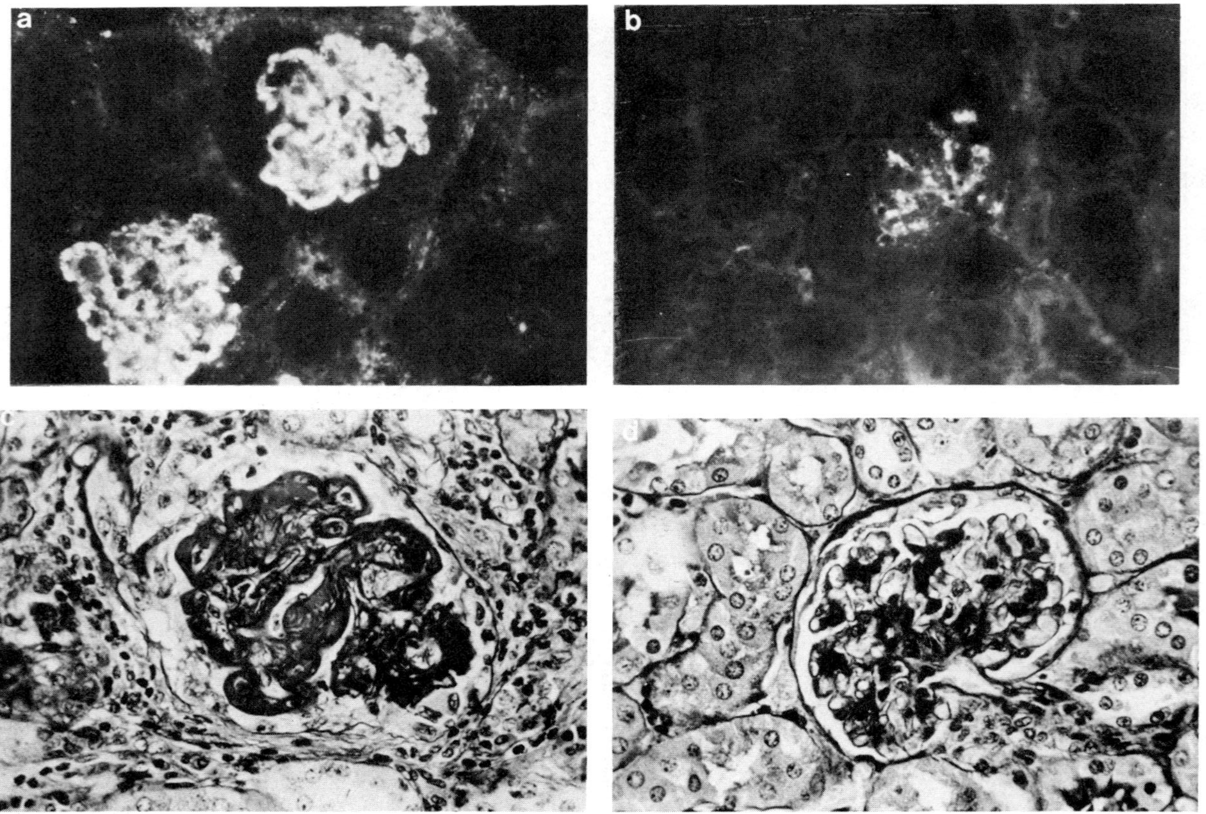

FIGURE 2. IF staining with anti-IgG conjugate and PAS staining of kidney sections ($\times$400). IF of specimens from (1 → 1) mice (a) and (n → 1) mice (b). PAS staining of specimens from (1 → 1) mice (c) and (n → 1) mice (d).

disease at their older age, *lpr* gene alone cannot possibly be responsible for the onset of the morbid processes and, therefore, the role of the gene would be to accelerate or augment the predisposition inherent to both MRL/1 and MRL/n strains.[12] In agreement with the observations by other investigators[3,5,10] on NZB, BXSB, and MRL/1 strains, the results of this study confirmed that genetically determined potential for early onset of lymphoproliferation leading to autoimmunity resides in hemopoietic stem cells of MRL/1 strain.

Further studies on this experimental model would have to be performed to investigate not only other factors responsible for the induction of autoimmunity, but also the subtle obstacle encountered in the BMC transfer, since bone marrow transplantation has become a potential clinical approach for the treatments of immunological and hematological disorders.[13]

SUMMARY

Two-month old MRL/1 mice, which spontaneously develop lymphoproliferative syndrome and severe lupus nephritis at 4–5 months of age, were irradiated and reconstituted with bone marrow cells from a congenic strain MRL/n mice. In the chimeric (n → 1) mice, the early onset of the lupus nephritis was prevented as evidenced by decrease in the degree of proteinuria, diminished intensity of immune deposits in glomeruli, and milder histopathologic changes in kidneys. Lymph node swelling as well as generation of a large number of splenic IgSC was also prevented. Anti-ss DNA antibody response in the chimeric mice, however, varied tremendously from one animal to another.

ACKNOWLEDGMENT

The authors with to thank Ms. Ai Kariyone for her excellent technical assistance.

REFERENCES

1. ANDREWS, B. S., R. A. EISENBERG, A. N. THEOFILOPOULOS, S. IZUI, C. B. WILSON, P. J. MCCONAHEY, E. D. MURPHY & F. J. DIXON. 1978. Spontaneous murine lupus like syndromes. Clinical and immunopathological manifestations in several strains. J. Exp. Med. **148:** 1198–1215.
2. AOTSUKA, S., M. OKAWA, K. IKEBE, & R. YOKOHARI. 1979. Measurement of anti-double-stranded DNA antibodies in major immunoglobulin classes. J. Immunol. Methods **28:** 146–162.
3. BALDERAS, A. N., A. N. THEOFILOPOULOS & F. J. DIXON. 1981. Cell transfer studies among MRL/Mp-*lpr*/*lpr* and MRL/Mp-+/+ SLE-prone murine substrains. Fed. Proc. **40:** 973.
4. CUNNINGHAM, A. J. & A. SZENBERG. 1968. Further improvements in the plaque technique for detecting single antibody-forming cells. Immunology **14:** 599–601.
5. EISENBERG, R. A., S. IZUI, P. J. MCCONAHEY, L. HANG, C. J. PETERS, A. N. THEOFILOPOULOS & F. J. DIXON. 1980. Male determined accelerated autoimmune disease in BXSB mice; Transfer by bone marrow and spleen cells. J. Immunol. **125:** 1032–1036.
6. GRONOWICZ, E., A. COUTINHO & F. MELCHERES. 1976. A plaque assay for all cells secreting Ig of a given type or class. Eur. J. Immunol. **6:** 588–590.
7. KAWAMURA, A., JR. 1977. Fluorescent antibody techniques and their applications. 2nd edit. pp. 77–94. University of Tokyo Press. Tokyo.

8. KUHARA, T., M. FUJIWARA, K. SUDO, K. SUZUKI & A. KAWAMURA, JR. 1980. Immunological properties of athymic nude mice born from homozygous (nu/nu) parents. Laboratory Animals **14:** 167–172.

9. MATHIESEN, L. R., S. M. FEINSTONE, D. C. WONG, P. SKINHOEJ & R. H. PURCELL. 1978. Enzyme-linked immunosorbent assay for detection of hepatitis A antigen in stool and antibody to hepatitis A antigen in sera: Comparison with solid-phase radioimmunoassay, immune electron microscopy, and immune adherence hemagglutination assay. J. Clin. Microbiol. **7:** 184–193.

10. MORTON, J. I. & V. SIEGEL. 1974. Transplantation of autoimmune potential. I. Development of anti-nuclear antibodies in H-2 histocompatible recipients of bone marrow from New Zealand Black mice. Proc. Natl. Acad. Sci. USA **71:** 2162–2165.

11. MURPHY, E. D. & J. B. ROTHS. 1978. Autoimmunity and lymphoproliferation: Induction by mutant gene *lpr*, and acceleration by a male-associated factor in strain BXSB mice. *In* Genetic Control of Autoimmune Disease. N. R. Rose, P. E. Bigazzi & N. L. Warner, Eds.: 207–221. Academic Press. New York.

12. PISETSKY, D. S., S. A. CASTER, J. B. ROTH & E. D. MURPHY. 1982. *lpr* gene control of the anti-DNA antibody response. J. Immunol. **128:** 2322–2325.

13. STORB, R. & E. D. THOMAS. 1979. Human bone marrow transplantation. Transplantation **28:** 1–3.

14. THEOFILOPOULOS, A. N. & F. J. DIXON. 1981. Etiopathogenesis of murine SLE. Immunol. Rev. **55:** 179–216.

15. THEOFILOPOULOS, A. N., R. A. EISENBERG, M. BOURDON, J. S. CROWELL, JR. & F. J. DIXON. 1979. Distribution of lymphocytes identified by surface markers in murine strains with systemic lupus erythematosus-like syndromes. J. Exp. Med. **149:** 516–534.

16. THEOFILOPOULOS, A. N., P. J. McCONAHEY, S. IZUI, R. A. EISENBERG, A. B. PEREIRA & W. D. CREIGHTON. 1980. A comparative immunological analysis of several murine strains with autoimmune manifestations. Clin. Immunol. Immunopathol. **15:** 258–278.

Methodological Aspects of Immunofluorescence Applied to Nephrology

F. SHIMIZU,[a] S. SAEGUSA,[b] AND A. KAWAMURA, JR.[b]

[a]Department of Immunology
Institute of Nephrology
School of Medicine
Niigata University
Niigata, Japan

[b]Department of Immunology
Institute of Medical Science
University of Tokyo
Tokyo, Japan

INTRODUCTION

Immunofluorescence has been generally applied to the immunopathological studies of renal diseases and has played an important role in establishing that many human renal diseases have an immunological basis.[2] However, it seems that no systematic study of its application to kidneys has been performed and that no effort to standardize it in its application to kidneys has been made.

The purpose of this report is to reexamine some basic problems in immunofluorescence as applied to nephrology, especially on the problems of pretreatments so that more detailed and precise information can be obtained from the fluorescent patterns of kidneys.

MATERIALS AND METHODS

Kidneys with positive staining for immunoglobulins were obtained from (1) NZB/WF1 mice, (2) WKA and SHR rats with Masugi nephritis or with experimental immune complex nephritis (bovine serum albumin [BSA] or human IgG [HGG]-induced nephritis, active and passive Heymann nephritis [AHN;PHN]), (3) WKA rats with aminonucleoside [AN] nephrosis and with alloxan diabetes, (4) normal mice of various strains, (5) uremic cats, (6) normal mice into which HGG was injected intravenously four hours before sacrifice, and (7) human renal biopsy specimens. In order to examine the direct effect of solvents used for pretreatments, fluorescein isothiocyanate (FITC)-labeled HGG was also injected and stained further with tetramethylrhodamine isothiocyanate (TRITC)-labeled anti-HGG *in vitro*.

Antisera were prepared as previously described[2,4] and labeled with FITC[1] (Interna-

Abbrevations: BSA = bovine serum albumin, HGG = human IgG, AHN = active Heymann nephritis, PHN = passive Heymann nephritis, AN = aminonucleoside, PBS = phosphate-buffered saline, FITC = fluorescein isothiocyanate, and F/P = fluorescein/protein.

tional Biological Supplies Inc., Florida) or were obtained commercially. All procedures in immunofluorescence were performed, in principle, according to the method of Kawamura *et al.*[1] Briefly, kidneys were frozen quickly in a large test tube containing *n*-hexane precooled in dry-ice–acetone mixture. Sections of 4 μm thickness were made with a cryostat (Sakura, Tokyo), and pretreated at room temperature for 5 min. Sections were stained with FITC-labeled antibody at 37°C for 30 min, then washed under continuous shaking in three changes of PBS for 15 min, air-dried, and mounted with buffered glycerol. The preparations were observed by a transmitted-light (Tiyoda) or an incident-light fluorescence microscope (Zeiss) under UV excitation, or with an interference filter system for FITC. Fujichrome (ASA 100) was used for the photography.

The following variations of the conditions were performed in order to examine the influence of the change on the fluorescence patterns.

Specimens (Antigens)

Conditions of Storage Before Freezing of Renal Tissue

Kidneys from rats with Masugi nephritis or from NZB/WF1 mice with naturally occurring immune complex nephritis were kept at 4°C, room temperature, or 37°C. After 30 min, 1, 6, 12, 24, and 72 hr, a block of each renal tissue was frozen.

Conditions of Freezing

Instead of quick-freezing, specimens were left in a deep-freezer at −70°C or −20°C.

Thickness

Sections 2 to 8 μm thick were cut with cryostat.

Conditions of Storing Before the Staining of Sections

Sections were kept at −20°C, 4°C, and at room temperature with or without acetone pretreatment for a week or a month before staining.

Solvents for Pretreatment

In addition to acetone, carbon tetrachloride, ethanol, distilled water, detergent, and phosphate-buffered saline (PBS) were used for comparison at room temperature for 5 min.

Pretreatments with PBS for 30, 60, 120 min, and 24 hr, with 0.5 M, 1.0 M, and 3.0 M KSCN in 0.1 M phosphate buffer (pH 7.0) and 0.1 M glycine-HCl buffer (pH 3.0), or 0.02 M citrate buffer (pH 3.4) and 0.1 M glycine-NaOH buffer (pH 10.0) each for 20 min at room temperature were also carried out.

FITC-labeled Antibodies

The Molar Ratio of Fluorochrome to Protein

F/P molar ratio was determined by the optical densities at 495 and 280 nm as described by Kawamura *et al.*[1]

FITC-labeled anti-mouse IgG with a F/P molar ratio of 1.4 was eluted from DEAE cellulose with 0.1 M NaCl containing 0.005 M phosphate buffer, pH 7.0. Then, anti-mouse IgG with a higher F/P molar ratio (F/P = 3.0) was obtained by elution with 0.3 M NaCl containing buffer.

Either the amount of FITC or protein was matched for the comparison of patterns stained with conjugated anti-mouse IgG with a different F/P molar ratio. Normal rabbit IgG was also labeled with FITC.

Tissue Interactions of Fc Portions of Labeled Antibodies

The fluorescent pattern of kidneys stained with commercially available FITC-labeled IgG/F(ab)$_2$-fragment of each antibody (Hoechst, Marburg) was compared to that with FITC-labeled IgG whole molecule.

RESULTS AND DISCUSSION

Conditions of Storage Before Freezing

The antigenicity of immunoglobulins detected in glomeruli of Masugi nephritis could be detected for more than 72 hr, if kept at 4°C. When the glomeruli were kept at 37°C, a remarkable decrease in fluorescence and destructive changes of tissues, presumably due to autolysis, were observed within 12 hr. These results are of importance with regard to autopsy materials.

Conditions of Freezing

A remarkable distortion of the tissue, presumably induced by ice-crystal formation was observed, although the antigenicity of immunoglobulins was not changed.

Thickness of Sections

The fluorescent patterns could be analyzed in more detail in the specimen of 2 μm thickness than in that of 8 μm thickness. But the intensity of fluorescence apparently decreased in the specimen of 2 μm thickness. It is generally accepted that 4 μm is most adequate.[1] This was also demonstrated to be the case for the kidney. If a detailed examination of the fluorescent pattern is desired, 2 μm sections may be used if the fluorescence is strong enough.

Conditions of Storage Before the Staining of Sections

Sections could be preserved without significant loss of antigenicity for more than one month if kept at −20°C regardless of pretreatment with acetone. When kept at room temperature, however, they showed remarkable decrease of specific fluorescence and increase of blue autofluorescence.

Fixation of Tissue

Fixatives, generally used for pretreatment in immunofluorescence, could also be applied to kidneys. But fine, nonspecific granules of FITC, which might disturb the analysis of the pattern to some degrees, were observed in specimens pretreated with ethanol. There was no remarkable difference between the patterns of acetone-treated and untreated specimens. Fine to coarse fluorescent granules, noted in some specimens, could be removed by prewashing with PBS. In some cases, the pretreatment of PBS, distilled water, or detergent changed negative into positive results. Such an interesting phenomenon could be observed in almost all cases of kidneys from uremic cats, and in some cases of mice and rat specimens. The mechanism of this interesting phenomenon is still unknown. A blocking substance might exist and be removed from renal tissues by washing in PBS, disclosing covered antigenic determinants. Taking the clean-up effect mentioned above into consideration, washing with PBS before staining is recommended. The examination of a normal mouse kidney injected with FITC-labeled HGG revealed that soluble antigen in the specimen can be retained best by ethanol pretreatment or direct reaction with the corresponding antibody.

From these results, pretreatment would be expected to change fluorescence patterns and further comprehensive studies on pretreatments are in progress in our laboratories, including attempts not only to clarify the mechanism of the change in fluorescent findings from negative to positive by PBS, but also to identify the nature of immunoglobulins in kidneys, according to its mode of changes in fluorescent patterns caused by various pretreatments of specimens. For example, immunoglobulins detected in Masugi nephritis and immune complex nephritis at various intervals after the last injection of antigen remained positive in spite of washing with PBS for 24 hr.

On the other hand, the fluorescence of IgG of unknown origin,[3] as typically observed in germ-free mice was easily lost by prewashing with PBS. The same tendency could be observed in cases of rats with AN nephrosis or alloxan diabetes and normal mouse injected with HGG.

The remarkable decrease of the fluorescence by prewashing with PBS for 24 hours was seen in the case of PHN rat, injected once with the rabbit anti-rat Fx1A antibody 5 min before sacrifice. The mechanism of this immune complex formation is now being discussed as to whether it occurs in circulation or *in situ*.

The effect of acid elution was also examined on specimens from individuals with BSA or HGG immune complex nephritis. Positive staining for the antigen (BSA or HGG) could be observed even without acid elution. Acid elution procedures did not change the fluorescence of the antigen and decreased the fluorescence of the antibody-antigen complexes.

Results of the various pretreatments are summarized in TABLE 1. Prewashing with PBS for 30 min, for example, allows for distinction of the immunologically significant IgG from non-specifically adsorbed IgG.

TABLE 1. Effect of Various Pretreatments on Fluorescence Patterns in Kidneys

| Origin of Immunoglobulins | Fluorescent Antibody | Treatment (at Room Temperature) | | | | | | | | | |
| | | PBS | | | | | KSCN (20 min) | | | pH (20 min) | |
		5 min (control)	30 min	60 min	120 min	24 hr	0.5M	1.0M	3.0M	3.0	10.0
Masugi nephritis rat	Anti-rabbit IgG	++	++	++	++	++	++	+	±	±	++
BSA immune complex nephritis rat	Anti-rat IgG	++	++	++	++	++	++	NT	±	±	++
	Anti-BSA	++	++	++	++	++	++	NT	±	±	++
HGG immune complex nephritis rat	Anti-rat IgG	++	++	++	++	++	++	NT	±	±	++
	Anti-HGG	++	++	++	++	++	++	NT	±	±	++
AHN rat	Anti-rat IgG	++	++	++	++	++	++	NT	+	NT	++
PHN rat 5 min after injection	Anti-rabbit IgG	++	+	+	+	±	NT	NT	NT	NT	NT
PHN rat 2 weeks after injection	Anti-rabbit IgG	++	++	++	++	++	NT	NT	NT	NT	NT
AN nephrosis rat	Anti-rat IgG	++	±	±	−	−	±	−	−	−	NT
Alloxan diabetes rat	Anti-rat IgG	++	+	NT	NT	±	NT	NT	NT	NT	NT
NZB/W F1 mouse	Anti-mouse IgG	++	++	++	++	++	++	+	±	±	NT
	Anti-mouse IgM	++	+	+	±	±	++	+	±	−	NT
C3H mouse	Anti-mouse IgG	+	±	±	−	−	±	−	−	−	NT
	Anti-mouse IgM	+	±	±	−	−	±	−	−	−	NT
Germ-free CF #1 mouse	Anti-mouse IgG	+	−	−	−	−	−	−	−	−	NT
HGG-injected mouse	Anti-HGG	+	±	−	−	−	NT	NT	NT	NT	NT
Uremic cat	Anti-Cat IgG	++	++	++	++	++	+	NT	±	±	NT

++: strong positive, +: positive, ±: trace, −: negative, and NT: not tested.

F/P Molar Ratio of Conjugates

Fluorescent antibody with a high F/P molar ratio (F/P = 3.0) stained Bowman's capsule and tubular basement membrane with slight staining of background. The same amount of antibody of identical lot with lower molar ratio (F/P = 1.4) stained only glomerulus.

The specimens were strongly and probably nonspecifically stained by fluorescent antibodies with a high F/P molar ratio but not a low molar ratio, when the same amount of FITC was applied. Thus a "false positive" result may be caused by fluorescent antibodies with a high F/P molar ratio. Moreover, FITC-labeled normal rabbit IgG with a high F/P molar ratio stained mouse kidney specimens, as observed with anti-mouse IgG of high F/P molar ratio. This suggests non-specificity of staining of the fluorescent antibody with a high molar ratio. The fact seems to be very important, considering that biopsy specimens are investigated generally by commercially obtained fluorescent antibodies, the F/P ratio of which is, in our experiences, sometimes high.

There was no remarkable difference between the fluorescence patterns with FITC-labeled IgG/F(ab)$_2$-fragment and with FITC labeled "whole" IgG.

SUMMARY

The influences of different conditions of specimens and conjugates, especially of pretreatments on the fluorescent patterns of kidneys with positive staining for immunoglobulins were examined. The nature of immunoglobulins might be differentiated by the proper combination of pretreatments. It was also found that fluorescent patterns of kidneys were remarkably changed by different methodological conditions.

REFERENCES

1. KAWAMURA, A., JR., Ed. 1977. Fluorescent antibody techniques and their application. 2nd edit. University of Tokyo Press. Tokyo.
2. SHIMIZU, F. 1970. Histopathologic and immunological studies on chronic renal lesions in rats induced by aminonucleoside of Puromycin. Japan. J Exp. Med. **40:** 227–242.
3. SHIMIZU, F., F. ABE, K. ITO & S. KAWAMURA. 1977. On the age-associated presence of immunoglobulin and complement in the renal glomeruli of mice. Contr. Nephrol. **6:** 79–93.
4. SHIMIZU, F., K. ITO, S. SAEGUSA, M. FUKUI & F. ABE. 1976. The application of immunofluorescence to the study of renal disease. I. Influences of various conditions of specimens and conjugates on the fluorescent patterns of kidney. Japan. J. Exp. Med. **46:** 37–43.

The Brush Border of Proximal Tubules of Normal Human Kidney Activates the Alternative Pathway of the Complement System *In Vitro*[a]

GIOVANNI CAMUSSI, CIRO TETTA, GIANNA
MAZZUCCO, AND ANTONIO VERCELLONE

Laboratorio Immunopatologia
Istituto di Anatomia Patologica
Cattedra di Nefrologia
Universita di Torino
Ospedale Maggiore S.G. Battista
Torino, 10126 Italy

INTRODUCTION

It is the purpose of this study to investigate whether any of the structures of the normal human nephron is capable of directly activating the complement (C) system *in vitro*. It was found that the brush border of proximal tubules of human kidney fixes C by activating the alternative pathway. The observation confirms and extends the results of an earlier study in which normal rat kidney was used as substrate.[2]

MATERIALS AND METHODS

Fixation of Complement by Cryostat Sections of Human Kidney

Histologically normal specimens were obtained from kidneys with tumor growth. To preserve the brush border of proximal tubules it appeared essential to snap-freeze the kidney specimens immediately after removal (without embedding in OCT compound) in liquid nitrogen, and to use cryostat sections fixed in acetone for 20 min at 4°C. After washing in PBS, the sections were incubated with the following reagents: three different fresh normal human sera (NHS); EDTA-treated NHS; Mg-EGTA NHS; C-inactivated NHS (heating at 56°C for 60 min); the enzyme inhibitor Aprotin®, followed by NHS; and NHS preincubated with FxlA (see below).[2] After washing with PBS, the sections were incubated with monospecific FITC-conjugated antisera to human IgG, IgA, IgM, C3, Clq, C4, and properdin.[2] The intensity of fluorescence staining was evaluated semiquantitatively as follows: 0, negative; +, slight; + +, moderate; + + +, strong.

Preparation of a Glomerular and Tubular Kidney Fraction

Fx1A, a cortical fraction containing brush border of proximal tubules, was prepared from human and rat kidney tissue according to the method described by

[a]Supported by the Consiglio Nazionale della Ricerche (CNR) Contract 82.02214.04.

Edgington *et al.*[3] A glomeruli-rich fraction obtained during the preparation of human Fx1A was further purified by repeated centrifugation at 400 $\times g$ for 5 min. Tubular and glomerular preparations were lyophilized. Endotoxin was not detectable in the preparations when tested by the limulus assay.[5]

Complement Assays

C consumption in NHS after incubation for 1 hr at 37°C with tubular (20 mg Fx1A/ml of serum) or glomerular preparation (20 mg/ml of serum) was determined by measuring total hemolytic C activity (CH50). To differentiate classic from alternative C pathway activation, Mg-EGTA- and EDTA-treated sera were used.[4] Effective inhibition of either pathway in these Mg-EGTA- or EDTA-treated sera was shown by measuring CH50 values after preincubation of these sera with sheep red blood cells sensitized with rabbit antiserum (EA cells; 5×10^8/ml of serum) or with inulin (25 mg/ml of serum).

C3 activation in serum was studied by a modification of the bidirectional immunoelectrophoresis technique.[2]

Generation and Purification of C5a

Incubation of epsilon-aminocaproic acid–treated NHS with human Fx1A generated a component with anaphylatoxic activity. This component was purified and its properties determined. A comparison was made with properties of a reference C3a and C5a preparation. For details of this part of the study, see Camussi *et al.*[2]

RESULTS AND DISCUSSION

The results of the present study indicate that the brush border of proximal tubules of normal human kidney is capable of directly activating the C system (TABLE 1). Activation by the wall of renal vessels was also seen.[8] C was not fixed by any other kidney structure, including the glomeruli. Similar results have been obtained in an earlier investigation on the C-fixing properties of rat kidney tissue.[2]

The following observations strongly suggest that the brush border of proximal tubules of human kidney activate the C system via the alternative pathway: (1) After

TABLE 1. *In Vitro* Fixation of Human Complement Components along the Luminal Side of Proximal Tubules of Normal Human Kidney

Treatment of Normal Fresh Human Serum	C3	Properdin	C1q	C4	Ig
None	+ +	+ +	0	0	0
56°C, 30 min	0	0	0	0	0
EDTA	0	0	0	0	0
Mg-EGTA	+ +	+ +	0	0	0
Aprotin®	+ +	+ +	0	0	0

TABLE 2. Complement Consumption Measured by Determining Total Hemolytic Complement Activity (CH50) in Normal Fresh Human Serum (NHS) or Mg-EGTA-treated NHS Incubated with Various Preparations[a]

Preparation	NHS	Mg-EGTA NHS
Human Fx1A	56–72	50–66
Rat Fx1A	90–100	80–90
Human glomerular preparation	10–15	8–15
Inulin	100	90–100
Sensitized sheep red blood cells	70–80	0–6+

[a]The complement consumption is expressed in percentages; indicated are the extreme values obtained with three different normal fresh human sera.

incubation of kidney sections with NHS, C3 and properdin, but no Clq and C4, could be demonstrated along the brush border of proximal tubules (TABLE 1). (2) Treatment of NHS with Mg-EGTA, which blocks the classic but not the alternative pathway, did not influence the binding of C to the tubules (TABLE 1). Although an enzyme inhibitor, Aprotin®, failed to prevent fixation of C to the proximal tubules, the possibility that enzymes activate the C system cannot be completely discounted. For example, angiotensin converting enzyme, present along the luminal side of proximal tubules, appears able to activate the C system *in vitro*.[1] (3) Experiments involving incubation of NHS with partially purified human brush border (a kidney fraction, called FxlA) support the studies performed by the indirect immunofluorescence technique. Incubation of Fx1A with NHS blocked the binding of C to cryostat sections of kidney, promoted the consumption of total complement hemolytic activity (TABLE 2), and generated C3 breakdown products and an anaphylatoxin with the characteristics of C5a. Incubation of human Fx1A with Mg-EGTA serum still resulted in C consumption. For reasons unknown, rat Fx1A was, on a weight base, more effective in consuming C than human Fx1A (TABLE 2). The addition of EDTA to NHS completely inhibited the consumption of C by human or rat FxlA.

While under normal conditions C components are not filtered by glomeruli, their presence has been demonstrated in the urine of patients with certain forms of the nephrotic syndrome.[6,7] It is conceivable that in these patients with a non-selective proteinuria, the C system is activated by the brush border resulting in damage of the proximal tubules. This possibility is supported by micropuncture studies in the rat by Sato and Ullrich, demonstrating functional as well as morphological lesions of the proximal tubules induced by intraluminal perfusion with fresh serum. Inactivation of the C system prevented the development of proximal tubular lesions.[6,9]

SUMMARY

The aim of this investigation was to study complement fixation by normal human kidney tissue. C fixation was assessed on acetone-fixed sections of frozen human kidney. In addition, C consumption following incubation of normal fresh human serum with tubular or glomerular fractions of human kidney was measured. The results are consistent with the interpretation that the brush border of proximal tubules of human kidney activates the C system via the alternative pathway. It is suggested that this activation may occur *in vivo* in patients with a non-selective proteinuria.

REFERENCES

1. CALDWELL, P. R. B., H. J. WIGGER, L. T. FERNANDEZ, R. M. D'ALISA, D. TSE-ENG, V. P. BUTLER, Jr. & I. GIGLI. 1981. Lung injury induced by antibody fragments to angiotensin-converting enzyme. Am. J. Pathol. **105:** 54–63.
2. CAMUSSI, G., M. ROTUNNO, G. SEGOLONI, J. R. BRENTJENS & G. A. ANDRES. 1982. *In vitro* alternative pathway activation of complement by the brush border of proximal tubules of normal rat kidney. J. Immunol. **128:** 1659–1663.
3. EDGINGTON, G. S., R. J. GLASSOCK & F. J. DIXON. 1968. Autologous immune complex nephritis induced with renal tubular antigens. I. Identification and isolation of the pathogenic antigen. J. Exp. Med. **127:** 555–571.
4. FINE, D. P., S. R. MARNEY, D. G. COLLEY, J. S. SERGENT & R. M. DEZPREZ. 1972. C3 shunt activation in human serum chelated with EGTA. J. Immunol. **109:** 807–809.
5. JORGENSEN, J. H., H. F. CARVAGEL, B. E. CHIPPS & R. F. SMITH. 1973. Rapid detection of gram negative bacteriuria by use of the *Limulus* endotoxin assay. Appl. Microbiol. **26:** 38–42.
6. KJALMAN, A., A. AVITAL & B. D. MYERS. 1976. Renal handling of the third (C3) and fourth (C4) components of the complement system in the nephrotic syndrome. Nephron **16:** 333–343.
7. LANGE, K. & E. J. WENK. 1954. Complement components in the sera and urine of patients with severe proteinurias. Am. J. Med. Sci. **228:**448–453.
8. LINDER, E. 1981. Binding of Clq and complement activation of vascular endothelium. J. Immunol. **126:** 648–658.
9. SATO, K. & K. J. ULLRICH. 1974. Serum-induced inhibition of isotonic fluid absorption by the kidney proximal tubule. I. Mechanism of inhibition. Biochim. Biophys. Acta **343:** 609–614.

IgA Class Endomysium Antibodies in Dermatitis Herpetiformis and Coeliac Disease[a]

TADEUSZ P. CHORZELSKI,[b] JADWIGA SULEJ, HANNA TCHORZEWSKA, STEFANIA JABLONSKA, ERNST H. BEUTNER, AND VIJAY KUMAR

Department of Dermatology
Warsaw Academy of Medicine
Warsaw, Poland

Department of Microbiology
State University of New York at Buffalo
Buffalo, New York 14214

IF Testing Service
Buffalo, New York 14223

INTRODUCTION

Dermatitis herpetiformis (DH) is a vesicular skin disorder with characteristic histologic features (PMNs, microabscesses in dermal papillae adjacent to the vesicles) and disease-specific immunopathological findings (granular or fibrillar IgA deposits most commonly found in dermal papillae) usually associated with gluten-sensitive enteropathy.[2,3,8] Cases with linear IgA deposits in the basement membrane zone (BMZ) have been referred to by various names: "DH with linear IgA deposits,"[5,8] "IgA bullous pemphigoid,"[10] or "polymorphic variant of bullous pemphigoid."[5] It is becoming increasingly evident that this disease is a distinct entity.[2,3,6] We refer to it here as "linear IgA bullous dermatosis" or LABD. Detailed reviews appear elsewhere.[2,6]

Two types of antibodies have been detected in sera of DH patients, notably anti-reticulin antibodies (ARA) with reported frequencies of 5–18% and anti-gliadin antibodies (AGA) in about 40% of the cases. However, these antibodies occur predominantly in coeliac disease and are chiefly of IgG class: ARA have been reported in 38% of adults and in 59% of children; AGA in 40% of adults[7] and over 90% of children.[13] Both types of antibodies are occasionally detected in diseases other than with gluten-sensitive enteropathy.[11]

This report sets forth the first findings on IgA class antibody reacting with endomysium IgA-EmA of smooth muscle, particularly of monkey esophagus. We have found these IgA-EmA in a significant proportion of patients with DH, coeliac disease, and other gut diseases but not in cases of LABD. We have also shown that both IgA-EmA and IgG-AGA can be detected on monkey esophagus in the same serum sample by the indirect IF test.

[a]Supported in part by the Polish Academy of Sciences and in part by the Summerhill Foundation.

[b]Reprint requests to: Tadeusz P. Chorzelski, M.D., Department of Dematology, Warsaw Academy of Medicine, ul. Koszykowa 82a, 02-008 Warsaw, Poland.

TABLE 1. Characteristics of Conjugates

Specificity (Source)	Molar F/P Ratio[a]	Antibody Conc.	Dilution for Use
IgG (Hyland)	2.2	1.8 mg/ml	1:32
IgA (Hyland)	2.3	1.6 mg/ml	1:16

[a]F/P ratio = fluorescein-to-protein ratio.

MATERIALS AND METHODS

Sera from DH patients with granular IgA deposits in dermal papillae, from LABD patients with linear IgA deposits along the BMZ, from coeliac disease cases, from other gut diseases, as well as from control subjects have been studied using monkey esophagus as the substrate for a defined indirect immunofluorescence (IF) technique as described by Beutner *et al.*[1] TABLE 1 lists the characteristics of the conjugates. For the detection of anti-gliadin antibodies, a modification of the method of Unsworth *et*

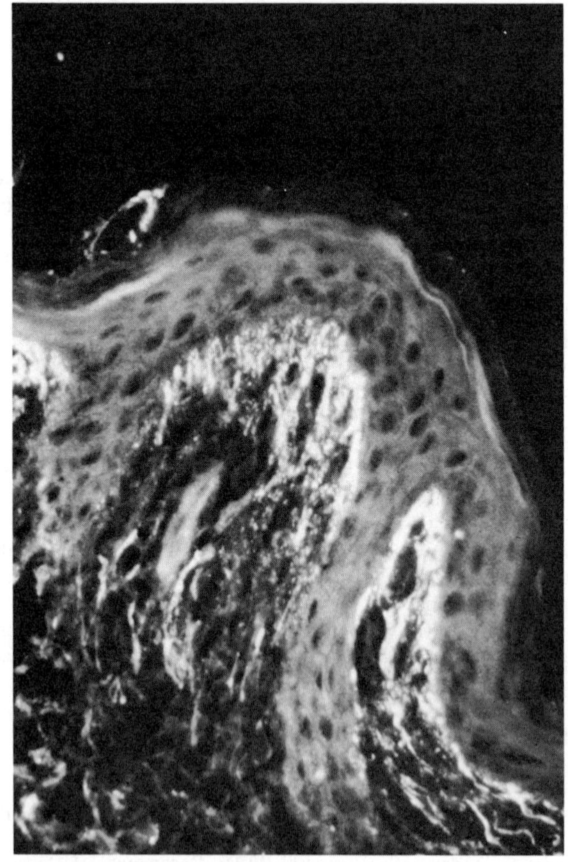

FIGURE 1. Direct IF staining with anti-IgA FITC conjugate on 4 μm sections of skin of patient with DH. Note fibrillar granular IgA deposits in dermal papillae. ×100.

al.[12] was used. Briefly, cryostat sections of monkey esophagus cut at 4 μm were incubated with 0.1 mg/ml of gliadin (gliadin—Catalog No. G 3375, Sigma Chemical Co., St. Louis, MO) solution in distilled H_2O, washed in phosphate-buffered saline (PBS) and then treated with patients' sera of 1:2.5 dilution. After serum treatment, sections were washed in PBS and treated with anti-human IgG or anti-IgA conjugates (TABLE 1). Most that were positive at 1:2.5 were titrated to an endpoint.

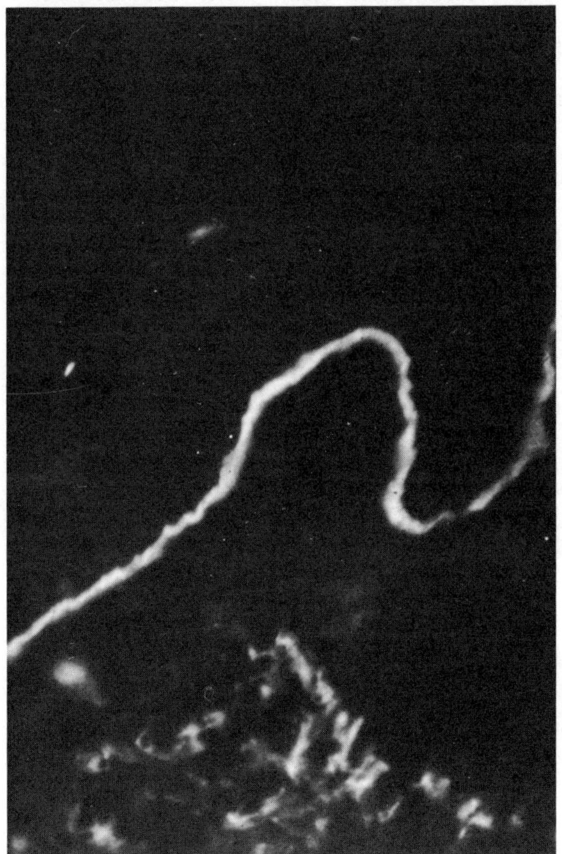

FIGURE 2. Direct IF staining as for FIGURE 1 on skin section of patient with LABD. Note the linear IgA deposits along the BMZ. ×112.

Slides were examined with Zeiss "Opton" epi-illumination system with an HBO 50 light source.

RESULTS

FIGURE 1 shows fibrillar granular IgA deposits seen by direct IF staining in the dermal papillae in a skin biopsy specimen taken from a patient with DH associated with gluten-sensitive enteropathy. The linear pattern of IgA deposits demonstrated by direct IF in another case (FIGURE 2) characterizes immunopathological findings in LABD. FIGURE 3 depicts the indirect IF pattern of IgA-EmA found in the serum of a DH patient. The reactive antigen appears as a network of thin, irregular lines around the sarcolemma of the individual smooth muscle myofibrils. This stands in sharp contrast to the indirect IF pattern of the anti-smooth muscle antibodies (ASMA) shown in FIGURE 4. The latter react only with the sarcoplasm. The ASMA, which have been reported to be directed against actin,[4,9] cross-react weakly with skeletal muscle while the antigen reactive with IgA-EmA has only been detected in smooth muscle. Usually the strongest reactivity of IgA-EmA occurs in the smooth muscle layer adjacent to the epithelium.

Silver impregnation (Belschowsky's method) stains structures shown in FIGURE 5 similar to those reactive with IgA-EmA (FIGURE 5a) thus suggesting that the endomysial antigenic component may be in the reticulin fibrils. They may also be surface components of myofibrils.

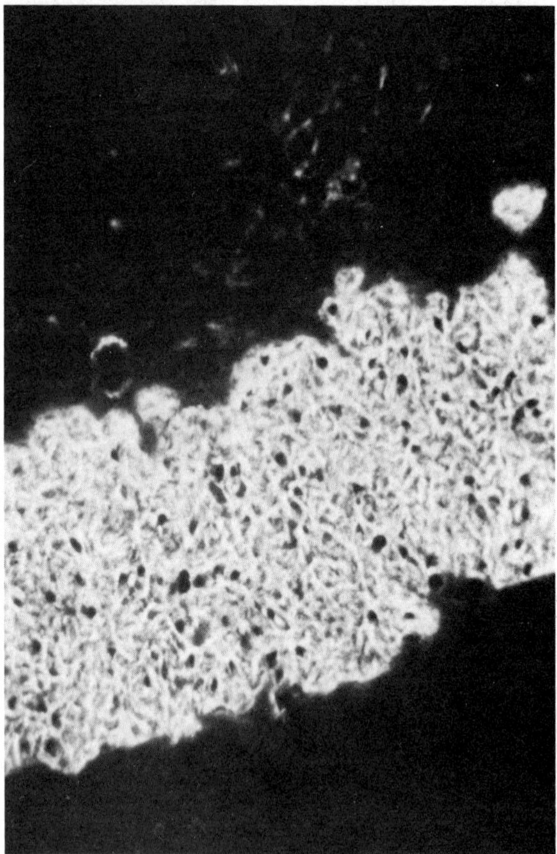

FIGURE 3. Indirect IF staining with DH serum containing IgA-EmA and anti-IgA FITC conjugate on monkey esophagus section. Note the reticulin-like pattern on smooth muscle. ×100.

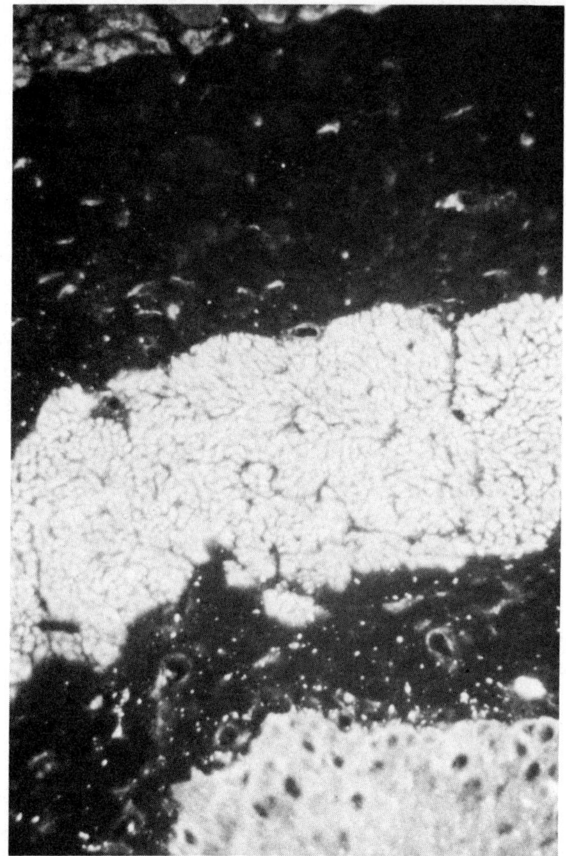

FIGURE 4. Indirect IF staining with serum containing ASMA and with anti-IgA FITC conjugate on monkey esophagus section. Note homogeneous cytoplasmic pattern in smooth muscles. ×100.

Pretreatment of monkey esophagus sections with gliadin followed by incubation first with DH serum containing AGA and then with anti-IgG conjugate gives an IF pattern of intercellular staining similar to that of pemphigus antibodies (FIGURE 6). Negative IF staining (FIGURE 6a) when gliadin is omitted proves that the intercellular IF pattern does depend on the presence of AGA in the patient's serum.[1] TABLE 2 lists some typical IF findings with four sera on monkey esophagus sections. As can be seen, the "three compartment test" format used in these studies serves to reveal three types of antibodies, IgA-EmA, AGA, and anti-intercellular antibodies (AICA), in varying combinations. The first two appear to be frequent findings in DH. The tests for IgG class antibodies (which reveal the AICA) serve as controls for both the IgA-EmA and the AGA; these AICA are the same as the pemphigus-like antibodies.

TABLE 3 summarizes the frequencies with which IgA-EmA and AGA occurred at titers of 2.5 to 160 in selected diseases. About 70 to 80% of DH and coeliac or suspected coeliac disease patients had at least one of these antibodies and about 40 to 50% had

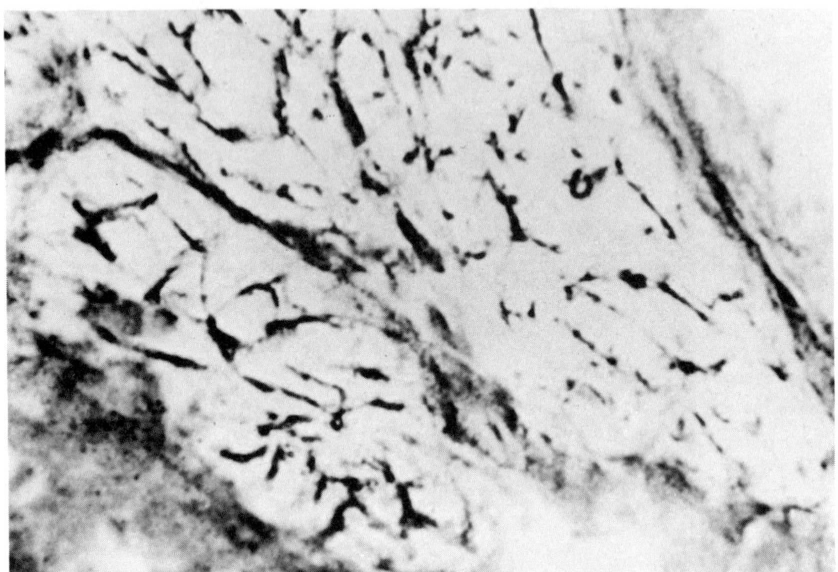

FIGURE 5. Silver impregnation (Belschowsky's method). Note reticulin-like pattern. × 250.

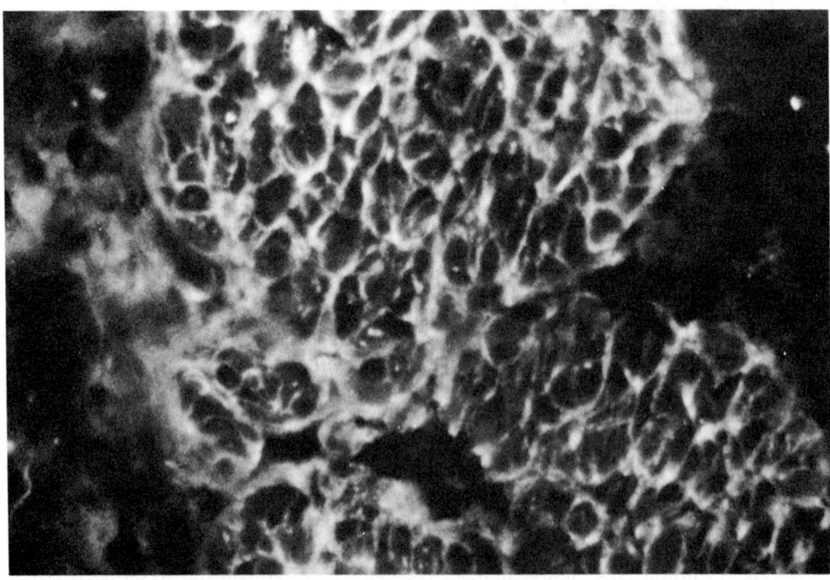

FIGURE 5a. The same as in FIGURE 3, at higher magnification. Note similar reticulin-like pattern as in FIGURE 5. ×250.

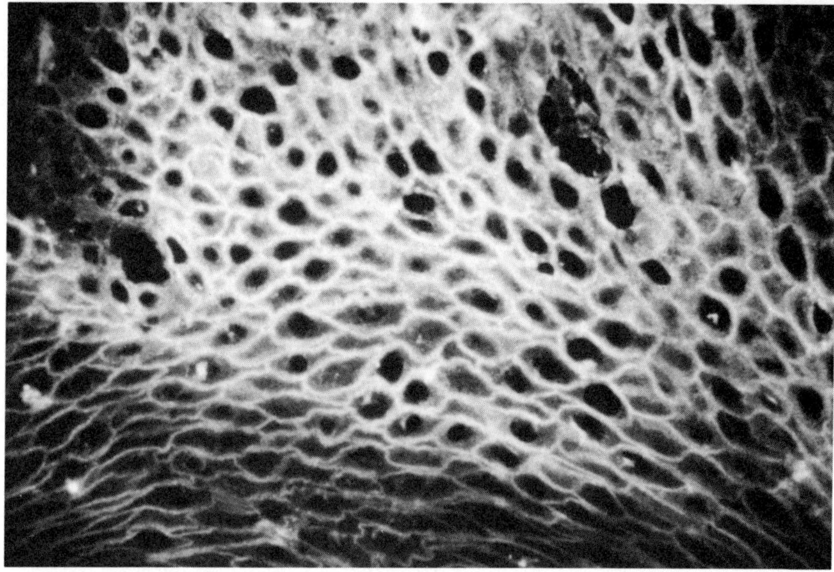

FIGURE 6. Indirect IF with DH serum containing AGA and with anti-IgG FITC conjugate on monkey esophagus section (pretreated with gliadin). Note intercellular pemphigus-like pattern in the epithelium. ×100.

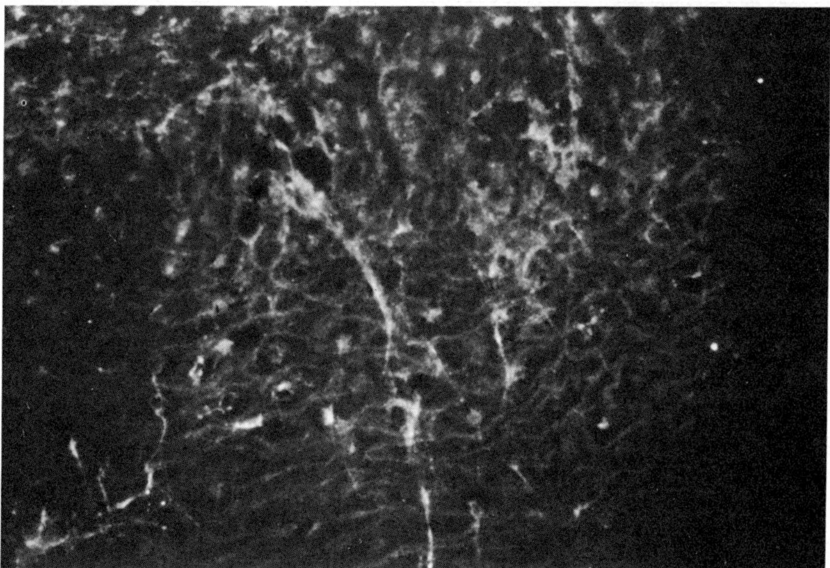

FIGURE 6a. The same procedure as in FIGURE 6 but without gliadin. Note negative reaction. ×100.

both. Importantly, both types of antibodies also occur in a rather high frequency in other gut diseases but neither was found in the 13 cases of LABD examined in these studies. Occasionally, EmA is not only of the IgA class but also of the IgG class though the latter appear in lower titers.

DISCUSSION

The newly described IgA-EmA has been found only in cases of DH associated with granular IgA deposits in dermal papillae, in cases of coeliac disease, and other gut diseases, but not in linear IgA bullous dermatosis (LABD). This points to the relation of this immunologic response to jejunal involvement and provides additional evidence that enteropathy in DH is in essence a coeliac disease and that LABD is not related to

TABLE 2. Examples of Findings with Sera of Patients with DH and Coeliac Disease in Indirect IF Tests on Monkey Esophagus

		IF Findings on Monkey Esophagus Sections		
		Epithelium (ICA) Anti-IgG Conjugate		
Case No.	Smooth Muscle (Em) Anti-IgA Conjugate	Unpretreated	Gliadin Pretreated	Interpretation of IF Findings
#1	positive	—	—	IgA-EmA
#2	—	—	positive	AGA
#3	—	positive	positive	pemphigus-like antibodies or pemphigus antibodies[a]

DH = dermatitis herpetiformis, CD = coeliac disease, ICA = intercellular area, Em = endomysium, IgA-EmA = IgA class anti-endomysial antibodies, and AGA = anti-gliadin antibodies.

[a] If indirect IF titer is high and complement IF is negative.

these as has been claimed in the past. IgA-EmA reacts with reticulin of endomysium of smooth muscle but not with reticulin fibrils of other organs. This suggests that the antigenic composition of reticulin fibrils or cell surface antigens differs in various tissues and that the antigenic stimulus may originate in the gut. It should be stressed that the IgA deposits found in the skin of DH patients probably originate in the gut since they contain secretory (dimeric) IgA.[14]

It is interesting to note that gliadin does not bind to the endomysial structure in smooth muscles or other reticulin-like structures in monkey esophagus sections as it does to reticulin in other organs. Instead, it binds to an intercellular component of the esophageal epithelium, giving the AGA an IF staining pattern similar to that of pemphigus antibodies. Thus, using monkey esophagus sections as a substrate, one can detect two types of antibodies in the serum: IgA-EmA with anti-IgA conjugate and AGA with gliadin pretreatment and anti-IgG conjugate. In addition, esophagus has the advantage that anti-reticulin antibodies, which interfere in the detection of AGA on rodent tissue, do not react on esophagus, thus making it easy for detecting the AGA antibodies.

TABLE 3. Survey of Frequencies of IgA-EmA[a] and AGA[a] in Selected Diseases

Clinical Diagnoses	Ab type[a]	<2.5 "neg"	2.5 or ≥2.5	5	10	20	40	80	≥160	Total Positive (Percent pos.) IgA-EmA	AGA	Positive IgA-EmA and AGA	Positive IgA-EmA or AGA or Both
Dermatitis herpetiformis	IgA-EmA	12	6	1	6	7	3	3		26/38 (68.4%)		14/38 (36.8%)	28/38 (73.7%)
	AGA	22	4		4	4	2	1	1		16/38 (42.1%)		
Coeliac disease	IgA-EmA	4		2		1	1		2	6/10 (60%)		5/10 (50%)	7/10 (70%)
	AGA	4	1			1	2		2		6/10 (60%)		
Suspected coeliac disease	IgA-EmA	6	1	2	1	2	1			7/13 (53.8%)		6/13 (46.2%)	10/13 (76.9%)
	AGA	4	4		1	1		1	2		9/13 (69.2%)		
Other gut diseases[b]	IgA-EmA	8		1			1	1		3/11 (27.3%)		2/11 (18.2%)	6/11 (54.5%)
	AGA	6	1			3			1		5/11 (45.5%)		
Linear IgA bullous dermatosis	IgA-EmA	13								0/13 (0%)		0/13 (0%)	0/13 (0%)
	AGA										0/13 (0%)		
Controls[c]	IgA-EmA									0/63 (0%)		0/63 (0%)	0/63 (0%)
	AGA	63									0/63 (0%)		

[a]Ab types: IgA-EmA or IgA class anti-endomysial antibodies; IgG class AGA or anti-gliadin antibodies.

[b]Other gut disease: 1 case sucrose intolerance (positive for both IgA-EmA and AGA); 3 cases cow milk–sensitive enteropathy (all AGA pos.); 6 cases malabsorption syndrome; and 1 case undiagnosed gut disease (patient refused examination).

[c]The controls include 5 pemphigus, 5 bullous pemphigoid, 7 systemic HLE (ANA pos.), 6 systemic sclerosis (ANA pos.), 5 psoriasis, and 36 other skin diseases.

REFERENCES

1. BEUTNER, E. H., R. J. NISENGARD & V. KUMAR. 1979. Defined immunofluorescence. Basic concepts and their application to clinical immunodermatology. *In* Immunopathology of the Skin. 2nd edit. E. H. Beutner, T. P. Chorzelski & S. F. Bean, Eds.: 29–75. John Wiley and Sons. New York.
2. CHORZELSKI, T. P., S. JABLONSKA, E. H. BEUTNER, S. F. BEAN & N. L. FUREY. 1979. Linear IgA bullous dermatosis. *In* Immunopathology of the Skin. 2nd edit. E. H. Beutner, T. P. Chorzelski & S. F. Bean, Eds.: 315–319. John Wiley and Sons. New York.
3. CHORZELSKI, T. P., S. JABLONSKA, E. H. BEUTNER & M. JARZABEK-CHORZELSKA. 1981. Linear IgA bullous dermatosis. *In* Epidermis in Disease. R. Marks & E. Christophers, Eds.: 577–583. MTP Press, Ltd. Lancaster, England.
4. GABBIANI, G., C. B. RYAN, J. P. LAMELIN, P. VASSALI, G. MAJNO, C. A. BOUVIER, A. CRUCHAND & E. F. LUSCHER. 1973. Human smooth muscle autoantibody. Its identification as antiactin antibody and a study of its binding to "nonmuscular" cells. Am. J. Pathol. **3:** 473–486.
5. HONEYMAN, J. F., A. R. HONEYMAN, M. A. DE LA PARRA, A. PINTO & G. J. EGUIGUREN. 1979. Polymorphic pemphigoid. Arch. Dermatol. **115:** 423–427.
6. KATZ, S. I. & W. STROBER. 1978. The pathogenesis of dematitis herpetiformis. J. Invest. Dermatol. **70:** 63–75.
7. LEONARD, J. N., G. P. HAFFENDEN, A. F. SWAIN, N. P. RING, E. J. HOLBOROW & L. FRY. 1981. Dematitis herpetiformis—A comparison between patients with linear and papillary IgA deposits (based on multicentric European study). Br. J. Dermatol. **105** (suppl.) **19:** 14–15.
8. LAWLEY, T. J., W. STROBER, H. YAOITA & S. I. KATZ. 1980. Small intestinal biopsies and HLA types in dermatitis herpetiformis patients with granular and linear IgA skin deposits. J. Invest. Dermatol. **74:** 9–12.
9. NORBERG, R., G. BIBERFELD, A. FAGRAEUS, K. LIDMAN, R. THORSTENSSON & G. UTTER. 1977. The reaction of cells with anti-actin sera in relation to the amount of cellular actin. Clin. Exp. Immunol. **28:** 512–516.
10. PROVOST, T., J. C. MAIZE, J. S. AHMED, J. S. STRAUSS & R. L. DOBSON. 1979. Unusual subepidermal bullous diseases with immunologic features of bullous pemphigoid. Arch. Dermatol. **115:** 156–160.
11. SEAH, P. P., L. FRY, E. J. HOLBOROW, M. A. ROSSITER, W. F. DOE, A. F. MAGALHAES & A. V. HOFFBRAND. 1973. Antireticulin antibody: Incidence and diagnostic significance. Gut **14:** 311–315.
12. UNSWORTH, D. H., G. D. JOHNSON, G. HAFFENDEN, L. FRY & E. J. HOLBOROW. 1981. Binding of wheat gliadin in vitro to reticulin in normal and dermatitis herpetiformis skin. J. Invest. Dermatol. **76:** 88–93.
13. UNSWORTH, D. J., P. D. MANUEL, J. A. WALKER-SMITH, C. A. CAMPBELL, G. D. JOHNSON & E. J. HOLBOROW. 1981. New immunofluorescent blood test for gluten sensitivity. Arch. Dis. Child. **56:** 864–868.
14. UNSWORTH, D. J., J. N. LEONARD, A. W. PAYNE, L. FRY & E. J. HOLBOROW. 1982. IgA in dermatitis herpetiformis is dimeric. The Lancet (Feb. 27): 478–479.

A Functional Assay for Complement-Activating Antibodies to the Cutaneous Basement Membrane Zone

W. RAY GAMMON AND CAROLYN C. MERRITT

Department of Dermatology
School of Medicine
The University of North Carolina at Chapel Hill
Chapel Hill, North Carolina 27514

Using a qualitative leukocyte attachment assay, we previously reported that complement-binding bullous pemphigoid antibodies could mediate the attachment of leukocytes to the cutaneous basement membrane zone in the presence of serum complement. Although it was shown that complement-mediated leukocyte chemotaxis was a major mechanism in attachment, the role of immune adherence to basement membrane–bound IgG (Fc) and complement (C3b) was not determined and the relationship between leukocyte attachment and complement-binding antibody titers was not defined. In this study we have examined these questions using a quantitative leukocyte attachment assay, bullous pemphigoid sera of varying complement binding antibody titers, and basement membrane IgG (Fc) and complement (C3b) prepared with skin sections, bullous pemphigoid sera, and serum complement. The results confirm the ability of complement-binding bullous pemphigoid antibodies to mediate leukocyte attachment to the basement membrane, demonstrate a correlation between leukocyte attachment and antibody titers, and show that IgG (Fc)- and C3b-mediated immune adherence does not account for the accumulation of leukocytes at the basement membrane.

INTRODUCTION

Indirect immunofluorescence studies have shown complement-binding antibodies to the cutaneous basement membrane zone in sera from patients with bullous pemphigoid (BP).[5] Using a qualitative leukocyte attachment assay, we reported that these antibodies could mediate the accumulation and nonrandom side-by-side attachment of normal human peripheral blood leukocytes to the basement membrane zone of cryostat sections of fresh-frozen normal human skin.[2] A requirement for fresh but not heat-inactivated normal human serum during the attachment reaction supported a role for complement as well as antibody. Based on current concepts concerning the functional interactions between leukocytes and complement-activating immune complexes, it was assumed the accumulation and attachment of cells to the basement membrane zone was due to chemotaxis in response to complement-derived chemotactic

---

List of abbreviations: C = complement, BMZ = basement membrane zone, BP = bullous pemphigoid, BPS = BP serum, BPS$_i$ = BPS heat-inactivated, LA = leukocyte attachment, PBL = peripheral blood leukocytes, NHS = normal human serum, NHS$_i$ = NHS heat-inactivated, PBS = phosphate-buffered saline, GBSS = Geys Balanced Salt Solution, FITC = fluorescein isothiocyanate, and H&E = hematoxylin and eosin.

335

factors and/or immune adherence via leukocyte membrane receptors for basement membrane zone–bound IgG Fc or complement components. In an additional study, we showed that a major mechanism responsible for the accumulation of peripheral blood leukocytes at the basement membrane zone was chemotaxis, however, a role for leukocyte receptors for basement membrane zone–bound IgG Fc and complement components was not excluded.[3]

In this study, the leukocyte attachment assay has been quantitated by counting numbers of peripheral blood leukocytes attached to the basement membrane zone per unit length of basement membrane zone. Using the method, we have reexamined the ability of bullous pemphigoid sera (BPS) to mediate leukocyte attachment in the presence of a complement source and have examined the effect of anti–basement membrane zone antibody titers on leukocyte attachment. In addition, the method has been used to determine if the accumulation of peripheral blood leukocyte at the basement membrane zone occurs as a result of cell attachment via receptors for IgG Fc and complement. The results confirmed the ability of complement-binding IgG anti–basement membrane zone antibodies to mediate leukocyte attachment, showed a correlation between the concentration of anti–basement membrane zone antibody and leukocyte attachment, and failed to demonstrate a role for basement membrane zone–bound IgG Fc and complement components in the accumulation of peripheral blood leukocytes at the basement membrane zone.

MATERIALS AND METHODS

Skin

Neonatal human foreskin was obtained immediately following routine circumcision of healthy neonates, trimmed of excess fat, washed 5 min in sterile 0.15 M NaCl buffered with 0.01 M NaH_2PO_4 and $NaHPO_4$, pH 7.2 (PBS), and excess moisture removed on filter paper. Skin was then snap-frozen in liquid N_2, embedded in Ames O.C.T. compound (Ames Co., Elkhart, IN) and stored frozen at $-70°C$. Immediately prior to use, 6 μm thick sections were cut on a cryostat and allowed to dry briefly on glass microscope slides.

Bullous Pemphigoid Sera

Four bullous pemphigoid sera (BPS 1–4) were obtained from patients with active disease diagnosed by established clinical, histologic, immunohistologic, and immunoelectron microscopic criteria.[8] Titers of IgG and C3-binding anti–basement membrane zone antibodies were determined using standard indirect IgG- and complement-binding immunofluorescent procedures.[1,5] All sera were stored in 1.0 ml aliquots at $-70°C$. Prior to use, sera were heat-inactivated (BPS_i) at 56°C for 30 min and working dilutions prepared with PBS.

Complement

Fresh-frozen, platelet-poor normal human sera from a single blood group AB, type Rh-positive donor was used as a source of serum complement and normal human serum in all studies. Blood was collected under sterile conditions and platelet-poor normal human sera prepared as previously described.[2] Aliquots (0.5 ml) were stored frozen at $-70°C$. Some aliquots were heat-activated ($56°C \times 30$ min) for use as a complement and normal human sera control.

Peripheral Blood Leukocytes

Peripheral blood leukocytes from normal human donors were prepared by dextran sedimentation of heparinized whole blood as previously described.[3] Following dextran sedimentation for 30 min at 37°C, the leukocyte-rich supernatant was recovered and cells washed twice in 10.0 ml Geys Balanced Salt Solution (GBSS) (Flow Labs, McLean, VA) containing 2% bovine serum albumin. Total leukocyte counts and percentage of granulocytes were determined by counting methylene-blue–stained cells in a hemocytometer. The percentage of granulocytes was consistently 70–85%. Cell viability determined by trypan blue exclusion was 98% or more. Prior to use, peripheral blood leukocytes were suspended at a concentration of 10×10^6 peripheral blood leukocytes/ml in either 10% fresh normal human sera, 10% heat-inactivated normal human sera (NHS_i), or GBSS.

Preparation of Skin Sections with Anti–Basement Membrane Zone Antibody and Complement

Six μm thick sections were incubated at $25°C \times 20$ min with each of the four BPS_i diluted 1:20, 1:80, or 1:160. As a control, some sections were incubated under the same conditions with either GBSS or NHS_i diluted 1:20 in PBS. Following incubation, sections were washed for 15 min at 25°C in three changes of PBS and kept in a humidity chamber at 25°C until used.

For the preparation of skin sections with basement membrane zone (BMZ)–bound anti-BMZ antibody and C3b, sections were incubated with BPS_i 1:20 as above, washed 5 min at 25°C in GBSS and incubated with 25% fresh normal human sera/GBSS at $37°C \times 30$ min. Sections were then rinsed for 15 min in three changes of GBSS and placed in a humidity chamber at 25°C.

To demonstrate binding of IgG and C3b to the basement membrane zone, representative sections were processed for immunofluorescence using fluorescein isothiocyanate (FITC)–conjugated goat anti-human IgG (Cappel Labs, Cochranville, PA) or mouse monoclonal anti-human C3b (Bethesda Research Labs, Inc., Rockville, MD) followed by FITC-conjugated, species-specific, sheep anti-mouse IgG (Cappel Labs). Total protein concentrations, molar fluorescein-to-protein ratios, and working dilutions were: 6.1 mg/ml, 3.2, and 1:10 for FITC-conjugated goat anti-human IgG; 7.0 mg/ml, 3.4, and 1:10 for FITC-conjugated sheep anti-mouse IgG; and 11.0 mg/ml, 3.0, and 1:50 for mouse monoclonal anti-human C3b.

Leukocyte Attachment Assay

The leukocyte attachment assay was performed as previously described using a modification of the method of Yamamoto et al.[2,10] The assay was performed on skin sections pretreated with BPS_i, NHS_i, buffer, or BPS_i and 25% NHS. Briefly, slides with treated sections were covered with a second microscope slide (covering slide) to which two thicknesses of #88 vinyl electrical tape (3M Co, St. Paul, MN) had been attached to each end. The tape prevented contact between the covering slide and skin sections and created a chamber approximately 0.2 mm thick with a volume of 0.4 ml. The ends of the slides were secured with clamps and the chambers injected with 10 × 10^6 peripheral blood leukocytes suspended in 10% fresh NHS, 10% NHS_i, or GBSS. Sections that had been pretreated with BPS_i 1–4 at dilutions of 1:20, 1:80, or 1:160 received peripheral blood leukocytes in 10% fresh NHS, in GBSS, or in 10% NHS_i. Sections pretreated with PBS or NHS_i received peripheral blood leukocytes in 10% NHS. Sections pretreated with BPS 1:20 and 25% NHS received peripheral blood leukocytes in GBSS.

Following injections of peripheral blood leukocytes, chambers were incubated for 30 min at 37°C in a humid air atmosphere. Following incubation, chambers were disassembled and tissue slides rinsed for 10 min at 25°C in PBS to remove nonadherent cells. Slides were then briefly dried at 25°C, fixed in 100% ethanol, stained with hematoxylin and eosin (H&E) and examined at 250× magnification using a light microscope equipped with a 10× filar linear micrometer eyepiece (American Optical Corp., Buffalo, NY). The number of peripheral blood leukocytes attached to the BMZ per mm BMZ was counted in 3–7 random fields per slide and averaged. Each

TABLE 1. IgG and C3 Binding Anti-BMZ Antibody Titers in Bullous Pemphigoid Sera

BPS_i	IgG Anti-BMZ Antibody	C3 Binding Anti-BMZ Antibody
1	1280	320
2	1280	320
3	2560	640
4	2560	640

experiment was performed in triplicate and results of each treatment were expressed as a mean of triplicate experiments ± standard error. Statistical analysis of results was performed by t-test.

RESULTS

Anti-BMZ Antibody Titers in BPS

IgG- and C3-binding anti-BMZ antibody titers for BPS_i 1–4 are shown in TABLE 1. BPS_i 1 and 2 had anti-BMZ and C3-binding antibody titers of 1:1,280 and 1:320, respectively. These titers in BPS_i 3 and 4 were 1:2,560 and 1:640, respectively. All four BPS_i had C3-binding anti-BMZ antibody titers greater than the highest titer (1:160) used for the leukocyte attachment assay.

Formation of BMZ-Bound C3b

Indirect immunofluorescence of skin sections treated with BPS 1–4 at dilutions of 1:20, and subsequently incubated with 25% NHS at 37°C showed deposition of C3b. The intensity of staining for C3b was 3+ (maximal staining).

TABLE 2. Leukocyte Attachment to the BMZ of Skin Sections Treated with BPS_i, NHS_i, and GBSS

BPS_i	Mean LA ± SEM (#PBL/mm BMZ)	Signifiance[a]
1	160.8 ± 9.2	p > .001
2	113.4 ± 38.2	p > .05
3	193.0 ± 51.5	p > .05
4	158.3 ± 7.5	p > .001
NHS_i	20.9 ± 4.9	p < .05
GBSS	22.1 ± 2.3	

[a]Significance of leukocyte attachment (LA) determined by t-test comparison between BPS_i and buffer control and NHS_i and buffer control.

Leukocyte Attachment to Skin Sections Treated with Anti-BMZ Antibody

To demonstrate a requirement for anti-BMZ antibody in leukocyte attachment, sections treated with BPS_i 1–4 at a dilution of 1:20 and subsequently incubated with peripheral blood leukocytes in 10% NHS were compared to sections treated with NHS_i 1.20 or GBSS followed by peripheral blood leukocytes in 10% NHS. The results (TABLE 2) show significantly greater leukocyte attachment for all four BPS compared to NHS_1 or buffer-treated sections (p < .05–.001). No differences in leukocyte attachment between sections treated with buffer or NHS_i were observed. Furthermore, no differences in leukocyte attachment among the four BPS were observed at a dilution of 1:20. Although the exact percentage of each cell type attached to the BMZ was not determined, most were neutrophils.

Requirement for Serum C in Leukocyte Attachment

To demonstrate a requirement for serum C in leukocyte attachment, skin sections were pretreated with BPS_i 1–4 diluted 1:20 and subsequently incubated with peripheral blood leukocytes suspended in 10% NHS, 10% NHS_i, or GBSS. The results (TABLE 3) show significantly greater leukocyte attachment in the presence of NHS compared to NHS_i or GBSS.

Effect of C3 Binding Titer on Leukocyte Attachment

To determine if C3 binding titer affected leukocyte attachment, skin sections were treated with each of the four BPS_i at dilutions of 1:20, 1:80, and 1:160. The results

TABLE 3. Leukocyte Attachment to the BMZ in Presence of NHS, NHS_i, and GBSS

	Mean LA ± SEM (#PBL/mm BMZ)				
BPS	10% NHS	10% NHS_i	Significance[a]	GBSS	Significance[b]
1	160.8 ± 9.2	27.6 ± 9.4	p < .001	43.1 ± 10.4	p < .05
2	113.4 ± 38.2	23.7 ± 2.6	p < .05	23.5 ± 2.9	p < .05
3	193.0 ± 51.5	12.8 ± 0.2	p < .05	21.2 ± 6.5	p < .05
4	158.3 ± 7.5	30.3 ± 3.2	p < .001	ND	

[a]Significance determined by t-test comparison between 10% NHS and 10% NHS_i.
[b]Significance determined by t-test comparison between 10% NHS and buffer.

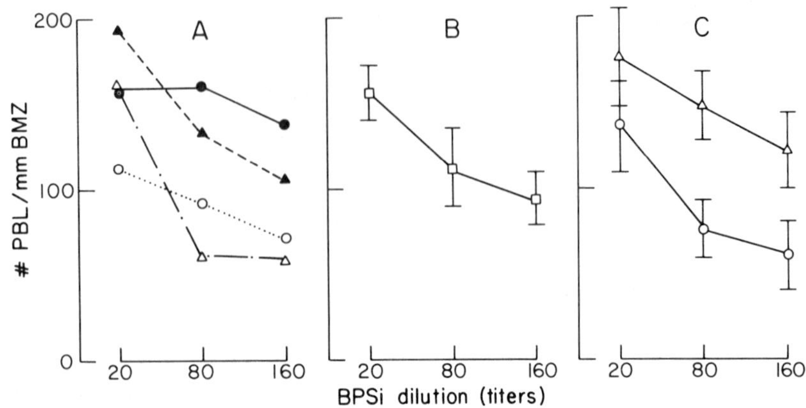

FIGURE 1. (A) Effect of serum titer on LA for BPS 1 ▲ – – – ▲; BPS 2 ●——●; BPS 3 Δ– · –Δ, and BPS 4, O· · · · ·O. (B) Mean LA for BPS 1–4 at titers of 1:20–1:160. (C) Comparison of mean LA between BPS 1 and 2 and BPS 3 and 4 dilutions of 1:20–1:160. Mean LA for BPS 1 and 2 = Δ——Δ and Mean LA for BPS 3 and 4 = O——O.

(FIGURE 1a) showed a decrease in leukocyte attachment with increasing dilutions of each BPS_i. The mean leukocyte attachment for all four BPS at dilutions of 1:20, 1:80, and 1:160 is shown in FIGURE 1b. A comparison of leukocyte attachment between BPS with a C3 binding titer of 1:320 (BPS_i 1 and 2) and a C3 binding titer of 1:640 (BPS 3 and 4) also showed differences in leukocyte attachment at dilutions of 1:80 and 1:160 (FIGURE 1c).

Leukocyte Attachment to BMZ-bound IgG Fc and C3b

To determine if PBL would bind to BMZ IgG, Fc, and C3b skin sections were treated with BPS_i 1–4, diluted 1:20, followed by PBS (BMZ, IgG, Fc) or 25% NHS (BMZ, IgG Fc, C3b). Following these treatments, the leukocyte attachment assay was performed with peripheral blood leukocytes suspended in GBSS. The results (TABLE 4)

TABLE 4. LA to BMZ-Bound IgG Fc and IgG Fc + C3b

Treatment	Mean LA ± SEM (#PBL/mm BMZ)	Significance[a]
IgG Fc		
BPS 1	43.1 ± 10.4	p > .1
BPS 2	23.5 ± 2.9	p > .1
BPS 3	21.5 ± 6.5	p > .1
BPS 4	ND	
IgG Fc + C3b		
BPS 1	41.2 ± 8.4	p > .1
BPS 2	36.1 ± 3.9	p > .1
BPS 3	36.0 ± 1.5	p > .1
BPS 4	29.3 ± 5.3	p > .1
Control Treatment	22.1 ± 2.3	

[a]Significance determined by *t*-test comparison between experimental and control treatment.

showed no significant attachment to BMZ-bound IgG Fc or IgG Fc C3b using any of the four BPS_i.

DISCUSSION

In this study we have quantitated leukocyte attachment by counting numbers of peripheral blood leukocytes attached to basement membrane zone (BMZ). Using this method, we have confirmed that leukocyte attachment occurs in the presence of complement-binding anti-BMZ antibodies and fresh normal human sera (NHS) and that fresh NHS (serum C) must be present for peripheral blood leukocytes to accumulate at the BMZ. When skin sections were treated with BPS alone or BPS and NHS and subsequently incubated with PBL suspended in NHS_i or buffer, no significant attachment occurred. This finding suggests that an interaction between leukocytes and BMZ-bound IgG Fc and C3b is not sufficient for significant leukocyte attachment and further suggests that chemotaxis, not immune adherence, is the mechanism for cell accumulation at the BMZ. Undoubtedly, once cells have reached the BMZ, they do interact with BMZ-bound IgG and C3. Although immune adherence is not primarily responsible for cell accumulation, it may be responsible for keeping cells at the BMZ once they arrive.

Leukocytes observed at the BMZ consisted mainly of neutrophil granulocytes and a few mononuclear cells (probably monocytes) and eosinophils. These cells are known to possess membrane receptors for complement-derived chemotactic factors (C5a) and for surface-bound IgG Fc and complement components, mainly C3b[4,6,7,9] In this study, the presence of IgG and C3b at the BMZ was confirmed by immunofluorescence. Functional evidence for BMZ-bound IgG Fc was provided by the findings that complement was activated and deposited at the site of IgG deposition. Thus the explanation for why peripheral blood leukocytes did not accumulate at the BMZ via immune adherence probably lies in the methodology. As has been previously shown, when peripheral blood leukocytes are injected into leukocyte attachment chambers, they tend to settle randomly on skin sections[3] In order for cells to accumulate at the BMZ, they must migrate to it. In the absence of a stimulus for directed migration, such as complement-derived chemotactic factor (C5a), migration and attachment would be expected to be a more or less random event. It is recognized that the presence of IgG and C3b at the BMZ might function to "capture" cells that had randomly migrated to it. However, this effect was apparently not sufficient to significantly increase cell accumulation at the BMZ.

These studies also suggest that the degree of leukocyte attachment is dependent on the amount of complement-binding anti-BMZ antibody used to sensitize skin sections. There was a significant decrease in leukocyte attachment when BPS were diluted (titered from 1:20 to 1:160). Furthermore, when skin sections were treated with equivalent dilutions of BPS_i (1:40 and 1:160), leukocyte attachment was significantly greater on sections treated with BPS with complement-binding anti-BMZ antibody titers of 1:640 compared to the sera with complement-binding titers of 1:320. Presumably this effect is due to decreasing chemotactic factor (C5a) production at the BMZ resulting from a decrease in immune complex formation.

The leukocyte attachment assay is a useful adjunct to indirect immunofluorescence for detecting complement-binding anti-BMZ antibodies and has the advantage of detecting functional interactions among BMZ immune complexes, complement, and leukocytes. The method should be of value for examining a number of questions concerning immune complex function in skin and immune-mediated interactions between leukocytes and the BMZ.

REFERENCES

1. BEUTNER, E. H. & R. J. NISENGARD. 1973. Defined immunofluorescence in clinical immunopathology. *In* Immunopathology of the Skin: Labeled Antibody Studies. E. H. Beutner, T. P. Chorzelski, S. F. Bean & R. E. Jordan, Eds.: 197–247. Dowden, Hutchinson and Ross. Stroudsberg, PA.
2. GAMMON, W. R., D. M. LEWIS, J. R. CARLO, W. M. SAMS, JR. & C. E. WHEELER, JR. 1980. Pemphigoid antibody mediated attachment of peripheral blood leukocytes at the dermal-epidermal junction of human skin. J. Invest. Dermatol. **75:** 334–339.
3. GAMMON, W. R., C. C. MERRITT, D. M. LEWIS, W. M. SAMS, JR., C. E. WHEELER, JR. & J. R. CARLO. 1981. Leukocyte chemotaxis to the dermal-epidermal junction of human skin mediated by pemphigoid antibody and complement: Mechanisms of cell attachment in the in vitro leukocyte attachment method. J. Invest. Dermatol. **76:** 514–522.
4. HUBER, H., M. J. POLLEY, W. D. LINSCOTT & H. G. MULLER-EBERHARD. 1968. Human monocytes: Distinct receptors for the third component of complement and for immunoglobulin G. Science **162:** 1281–1283.
5. JORDON, R. E., W. M. SAMS, JR. & E. H. BEUTNER. 1969. Complement immunofluorescence in bullous pemphigoid. J. Lab. Clin. Med **74:** 548–556.
6. NEWMAN, S. L. & R. B. JOHNSTON, JR. 1979. Role of binding through C3b and IgG in polymorphonuclear neutrophil function: Studies with trypsin-generated C3b. J. Immunol. **123:** 1839–1846.
7. SAMS, W. M., JR. & W. R. GAMMON. 1982. The pathogenesis of blisters in pemphigus and bullous pemphigoid. JAAD
8. SHER, R. & A. GLOVER. 1976. Isolation of human eosinophils and their lymphocyte-like rosetting properties. Immunology **31:** 337–341.
9. WILLIAMS, L. T., R. SNYDERMAN, M. C. PIKE, & R. J. LEFKOWITZ. 1977. Specific receptor sites for chemotactic peptides in human polymorphonuclear leukocytes. Proc. Natl. Acad. Sci. USA **74:** 1204–1208.
10. YAMAMOTO, T., I. KIHARI, T. MORITA & T. OITE. 1979. Attachment of polymorphonuclear leukocytes to glomeruli with immune deposits. J. Immunol. Meth. **26:** 315–323.

Experimental Production of Intercellular Antibodies in Monkeys[a]

SUSAN A. KRASNY AND ERNST H. BEUTNER

Department of Microbiology
School of Medicine
State University of New York at Buffalo
Buffalo, New York 14214

INTRODUCTION

Efforts to induce the formation of pemphigus antibodies have led several research groups to conduct rabbit immunization experiments. These have induced antibodies to the intercellular areas of stratified epithelia as detected by the indirect immunofluorescence (IF) test.[1,5,7,10] However, none of these demonstrated *in vivo* bound intercellular antibodies, thus suggesting that antibodies produced were pemphigus-like. These antibodies yield indirect IF staining patterns resembling those of pemphigus antibodies. From the standpoint of the antibody producer, the most important difference is that true pemphigus antibodies bind *in vivo* to their antigens in the normal skin and mucosa and produce epithelial damage whereas pemphigus-like antibodies fail to do so.

The reports of Michel, Schiltz, and their co-workers[8,9] on the binding of pemphigus antibodies to skin explants with subsequent development of intraepithelial lesions in the absence of complement are the most convincing experimental evidence for the pathogenic potential of pemphigus antibodies. Recently, we have adapted this organ culture system for use with mucosal explants.

The purpose of this study was to elicit antibodies to the intercellular antigens of stratified epithelia by immunization of monkeys with a saline extract of human esophageal mucosa containing pemphigus antigen. We also characterized these antibodies with regard to their pathogenic potential in organ culture. Because a number of reports describe cases of human pemphigus provoked by D-penicillamine, some of the animals were pretreated with this drug.[6,12]

MATERIALS AND METHODS

Immunofluorescence Staining Procedures

Defined IF tests were performed as described previously.[2] The fluorescein isothiocyanate–labeled conjugates used for all immunofluorescence studies were produced at the Department of Microbiology at SUNY at Buffalo and had the following characteristics: goat anti-human IgG, molar $F/P = 4.0$, Ab/ml = 5.5 mg, protein/ml = 12.9 mg; goat anti-human IgM, molar $F/P = 2.2$, protein/ml = 14.3 mg; goat anti-human IgA, molar $F/P = 2.1$, Ab/ml = 0.3 mg, protein/ml = 12.0 mg; goat anti-human C3, molar $F/P = 2.4$, Ab/ml = 0.7 mg, protein/ml = 11.2 mg. The

[a]This work was supported in part by a grant from the Summerhill Foundation and by grant AI070880 from the U.S. Public Health Service.

343

dilutions used contained approximately 50–100 μg/ml antibody protein for detecting human antibody and 50–200 μg/ml antibody protein for detecting monkey antibody.

Extraction of Human Esophagus

Crude extracts of human esophagus mucosa containing pemphigus antigen were prepared by a modification of the method of Shu and Beutner.[11] All procedures were performed in the cold. Briefly, an unfixed esophagus was removed at autopsy within 24 hr postmortem. The mucosal layer was separated from the adventitia, cut into strips, fast-frozen in liquid nitrogen, and ground with a mortar and pestle to a fine powder. This powder was suspended in 2 ml of PBS/g tissue. The slurry was sonicated using short 10-sec bursts until the mucosa appeared homogenized. The homogenate was centrifuged at 12,000 × g for 30 min and the supernatant fluid then dialyzed against distilled water for 18–24 hr. The retentate was centrifuged as before to remove the heavy precipitate.

Prior to use for immunization of animals, the protein and pemphigus antigen content of the esophagus preparations were determined by the biuret method and inhibition of indirect immunofluorescence staining of pemphigus sera, respectively. To determine the latter, doubling dilutions of pemphigus sera in PBS were each incubated with equal volumes of undiluted and doubling dilutions of esophagus extract for 30 min at room temperature.

Immunization of Monkeys

Seven monkeys, one stumptail (*Macaca arctoides*) and six pigtails (*Macaca nemestrina*), were divided into three groups. Two groups were pretreated with D-penicillamine intramuscularly or *per os* and all three received intradermal injections of 1 mg protein/ml esophagus extract in Freund's complete adjuvant as indicated by arrows in FIGURES 1–3.

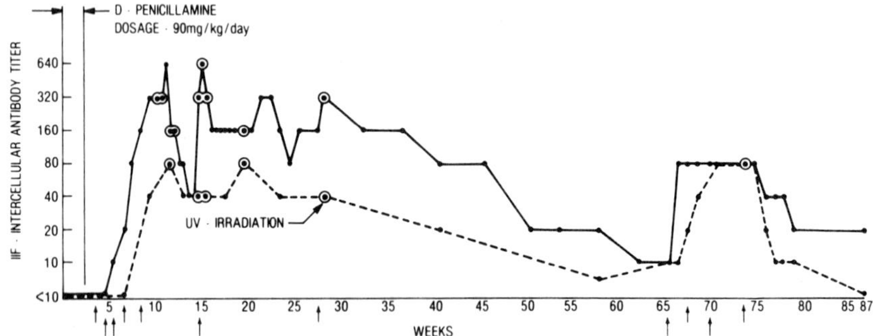

FIGURE 1. Group 1: monkey 7394. Antibody titers to epithelial intercellular antigens from heterologous (—) or autologous (---) monkey tissue following injection with an extract of human esophagus. Prior to immunization this monkey received D-penicillamine i.m. for 2.5 wk. Intercellular antibody titers were determined by an indirect immunofluorescence (IF) test on African Green monkey esophagus or autologous lip mucosa using a goat anti-human IgG conjugate. (☉) Indicates sera used for maintaining explants of monkey lip mucosa in organ culture.

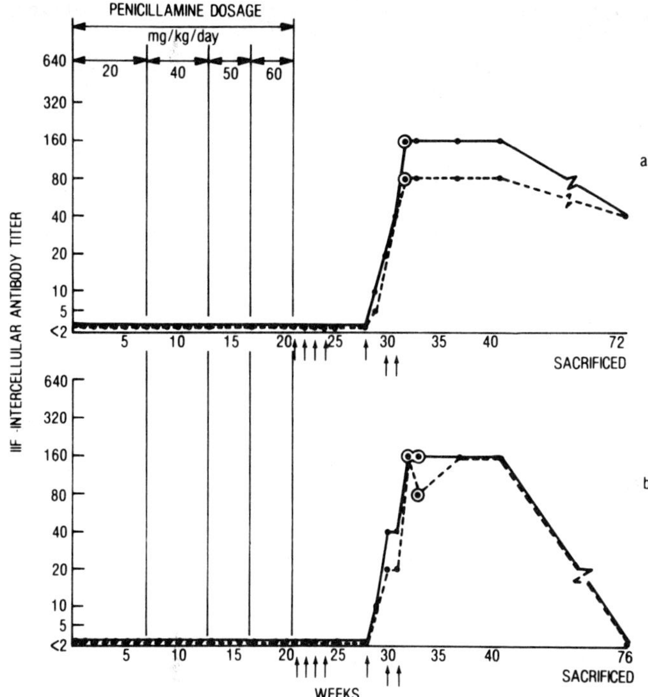

FIGURE 2. Group 2: monkey 7902 (a) and 7911 (b). Intercellular antibody titers determined as in FIGURE 1 to heterologous (—) and autologous (---) monkey tissue. Prior to immunization these monkeys received increasing doses of D-penicillamine *per os* for 21 wk. (⊙) Same as in FIGURE 1.

Organ Culture

A modification of the method of Michel and Ko was employed.[8] Sheets of monkey lip mucosa were obtained from a biopsy or from animals immediately after sacrifice. The mucosal sheets were washed three times with an antibiotic/antimycotic (Grand Island Biological Co.) and cut into 2 × 2 mm explants. These were cultured on sterilized discs of Whatman filter paper supported by stainless steel grids and placed in plastic tissue culture dishes. Explants were cultured in 0.2 ml of whole human or monkey sera to which 40 units of antibiotic/antimycotic had been added. Explants were incubated in a moist chamber at 31°C. At 8, 24, 48, and 72 hr explants were removed and snap-frozen in liquid nitrogen. These were sectioned and examined by direct IF for IgG, IgM, IgA, and C3 staining in the intercellular areas. Additional sections were stained with hematoxylin and eosin for histologic examination.

Sera

Sera from three patients with pemphigus vulgaris were employed as positive controls for indirect IF tests or for organ culture experiments. All pemphigus sera yielded characteristic intercellular staining on sections of monkey epithelial tissue using a goat anti-human IgG conjugate. Indirect IF tests of three additional human sera also revealed typical intercellular staining. However, clinical evaluation, and/or direct IF and histologic examination of biopsies from these three patients ruled out a diagnosis of pemphigus confirming these as pemphigus-like antibody–containing sera.

RESULTS

FIGURES 1–3 summarize the titers of intercellular antibodies produced by five of seven monkeys upon immunization with human esophageal extract as seen in IF tests

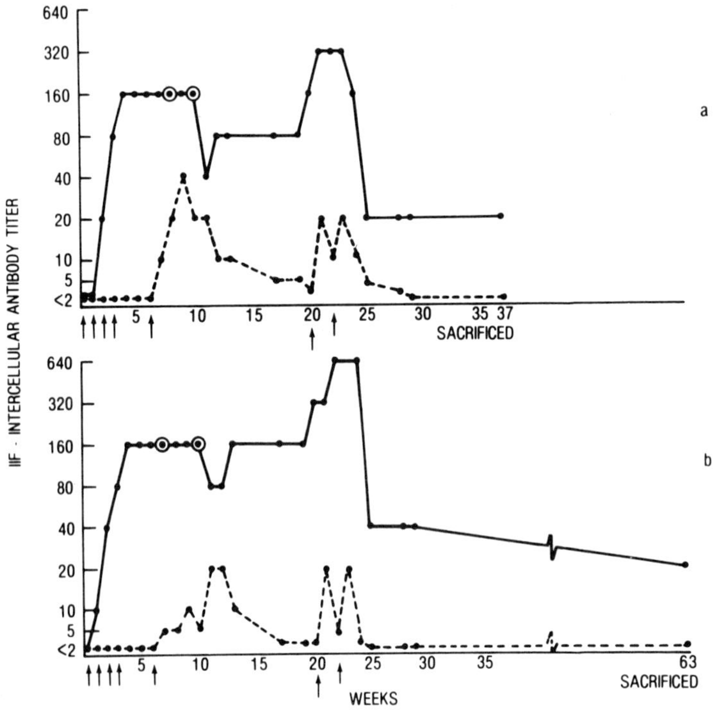

FIGURE 3. Group 3: monkey 7910 (a) and 7913 (b). Intercellular antibody titers determined as in FIGURE 1 to heterologous (—) and autologous (---) monkey tissue. No D-penicillamine administered before immunization. (⊙) Same as FIGURE 1.

with an anti-IgG conjugate on heterologous (solid line) monkey esophagus and autologous (dashed line) monkey lip mucosa as the substrate. There was one non-responder in Group 2 (D-penicillamine *per os*) and in group 3 (no D-penicillamine). No intercellular antibodies were detected in either group 1 or group 2 during penicillamine administration.

Three differences were observed in the intercellular antibody responses of monkeys receiving D-penicillamine and those not administered the drug. Animals given

D-penicillamine prior to immunization exhibited a delay of 3–7 weeks in the production of antibodies to heterologous intercellular antigens, compared to monkeys not given the drug, which responded within one week. However, irrespective of D-penicillamine administration, a delay of approximately six weeks occurred in autoantibody production in all monkeys that gave detectable immune responses. The maximum antibody titers to heterologous intercellular antigens essentially were the same in all three groups of monkeys studied. However, the autoantibody titers achieved by the groups given D-penicillamine were as much as 16-fold greater than that of monkeys not given the drug.

The IF staining pattern of the intercellular antibodies found in the sera of monkeys in groups 1 and 2 (D-penicillamine) resembled true pemphigus antibodies in that they failed to stain the basal cell areas or the upper epithelium of an esophagus section. The sera from group 3 (no penicillamine) stained the intercellular areas throughout the esophagus epithelium.

Routine examination throughout D-penicillamine administration and immunization failed to reveal skin or mucous membrane lesions suggestive of pemphigus. When examined with anti-IgG, anti-IgM, anti-IgA, and anti-C3 conjugates, skin and mucosal biopsies and tissues taken at necropsy failed to reveal *in vivo* deposits in the epithelial intercellular substance irrespective of circulating antibodies present in sera drawn simultaneously.

In order to assess the pathogenic potential of monkey intercellular antibodies, we compared them to pemphigus antibodies in organ culture. Because some pemphigus sera and all monkey sera reacted only with mucosal epithelium and not skin, explants of monkey lip mucosa were used. Direct IF and H & E studies of mucosal explants cultured in monkey and human sera containing intercellular antibodies are summarized in TABLE 1. Binding of intercellular antibodies occurred in 19/42 (45%) heterologous explants incubated in sera from immunized monkeys, whereas 2/32 (6.3%) explants exhibited acantholysis. No intercellular binding or acantholytic clefts were observed in explants of autologous lip mucosa. A granular deposition (FIGURE 4) was seen in the intercellular areas of mucosal explants with sera of three of the five responder monkeys (group 1–penicillamine and group 3–no penicillamine). The intercellular staining appeared at 8–16 hr and gradually increased in intensity to a maximum at 24 hr. The intensity then decreased over a period of 48 to 72 hr. When present, acantholysis was observed at 48–72 hr (FIGURE 5). Not all serum samples from the three monkeys were consistently positive. Moreover, of the sera from monkeys with intercellular antibodies that bound in organ culture, only sera of monkey 7394 produced acantholysis.

The behavior of pemphigus antibodies in mucosal explants was very similar to that reported in skin explants.[8,9] Binding of intercellular antibodies was observed in 26/29 (90%) of mucosal explants. In contrast to the granular staining observed with monkey antibodies, pemphigus antibodies produced a strong linear pattern (FIGURE 6). Acantholysis occurred in 16/25 (64%) of these explants (FIGURE 7).

Direct IF and H & E staining of mucosal explants in sera containing pemphigus-like antibodies at titers equal to or greater than that of pemphigus or monkey intercellular antibodies failed to yield detectable binding or acantholysis. Moreover, indirect IF tests with the sera employed in organ culture studies always demonstrated antigen reactivity in explants negative by direct IF.

Examination by direct immunofluorescence using anti-IgM, anti-IgA, or anti-C3 conjugates failed to reveal staining in explants cultured in monkey or human sera. All explants incubated in normal human or monkey control sera were negative by direct immunofluorescence and revealed no acantholytic changes. Hematoxylin and eosin staining procedures on both control and test explants demonstrated varying degrees of

TABLE 1. Immunofluorescence and Histologic Findings in Mucosal Explants Incubated in Monkey or Human Sera that Contain Intercellular Antibodies

| Intercellular Antibodies | No. of Serum Samples Tested | Ranges of Indirect IF Intercellular Antibody Titers on Explant Mucosa | | Organ Culture Reactions | | | |
| | | | | Direct IF[c]–Intercellular Binding (No. Pos./No. Tests) | | Acantholysis (No. Pos./No. Tests) | |
		Heterologous[b]	Autologous[b]	Heterologous	Autologous	Heterologous	Autologous
Monkey							
7394[a]	10	80–320	40–80	14/32 (45%)	0/4	2/24 (8.3%)	0/4
7902[a]	1	160	80	0/2	0/2	0/2	0/2
7911[a]	2	160	80–160	0/3	0/3	0/3	0/3
7910	2	160	20	2/2	ND	0/1	ND
7913	2	160	5	3/4	ND	0/2	ND
Total	17			19/42 (45%)	0/9	2/32[d] (6.3%)	0/9
Human							
Pemphigus	1	160	ND	26/29 (90%)	ND	16/25 (64%)	ND
Pemphigus-like	3	160–640	ND	0/8	ND	0/8	ND

[a]Received D-penicillamine.
[b]Lip mucosa from African Green monkey (heterologous) or from the antibody producer (autologous).
[c]Goat anti-human IgG conjugate.
[d]Does not include four experiments (10 monkey sera, 4 human sera) in which explants were spongiotic after 24 hr.

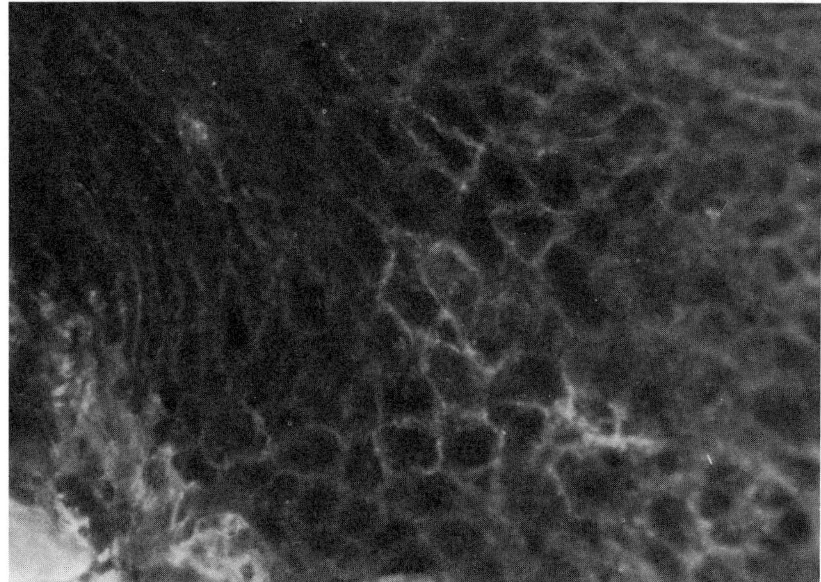

FIGURE 4. Direct immunofluorescence staining for IgG on explant of monkey lip mucosa incubated for 24 hr in antiserum from monkey 7394 with intercellular antibodies to a titer of 160. Note the intercellular staining has a granular appearance. (×320)

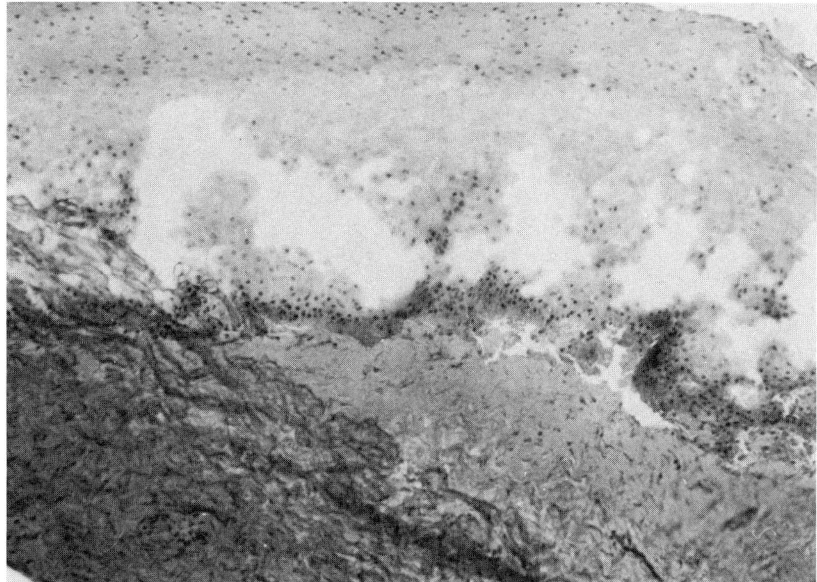

FIGURE 5. H & E stained section of mucosa of a monkey lip explant incubated 72 hr in serum from monkey 7394 containing intercellular antibodies. Note the marked acantholysis with few acantholytic cells. (×80)

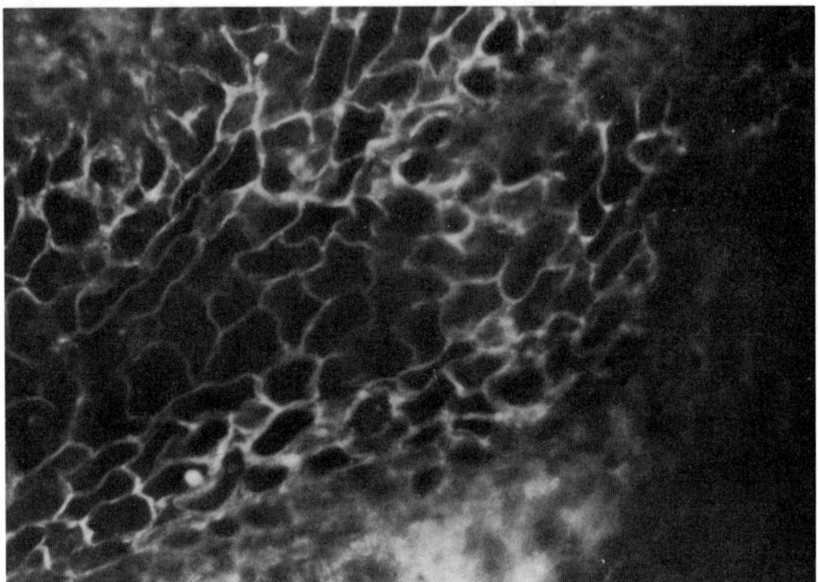

FIGURE 6. Direct immunofluorescence staining for IgG of the mucosa of a monkey lip explant after incubation for 24 hr in human pemphigus serum. Note strong linear intercellular staining. ($\times 360$)

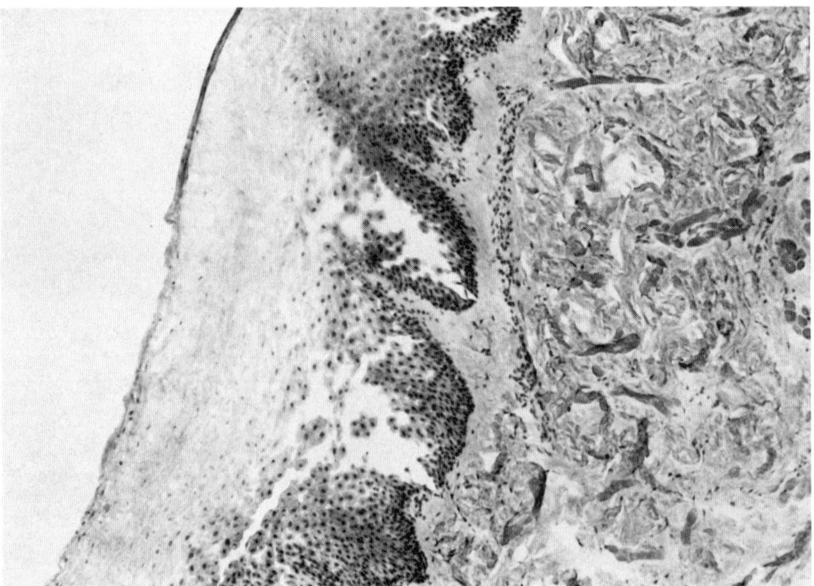

FIGURE 7. H & E stained section of an explant of monkey lip mucosa incubated for 48 hr in human pemphigus serum. Note the intraepithelial cleft with numerous acantholytic cells. ($\times 80$)

nonspecific spongiosis and vacuolation, which were attributable to the culture conditions.

DISCUSSION

We have demonstrated that upon heterostimulation by injection of a crude saline by extract of human esophageal mucosa, one stumptail and four of six pigtail monkeys produced antibodies to intercellular antigens as detected by indirect IF. Previous attempts to produce pemphigus antibodies in rabbits resulted in intercellular antibody titers comparable to those observed in the sera of our monkeys.[1,5,7,10]

We also examined the effect of D-penicillamine on the antibody response of these monkeys to injection with pemphigus antigen. However, due to the small number of monkeys employed in our study, the differences in the antibody responses of groups 1 (penicillamine–i.m.), 2 (penicillamine–*per os*), and 3 (no penicillamine) must be viewed with some caution and no conclusions may be drawn. In addition, the experimental conditions involving doses, route, and length of time of administration may not have been optimal to observe the effects of this drug on the immune response. Further studies are indicated.

The binding of intercellular antibodies and acantholytic changes observed in mucosal explants when cultured in certain monkey sera indicates their pathogenic potential. This is the first study to equate experimentally induced intercellular antibodies to pemphigus antibodies. Moreover, negative IF staining for C3 in the intercellular substance of these same explants positive for IgG, provided further support for the tenet of Schiltz and Michel[9] that complement is not involved in the primary mechanism of acantholysis in pemphigus.

The selectivity in binding to intercellular antigens and acantholytic changes displayed by antibodies produced in monkeys suggested that a variety of specificities existed in the intercellular antibodies produced in immunized monkeys. *In vitro,* some monkey intercellular antibodies had features of true pemphigus antibodies. However, the failure of responder monkeys to develop skin or mucosal lesions, the absence of *in vivo* deposition of intercellular antibodies, and the failure to reveal immunoglobulin binding and acantholysis in autologous explants of mucosa cultured in sera from the donor monkeys suggest that these autoantibodies were pemphigus-like. These results prompted the study of the *in vitro* reactivity of human pemphigus-like antibodies. These are differentiated from true pemphigus antibodies by several characteristics.[4] Most importantly, pemphigus antibodies bind *in vivo* to their antigens in the normal skin and mucosa whereas pemphigus-like antibodies do not bind. The finding that true pemphigus antibodies bind in organ culture and that pemphigus-like antibodies do not, now serves as an additional criterion for the differentiation of these intercellular antibodies. This failure to bind *in vitro* was not due to the loss of pemphigus antigen. Moreover, all sera contained titers of intercellular antibodies at concentrations (≥ 80) sufficient to produce acantholysis.[3]

The failure of monkey antibodies directed to intercellular autoantigens and human pemphigus-like antibodies to bind in organ culture, suggests that these antibodies may be directed to inaccessible antigen determinants. Conversely, pemphigus antibodies and some monkey antibodies that bind to explants *in vitro* are directed to accessible intercellular antigens that share the same histologic location but are different from hidden intercellular antigens.

In conclusion, the demonstration in the serum from an immunized monkey of antibodies that bind to intercellular antigens in organ culture and produce acantholytic changes in normal monkey epithelia encourages continued efforts to produce an experimental model of pemphigus.

ACKNOWLEDGMENTS

We thank Dr. T. Chorzelski (Warsaw Medical School, Warsaw, Poland) for providing a serum with a high titer of pemphigus-like antibodies and for advice throughout this research. The gift of D-penicillamine (Cuprimine) from Merck, Sharpe and Dohme (West Point, PA) is gratefully acknowledged.

REFERENCES

1. ABLIN, R. J. & E. H. BEUTNER. 1969. Experimental production of pemphigus-like antibodies. Clin. Exp. Immunol. **4:** 283–291.
2. GROB, P. J. & T. M. INDERBITZIN. 1967. Pemphigus antigen and blood group substances A and B. J. Invest Dermatol. **49:** 637–641.
3. INDERBITZIN, T. M. & P. J. GROB. 1967. Destruction of epithelial cells *in vivo* by anti-epithelial autoantibodies. J. Invest. Dermatol. **49:** 642–645.
4. SHU, S. & E. H. BEUTNER. 1970. Experimental production of pemphigus-like autoantibodies by isoimmunization of rabbits. Fed. Proc. **29:** 722 (Abstract).
5. MICHEL, B. & C. S. KO. 1977. An organ culture model for the study of pemphigus acantholysis. Br. J. Dermatol. **96:** 295–302.
6. SCHILTZ, J. R. & B. MICHEL. 1976. Production of epidermal acantholysis in normal human skin *in vitro* by the IgG fraction from pemphigus serm. J. Invest. Dermatol. **67:** 254–260.
7. HAY, K. D., H. K. MULLER & P. C. READE. 1978. D-penicillamine-induced mucocutaneous lesions with features of pemphigus. Oral Surg. **45:** 385–395.
8. YUNG, C. W. & G. W. HAMBRICK, JR. 1982. D-penicillamine-induced pemphigus syndrome. J. Am. Acad. Dermatol. **6:** 317–324.
9. BEUTNER, E. H., R. J. NISENGARD & V. KUMAR. 1979. Defined immunofluorescence: Basic concepts and their application to clinical immunodermatology. *In* Immunopathology of the Skin. 2nd edit. E. H. Beutner, T. P. Chorzelski & S. F. Bean, Eds.: 29–75. John Wiley and Sons. New York.
10. SHU, S. & E. H. BEUTNER. 1973. Isolation and characterization of antigens reactive with pemphigus antibodies. J. Invest. Dermatol. **61:** 270–276.
11. CHORZELSKI, T. P., E. H. BEUTNER & S. JABLONSKA. 1979. The role and nature of autoimmunity in pemphigus. *In* Immunopathology of the Skin. 2nd edit. E. H. Beutner, T. P. Chorzelski & S. F. Bean, Eds.: 183–229, John Wiley and Sons. New York.
12. BINDER, W. L., E. H. BEUTNER & T. P. CHORZELSKI. 1978. In vitro effects of pemphigus antibodies on skin. Br. J. Dermatol. **99:** 39–42.

Pemphigus and Pemphigoid in Dogs, Cats, and Horses

D. W. SCOTT, T. O. MANNING, C. A. SMITH, AND
R. M. LEWIS

Departments of Clinical Sciences and Pathology
New York State College of Veterinary Medicine
Cornell University
Ithaca, New York 14853

INTRODUCTION

Pemphigus (from the Greek "blister") and pemphigoid (from the Greek "resembling pemphigus") have been recognized in humans for many years. These disorders are of presumed autoimmune origin, but their exact pathogenesis remains elusive. Reports on pemphigus and pemphigoid in domestic animals are of much more recent vintage. In the dog, pemphigus vulgaris was first reported in 1975,[2,16] pemphigus vegetans in 1977,[9] pemphigus foliaceus in 1977,[1] bullous pemphigoid in 1978,[4] and pemphigus erythematosus in 1980.[10,11] In the cat, pemphigus vulgaris, pemphigus foliaceus, and pemphigus erythematosus were reported in 1980.[10,12] Equine pemphigus foliaceus and equine bullous pemphigoid were reported in 1981.[3,6,7]

The purpose of this article is to review the clinical, pathological, immunopathological, and therapeutic aspects of pemphigus and pemphigoid as studied in 34 dogs, 10 cats, and 9 horses at the New York State College of Veterinary Medicine.

MATERIALS AND METHODS

Thirty-four dogs, 10 cats, and 9 horses with various forms of pemphigus and pemphigoid were studied. Skin and mucosal biopsies were fixed in 10% neutral formalin and stained with hematoxylin and eosin (H&E) and acid orcein giemsa (AOG) for histologic examination.

Direct immunofluorescence testing was performed as previously reported.[10] In dogs, anti-IgG, -IgM, -IgA, -whole Ig, and -C_3 were used. In cats and horses, only anti-whole Ig and anti-C_3 were used. Indirect immunofluorescence testing, antinuclear antibody tests, lupus erythematosus cell tests, and cellulose acetate serum protein electrophoresis were also performed as previously reported.[10,14]

Various chemotherapeutic regimens were used to treat pemphigus and pemphigoid in domestic animals. In general, the chemotherapeutic regimens were used in the following sequence: (1) prednisolone, 1.1 mg/kg, orally twice daily; (2) prednisolone, 3.3 mg/kg, orally twice daily; (3) prednisolone, 3.3 mg/kg, orally twice daily in conjunction with cyclophosphamide (Cytoxan[R], Mead-Johnson), given orally at the rate of 50 mg/m^2 of body surface every other day; and (4) aurothioglucose (Solganol[R], Schering). Aurothioglucose was administered intramuscularly at weekly test doses of 1 and 2 mg for cats and dogs weighing less than 10 kg; 5 and 10 mg for dogs weighing over 10 kg, and 20 and 40 mg for horses. Aurothioglucose was then given intramuscularly at 1 mg/kg/week until good clinical response. The decision to switch to the next therapeutic regimen in the sequence was made on the basis of lack of therapeutic efficacy or unacceptable side effects.

RESULTS

Clinical Findings

Pemphigus vulgaris was recognized in 12 dogs and 4 cats. Affected dogs ranged from 1 to 11 years of age with no breed or sex predilections. Cats ranged from 1 to 14 years of age with no breed or sex predilections. Clinically, pemphigus vulgaris was characterized by a vesicobullous, deeply erosive disorder that may affect the oral cavity, mucocutaneous junctions (lips, nostrils, eyelids, prepuce, vulva, anus), skin, or any combination thereof. Oral involvement was the initial sign in about 50% of the patients, and was present in over 90% of the patients at the time of diagnosis. Pemphigus vulgaris limited to the skin is rare in dogs[15] and unreported in cats. Severely affected animals may be anorectic, depressed, and febrile.

Pemphigus foliaceus was recognized in 7 dogs, 5 cats, and 8 horses. Affected dogs ranged from 1 to 12 years of age with no breed or sex predilections. Affected cats ranged from 6 months to 11 years of age with no breed or sex predilections. Affected horses ranged from 4 months to 10 years of age with no breed or sex predilections. Clinically, pemphigus foliaceus was characterized by a vesicobullous, superficially erosive skin disease. The face, feet, and ventrum appear to be predilectd areas. Affected dogs and cats frequently had marked hyperkeratosis of the footpads. Severely affected animals may be anorectic, depressed, and febrile. Horses usually have pitting edema of all four limbs and the ventrum.

Pemphigus erythematosus was recognized in 5 dogs and 1 cat. Affected dogs ranged from 2 to 9 years of age with no breed or sex predilections. The cat was an 8-year-old, male American shorthaired tabby. Clinically, pemphigus erythematosus was characterized by a vesicobullous, superficially erosive skin disease, with marked scaling and crusting, that was limited to the head and ears. Affected animals were usually otherwise healthy.

Pemphigus vegetans was recognized in a 15-year-old terrier. Clinically, it was characterized by a generalized vesicopustular dermatitis that evolved into verrucous vegetations and papillomatous proliferations. The dog was otherwise healthy.

Bullous pemphigoid was recognized in 9 dogs ranging from 2 to 10 years of age. There was no apparent sex predilection, but 6 of the 9 dogs were collies. Bullous pemphigoid was also recognized in a 14-year-old female quarter horse. Clinically, bullous pemphigoid was characterized by a vesicobullous, ulcerative disorder affecting the oral cavity, mucocutaneous junctions, skin, or any combination thereof. About 80% of the patients had oral cavity involvement at the time of diagnosis. Severely affected animals may be anorectic, depressed, and febrile.

Clinical Pathology

Animals with pemphigus vulgaris, pemphigus foliaceus, and bullous pemphigoid often had mild to moderate leukocytosis and neutrophilia, mild nonregenerative anemia, mild hypoalbuminemia, and moderate elevations of alpha-2, beta, and gamma globulins.

Pemphigus vulgaris was characterized by suprabasilar acantholysis and cleft formation. Pemphigus foliaceus and erythematosus were characterized by subcorneal

or intragranular acantholysis and cleft formation. Pemphigus vegetans was characterized by papillomatosis and intraepidermal abscesses containing eosinophils and
acantholytic keratinocytes. Bullous pemphigoid was characterized by subepidermal
cleft and vesicle formation. In all forms of pemphigus and pemphigoid, hair follicles
were often involved in the pathology and a lichenoid band of inflammatory cells was
common.

Immunopathology

Results of the immunopathologic studies are summarized in TABLES 1–3. Direct
immunofluorescence testing was positive in all dogs and cats, and in 6 of 9 horses. All
forms of pemphigus were characterized by the diffuse deposition of immunoglobulin,
and rarely C3, in the intercellular spaces. In addition, the 5 dogs and 1 cat with
pemphigus erythematosus had immunoglobulin, and occasionally C3, deposited at the
basement membrane zone. Bullous pemphigoid was characterized by the deposition of
immunoglobulin, and usually C3, at the basement membrane zone.

Indirect immunofluorescence testing was positive in only 1 of 23 dogs with
pemphigus, 1 of 6 dogs with pemphigoid, and 0 of 10 cats with pemphigus. However, in
horses, indirect immunofluorescence testing was positive in 5 of 8 with pemphigus, and
the 1 horse with pemphigoid.

Antinuclear antibody tests were positive in 4 of 5 dogs and the 1 cat with
pemphigus erythematosus. Lupus erythematosus cell tests were negative in every
case.

Therapy

The first chemotherapeutic regimen failed to control the disease in any of the 22
dogs and 5 cats in which it was used. It was successful in all 3 horses with pemphigus
foliaceus in which it was used. One of these horses has been maintained on
alternate-day prednisolone therapy for almost two years.

The second chemotherapeutic regimen was effective in inducing remission in 26 of
the 30 dogs (94%) in which it was used. However, in 16 dogs (52%) this regimen was
unsatisfactory because alternate-day prednisolone therapy sufficient to control the
disease but not cause unacceptable side effects could not be achieved in 13 dogs, and 3
dogs suffered attacks of acute pancreatitis within 7 to 10 days of the initiation of
therapy. Fifteen dogs were put in remission with this regimen and successfully
maintained with alternate-day prednisolone (1.1 to 2.2 mg/kg alternate mornings) for
one to six years. This regimen was successful in 5 of the 10 cats in which it was used,
with the 5 cats being maintained on alternate-day prednisolone (2.2 mg/kg alternate
evenings) for up to two years.

The third chemotherapeutic regimen was used for 5 dogs. These dogs were then
successfully maintained on alternate-day prednisolone (1.1 to 2.2 mg/kg) and cyclophosphamide for 4 months to 4 years.

The fourth chemotherapeutic regimen was used in 6 dogs, 4 cats, and 2 horses
(TABLE 4). All animals have been successfully managed for 6 months to 3 years.

Two animals underwent spontaneous remissions. A 7-year-old, spayed female
Scottish Terrier with a 10-month course of oral pemphigus vulgaris underwent a
spontaneous remission that has lasted 6 months. A 10-year-old female Arabian horse
with pemphigus foliaceus was also 7 months pregnant. When the mare was aborted,
the pemphigus cleared within 2 months and has not recurred for 6 months.

TABLE 1. Immunopathologic Findings in Canine Pemphigus and Pemphigoid

| Disease | Number of Patients | Direct Immunofluorescence Testing | | | | Indirect Immunofluorescence Testing | Antinuclear Antibody[a] | Lupus Erythematosus Cell Test |
		IgG	IgA	IgM	C3			
Pemphigus vulgaris								
ICS[b]	11	11	1	1	0	0	1(1:20)	0
Pemphigus foliaceus								
ICS	7	7	1	1	1	1(1:16)	0	0
Pemphigus erythematosus								
ICS	5	5	2	2	2	0	4(1:10 to 1:20)	0
BMZ[c]		2	3	2	2			
Bullous pemphigoid								
BMZ	6	3	4	2	4	1(1:2048)	0	0

[a]Normal dogs in this laboratory will have an ANA of 0.
[b]Intercellular spaces.
[c]Basement membrane zone.

TABLE 2. Immunopathologic Findings in Feline Pemphigus

| Disease | Number of Patients | Direct Immunofluorescence Testing | | | Indirect Immunofluorescence Testing | Antinuclear Antibody[a] | Lupus Erythematosus Cell Test |
		Whole	Ig	C3			
Pemphigus vulgaris							
ICS[b]	4	4		1	0	0	0
Pemphigus foliaceus							
ICS	5	5		0	0	0	0
Pemphigus erythematosus	1						
ICS		1		1	0	1(1:5)	0
BMZ[c]		1		1			

[a]Normal cats in this laboratory will have an ANA of 0.
[b]Intercellular spaces.
[c]Basement membrane zone.

TABLE 3. Immunopathologic Findings in Equine Pemphigus and Pemphigoid

| Disease | Number of Patients | Direct Immunofluorescence Testing | | | Indirect Immunofluoresence Testing | Antinuclear Antibody[a] | Lupus Erythematosus Cell Test |
		Whole	Ig	C3			
Pemphigus foliaceus							
ICS[b]	8	6		0	5(1:10 to 1:8000)	2 (1:4)	0
Bullous pemphigoid							
BMZ[c]		0		0	1(1:80)	0	0

[a]Normal horses in this laboratory can have an ANA of 1:5.
[b]Intercellular spaces.
[c]Basement membrane zone.

DISCUSSION

Pemphigus and pemphigoid in dogs, cats, and horses are very similar, clinically and pathologically, to the diseases in humans.[13] Prognostically, pemphigus vulgaris, pemphigus foliaceus, and bullous pemphigoid appear to be severe diseases in domestic animals, often resulting in death or euthanasia unless effectively treated.

Immunopathologically, pemphigus and pemphigoid in dogs, cats, and horses have many similarities to the situation in humans.[13,14] Direct immunofluorescence testing is very reliable for diagnosis. However, indirect immunofluorescence testing, contrary to the situation in humans, is quite unreliable for the diagnosis of canine and feline pemphigus and pemphgoid. Indirect immunofluorescence testing was positive in only 1 of 23 dogs (4.3%) with pemphigus, 1 of 6 dogs (16.7%) with pemphigoid, and 0 of 10 cats (0%) with pemphigus. When pemphigus antibody titers have been reported in dogs,[1,5] they have been very low (1:5 to 1:16) compared to humans. In horses, indirect immunofluorescence testing was positive in 5 of 8 cases (62.5%) of equine pemphigus foliaceus, with titers ranging from 1:10 to 1:8000, and in the 1 case of equine bullous pemphigoid (titer 1:80).

TABLE 4. Results of Chrysotherapy in Dogs, Cats, and Horses with Pemphigus

Disease	Species	Results
Pemphigus foliaceus	Horse	Received gold for 6 months: in remission for 1 year
Pemphigus foliaceus	Horse	Received gold for 6 months: in remission for 8 months
Pemphigus foliaceus	Cat	Received gold for 1.5 years: in remission for 1.5 years
Pemphigus foliaceus	Cat	Received gold for 6 months: in remission for 6 months
Pemphigus erythematosus	Cat	Received for gold 1 year: in remission for 1 year
Pemphigus vulgaris	Cat	Received gold for 2 years: in remission for 2 years
Pemphigus foliaceus	Dog	Received gold for 7 months: in remission for 3 years
Pemphigus foliaceus	Dog	Received gold for 8 months: in remission for 3 years
Pemphigus foliaceus	Dog	Received gold for 1 year: in remission for 1 year
Pemphigus erythematosus	Dog	Received gold for 2 years: in remission for 2 years
Pemphigus erythematosus	Dog	Received gold for 6 months: in remission for 6 months
Pemphigus vulgaris	Dog	Received gold for 1 year: in remission for 1 year

Massive doses of glucocorticoids were initially considered to be the preferred treatment for pemphigus and pemphigoid.[13,14] However, although glucocorticoids reduced the death rate from these diseases, the therapeutic results were unsatisfactory in many instances. In addition, at present the major causes of mortality and morbidity in pemphigus and pemphigoid are side effects of glucocorticoid therapy.[13,14] Mortality and morbidity increase dramatically as the dose of required glucocorticoid increases.

Recently, the use of various immunomodulating drugs has been advocated for pemphigus and pemphigoid.[13,14] Drugs such as cyclophosphamide, azathioprine, methotrexate, and gold compounds have greatly reduced the doses of glucocorticoids needed.

The therapeutic studies reported here allow interesting preliminary conclusions. (1) A commonly used "immunosuppressive" dose of prednisolone (2.2 mg/kg/day) was ineffective in all 22 dogs and 5 cats with pemphigus and pemphigoid in which it was used; however, it was effective in the 3 horses with pemphigus foliaceus in which it was used. (2) The larger dose of prednisolone (6.6 mg/kg/day) was effective for inducing remission in 26 of 31 dogs (94%) and 5 of 10 cats (50%), but in 16 of the 31 dogs (52%), therapy had to be stopped because of unsatisfactory response or unacceptable side effects. (3) Combination chemotherapy with prednisolone and cyclophosphamide was successful in all 5 dogs in which it was used. (4) Aurothioglucose was successful in all 6 dogs, 4 cats, and 2 horses in which it was used (TABLE 4). Encouraging therapeutic results with gold compounds have been previously reported in human,[13] canine,[5,11] feline,[12] and equine[8] pemphigus.

REFERENCES

1. HALLIWELL, R. E. W. & M. H. GOLDSCHMIDT. 1977. Pemphigus foliaceus in the canine: A case report and discussion. J. Am. Anim. Hosp. Assoc. **13:** 431–436.
2. HURVITZ, A. I. & E. FELDMAN. 1975. A disease in dogs resembling human pemphigus vulgaris: Case reports. J. Am. Vet. Med. Assoc. **166:** 585–590.
3. JOHNSON, M. E., D. W. SCOTT, T. O. MANNING *et al.* 1981. Pemphigus foliaceus in the horse. Equine Pract. **3:** 40–45.
4. KUNKLE, G., M. H. GOLDSCHMIDT & R. E. W. HALLIWELL. 1978. Bullous pemphigoid in a dog: A case report with immunofluorescent findings. J. Am. Anim. Hosp. Assoc. **14:** 52–57.
5. MANNING, T. O., D. W. SCOTT, S. A. KRUTH *et al.* 1980. Three cases of canine pemphigus foliaceus, and observations on chrysotherapy. J. Am. Anim. Hosp. Assoc. **16:** 189–202.
6. MANNING, T. O., D. W. SCOTT, W. C. REBHUN *et al.* 1981. Pemphigus-pemphigoid in a horse. Equine Pract. **3:** 38–44.
7. PETER, J. E., P. G. MORRIS & B. J. GORDON. 1981. Pemphigus in a thoroughbred. Vet. Med. Small Anim. Clin. **76:** 1203–1206.
8. POWER, H. T., E. O. MCEVOY & T. O. MANNING. 1982. Use of a gold compound for the treatment of pemphigus foliaceus in a foal. J. Am. Vet. Med. Assoc. **180:** 400–403.
9. SCOTT, D. W. 1977. Pemphigus vegetans in a dog. Cornell Vet. **67:** 374–384.
10. SCOTT, D. W., M. J. WOLFE, C. A. SMITH *et al.* 1980. The comparative pathology of nonviral bullous skin diseases in domestic animals. Vet. Pathol. **17:** 257–281.
11. SCOTT, D. W., W. H. MILLER, R. M. LEWIS *et al.* 1980. Pemphigus erythematosus in the dog and cat. J. Am. Anim. Hosp. Assoc. **16:** 815–823.
12. SCOTT, D. W. 1980. Feline dermatology 1900 to 1978: A monograph. J. Am. Anim. Hosp. Assoc. **16:** 331–459.
13. SCOTT, D. W. & R. M. LEWIS. 1981. Pemphigus and pemphigoid in dog and man: Comparative aspects. J. Am. Acad. Dermatol. **5:** 148–167.
14. SCOTT, D. W., T. O. MANNING, C. A. SMITH *et al.* 1982. Observations on the immunopathology and therapy of canine pemphigus and pemphigoid. J. Am. Vet. Med. Assoc. **180:** 48–52.
15. SCOTT, D. W., T. O. MANNING, C. A. SMITH *et al.* 1982. Pemphigus vulgaris without mucosal or mucocutaneous involvement in two dogs. J. Am. Anim. Hosp. Assoc. **18:** 401–404.
16. STANNARD, A. A., D. H. GRIBBLE & B. B. BAKER. 1975. A mucocutaneous disease in the dog, resembling pemphigus vulgaris in man. J. Am. Vet. Med. Assoc. **166:** 575–582.

Immunofluorescence Studies in Psoriasis: Detection of Antibodies to Stratum Corneum in Psoriatic Scales[a]

VIJAY KUMAR, PAMELA JONES, ERNST H. BEUTNER,
AND STEFANIA JABLONSKA

Department of Microbiology
State University of New York at Buffalo
Buffalo, New York 14214

Department of Dermatology
Warsaw Academy of Medicine
Warsaw, Poland

INTRODUCTION

There is considerable evidence available in the literature in regard to the role of immunological phenomena in the pathogenesis of psoriasis.[1,2,8,9,11,13,24] The presence of circulating anti-stratum corneum (SC) antibodies and the *in vivo* binding of IgG, IgA, IgM, as well as fibrin and complement C3, to lesional, but not to normal, SC of psoriatic patients is a prime example of this evidence.[5–7,13–16]

These circulating anti-SC antibodies are found in sera of normal as well as psoriatic individuals, which raises doubt as to their pathogenic role in psoriasis. Beutner *et al.*[3] have shown that SC antigen is inaccessible for reaction with SC antibodies in normal skin but becomes reactive either by physical trauma to the skin or by certain biochemical changes.[22] These may also occur in lesional skin of psoriatics and the inflammatory response and increased mitotic rates of epithelial cells may be sequelae to SC antigen-antibody reaction. This may imply that immune deposits detected by immunofluorescence in the SC reflect on antigen-antibody interaction and may be a primary event in the pathogenesis of psoriasis.[3]

The antibody nature of these immune deposits is in question.[23,25] Kimura and Nishikawa concluded that immune deposits in psoriatic scales depend mostly on nonimmunologic processes.[20] Other investigators agree that immune deposits found in the SC are not disease specific and may be found in many dermatoses.[23,25]

In this paper, we reexamine the question of the specificity of immune deposits in psoriatic scales by sequential acid-buffer elution studies and by determining the SC antibody and IgG content of psoriatic scale eluates.

MATERIALS AND METHODS

Psoriatic scales were kindly provided by Professor Jablonska from eight patients with generalized psoriasis. Human callus of the mechanical hyperkeratosis type were kindly supplied by Dr. J. M. Carrell.

[a]Supported in part by a Biomedical Research Support Grant from State University of New York at Buffalo.

Elution

Psoriatic scales and callus were ground to a fine powder after freezing in liquid nitrogen. 50 mg of the powder was treated with 1 ml of PBS or with citrate buffer adjusted to different pH (TABLE 1), centrifuged, and the eluates removed as described elsewhere.[18] The eluates were neutralized immediately and kept frozen until used.

Protein content of these eluates was determined by a modified Lowry method.[12] IgG content was determined by the solid-phase chemiluminescent assay system.[17]

Anti-SC antibody titers were determined both by indirect IF (see below) and by passive hemagglutination.[4,21]

Cryostat-cut sections of scales and callus employed in the study were similarly treated with buffers of varying pH for 10 min and washed in PBS before being tested by direct IF.

Immunofluorescence

Direct IF studies were performed on 4 μm thick sections by the standard IF method.[4] All the readings were made on the AO epi-illumination microscope equipped

TABLE 1. Steps Involved in Treating Callus and Psoriatic Scales and Function of Such Treatments

Treatment	Function
pH 7 (PBS) ↓	Removes unbound proteins
pH 5 ↓	Removes loosely bound proteins
pH 3 ↓	Dissociates antigen-antibody complexes (especially of low avidity)
pH 2	Dissociates antigen-antibody complexes

with a quartz halogen lamp and narrow band FITC filters. The conjugates employed in this study had the following characteristics. IgG–protein 10.65 mg/ml, molar F/P[a] 1.47, unitage[a] 8, use dilution 1:32; C3–protein 13.6 mg.ml, molar F/P 3.03, unitage 16, use dilution 1:64; and fibrin–protein 11.6 mg/ml, molar F/P 2.6, unitage 4, use dilution 1:16.

Eluates were tested for SC antibodies by indirect IF on 4 μm frozen human callus sections by indirect IF methods[4] using the IgG conjugate with the above-mentioned characteristics. A normal human serum with a SC antibody titer of 160 was used as a control.

Organ Culture

For studying the question of accessibility of the SCAg, part of the normal human skin samples obtained from radical mastectomy was stripped 20 times with cellophane

[a]F/P = fluorescein-to-protein molar ratio. Unitage = the reciprocal of the conjugate dilution that gives a visible line of precipitation by gel diffusion when tested against 1 mg/ml of the antigen.

TABLE 2. Frequency of Deposits of IgG, Fibrin, and Complement C3 in Psoriatic Scales and Hyperkeratotic Callus

	IgG	C3	Fibrin
		(No. Positive/No. Studied)	
Psoriatic scales	8/8	8/8	8/8
Callus	0/2	0/2	0/2

tape. Skin explants (4 × 4 mm) were cut with a keratome and were cultured with serum containing SC antibodies to a titer of 40 for 4 hr at 37°C.[22] The explants were then washed in PBS to remove unbound proteins and quick-frozen in liquid nitrogen for IF studies.

RESULTS

Direct IF studies of untreated psoriatic scales and callus gave expected results for *in vivo* bound IgG, C3, and fibrin (TABLE 2). Eight of eight psoriatic scales gave positive IF findings of linear intercellular staining as shown in FIGURE 1.

Direct IF staining of skin explants cultured in normal human serum containing anti-SC antibodies before and after tape stripping are shown in FIGURE 2. Direct IF studies of untreated skin explants revealed no binding to SC antigen (FIGURE 2a). Explants of skin, when cultured after stripping with tape, gave positive SC staining (FIGURE 2b).

The sequence of steps followed in eluting proteins unbound, loosely bound, or those bound with high affinity from ground psoriatic scales or callus are summarized in TABLE 1. Following the above protocol, eluates obtained of ground scales or callus were

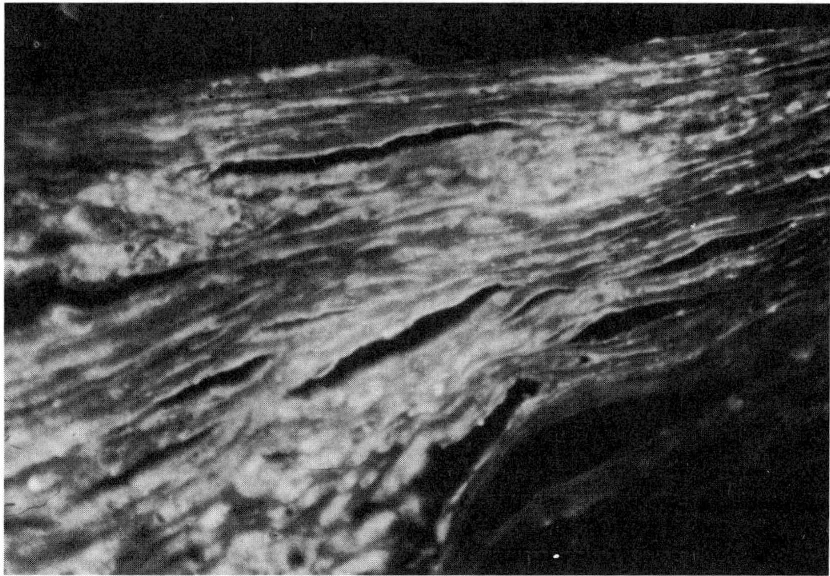

FIGURE 1. Direct IF staining for IgG of psoriatic scales from a patient with active lesion. Note IF staining of the stratum corneum is primarily linear. 336×.

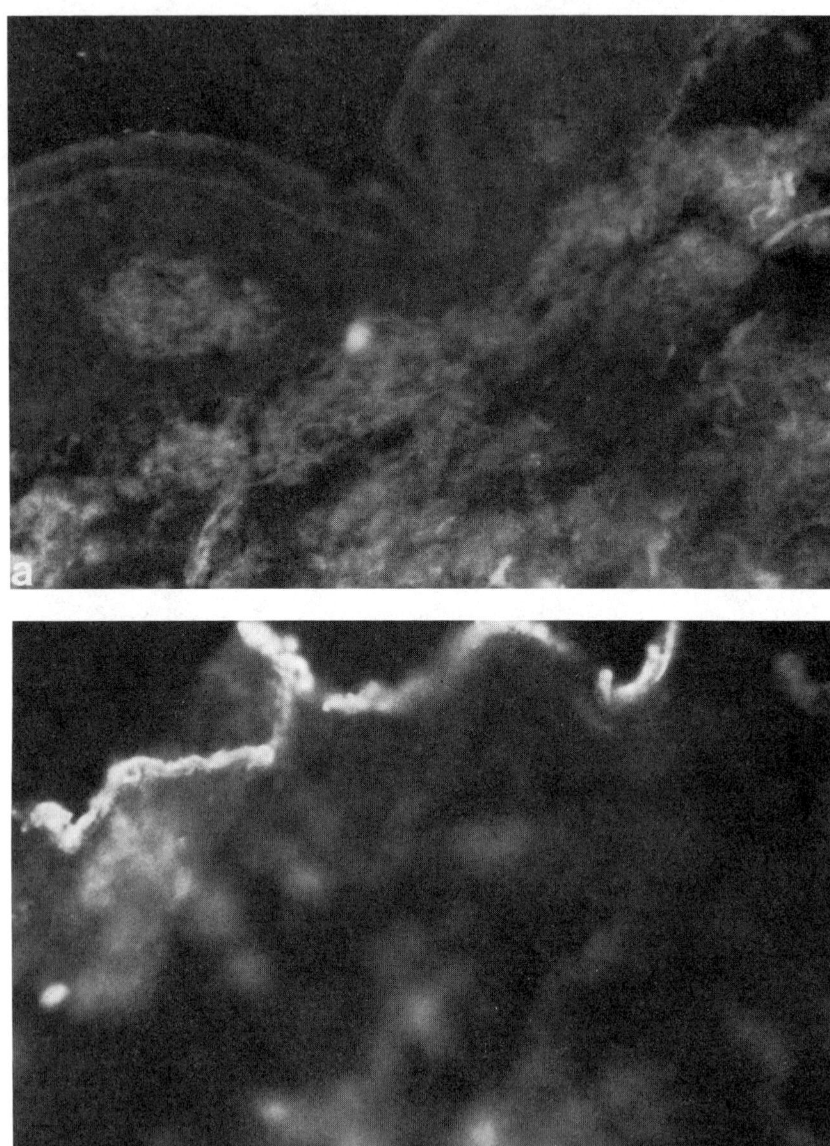

FIGURE 2. Direct IF staining of an organ culture maintained in normal human serum having SC antibody titers of 40. (a) Without stripping, showing a negative reaction and (b) after stripping, strong staining of SC. 336×.

TABLE 3. Anti-stratum Corneum Antibodies in Eluates of Psoriatic Scale and Callus

| | Scale (No. Positive/No. Tested) | | Callus (No. Positive/No. Tested) | |
Treatment	IIF[a]	PHA[a]	IIF	PHA
pH 7	6/10	2/5	0/2	0/2
pH 5	2/10	1/5	0/2	1/2
pH 3	0/10	0/5	0/2	0/2
pH 2	1/10	5/5	0/2	0/2

[a]IIF = indirect immunofluorescence and PHA = passive hemagglutination.

tested for anti-SC antibodies and the results are summarized in TABLE 3. Five of the five eluates obtained after pH 2 treatment were positive for anti-SC antibodies by passive hemagglutination (PHA), whereas only one of ten were positive by indirect IF. No SC antibody activity, either to the carbohydrate antigen as detected by PHA or to the protein antigen as detected by indirect IF was found in pH 3 eluates. No anti-SC antibody activity was detected in eluates of normal callus either by indirect IF or by PHA. However, one of the callus specimens examined contained weakly positive hemagglutinating SC antibodies in pH 5 eluate. Total protein and IgG content were highest in pH 7 eluates (TABLE 4) and were lowest in pH 3; pH 2.0 eluates had higher concentrations of both protein and IgG as compared to pH 3 eluates. To further confirm the above findings, direct IF was performed on psoriatic scales after treatment with buffers of varying pH. As shown in FIGURE 3, virtually all the scales showed either a decrease or absence of staining activity after pH 2 treatment as compared to pH 3. Only 37% of the scales decreased in staining intensity at pH 3.

DISCUSSION

Stratum corneum antibodies occur both in normal individuals and in psoriatic patients. In the normal situation these antibodies may play a role as "physiological antibodies" in removing the damaged stratum corneum as a normal "autoimmune repair mechanism."[2,3] Tissue culture studies have shown that SC antigens are inaccessible under normal conditions. The same appears to be true *in vivo*. However, damage to SC by mechanical stripping makes this antigen accessible to stratum corneum antibodies.[3] The same SC antibodies have been implicated in triggering the development of psoriatic lesions by extensive, self-perpetuating SC antigen-antibody

TABLE 4. Total Protein, IgG Content, and Anti-SC Activity in the Eluates of Psoriatic Scales from Five Patients

Treatment	Total Protein[a] (mg/ml)	Total IgG[a] (mg/ml)	Units of Specific Activity (PHA titer/mg IgG)
pH 7	4.9 (3.5–6.6)	0.59 (0.4–0.9)	1
pH 5	0.95 (0.68–1.5)	0.045 (0.03–0.07)	7
pH 3	0.53 (0.16–0.95)	0.024 (0.005–0.05)	ND[a]
pH 2	0.97 (0.51–1.9)	0.058 (0.01–0.12)	19

[a]Values are mean (range). ND = not detectable.

reactions. That is, the "SC antigen conversion" phenomenon appears to extend beyond the site of trauma.[3,13]

Deposits of immunoglobulins and complement in the SC seem to be highly characteristic but not specific of psoriasis since they have been shown to be present in other nonpsoriatic parakeratotic lesions.[10,14,16,19,23,25] Just what causes some of these apparent "physiologic" *in vivo* reactions of SC antibodies to become self-perpetuating is not clear as yet. The question under consideration in these studies is whether the immune deposits seen in psoriatic scales are in fact caused by *in vivo* bound SC antibodies. Kimura and Nishikawa[20] concluded from their observations on the reduction of SC staining only upon fibrinolysin treatment and not pH 3 acid citrate

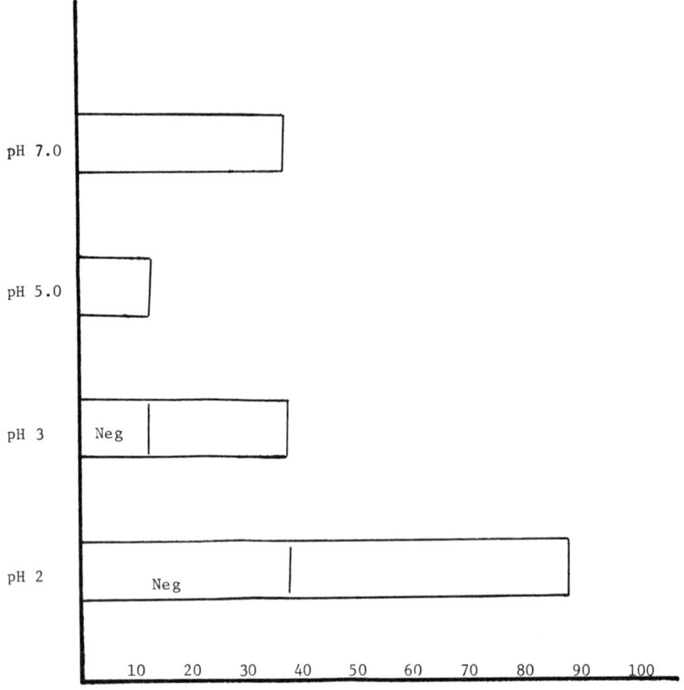

FIGURE 3. Loss of direct IF for IgG from psoriatic scales upon sequential acid citrate buffer elution.

buffer treatment, that immunoglobulin deposits in the horny layer are derived from extravasation of plasma proteins rather than antigen-antibody reaction. They failed, however, to recognize that immunoglobulins present in eluates may or may not have SC antibody activity. We have found that fibrinolysin treatment of psoriatic scales did not completely remove all the *in vitro* bound SC reactivity.[18] Also, while there was a decrease in staining intensity of psoriatic scales for IgG upon pH 3.0 acid buffer treatment, this decrease in staining intensity was not as significant as with pH 2 treatment. These findings are further corroborated by the presence of higher amounts of total protein, IgG, and importantly, of anti-SC antibody in pH 2 eluates compared to pH 3.0 eluates.

Thus, it can be said that in psoriatic scales from patients with fully developed lesions, some but not all, immunoglobulin deposits are due to the binding of SC antibodies to SC antigen–reactive sites.

REFERENCES

1. BERGSTRESSER, P. R. & J. N. GILLIAM. 1981. The immunology of psoriasis. Pharmocol. Ther. **14:** 345–354.
2. BEUTNER, E. H., Ed. 1982. Autoimmunity in Psoriasis. CRC Press. Boca Raton, FL.
3. BEUTNER, E. H., W L. BINDER, S. JABLONSKA & V. KUMAR. 1979. Nature of stratum corneum autoantibodies, antigens and antigen conversion and their role in healing and in psoriasis. *In* The Epidermis in Disease. R. Marks & E. Christophers, Eds.: 333–357. MTP Press, Ltd. Lancaster, England.
4. BEUTNER, E. H., W. L. BINDER & V. KUMAR. 1982. Methods for immunofluorescence studies of stratum corneum antibodies. In Autoimmunity in Psoriasis. E. H. Beutner, Ed.: 267–281. CRC Press. Boca Raton, FL.
5. BEUTNER, E. H., S. JABLONSKA, V. KUMAR, W. L. BINDER & T. P. CHORZELSKI. 1981. Letter to Editor on "A further study on the mechanism of deposition of plasma proteins in psoriasis scales." Arch. Dermatol. Res. **270:** 217–219.
6. BEUTNER, E. H., M. JARZABEK–CHORZELSKA, S. JABLONSKA, T. P. CHORZELSKI & G. RZESA. 1978. Autoimmunity in psoriasis. A complement immunofluorescence study. Arch. Dermatol. Res. **261:** 123–134.
7. BINDER, W. L. 1978. Studies on the role of the immune system in the pathogenesis of psoriasis. Ph.D. Thesis. State University of New York at Buffalo.
8. CORMANE, R. H. 1981. Immunopathology of psoriasis. Arch. Dermatol. Res. **270:** 201–215.
9. CRAM, D. L. 1981. Psoriasis: Current advances in etiology and treatment. J. Am. Acad. Dermatol. **4:** 1–14.
10. DOYLE, J. A., S. A. MULLER, R. S. ROGERS & A. L. SCHROETER. 1981. Immunofluorescence in psoriasis: A clinical study of sixty patients. J. Am. Acad. Dermatol. **5:** 655–660.
11. GUILHOU, J. J., J. MEYANDIER & J. CLOT. 1978. New concepts in the pathogenesis of psoriasis. Br J. Dermtaol. **98:** 585–592.
12. HARTREE, E. F. 1972. Determination of protein: A modification of the Lowry method that gives a linear photometric response. Anal. Biochem. **48:** 122–127.
13. JABLONSKA, S., E. H. BEUTNER, W. L. BINDER, M. JARZABEK-CHORZELSKA, G. RZESA & O. CHOWANIEC. 1979. Immunopathology of psoriasis. Arch. Dermatol. Res. **265:** 65–71.
14. JABLONSKA, S., T. P. CHORZELSKI, E. H. BEUTNER, E. MACIEJOWSKA, M. JARZABEK-CHORZELSKA & G. RZESA. 1978. Autoimmunity in psoriasis. Relation of disease activity and forms of psoriasis to immunofluorescence findings. Arch. Dermatol. Res. **261:** 135–146.
15. JABLONSKA, S., T. P. CHORZELSKI, M. JARZABEK-CHORZELSKA & E. H. BEUTNER. 1975. Studies in immunodermatology. VII. Four compartment system studies of IgG in stratum corneum and of stratum corneum antigen in biopsies of psoriasis and control dermatoses. Int. Arch. Allergy **48:** 324–340.
16. JABLONSKA, S., M. JARZABEK-CHORZELSKA, E. H. BEUTNER, G. RZESA, E. MACIEJOWSKA & T. P. CHORZELSKI. 1979. Clinical relevance of stratum corneum antibodies in psoriasis. *In* Immunopathology of the Skin. 2nd edit. E. H. Beutner, T. P. Chorzelski & S. F. Bean, Eds.: 427–443. John Wiley and Sons. New York.
17. JONES, P. & V. KUMAR. 1984. Solid-phase chemiluminescent immunoassay for determination of IgG in eluates of psoriatic scales. J. Immunol. Methods. (In press.)
18. JONES, P., V. KUMAR, E. H. BEUTNER & S. JABLONSKA. 1984. Detection of high avidity SC antibodies in eluates of psoriatic scales. (In preparation.)
19. KANEKO, F., H. GUSHIKEN, I. KAWAGISHI, Y. MIURA, K. KOBAYASHI & T. KONNO. 1980.

Analysis of immunological responses in psoriatic lesions. Immunological studies on psoriatic lesions. J. Invest. Dermatol. **75:** 436–439.

20. KIMURA, S. & T. NISHIKAWA. 1980. A further study on the deposition of plasma protein in psoriasis scale. Arch. Dermatol. Res. **268:** 79–84.
21. KROGH, H. K., J. A. MAELAND & O. TONDER. 1972. Indirect haemmagglutination for demonstration of antibodies to stratum corneum of skin. Int. Arch. Allergy **42:** 493–502.
22. KUMAR, V., T. ROGOZINSKI, E. H. BEUTNER & S. JABLONSKA. 1983. Studies in immuno-dermatology. IX. Effect of enzyme and organic solvents on the reactivity of stratum corneum antigens. Int. Arch. Allergy. **71:** 112–116.
23. TOSCA, A., Z. CHRISOBOLI, J. HATZIS, A. VARELZIDIS & J. STRATIGOS. 1981. Secondary autoimmune phenomenon in the pathogenesis of psoriasis. Arch. Dermatol. Res. **271:** 351–356.
24. WAHBA, A. 1980. Immunological alterations in psoriasis. Int. J. Dermatol. **19:** 124–129.
25. WEISS, V. C., H. VAN DEN BROEK, S. BARRETT & D. P. WEST. 1982. Immunopathology of psoriasis: A comparison with other parakeratotic lesions. J. Invest. Dermatol. **78:** 256–260.

Use of Immunofluorescence in the Study of the Pathogenesis of Respiratory Syncytial Virus Infection[a]

ROBERT C. WELLIVER AND PEARAY L. OGRA

Departments of Pediatrics and Microbiology
State University of New York at Buffalo

Division of Infectious Diseases
Children's Hospital
Buffalo, New York 14222

INTRODUCTION

Previous studies of respiratory syncytial virus (RSV) infection have been limited by the difficulty in recovering RSV in tissue culture from infected patients[3] and by the lack of a sensitive, technically facile assay to determine the antibody response to RSV infection.[2,4] Using indirect immunofluorescence techniques, we have developed the capability to document RSV infection in infants and children within 24 hours of the time of hospitalization by identification of viral antigen in exfoliated nasopharyngeal epithelial cells (NPEC). In addition, use of immunofluorescent antibody techniques have also permitted the study of the antibody response in nasopharyngeal secretions (NPS) in IgG, IgM, IgA, and IgE immunoglobulin classes following primary and secondary infection with RSV.[2,7] This report summarizes the results of the above investigations.

MATERIALS AND METHODS

Study Population

The study population consisted of infants and children hospitalized in the first year of life with infection due to RSV, and subsequently followed for each episode of respiratory illness as part of an ongoing study of respiratory illness in early life. Patients were classified as having upper respiratory tract illness only, pneumonia with minimal or no wheezing, or bronchiolitis-asthma on the basis of criteria described previously.[2]

Collection and Processing of Specimens

Samples of NPS for initial diagnosis of RSV infection and for determination of secretory antibody responses were obtained at the onset of illness and at intervals up to

[a]The original studies were supported in part by grants from the National Heart, Lung and Blood Institute (HL-21829) and National Institute of Allergy and Infectious Diseases (AI-15939).

12 weeks later. Samples were aspirated into polyethylene tubes and rinsed into traps using 2.5 ml of a phosphate-buffered saline solution. NPEC were sedimented by low-speed centrifugation and processed for detection of RSV antigen[1] and for cell-bound IgE[7] as described previously and below. Aliquots of supernatants were inoculated into tissue culture using human epithelial (HEp-2) cell lines[3] for detection of RSV.

Detection of RSV Antigen in NPEC

The slides of NPEC were incubated in a moist chamber with a 1:5 dilution of bovine anti–RS virus serum for 30 minutes at 37°C. The NPEC smears were then incubated with a 1:10 dilution of fluorescein isothiocyanate (FITC)–conjugated anti-bovine immunoglobulin in 0.005% of Evans blue counterstain for 30 mintues at 37°C. Extensive washing procedures in PBS were used after each incubation. A final wash in distilled water was used before the slides were air-dried and examined. A negative control using normal bovine serum in place of the bovine anti–RS virus serum was included with every specimen. A positive control using RS virus–infected HEp-2 cells was used with each experiment. The stained slides were examined using an epi-illumination system.

Before the test system could be used in a clinical setting, several control experiments were done to assess the sensitivity and the specificity of the technique. The bovine anti–RS virus serum and anti-bovine FITC conjugate were titrated using immunofluorescent antibody test (IFAT) to determine the optimal dilution of both reagents that gave maximum specific fluorescence for RS virus–infected HEp-2 cells and minimum nonspecific fluorescence with uninfected HEp-2 cells. At the optimal dilution of the antisera, at least 50% of the RS virus–infected HEp-2 cells showed a positive fluorescence of 3^+ to 4^+ intensity. Each new batch of reagents required such calibration before use. HEp-2, rhesus monkey kidney, and WI-38 cells infected with a wide range of viruses, including parainfluenza 1, 2, and 3, adenovirus, influenza, and RS virus, were tested in the RS virus–IFAT system to ensure the absence of cross-reactivity of these reagents with the other viral agents. Positive fluorescence was seen only with the RS virus–infected cells. The staining of RS virus could be effectively blocked by treating the NPEC smears with human antibody to RS virus before staining the slides with FITC-labeled antisera to bovine immunoglobulin. As additional controls, antisera to adenovirus and parainfluenza 1, 2, and 3 gave no staining reactions with RS virus-positive smears. No staining reactions were seen in uninfected tissue cultures used as negative controls.

Demonstration of Cell-bound IgE on Respiratory Epithelial Cells

After acetone fixation to glass microscope slides, RSV-infected cells and cell-bound IgE were simultaneously detected by double-immunofluorescent–staining techniques as previously described.[7] Briefly, RSV-infected cells were first indentified by incubation of infected cells with rabbit anti-RSV antibody (Wellcome, Beckenham, England) and rhodamine-conjugated goat anti-rabbit globulin (Cappel Laboratories, Cochranville, PA). The same cells were then studied for the presence of cell-bound IgE by staining with fluorescein-conjugated goat anti-human IgE (Meloy, Springfield, VA). Slides were then examined under a fluorescence microscope. Rhodamine staining was demonstrated with a 4-mm UG5 primary and BG38 barrier filter under a Leitz Orthoflux fluorescence microscope. Fluorescent staining in the same field was reexamined after conversion to a 3-mm BG12 primary and OG1-GG4 barrier filter. By viewing the same field through the second filter system, we could detect cell-bound IgE by yellow-green fluorescence of the fluorescein conjugate. Cell-bound IgE was observed essentially only on RSV-infected cells. Controls consisted of uninfected human epithelial tissue culture cells and samples of NPECs taken several days after the onset of illness in patients with upper-respiratory-tract illness due to parainfluenza virus Types 1 and 3. These cells were examined by the microscopist, who was unaware of whether or not they were controls. All controls gave negative results when tested by this method. Appropriate blocking procedures were employed in all experiments to rule out nonspecific staining. Preincubation of anti-RSV serum with high titers of virus and preincubation of anti-IgE conjugates with human IgE before testing resulted in blocking of fluorescence.

Free Secretory Antibody Determinations

Other aliquots of supernatant fluids were processed for determination of free secretory IgG, IgM, and IgA antibody to RSV as previously described.[2] After dialysis against distilled water at 4°C for 48 hours, secretions were lyophilized and held at -20°C until reconstitution in 1 ml of buffer for determination of RSV antibody content. Briefly, HEp-2 tissue culture cells were infected with a stock strain of RSV with a titer of 10^5 tissue culture infective dose $(TCID_{50})$/ml. After three days of incubation, the infected cells were trypsinized, washed, and resuspended in phosphate-buffered saline solution (5×10^6 cells/ml). Cells infected with RSV and uninfected cells were fixed with acetone on glass slides and incubated with twofold serial dilutions of NPS specimens for one hour at 37°C. For RSV-IgG and RSV-IgM determinations, the slides were next incubated with fluorescein-labeled goat anti-human IgG and IgM for 30 minutes at 37°C. For RSV-secretory IgA determinations, slides were incubated with rabbit anti-human secretory component and then fluorescein-conjugated goat anti-rabbit globulin. Observed titers of RSV-specific IgG, IgM, and secretory IgA were adjusted to a concentration of 10 mg/dL of the respective immunoglobulin. Procedures consisting of the use of uninfected cells or substitution of phosphate-buffered saline solution for secretions or reagents did not exhibit fluorescence. Secretions from infants with infection caused by parainfluenza virus were available from a separate study, and these secretions did not show fluorescence when tested for RSV antibody. Specific fluorescence could be blocked by absorbing secretions with an equal volume of fluid that contained live RSV at a titer of 10^5 $TCID_{50}$/ml.

Total Immunoglobulin Concentration in NPS

Total immunoglobulin concentrations were measured in radial immunodiffusion plates.

RESULTS

Characteristics of IFAT Staining

The immunofluorescence seen in NPEC was entirely intracytoplasmic in the form of particulate material that varied in size from fine particles to large, inclusion-like bodies. The nucleus was unstained. The presence of as few as two to three cells with such characteristic fluorescence on one smear was sufficient for diagnosis. However, most positive specimens contained at least one infected cell per microscopic field. An experienced observer could screen one specimen in 10 to 15 minutes. The entire procedure of staining and examining the smear could be easily completed within four to five hours.

Comparative Evaluation of IFAT and Tissue Culture Infection

Using both IFAT and tissue culture infection (TCI), RS virus was detected in 123 NPS specimens. The RS virus was isolated by TCI in 106 (27%) NPS specimens. The remaining 281 (73%) NPS specimens failed to yield RS virus by TCI (TABLE 1). Of the 106 NPS specimens positive for RS virus by TCI, 98 (92%) were also found to be positive by IFAT. However, eight TCI-positive specimens were negative for RS virus by IFAT. It should be pointed out that three of the IFAT-negative NPS smears were found to have an insufficient number (less than two/high-power field) of exfoliated cells. Only five (4.7%) of the NPS specimens that were positive on TCI and contained adequate cells were negative for RS virus by IFAT (TABLE 1). Of the 387 specimens available for IFAT, only ten (2.5%) were considered to have an inadequate number of cells.

Ninety-four percent of the NPS specimens negative for RS virus by TCI were also negative by IFAT. However, it is of interest to note that 17 (6%) of the specimens negative by tissue culture were positive for RS virus antigen when tested by IFAT (TABLE 1). All such specimens were obtained on follow-up testing from subjects whose initial specimens were RS virus–positive by both TCI and IFAT.

The cytopathogenic effect of RS virus in TCI was manifested as early as six days and as late as three weeks after inoculation. The average duration of time required for detection of RS virus by TCI was about ten days.

Kinetics of Secretory RSV-Antibody Response

The temporal kinetics of the secretory antibody responses to primary infection with RSV are shown in FIGURE 1. The RSV-IgG response was present in the secretions of 9 of 15 (60%) patients during the first three days of illness. Peak titers were attained 8 to 13 days after the onset of illness, at which time RSV-IgG was present in 100% of NPS specimens. At 57 to 90 days after illness, the geometric mean titer (GMT) of RSV-IgG

TABLE 1. Comparative Evaluation of the Identification of Respiratory Syncytial (RS) Virus in Nasopharyngeal Secretions (NPS)

Tissue Culture Infectivity of NPS Specimens		Indirect Immunofluorescence of NPS Specimens			
		Positive for RS Virus		Negative for RS Virus	
		No.	%	No.	%
Total no. tested	387	115	29	272	70
No. positive for RS virus	106	98	92	5[a]	4.7
No. negative for RS virus	281	17	6	264	94.0

[a]Three other specimens were considered unsuitable for IFAT because of insufficient number of exfoliated cells.

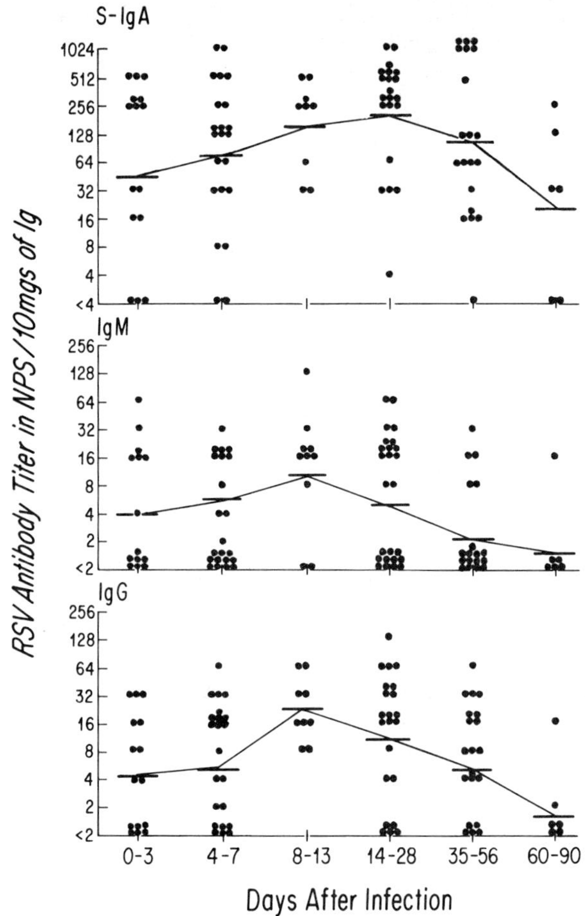

FIGURE 1. Temporal pattern of secretory antibody response to primary infection with respiratory syncytial virus (RSV). Horizontal bars indicate geometric mean titers; NPS, nasopharyngeal secretions; Ig, immunoglobulin; S-IgA, secretory IgA. (From Kaul *et al.*[2] By permission from *American Journal of Diseases of Children.* Copyright 1981, American Medical Association.)

in NPS had fallen to less than 2, and four of six specimens had no detectable RSV-IgG. The RSV-IgM was present in 7 of 14 (50%) of NPS specimens during the first three days of illness with a GMT of four. Peak titers of RSV-IgM were also attained at 8 to 13 days after the onset of illness (GMT = 11.3). Only one of six specimens still contained RSV-IgM at 57 to 90 days after illness, and the GMT had fallen to less than two. The predominant class of antibody in secretions was RSV-secretory IgA, being present in 12 of 15 (80%) NPS samples within the first three days of illness at a GMT of 50. Peak titers of RSV-IgA (GMT = 211) were attained at 14 to 28 days after the onset of illness, although a relatively stable titer of RSV-secretory IgA was maintained from 8 to 56 days after illness. During the latter interval, RSV-IgA antibody could be detected in 49 to 50 (98%) NPS specimens that were tested. In contrast, at 57 to 90 days after illness, RSV-IgA antibody was present in only four of six (67%) NPS specimens, and the GMT had fallen to less than 22.

Cell-bound IgE Analysis

The relation of the form of clinical illness experienced to the presence and persistence of cell-bound IgE is shown in FIGURE 2. Between 70 and 80 percent of patients had demonstrable cell-bound IgE in the first six days of illness, regardless of the form of clinical illness present. However, when samples taken 7 to 20 days after the onset of illness were studied, all patients with bronchiolitis or asthma and 33 percent of patients with upper-respiratory-tract illness or pneumonia had evidence of cell-bound IgE. In samples taken 21 to 34 days after the onset of illness, 50 percent of patients with bronchiolitis or asthma and 17 percent of patients with upper-respiratory-tract illness or pneumonia had cell-bound IgE present. When the results from samples taken seven to 34 days after the onset of illness were pooled, cell-bound IgE was found in 16 of 22 patients with bronchiolities or asthma (73 percent), but in only 4 of 15 patients

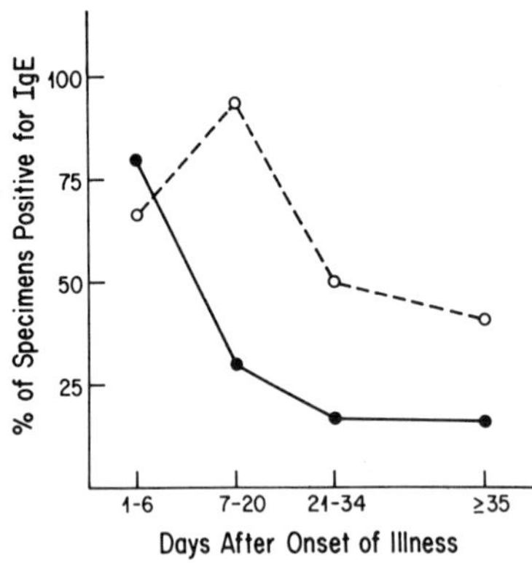

FIGURE 2. Relation of form of illness to persistence of cell-bound IgE in the respiratory tract after respiratory syncytial virus infection. The solid line indicates results in patients with upper respiratory tract illness or pneumonia; the broken line indicates results in patients with bronchiolitis or asthma. (From Welliver et al.[7] By permission from the *New England Journal of Medicine.*)

with upper-respiratory-tract illness or pneumonia (27 percent)—a difference that was statistically significant (chi-square with Yates' correction = 4.18; $p < 0.05$). When samples taken 35 to 55 days after the onset of illness were studied, cell-bound IgE was present in 45 percent of 22 subjects with bronchiolitis or asthma and in 20 percent of 10 patients with upper-respiratory-tract illness or pneumonia, but this difference was not statistically significant.

DISCUSSION

We found the IFAT to be a sensitive and specific test when applied in a clinical setting, with a 92% agreement with tissue culture infection procedures for positive specimens and 94% for the negative ones. For the past 18 months, we have been routinely using this approach for the rapid diagnosis of RS virus infection in our hospital. As a result, an increased awareness of the impact of RS virus in childhood respiratory infections has been established among physicians and nursing staff.

Preliminary data suggest that the rapid diagnosis of virus infection in children with respiratory diseases has modestly reduced the indiscriminate use of antibiotics, particularly in cases of acute respiratory infection. An attempt has also been made to reduce the incidence of nosocomial infections by rapidly identifying and separating the patients with RS virus–associated respiratory diseases.

The present study has shown that RSV-specific secretory IgA, as well as IgG and IgM, antibody appear in the respiratory tract as early as the first three days after the onset of illness caused by RSV infection. Peak titers were attained at 8 to 13 days after the onset of illness. Importantly, however, the antibody titers declined sharply at 90 days after primary infection.

This study demonstrates the appearance of IgE bound to exfoliated NPECs during the acute phase of illness in most patients with infection by RSV. Cell-bound IgE persisted for longer periods in patients with bronchiolitis or asthma due to RSV than in patients with upper-respiratory tract illness or pneumonia. One might expect to find cell-bound IgE in the acute phase of illness more commonly in patients with bronchiolitis or asthma than in patients with upper-respiratory-tract illness or pneumonia, but in our study cell-bound IgE was generally present up to six days after the onset of illness, regardless of the type of illness present. It is important to note that the assay described is only qualitative; quantitative differences in the amount or function of IgE may be present in different forms of RSV-related illness. In a previous study of RSV-induced pneumonia[5] hyperaeration was often demonstrated radiographically in the absence of clinical evidence of bronchospasm. In addition, other investigators have commented on the copious amounts of nasopharyngeal secretions that are often present in cases of RSV infection.[6] We have observed that patients with upper-respiratory-tract illness caused by RSV may have large amounts of clear or mucoid secretions similar to the secretions of patients with allergic rhinitis. Some clinical evidence therefore suggests that the production of IgE in response to RSV infection, presumably with subsequent release of such chemical mediators as histamine, may contribute to the pathogenesis of all forms of respiratory illness that are due to RSV. It is therefore possible that quantitative differences in the production of these compounds in the respiratory tract may determine the form and severity of disease at the time of initial infection.

REFERENCES

1. KAUL, A., R. SCOTT, M. GALLAGHER, J. CLEMENT & P. L. OGRA. 1978. Respiratory syncytial virus infection: Rapid diagnosis in children by use of indirect immunofluorescence. Am. J. Dis. Child. **132:** 1088–1090.
2. KAUL, T. N., R. C. WELLIVER, D. T. WONG, R. A. UDWADIA, K. RIDDLESBERGER & P. L. OGRA. 1981. Secretory antibody response to respiratory syncytial virus infection. Amer. J. Dis. Child. **135:** 1013–1016.
3. LENNETTE, E. H. & N. J. SCHMIDT. 1969. Diagnostic Procedures for Virus and Rickettsial Infections. American Public Health Assn. New York.
4. MCINTOSH, K., H. B. MASTERS, I. ORR *et al.* 1978. The immunologic response to infection with respiratory syncytial virus in infants. J. Infect. Dis. **138:** 24–32.
5. RICE, R. P. & F. LODA. 1966. A roentgenographic analysis of respiratory syncytial virus pneumonia in infants. Radiology **87:** 1021–1027.
6. REILLY, C. M., J. STOKES, JR., L. NEELAND *et al.* 1961. Studies of acute respiratory illness caused by respiratory syncytial virus. 3. Clinical and laboratory findings. N. Engl. Med. **264:** 1176–1182.
7. WELLIVER, R. C., T. N. KAUL & P. L. OGRA. 1980. The appearance of cell-bound IgE in respiratory tract epithelium after respiratory syncytial virus infection. N. Engl. J. Med. **303:** 1198–1202.

Immunofluorescence in Diagnostic Virology

KENNETH McINTOSH

Children's Hospital Medical Center
300 Longwood Avenue
Boston, Massachusetts 02115

Since the introduction of tissue culture methods for the growth and recognition of viruses in the 1940s, most viral diagnosis that depends on the detection of virus rather than measurement of antibody has used this sensitive and versatile, but expensive and cumbersome technique. It is ironic, therefore, that recognition of viral antigens in clinical specimens was first introduced long before the development of this tissue culture technology. In the 1920s infectious disease clinicians and pathologists were taking the lymph of smallpox lesions and showing that one could agglutinate it with specific antisera. In the 1930s Parker and Muckenfuss first developed the complement fixation reaction using lymph mixed with antibody and a fresh source of complement.[5] In fact, this became the procedure of choice for the laboratory diagnosis of smallpox before and even after the general use of tissue culture for viral isolation.

Immunofluorescence (IF) was also recognized early as a potentially valuable tool for detection of viral antigens,[3,7,8] but its use as a procedure that could supplement and, at times, replace tissue culture in diagnostic laboratories has been only gradually gaining acceptance. IF has, of course, been used at several stages in the diagnosis of viral diseases: the measurement of antibody, particularly when there is some highly specific antigen in question, as with Epstein-Barr virus or cytomegalovirus; the identification of viruses in tissue culture; and the direct detection of viruses in body fluids. It is this last application that I will elaborate on in this paper.

The detection of viruses by IF in specimens obtained directly from the patient takes advantage of two features of the method that are inherent in IF but are not widely recognized. First, the sensitivity of IF is enormously increased by the capacity (indeed, the necessity) to detect intracellular antigens. This natural concentration of antigens in specific compartments allows the detection of much smaller quantities than would be otherwise possible. Second, there is an enormous increase in the specificity of the method contributed by the morphologic features of the fluorescence produced by each specific virus in each specific sort of specimen. Not only can non-specific fluorescence, an unfortunate accompaniment of the IF technique with almost any tissue examined, be rapidly assessed and ignored; but staining of the structures other than the ones particularly sought (for example, bacteria) can be quickly discerned and appropriately discounted.

The rapidity of the technique is another advantage. Preparation of the specimens is crucial to the success of the method; it usually requires only a few minutes, and staining, particularly by the direct method, can be accomplished in an hour or two. Most solid-phase antigen detection systems require longer incubation periods than this. A second advantage, to be illustrated below, is that the antigens of clinical specimens are often more stable than their infectivity, and this allows the use of the method under circumstances where transportation of specimens is slow or difficult. And finally, because the antigens observed by IF are usually intracellular, and therefore in an immunologically protected position, they can often be detected even in the presence of antibody. This means that IF often remains positive later in the course of disease than in either culture[2] or tests for soluble antigen.[4]

In spite of these advantages, there are still many viral diagnostic laboratories that do not use IF for the detection of antigens in body fluids. There are several reasons for this, all of which should, with time, be overcome. First, skill and experience are necessary in obtaining of suitable specimens from patients, in the preparation of these specimens for staining, and in the examination and reading of the slides by the microscopist. Second, excellent reagents are absolutely essential to achieve the sensitivity and specificity that are possible and necessary to render the method competitive with culture techniques. Such reagents are now available in Europe and available to a limited extent in the United States or other countries in this hemisphere. The first of these problems, that of experience and training, should gradually be overcome as viral diagnostic tests are more widely used and the demand for rapid diagnosis increases. However, there are obstacles to this that will be discussed later in this paper. Reagents should become available both as the requirements are more widely recognized by the industries involved and as monoclonal antibodies for each of the important virus groups are made and produced by commercial firms. Some of the problems faced by those trying to adapt monoclonal antibodies to uses in solid-phase systems are fortunately not encountered in IF. Avidity, although important, is not such a crucial problem as it is when picogram quantities of soluble antigens are being measured. Sandwich methods are most efficient when two monoclonal antibodies recognizing two different determinants are used. This is not necessary for IF, since only a single antiviral antibody is used. If a wider spectrum of antibodies is needed, it may ultimately be possible to mix several monoclonal antibodies and achieve relatively quickly the spectrum desired.

Some time and effort should be spent in discussion of the proper method for obtaining specimens for IF diagnosis in human disease. Although this seems a trivial subject, an understanding of the techniques involved and the potential pitfalls is absolutely essential for success in diagnosis.

Since whole cells are needed, only certain specimens are adequate. For those viruses shed from the respiratory tract (a large proportion of the viruses of man), the chances of recovering whole cells shed from the respiratory epithelium should be maximized. Although swabs have been used with success, a far more reliable source of cells is a sample of respiratory secretion.[7] This is most easily obtained from the nasopharynx. Although tracheal epithelium may localize the virus more certainly in the lower respiratory tract, this is usually not necessary, and, indeed, the effort to obtain the specimen in infants and children often yields only a sample of esophageal epithelium (sometimes mixed with gastric secretion) and causes at the same time unnecessary discomfort. A No. 8 French plastic feeding tube is connected to a thumb valve, a trap, and then a source of gentle mechanical suction. The tube is introduced along the floor of the nose until it reaches the nasopharynx and is then slowly withdrawn as suction is applied. The resultant sample of secretion, which is usually rich in shed epithelial cells and virus, is then transported on ice to the laboratory where it is mixed with about 10 ml of buffered saline and centrifuged to deposit the cells. All steps are performed gently since the infected cells are fragile and easily disrupted. The supernatant is removed and the cell button again mixed with buffered saline. This process is repeated until there is no mucus found (two or three times). The cells are then deposited in 8 mm square spots on a clean microscope slide, allowed to dry in air, and fixed for ten minutes in acetone at 4°C. Some skill is required both in separating mucus from cells and in judging how many cells are required to make a slide that will render an accurate positive or negative diagnosis.

This method is applicable to the diagnosis of infections by influenza viruses, parainfluenza viruses, respiratory syncytial virus, adenovirus, measles, rubella virus, and coronaviruses. If antisera were available, the method might be applicable to enteroviruses and rhinoviruses as well, but the multiple serotypes involved have thus far impeded success with these groups. The choice of which antiserum to use in the laboratory requires a knowledge of the seasonal epidemiology of each of the viruses, some information about the clinical presentation of the patient sampled, and enough competence in the details of infectious disease to match intelligently the patient and the potential infecting viruses. In most instances several viruses should be tested for, if only as a source of controls for that particular sample. The proper and practical use of controls will be discussed more fully below.

For those viruses found in vesicular lesions on the skin, namely herpes simplex and varicella, the lesion should be gently scraped with a scalpel and the cells obtained should be deposited in two or more drops of saline on the surface of a clean side.[1] These will usually be fairly large chunks, which should immediately be broken up as finely as possible with two #26 needles or, preferably, dissecting needles. The slides are then allowed to dry, fixed as above, and stained appropriately.

The other major source of cells for IF is biopsy or autopsy. It has rarely been valuable to search for viral antigens in leukocytes, either in the peripheral blood, cerebrospinal fluid, or serosal fluids. Autopsy or biopsy specimens may be examined either in frozen sections or in touch preparations. The latter are often easy to prepare, of good quality for IF, and in some instances superior to frozen sections since they tend to be thin and well dispersed, and to have less non-specific staining.

Any good fluorescence microscope may be used for IF viral diagnosis, but the reading of multiple specimens is facilitated by the use of incident light and a 40× oil-immersion objective. The latter we have found to be essentially the only objective we need in clinical work. The medium magnifying power is adequate for observation of cellular detail, an absolute necessity for accuracy, and yet the field is large enough and the amount of light sufficient for rapid scanning of an entire specimen. Oil immersion, while not routinely used at this power, reduces the light scattering produced when specimens are examined directly and obviates the need for a coverslip. We find that glycerol, either buffered or used untreated from the container, has adequate optical properties and is much less irritating to mucous membranes than high performance immersion oils.

While the proper use of multiple controls is and should be the rule with fluorescence in the research laboratory, in the diagnostic setting the rigorous insertion of positives, negatives, preimmune sera, conjugate controls, blocking, and so on would become unnecessarily cumbersome. We depend heavily on the examination of large numbers of specimens, the repetitive use of the same sera and conjugates, and the accumulated experience of the observer. Moreover, replicate slides must always be made from each specimen, so that in situations where there is any question the required controls can be done, or the staining procedure can be repeated.

In general, before they are used in the clinical setting, all reagents should have been titrated in tissue culture, tested on frozen clinical slides of known positives and known negatives, shown to exhibit no cross-reactions with other species of virus, and examined with regard to non-specific reactions in multiple clinical specimens. Few antisera are perfect, and imperfect sera can be used with great accuracy if their characteristic imperfections are well known. Each clinical specimen should be judged as to its adequacy and rejected if there are not enough cells present of if the level of non-specific fluorescence is too high and irreducible. Finally, to make a positive reading on a given slide, certain criteria must be absolutely insisted on. The fluorescence must be in the proper cell type (i.e. not a squamous epithelial cell if respiratory viruses are being sought) and must be in the appropriate part of the cell for that virus (i.e. not in the nucleus for a parainfluenza virus). The morphology of the fluorescence must be right for that particular virus. Respiratory syncytial virus produces fine punctate fluorescence often denser around the limiting membrane of the cell, with the occasional bright spot of a cytoplasmic inclusion body. Herpes virus looks entirely different, often staining the nucleus or the entire cell densely with little morphology visible. The fluorescence must, as every microscopist knows, be of exactly the right color: if fluorescein is used, it must be apple green, not any other green or yellow. In virtually every case, at least two positive cells must be seen. Usually if two positive cells are found there are many other *almost* positive cells around, and one can often find material that looks as though a positive cell had been disrupted either during the obtaining or preparation of the specimen. A blatantly positive cell as an isolated finding in a blatantly negative background may mean that that cell was inadvertantly transferred to that slide by means of a pipet tip from another slide.

Whenever the microscopist is dealing with an even slightly unfamiliar situation, controls should be used. Positive and negative tissue culture controls should be stained. There should be a negative tissue control (e.g. a negative brain frozen section if virus is being sought in brain). The suspect specimen should be stained for several viruses, and if possible a preinoculation serum should be used for the particular virus most suspected. If the conjugate being used in an indirect staining procedure is in the least unfamiliar, the antiviral serum should be omitted as a conjugate control. Finally, if there is still doubt, or if the finding is unusual or unexpected, blocking tests should be done. The most important ingredient for specificity is, in the long run, however, the experience of the microscopist. There will always be judgement involved in the use of IF for viral diagnosis, and many years of experience and of comparison of culture and IF results are essential for the accurate reading of difficult slides.

How can a technician gain the confidence necessary to call a specimen positive or negative when tissue culture results either disagree or are unavailable? This is probably the most vexing problem in the use of immunofluorescence in a routine viral diagnostic laboratory, and undoubtedly the one which will continue for some time to inhibit its widespread application. There is unfortunately no substitute for experience: the repetitive comparison of IF results with culture in innumerable different samples, with constant application of irreproachable standards of honesty and quality control. Alternatively, the administrative framework can be made available for repetitive comparisons of readings with those in a laboratory where experience with the technique has been extensive. This type of system has been used with respiratory viruses in England,[6] and is used for rabies diagnosis in this country. For this reason, good, versatile IF microscopists are usually trained in tissue culture diagnostic laboratories. If they leave virus isolation behind, they must know and admit their limitations, and periodic return to comparative situations (either with culture or with more experienced readers) would seem to be an important part of the maintenance of high standards.

When all aspects of technique are satisfactorily cared for, the results of IF diagnosis come exceedingly close to those of tissue culture. In some situations they are better. One of these with which we have had recent experience is the testing of specimens for labile viruses, which are sent by taxi or bus from outlying hospitals. Our record in the past two years illustrates this well. In 1980–1981, during the respiratory syncytial virus (RSV) season, 90% of positive specimens came from patients within our own hospital. In that year, we recovered 94% of the total positive specimens in tissue culture, and detected 88% by IF. In 1981–1982, 46% of RSV-positive specimens were sent to us from other hospitals in Boston and the suburbs of Boston. In that season the sensitivity of culture fell to 75% while that of IF rose to 90%. Overall, specimens coming from outside the hospital were found to be IF-positive but culture-negative four times as often (36%) as those coming from within the hospital (9%).

Thus, now in our laboratory IF of clinical specimens has become a routine adjunct to our diagnostic armamentarium. We use it occasionally as the only diagnostic method, omitting culture: when specimens are old; when an unexpected load on the laboratory exhausts our supply of tissue culture; or sometimes in epidemiologic studies where a single virus is being sought during a known outbreak. It serves the irreplaceable function of giving the physician and patient an early answer, often allowing for timely withdrawal of antibiotics or intelligent infection control procedures. It has become a test much in demand by our medical staff and, by making virus recognition immediately available, has brought the laboratory in all its functions closer to the patient and the physician.

REFERENCES

1. GARDNER, P. S. & J. MCQUILLIN, Eds. 2nd Edit. 1980. Rapid Virus Diagnosis: Application of Immunofluorescence. Butterworths. London.
2. GARDNER, P. S., J. MCQUILLIN & R. MCGUCKIN. 1970. The late detection of respiratory syncytial virus in cells of the respiratory tract by immunofluorescence. J. Hyg. Camb. **68:** 575–580.
3. LIU, C. 1956. Rapid diagnosis of human influenza infection from nasal smears by means of fluorescent-labelled antibody. Proc. Soc. Exp. Biol. Med. **92:** 883–887.
4. MCINTOSH, K. & R. M. HENDRY. 1982. An enzyme-linked immunosorbent assay (ELISA) for detection of respiratory syncytial virus infection: Development and description of the method. J. Clin. Microbiol. **16:** 324–328.
5. PARKER, R. F. & R. S. MUCKENFUSS. 1932. Complement-fixation in variola and vaccinia. Proc. Soc. Exp. Biol. Med. **29:** 483–485.
6. Report to the M.R.C. Sub-committee on respiratory syncytial virus vaccines. 1978. Respiratory syncytial virus infections: Admissions to hospital in industrial, urban, and rural areas. Br. Med. J. **2:** 796–798.
7. STURDY, P. M., J. MCQUILLIN & P. S. GARDNER. 1969. A comparative study of methods for the diagnosis of respiratory infections in childhood. J. Hyg. Camb. **67:** 659–670.
8. WELLER, T. H. & A. H. COONS. 1954. Fluorescent antibody studies with agents of varicella and herpes zoster propagated *in vitro*. Proc. Soc. Exp. Biol. Med. **86:** 789–794.

Enzyme Immunoassays for the Diagnosis of Viral Infections[a]

ROBERT H. YOLKEN, FLORA LEISTER, LORA
WHITCOMB, DAWN DAVIS, AND MARY JANE MEARS

Eudowood Division of Infectious Diseases
Department of Pediatrics
School of Medicine
The Johns Hopkins University
Baltimore, Maryland 21218

Traditionally the diagnosis of infectious diseases has been accomplished by the isolation of the infecting microorganism in pure culture. However, it has long been realized that cultivation systems offer some disadvantages for the rapid diagnosis of infectious diseases. For example, many microorganisms, especially viruses and slower growing bacteria, require such a long period of time to accomplish cultivation that the results of the patient's culture are not available to the physician at a time when the result can alter therapy. This is especially true in the case of many viruses and mycobacteria, which can require two to five weeks for cultivation and identification.[10,15] Recently a number of viral agents have been identified that cause serious human disease, but cannot be cultivated in available tissue culture systems. Interestingly, many of these organisms infect the hepatic and gastrointestinal tract of humans. For example, hepatitis A and hepatitis B viruses are responsible for a large number of cases of clinical hepatitis yet cannot be cultivated in the tissue culture systems available in clinical virology laboratories.[1,13] While the agent of non-A and non-B hepatitis has not been conclusively identified, it is clear that this agent also is not cultivatable.[3] In addition, agents such as rotavirus, adenovirus, Norwalk virus, and astrovirus, which have been shown to cause the vast majority of nonbacterial diarrheas, cannot be cultivated in available systems.[4,5] Thus, if the clinical laboratory is to provide meaningful information to clinicians for many infectious diseases, more rapid and more sensitive means must be developed for detecting and identifying a wide range of infectious diseases.

One way of accomplishing an accurate diagnosis of an infectious disease is the direct identification of an infecting antigen in the clinical specimen. The fact that most infecting microorganisms contain protein or polysaccharide antigenic determinants, which can be readily distinguished from host cell determinants, allows for the development of immunoassay systems capable of identifying infectious antigens in body fluids.

Since there are a number of ways in which antigen-antibody reactions can be measured, there are a number of immunoassay systems that can be utilized for the immuno-detection of infectious antigen. However, in order to be successful for this purpose, an immunoassay system must be very sensitive, rapid, convenient, and sufficiently specific so as not to be interfered with by materials that might be present in body fluids. In addition, it would be preferable if the immunoassay system would allow

[a]This work was supported by the National Institute of Allergy and Infectious Diseases (Contract No. No1 AI 92616), the National Institutes of Health (Grant No. 1 RO1 AI 17604-01), and the Thrasher Research Fund, Salt Lake City, Utah.

for the quantitative measurement of the antigen-antibody reaction since the ability to determine the quantity of infecting antigens might allow for a more accurate assessment of the patients condition and a more rational means of assessing antimicrobial therapy. Initially, the assay system that we found most suited for this purpose was radioimmunoassay since it allows for the measurement of small amounts of antigen under a variety of reaction conditions.[14] However, radioimmunoassays (RIAs) are limited by the need for radioactive isotopes. The requirement for radioactivity limits the laboratory environments in which the assay systems can be utilized. In addition, the short half-life of the isotopes generally utilized in RIAs requires the frequent relabeling of reagents. This is not a significant problem in situations, such as the testing of banked blood for hepatitis B surface antigen, where a predictable number of assays will be performed in a certain period of time. However, in the case of the diagnosis of common infectious diseases, it is difficult to predict which infections will be prevalent during a particular period, and how many tests will need to be performed. It is thus preferable to have access to an assay system that utilizes stable reagents and would thus be available whenever needed without the need to constantly prepare and evaluate reagents.

Many of the disadvantages of radioimmunoassays can be avoided by the use of an assay system that makes use of the assay format of radioimmunoassay, but utilizes a sensitive, stable nonisotopic marker as the immunoglobulin label. One assay system that accomplishes this is that of the enzyme immunoassay also known as enzyme-linked immunosorbent assay (ELISA).[2,17,18] Enzyme immunoassays (EIA) are similar in design to radioimmunoassays except that an enzyme is utilized in place of the radioactive isotope as an immunoglobulin marker. The utility of the EIA is based on the inherent magnification of the enzyme substrate system. For example, a single molecule of alkaline phosphatase or horseradish peroxidase, enzymes that are commonly used in EIA systems, can react with up to 100,000 molecules of substrate per minute to generate a visible product. This magnification allows for the measurement of very small amounts of antigen—often in the picogram range—by the simple measurement of an enzyme substrate reaction. While there are a number of substrates that can be used, the use of ones that generate visible color are particularly advantageous since they can be quantitated by simple instrumentation or, alternatively, can be qualitatively determined by the unaided eye. The advantages of enzyme immunoassays are shown in TABLE 1. In theory, these assays would offer practical means of detecting a number of antigens in body fluids under different laboratory conditions ranging from centralized laboratories in universities or state health departments to small laboratories or doctors offices.

However, there are a number of potential pitfalls for EIA systems, as shown in TABLE 2. Some of these problems are common to all immunoassays systems; namely, the assays can be no better than the specificity and sensitivity of the immunoreactants that are utilized. Potential problems of immunoassays range from antibodies with excess cross-reactivity to immunoreactants that have too narrow a scope; that is, that

TABLE 1. Advantages of Enzymes Immunoassays

Inherent magnification of enzyme substrate reaction ($\times 10^6$/10 minutes)
Stability of labeled reagents
Results measured by visual interpretation or simple instrumentation
Small reaction volumes
Quantitative nature of results
Ability to perform control reactions
Low biohazard potential

TABLE 2. Problems with Enzyme Immunoassays

All Solid-Phase Immunoassays
    Dependence on antigen-antibody kinetics
      Cross-reactivity
      Over-specificity
      Limitations of antibody affinity
    Nonspecific reactivity due to antiglobulin
    Variation in solid-phase binding

Enzyme Immunoassays
    Loss of antibody affinity following conjugation
    Variability of enzyme-substrate reaction
    Sensitivity of enzymatic activity to inhibitors
    Presence of endogenous enzymatic activity
    Limits of substrate detection

react with one narrow serotype of antigen but that do not react with other closely related serotypes that might be present in a clinical specimen. In addition, there are a number of potential problems that are specifically related to the use of enzymes as a marker. These include the need to avoid non-specific reactions due to the endogenous enzymes that might be present in the clinical specimen as well as the need to devise practical methods for the efficient conjugation of enzymes to the immunoglobulins. However, a number of investigators have devised means of overcoming these problems and these methods have allowed the development of practical immunoassays in most cases where immunoreactants of sufficient quality and quantity are available.

There are a number of ways in which enzyme immunoassays can be formulated.[6,20,21] In the direct system (FIGURE 1) antibody is attached to the solid phase. While a number of solid phases can be utilized, we have found microtiter plates to be particularly convenient since they allow for the performance of a number of reactions during a single test run. The availability of microtiter plates in the form of strips of 8 or 12 wells has markedly increased the utility of microplate systems, since the steps allow for the performance of a variable number of tests. While the antibody can be bound to the solid phase by means of non-specific absorption, it is usually preferable to utilize a method of covalent linkage either through the free carboxyl or amino groups of the antibody molecules.[11] Previous problems with variation in the absorptive properties of microtiter wells appeared to be minimized by the use of covalent linkage and also by the use of some second-generation solid-phase supports, which are available from a number of commercial sources (Immulon II, Dynatech Laboratories, Alexandria, VA). Alternatively, test systems utilizing antibody-coated beads can be utilized in place of the microtiter plates.[9]

Following removal of the first antibody, the clinical specimen is added. Antigen present in the specimen will bind to the antibody linked to the solid phase, while unreacted antibody will be removed in the washing step. Next, enzyme-labeled antibody is added. This antibody will bind to reactive sites on the antigen attached to the solid phase. Following removal of unreacted enzyme-labeled antibody, substrate is added. Enzyme bound by the previous step will react with substrate to produce a measurable product. This can be quantitated in a simple spectrophotometer or in a microplate colorimeter capable of measuring multiple microtiter plate wells in a short period of time.

The direct form of the assay systems has the advantage of being simple and rapid and resulting in a minimum of nonspecific reactivity. However, it is somewhat inconvenient in that it requires the availability of distinct enzyme-labeled antibodies

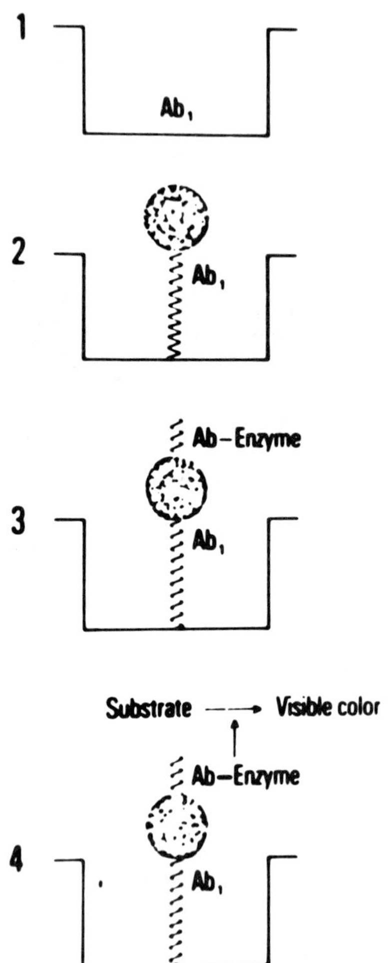

FIGURE 1. Direct ELISA for antigen measurement. (1) Antibody directed against the antigen to be measured is adhered to the well of a microtiter plate. (2) The test material is added. Any antigen to which the antibody is directed will adhere. (3) Antibody labeled with an enzyme is added. This will react with the antigen that is adhered to Ab_1. (4) A substrate is added. The enzyme adhered to the well will convert the substrate to a visible form. The amount of color measured is proportional to the amount of antigen in the test material.

directed at each virus or bacteria to be measured. For this reason, we often utilize the indirect form of assay system shown in FIGURE 2. In this system, antigen is bound to the solid phase in a manner identical to the direct assay. However, unlabeled second antibody (Ab_2 in FIGURE 2) is added in place of the enzyme-labeled antibody. Following the reaction of this antibody with the antigen, unreacted antibody is removed and the amount of Ab_2 is quantitated by the addition of an enzyme-labeled reactant directed at this antibody. This can either be an enzyme-labeled antiglobulin or enzyme-labeled staphylococcal protein A, a protein derived from staphylococci that has a high affinity for the Fc portion of many animal immunoglobulins.[23] The amount of enzyme-labeled antiglobulin or staphylococcal protein A bound to the solid phase is then measured as in the direct assay. Since unlabeled Ab_2 is utilized, the only enzyme-labeled reactant required is the antiglobulin or staphylococcal protein A directed at the second antibody. This allows for the use of a single enzyme-labeled reactant for the detection of any virus or bacteria provided that the antibody is derived from the appropriate animal species. In addition, we have found that the indirect systems are often somewhat more sensitive than the direct system, presumably because one molecule of Ab_2 is able to react with a number of molecules of antiglobulin. However, the indirect assay system has the disadvantage of requiring an extra incubation step. In addition, it is usually necessary that Ab_2 and the antibody used to coat the solid phase be prepared in different animal species to prevent the non-specific binding of the antiglobulin to the solid phase. However, in some cases different immunoglobulin subclasses of the same animal species can be utilized to avoid this non-specific interaction.

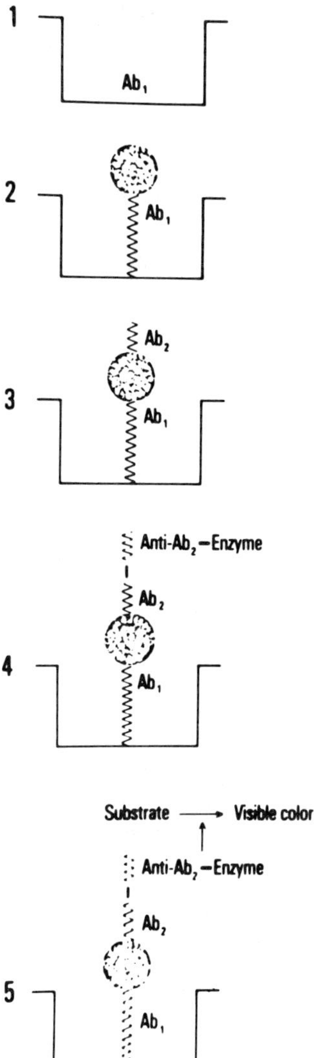

FIGURE 2. Indirect ELISA for antigen measurement. (1) Antibody directed against the antigen to be measured is adhered to the well of a microtiter plate. (2) The test material is added. Any antigen to which the antibody is directed will adhere. (3) Unlabeled antibody from a different animal than Ab_1 is added. This will react with any antigen that is adhered to Ab_1. (4) Enzyme-labeled antibody directed against the globulins of the animal source of Ab_2 is added. (5) A substrate is added. The enzyme adhered to the well will convert the substrate to a visible form. The amount of color measured is proportional to the amount of antigen in the test material.

Enzyme immunoassays can also be performed in a competitive manner as shown in FIGURE 3. Competitive immunoassays have the advantage of requiring fewer incubation steps and only a single immunoglobulin antibody reagent. However, they have the disadvantage of requiring the binding of antigen to the solid phase. In addition, we have found that the non-competitive assays are often less sensitive than the competitive assays.[22] However, in some cases, the competitive assays have proven to be useful for the overall measurement of antigens, which are present in the nanogram or microgram range.[16]

There are a number of determinants of the sensitivity and specificity of enzyme immunoassay systems. As discussed above, the most important is the sensitivity and specificity of the antibodies utilized. We have found that, since enzyme immunoassays are very sensitive, the reagents must be extremely specific to prevent the magnification

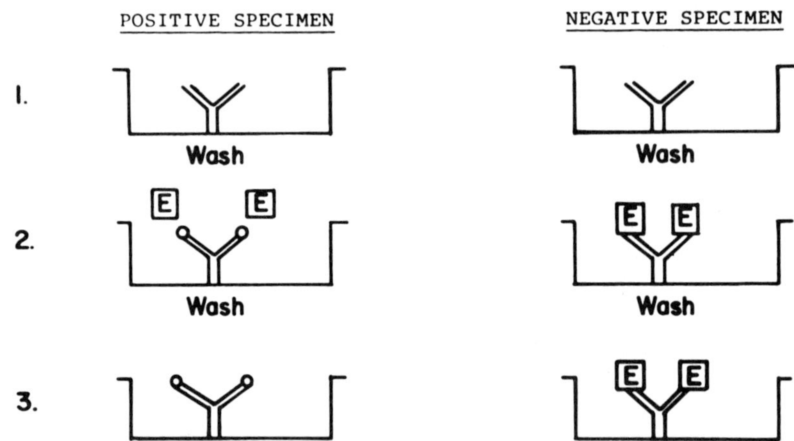

FIGURE 3. Competititve EIA for antigen measurement—labeled antigen method. (1) Antibody ($\curlyvee$) is bound to the solid phase. Unbound antibody is removed in the washing step. (2) The specimen is added. If it contains antigen (O) it will bind to the solid-phase antibody. Enzyme-labeled antigen (E) is then added. This will react with antibody sites not occupied by antigen from the specimen. (3) Unbound enzyme-labeled antigen is removed by washing and the amount of bound enzyme-labeled antigen is quantitated by the addition of the appropriate substrate. The amount of substrate product is inversely proportional to the amount of antigen in the test specimen.

of the non-specific responses and a high rate of background reactivity. We have found that we can prepare the most useful reagents by the immunization of experimental animals with the use of purified components of viruses or bacteria. In addition, we often further purify reagents by absorption with uninfected material or affinity chromatography. Also, we prefer to use the purified component of the antisera to minimize nonspecific interaction. One exciting new development in this area has been the ability to produce monoclonal antibodies by means of *in vitro* hybridoma systems.[8] The monoclonal antibody technology offers the possibility of producing unlimited amounts of immunoglobulins directed at a single antigenic determinant. In theory, monoclonal antibodies are virtually ideal for immunoassays directed at the direct detection of viral or bacterial antigens in body fluids. However, there are a number of pitfalls in monoclonal technology. One of the most important pitfalls to avoid is the possibility that monoclonal antibodies might be "too specific," that they would only react with a narrow range of antigenic determinants and not react with a majority of viruses or bacteria occurring in nature. However, it appears that many investigators have been able to select clones producing antibody directed at common viral antigens. These antibodies appear to react with most or all of the viruses that are present in clinical specimens. These monoclones might thus prove to be very useful reagents for enzyme immunoassays. An additional problem with the use of monoclonal antibodies is that many of them have relatively low affinity for the antigen. Many of the monoclones have affinity constants in the range of 10^{-8} to 10^{-9} as compared to polyclonal antibodies raised in animals that have affinity constants of 10^{-11}. Since the concentration of enzyme-labeled reactants is proportional to the affinity constant, the use of these monoclonal antibodies might result in the binding of less labeled antibody to the solid phase. It would thus be important to develop techniques for the production of monoclonal antibodies or mixtures of monoclonal antibodies with affinity constants comparable to those of standard antibodies.

An additional important factor in the performance characteristics of enzyme immunoassays is the minimization of nonspecific interactions between the enzyme-labeled antibodies in the solid phase. These reactions can be caused by the physical binding of the enzyme-labeled antibody to the solid phase or by immunological binding through the Fc portion of the antibody molecule. The physical binding can be reduced or eliminated by the use of detergents and immunologically neutral proteins, such as gelatin in the buffers and washing solutions as well as by the use of careful washing technique.

Nonspecific interaction can also occur when materials present in the clinical specimen are capable of binding to the Fc portion of the immunoglobulins used to coat the solid phase. Such interactions can be due to the presence of rheumatoid factor in the specimens or bacterial components with Fc binding capability. These reactions can be minimized by the addition of non-immune serum, IgG, or Fc fragments to the reaction mixture to neutralize the nonspecific reactants.[19] In addition, the nonspecific activity can be reduced by the use of the F (Ab) portion[7] of the immunoglobulin molecule for the immunoreactants. Rheumatoid factor can also be eliminated by the use of mild reducing agents such as *N*-acetyl-cysteine and other Fc binding materials can be denatured by the addition of a mild acidic buffer.[19] However, in all cases the presence of nonspecific interactions can be detected by the performance of control reactions in which non-immune IgG is utilized in place of the antiviral immunoglobulin to coat the solid phase (FIGURE 4). In this case, specific antigen is recognized by an increase in activity in the antiviral immunoglobulin wells as compared to the control wells. On the other hand, the nonspecific reaction will be manifested by equal amount of reactivity in both antigen and control wells. Quantitative measurements of specific activity are computed by subtracting the activity in the control well from that measured in the wells coated with antiviral antibody. The use of control reactions and the techniques described above mark an increase in the specificity of the EIA system and increase their general utility as diagnostic tools.

Our laboratory and others have devised enzyme immunoassay systems for the direct detection of a number of viral antigens in clinical specimens.[20,24] We have found EIA to be most useful for the detection of fastidious viruses found in stool and nasal-wash specimens obtained from patients with diarrhea and respiratory disease. In these cases the EIA systems not only provide a more rapid diagnosis but allow for the detection of fastidious antigens, which often cannot be cultivated in generally available tissue culture systems. Although specific antiviral therapy is not as yet available for most of these viruses, the ability to rapidly detect their presence in clinical specimens has been extremely useful both in the management of patients with suspected infections as well as in the early detection of epidemics. In addition, rapid diagnosis has proved useful for the prevention of the spread of infections among patients within hospitals and other institutions. In addition, the ability to test large numbers of specimens has been invaluable in studying the epidemiology of these infections and in documenting the varied clinical manifestations of viral infections.

Another area where enzyme immunoassays have been extremely useful has been in the rapid detection of bacterial antigens, especially those causing acute systemic disease in infants and children. These assays are particularly useful in cases where a patient has been treated with an antimicrobial agent, since such treatment might reduce the chances of recovering the organism by cultivation techniques.[12] One problem in the past has been that while immunoassays are more rapid than cultivation techniques the two to four hours required for the viral diagnostic assays is insufficiently rapid for the management of a child who is acutely ill. However, by making some simple modifications in the assay system, namely the incubation of plates on a plate shaker and the combination of the antigen and the conjugation incubation steps, we have been able to decrease the assay time to approximately 20 minutes (FIGURE 5). These rapid assays have great potential for the rapid diagnosis of serious bacterial and viral infections.[25]

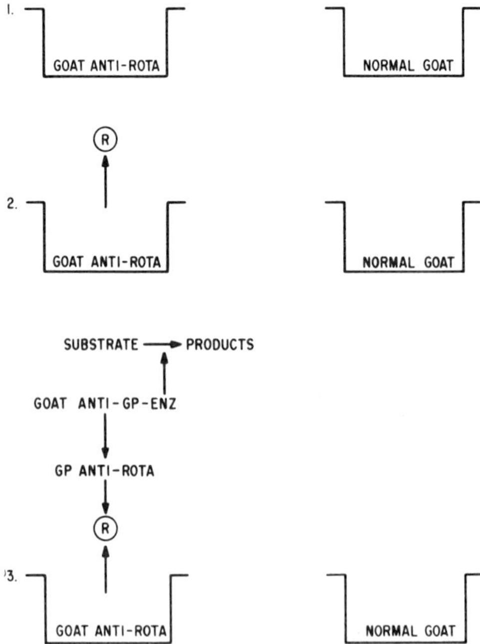

FIGURE 4. Confirmatory ELISA for rotavirus. (1) Alternate rows of microtiter plates are coated with goat anti-rotavirus (anti-rota) Ig and an equal concentration of Ig from the serum of a goat without demonstrable anti-rotavirus antibody. (2) The specimen is added. If it contains rotavirus it will react with the goat anti-rotavirus Ig. Nonspecific anti-Ig will react with both the goat anti-rotavirus Ig and the normal goat Ig. (3) Guinea pig anti-rotavirus, enzyme-linked goat anti-guinea pig Ig, and substrate are sequentially added with washing steps in between incubations. Specific rotavirus activity is manifested by a difference in color between the wells coated with goat anti-rotavirus serum and the wells coated with normal goat serum.

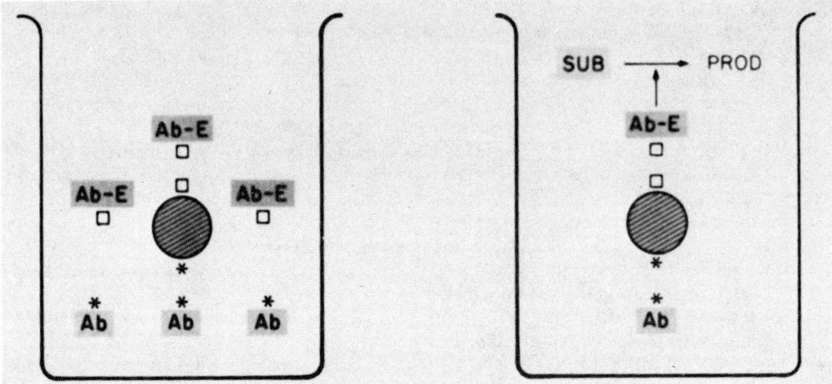

FIGURE 5. (Left) Antigen is added to a solid phase coated with antibody directed against one antigenic site (*Ab). Enzyme-labeled antibody directed against a different site on the antigen (□Ab-E) is then added. This will react with unbound sites on the antigen. (Right) Following a washing step to remove unreacted □Ab-E substrate is added. This will be converted by bound □Ab-E to a measurable product. The amount of product formed will be proportional to the concentration of antigen in the specimen.

Because enzyme immunoassays rely on the inherent magnification of enzyme substrate reactions, they have a great deal of versatility. While the current enzyme immunoassays are geared to use in large clinical laboratories, the stability of reagents and relative simplicity of the assays suggest that the assay systems might be utilized in a wide range of clinical situations. In addition, the inherent sensitivity to the enzyme substrate systems allows for the simple detection of small amounts of antigens in body fluids. Currently, the principal limitation of enzyme immunoassays is the limited availability of reagents with sufficient affinity and specificity to react with infectious antigens that are present in body fluids in small concentrations. Another limitation is that some intracellular antigens, such as chlamydia and herpes simplex virus, are difficult to detect until they can be extracted from the intracellular space. However, these problems are in the process of being overcome by a number of technical innovations, such as improved extraction techniques and improved methods of preparing monoclonal and polyclonal antibodies. While they have some limitations, enzyme immunoassays will undoubtedly prove to be useful tools for the detection and identification of infectious antigens in clinical specimens.

REFERENCES

1. ALMEIDA, J. D., D. RUBENSTEIN & E. J. STOTT. 1971. New antigen-antibody system in Australia antigen positive hepatitis. Lancet **2:** 1225–1227.
2. ENGVALL, E. & P. PERLMANN. 1972. Enzyme linked immunosorbent assay. ELISA. III. Quantitation of specific antibodies by enzyme-linked anti-immunoglobulin in antigen coated tubes. J. Immunol. **109:** 129–135.
3. FEINSTONE, S. M., L. F. BERKER & R. H. PURCELL. 1979. Hepatitis A and Ab. *In* Diagnostic Procedures for Viral, Rickettsial and Chlamydial Infections. E. H. Lennette & N. J. Schmidt, Eds.: 892–897. American Public Health Association. Washington, D.C.
4. GREENBERG, H. B., J. VALDESUSO, R. H. YOLKEN, E. GANGAROSA, W. GARY, R. G.

WYATT, T. KONNO, H. SUZUKI, R. M. CHANOCK & A. Z. KAPIKIAN. 1979. Role of Norwalk virus in outbreaks of nonbacterial gastroenteritis. J. Infect. Dis. **139:** 564–568.

5. KAPIKIAN, A. Z., H. W. KIM, R. G. WYATT, W. L. CLINE, J. O. ARROBIO, C. D. BRANDT, W. J. RODRIQUEZ, D. A. SACK, R. M. CHANOCK & R. H. PARROTT. 1976. Human reovirus-like agent as the major pathogen associated with "winter" gastroenteritis in hospitalized infants and young children. N. Engl. J. Med. **294:** 965–972.

6. KAPIKIAN, A. Z., R. H. YOLKEN, H. B. GREENBERG, R. G. WYATT, A. R. KALICA, R. M. CHANOCK & H. W. KIM. 1979. Gastroenteritis viruses. *In* Diagnostic Procedures for Viral, Rickettsial and Chlamydial Infections. E. H. Lennette & N. J. Schmidt, Eds.: 927–995. American Public Health Association. Washington, D.C.

7. KATO, K., Y. UMEDA, F. SUZUKI, D. HAYASHI & A. KOSAKA. 1979. Use of antibody Fab' fragments to remove interference by rheumatoid factors with the enzyme-linked sandwich immunoassay. FEBS Lett. **102:** 253–256.

8. KOHLER, G. & C. MILSTEIN. 1975. Continuous culture of fused cells secreting antibody of predefined specificity. Nature **256:** 495–497.

9. LEHTONEN, O. -P. & M. K. VILJANEN. 1980. Antigen attachment in ELISA. J. Immunol. Methods **34:** 61–70.

10. MCINTOSH, K., C. WILFERT, M. CHERNESKY, S. PLOTKIN & M. J. MATTHEIS. 1978. Summary of a workshop on new and useful methods in viral diagnosis. J. Infect. Dis. **138:** 414–419.

11. NEURATH, A. R. & N. STRICK. 1981. Enzyme-linked fluorescence immunoassays using beta-galactosidase and antibodies covalently bound to polystyrene plates. J. Virol. Methods **3:** 155–165.

12. PEPPLE, J., E. R. MOXON & R. H. YOLKEN. 1980. Indirect enzyme-linked immunosorbent assay for the quantitation of the type specific antigen of *Haemophilus influenzae* b: A preliminary report. J. Pediatr. **97:** 233–237.

13. PURCELL, R. H., J. L. DIENSTAG, S. M. FEINSTONE & A. Z. KAPIKIAN. 1975. Relationship of hepatitis A antigen to viral hapatitis. Am. J. Med. Sci. **270:** 61–71.

14. ROSENTHAL, J. D., K. HAUASHI & A. L. NOTKINS. 1973. Comparison of direct and indirect solid-phase microradioimmunoassay for the detection of viral antigens and antiviral antibody. Appl. Microbiol. **25:** 567–573.

15. RYTEL, M. 1979. Overview of newer diagnostic methods. *In* CRC Press. pp. 7–16. Boca Raton, FL.

16. SEGAL, E., R. A. BERG, P. A. PIZZO & J. E. BENNETT. 1979. Detection of candida antigen in sera of patients with candidiasis by an enzyme-linked immunosorbent assay inhibition technique. J. Clin. Microbiol. **10:** 116–118.

17. VOLLER, A., D. BIDWELL & A. BARTLETT. 1980. Enzyme-linked immunosorbent assay. *In* Manual of Clinical Immunology. 2nd Edit. N. Rose & H. Friedman, Eds.: 359–371. American Society for Microbiology. Washington, D.C.

18. WISDOM, G. B. 1976. Enzyme immunoassay. Clin. Chem. **22:** 1243–1255.

19. YOLKEN, R. H. & P. J. STOPA. 1979. Analysis of nonspecific reactions in enzyme-linked immunosorbent assay testing for human rotavirus. J. Clin. Microbiol. **10:** 703–707.

20. YOLKEN, R. H. 1980. Enzyme-linked immunosorbent assay (ELISA): A practical tool for rapid diagnosis of viruses and other infectious agents. Yale J. Biol. Med. **53:** 85–92.

21. YOLKEN, R. H., P. J. STOPA & C. C. HARRIS. 1980. Enzyme immunoassay for the detection of rotavirus antigen and antibody. *In* Manual of Clinical Immunology. 2nd Edit. N. Rose & H. Friedman, Eds.: 692–699. American Society for Microbiology. Washington, D.C.

22. YOLKEN, R. H. & P. J. STOPA. 1980. Comparison of seven enzyme immunoassay systems for measurement of cytomegalovirus. J. Clin. Microbiol. **11:** 436–551.

23. YOLKEN, R. H. & F. J. LEISTER. 1981. Staphylococcal protein A-enzyme immunoglobulin conjugates: Versatile tools for enzyme immunoassays. J. Immunol. Methods **43:** 209–218.

24. YOLKEN, R. H. 1982. Enzyme immunoassays for the detection of infectious agents in body fluids. Rev. Infect. Dis. **4:** 35–68.

25. YOLKEN, R. H. & F. J. LEISTER. 1982. Rapid multiple determinant enzyme immunoassay for the detection of human rotavirus. J. Infect. Dis. **146:** 1.

Negative Staining and Immune Electron Microscopy as Techniques for Rapid Diagnosis of Viral Agents

M. RIEPENHOFF-TALTY, H. J. BARRETT, B. A. SPADA,
AND P. L. OGRA

Children's Hospital
and
State University of New York at Buffalo
Buffalo, New York 14222

INTRODUCTION

In 1973, Bishop and her colleagues visualized reovirus-like particles in thin sections of duodenum biopsied from a young child with diarrhea. Shortly afterward, in England[7] and Canada,[15] similar particles were seen in feces using direct visualization with negative-stain electron microscopy. Now, nearly ten years later, the impact of rotaviruses on human disease is becoming apparent.

In 1972, an outbreak of gastroenteritis affected 50% of students and teachers in an elementary school in Norwalk, Ohio. Particles 27 nm in diameter were visualized in the infectious stool filtrate. This filtrate was mixed with convalescent serum from a patient who had had the disease. The resultant particles were coated with antibody and clumped together. This technique was termed immune electron microscopy and the virus-like particles were called Norwalk agent.[8]

It is estimated that these two agents account for 50% of hitherto undiagnosed gastroenteritis around the world. It is amazing that in this era of tissue culture, virology pathogens were diagnosed initially by electron microscopic techniques and that these techniques are still vital. Even in 1982 cell culture isolation is not an adequate diagnostic tool for rotaviruses. Norwalk agent and other antigenically related virus-like particles[1,20] are still not grown in cell culture systems.

Even more recently electron microscopy and immune electron microscopy have been shown to be critical techniques for detecting new virus-like particles associated with gastroenteritis in humans. These particles have two things in common. They are smaller than rotavirus (15–40 nm) and are round. They have logically been referred to by some investigators as small round virus-like particles (SRV). They are not antigenically related to Norwalk agent. Because these SRV have been detected by electron microscopy, their morphology is a distinguishing feature for some particles. Astrovirus, a 28 nm particle with a star-like surface was described by Madely and Cosgrove in 1975. Astrovirus has been detected in outbreaks of diarrhea in Scotland, England, and Canada.[2,11,13,16] Caliciviruses, also approximately 28 to 30 nm in size, have been associated with diarrhea in human infants according to several reports.[5,12,14,17,18]

Additional small round viruses (27 to 40 nm) have been distinguished by electron microscopy in fecal specimens from infants with nonbacterial gastroenteritis. Twenty-five to 28 nm particles, termed picorna-parvo have been reported to be associated with diarrhea.[16] Finally a group of slightly larger, 30–40 nm calici-like particles associated

<u>Negative staining for electron microscopy</u>

1. 400 mesh copper grid is covered with
    0.5% formvar and a thin layer of carbon.

2. Secure grid in forceps and suspend over petrie
    dish (as below)

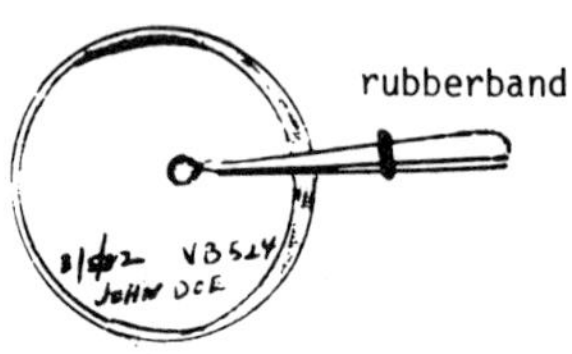

3. Make a suspension of specimen with 1% ammonium
    acetate.

4. Place a drop of liquid on suspended grid. Draw
    off excess by blotting the side of the
    liquid bubble with filter paper.

5. Place a drop of 2% phosphotungstate (pH 6.9-7.2)
    on grid. Wait approximately one minute.

6. Use filter paper to remove excess stain. Release
    grid to the dish.

7. Expose approximately 6 inches from UV light for
    3-5 minutes.

8. View with the aid of a transmission EM 30,000 X
    is an appropriate setting.

FIGURE 1. Procedure for preparing specimens for direct visualization by electron microscopy.

Centrifuge serum at 15,000 RPM for 1 hour
Prepare 2.3% stool suspension in transport
media (contains BSA)
↓
Centrifuge 4°C 1 hour 1,000 RPM
↓
Mix 0.8 ml supernatant fluid + 0.2 ml serum
(various dilutions)
↓
Incubate 1 hr at R.T.
↓
Centrifuge 90 min at 20,000 RPM
↓
Resuspend pellet DW + PTA

FIGURE 2. Stepwise preparation of fecal samples for immune electron microscopy of virus.

with diarrhea in several areas of the world may actually represent a single group. While they have various names: SRVL,[10] minireo,[16] fuzzy-wuzzy's,[6] minirota,[17] Otofuke agent,[19] and Sapporo agent,[9] they appear to have a similar size range and some common morphological features, like a spiky outer surface. Definitive etiological role and antigenic relatedness await good convalescent sera and serological reagents and appropriate immune electron microscopic testing.

The virology laboratory of Children's Hospital began diagnosis by electron microscopy near the end of 1978. Since then, rotavirus has been detected in approximately 500 children hospitalized with diarrhea. We have also found adenovirus, coronavirus-like particles, and representative samples of most of the SRV described above. We have begun using immune electron microscopy as a corollary diagnostic tool in an attempt to make a definitive diagnosis particularly in the case of the SRV.

```
Centrifuge serum at 15,000 RPM for 1 hour
Dilute serum with PBS in 4 fold serial dilutions
                        ↓
0.9 ml serum + 0.1 ml undiluted virus, mix well

                        ↓

            Incubate 4°C overnight

                        ↓

            Centrifuge 90 min.
               18,000 RPM

                        ↓

        Resuspend pellet in one drop
           distilled water + 3% PTA
```

FIGURE 3. Stepwise preparation of tissue culture–grown or purified virus for immune electron microscopy.

MATERIALS AND METHODS

Fecal Samples

Fecal samples were obtained either in a specimen bottle, soiled diaper, or rectal swab from children suffering from gastroenteritis and diarrhea or from age-matched control infants without diarrhea. The period of study under greatest concentration (especially for SRV) was October, 1979 to November, 1981. During that time fecal samples were obtained from 1260 infants. One thousand one hundred and sixty samples were taken from children with diarrhea. At least 90% of these were hospitalized. Fifty (50%) of the control infants were housed in the neonatal nursery during the winter season when rotavirus and SRV were prevalent. The remaining 50 were non-diarrheic infants either well, living at home, or attending a clinic with another complaint.

Negative-Stain Electron Microscopy

FIGURE 1 describes the protocol for negative-contrast stain electron microscopy. This technique was adapted from that of Szymanski and Middleton.[15]

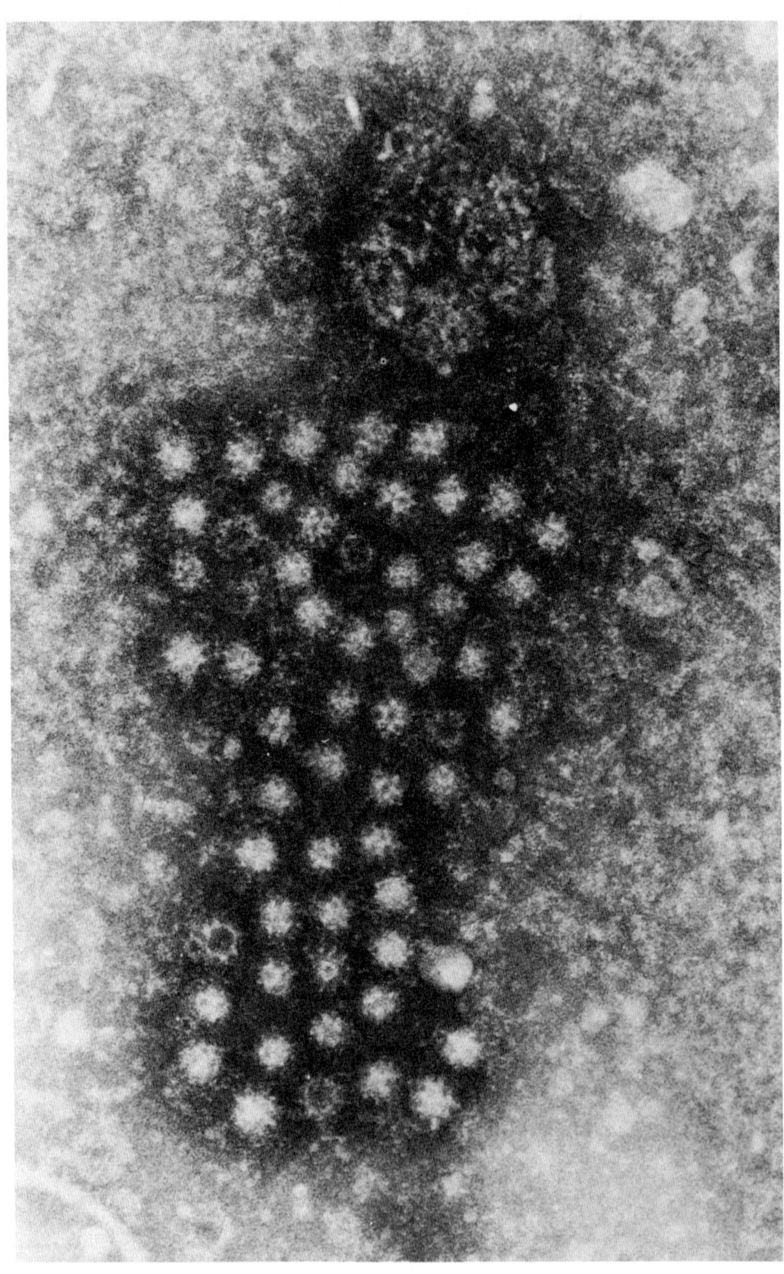

FIGURE 4. Electron micrograph of calicivirus-like agent (CVLA) mixed with porcine anti-CVLA serum at a dilution of 1:25. ×174,000 (Courtesy Dr. L. Saif, Ohio Research and Development Institute, Wooster, Ohio).

Immune Electron Microscopy

Two techniques were utilized in immune electron microscopy (IEM). The first described in FIGURE 2 was adapted from the method described by Brandt and colleagues[4] and was best suited for fecal samples. The second technique seen in FIGURE 3 was developed by Kapikian and is most useful for tissue culture grown and probably purified fecal virus.[8] Finally FIGURE 4 shows the result of IEM. Calicivirus-like agent

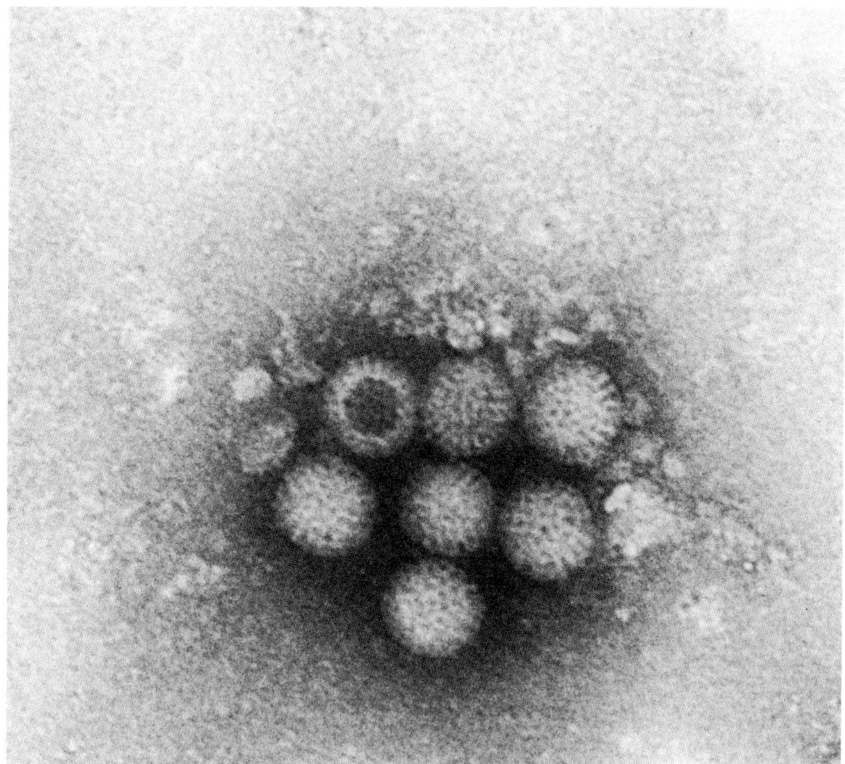

FIGURE 5. Electron micrograph of rotavirus ($\times$165,000).

(CVLA) was mixed with hyperimmune serum. Note the connecting bridges between the particles that have been produced by the virus-specific antibody.

RESULTS

Using the techniques just described we examined specimens from over 1,200 children during a two-year period. We found a variety of viruses and virus-like particles. FIGURES 5 through 7 are representative examples of the particles seen. In all cases the particles were seen in feces from children who had diarrhea at the time the samples were taken. Some of the agents, like rotavirus, are known pathogens and others are still awaiting a definite etiological role. In FIGURE 8b convalescent serum containing rotavirus-specific antibody had clumped particles in a fecal sample. FIGURE 8a is the control without the addition of virus-specific antiserum. When acute and convalescent serum samples are available, IEM can be used to document the cause of viral enteritis.

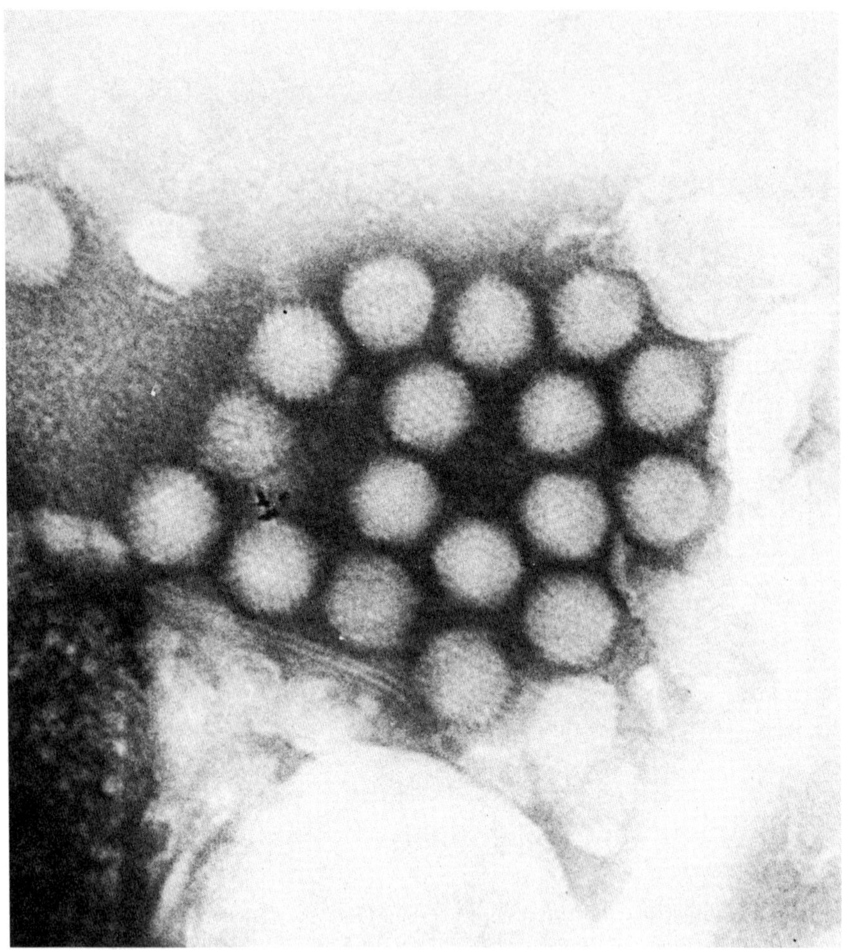

FIGURE 6. Electron micrograph of adenovirus (×165,000).

FIGURE 9 presents the number of the various viruses or virus-like particles seen. Of the 1,260 examined, 100 controls were negative. Of the 1,160 children with diarrhea, 314 contained particles. Of the 314, 230 or 73% were rotavirus. Forty-five (14%) of the specimens contained small round viruses. Of these 45, 25 could be given a tentative identification. Twenty appeared to be astrovirus and five had a calici-like appearance

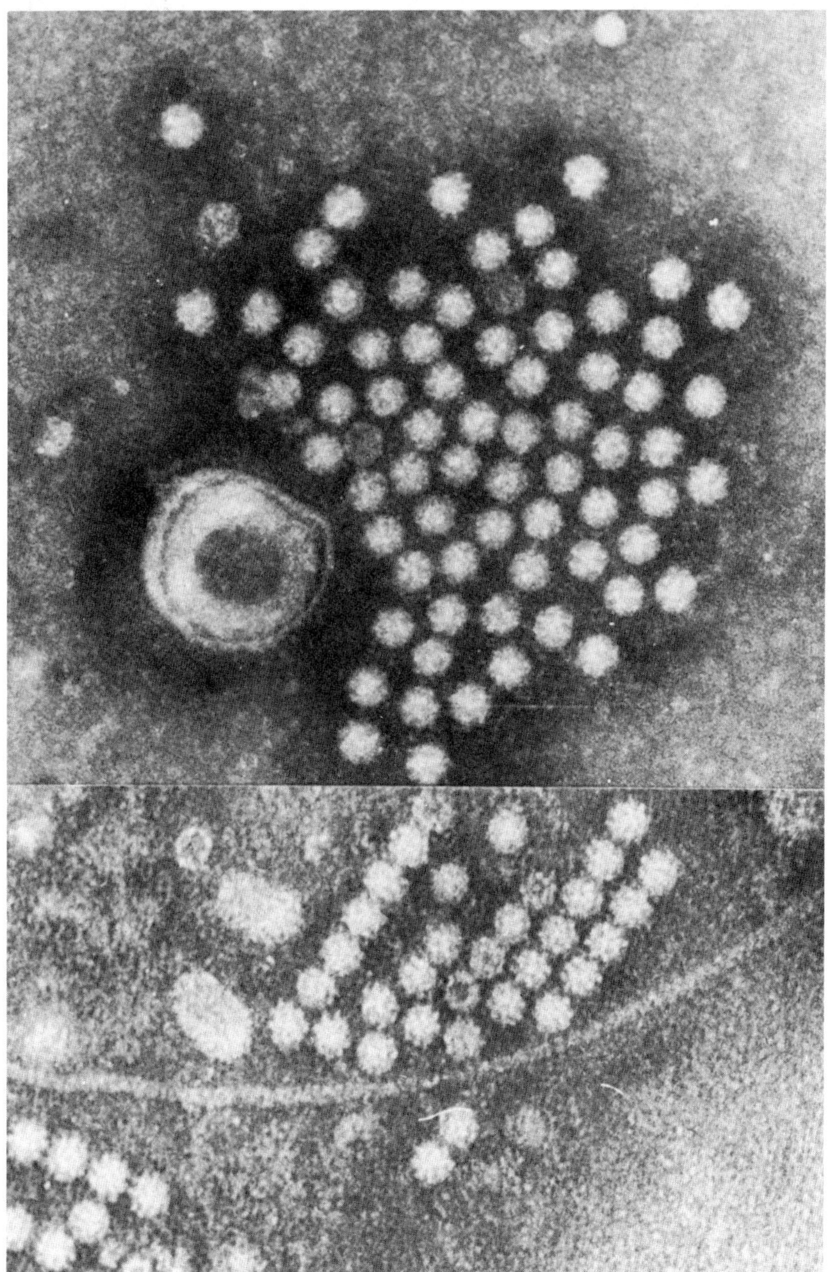

FIGURE 7. Electron micrograph of astrovirus (right frame) (×180,000) and calicivirus-like agent (left frame) (×165,000).

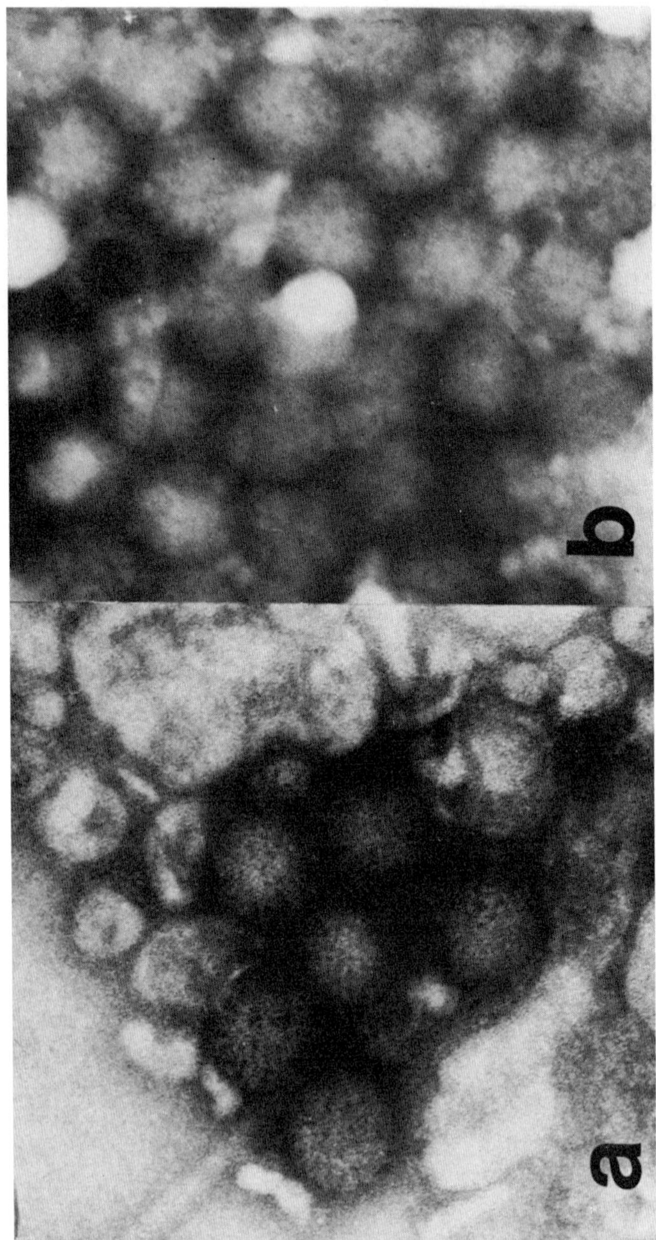

FIGURE 8. Electron micrograph of rotavirus. (a) Control without antiserum (b) Clumped virus after the addition of antirotavirus serum (1:1024).

and were large enough to be placed in the "minireo" or "minirota" category. The remaining SRV did not have a distinctive surface. They were negative in the radioimmunoassay for Norwalk virus as determined by Dr. H. Greenberg at the National Institute of Health. Therefore they may be left in the category described by Middleton and his associates[16] as picorna-parvo. The final 29 (9%) were identified as adenovirus and four particles appeared to be coronavirus-like.

We found that the mean age of children who were rotavirus positive was 11.5 months. This was significantly older than the 4.5 months, which was the mean for the infants who were positive for SRV.

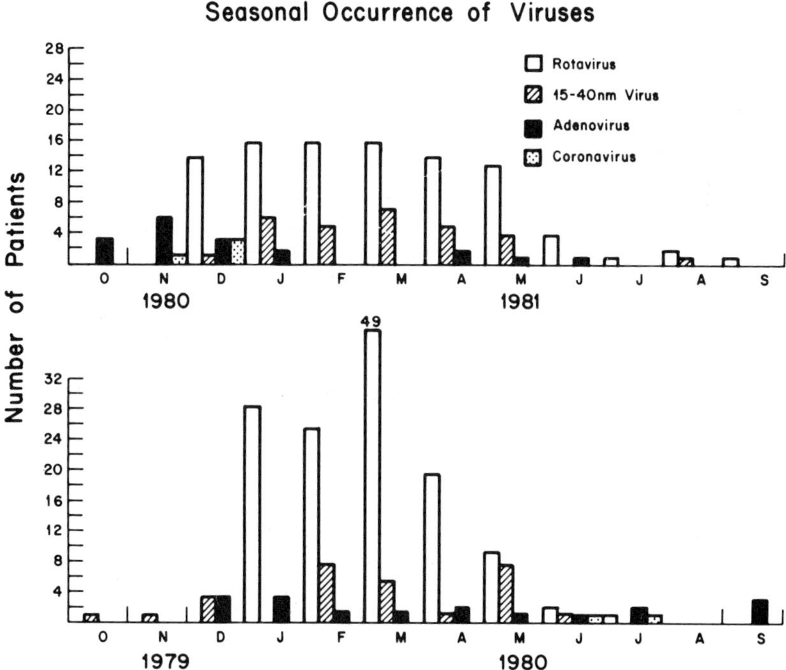

FIGURE 9. Distribution of the various viruses and viruslike particles during a two-year period at Buffalo Children's Hospital.

DISCUSSION

Negative-stain contrast electron microscopy can be held at least partially responsible for diagnosing 50% of the hitherto undiagnosed nonbacterial gastroenteritis in hospitalized children. In addition, IEM helped establish rotavirus as a major etiological agent of gastroenteritis in infants and young children. Finally, both techniques are being used to detect and to define the role of new virus-like particles associated with diarrhea.

Using these techniques, we were able to detect distinctive viruses or virus-like particles associated with diarrhea in 27% of all specimens examined. While the vast majority of these were rotavirus, the finding of astrovirus and other small round viruses associated with diarrhea in very young infants in the absence of any other pathogen is worthy of some attention.

REFERENCES

1. ADLER, J. L. & R. ZICKL. 1969. Winter vomiting disease. J. Infect. Dis. **119:** 668–673.
2. ASHLEY, C. R., E. O. CAUL & W. K. PAVER. 1979. Astrovirus-associated gastroenteritis in children. J. Clin. Path. **31:** 939–941.
3. BISHOP, R. F., G. P. DAVIDSON, I. H. HOLMES, & B. J. RUCK. 1973. Virus particles in epithelial cells of duodenal mucosa from children with acute nonbacterial gastroenteritis. Lancet **2:** 1281–1283.
4. BRANDT, C. D., H. W. KIM, W. J. RODRIQUEZ, L. THOMAS, R. H. YOLKEN, J. O. ARROBIO, A. Z. KAPIKIAN, R. H. PARROTT & R. M. CHANOCK. 1981. Comparison of direct electron microscopy, immune electron microscopy, and rotavirus enzyme-linked immunosorbent assay for detection of gastroenteritis. J. Clin. Micro. **13:** 976–981.
5. CHIBA, W., Y. SAKUMA, R. KOGASAKA, M. AKIHARA, H. TERASHIMA, K. HORINO & T. NAKAO. 1980. Fecal shedding of virus in relation to the days of illness in gastroenteritis due to calicivirus. **142:** 247–249.
6. FLEWETT, T. H. 1978. Electron microscopy in the diagnosis of infectious diarrhea **173:** 538–543.
7. FLEWETT, T. H., H. DAVIES, A. S. BRYDEN & M. J. ROBERTSON. 1974. Acute gastroenteritis associated with reovirus-like particles. J. Clin. Path. **27:** 608–614.
8. KAPIKIAN, A. Z., R. G. WYATT, R. DOLIN, T. S. THORNHILL, A. R. KALICA & R. M. CHANOCK. 1972. Visualization by immune electron microscopy of a 27-nm particle associated with acute infectious nonbacterial gastroenteritis. J. Virol. **10:** 1075–1081.
9. KOGASAKA, R., S. NAKAMURA, S. CHIBA, Y. SAKUMA, H. TERASHIMA, T. YOKOYAMA & T. NAKAO. 1981. The 33 to 39 nm virus-like particles tentatively designated as Sapporo agent associated with an outbreak of acute gastroenteritis. J. Med. Virol. **8:** 187–193.
10. KOGASAKA, R., Y. SAKUMA, S. CHIBA, M. AKIHARA, K. NORINO & T. NAKAO. 1980. Small round virus-like particles associated with acute gastroenteritis in Japanese children. J. Med. Virol. **5:** 151–160.
11. KURTZ, J. B., T. W. LEE, D. PICKERING. 1977. Astrovirus associated gastroenteritis in a children's ward. J. Clin. Path. **30:** 948–952.
12. MADELEY, C. R. & B. P. COSGROVE. 1976. Caliciviruses in man. Lancet **1:** 199–200.
13. MADELEY, C. R. & B. P. COSGROVE. 1975. Viruses in infantile gastroenteritis. Lancet **2:** 124.
14. McSWIGGAN, D. A., D. CUBITT & W. MOORE. 1978. Calicivirus associated with winter vomiting disease. Lancet **1:** 215.
15. MIDDLETON, P. J., M. T. SZYMANSKI, G. D. ABBOTT, R. BORTOLUSSI & J. R. HAMILTON. 1974. Orbivirus acute gastroenteritis of infancy. Lancet **1:** 1241–1244.
16. MIDDLETON, P. J., M. T. SZYMANSKI & M. PETRIC. 1977. Viruses associated with acute gastroenteritis in young children. Am. J. Dis. Child. **131:** 733–737.
17. SPRATT, H. C., M. I. MARKS, M. GOMERSALL, P. GILL & C. H. PAI. 1978. Nosocomial infantile gastroenteritis associated with minirotavirus and calicivirus. J. Peds. **93:** 922–926.
18. SUZUKI, H., T. KONNO, T. KATSUZAWA, A. IMAI, F. TAZAWA, N. ISHIDA, N. KATSUSHIMA & M. SALAMOTO. 1979. The occurrence of calicivirus in infants with acute gastroenteritis. J. Med. Virol. **4:** 321–326.
19. TANIGUCHI, K., S. URASAWA & T. URASAWA. 1979. Virus-like particle, 35 to 40 nm associated with an institutional outbreak of acute gastroenteritis in adults. J. Clin. Micro. **10:** 730–736.
20. THORNHILL, T. S., R. G. WYATT, A. R. KALICA, R. DOLIN, R. M. CHANOCK & A. Z. KAPIKIAN. 1977. Detection by immune electron microscopy of 26–27 nm virus-like particles associated with two family outbreaks of gastroenteritis. J. Infect. Dis. **135:** 20–27.

Immunologic Cross-reactivity Between
Streptococcus Mutans and
Mammalian Tissues[a]

RUSSELL J. NISENGARD, MURRAY W. STINSON, AND
LYNN PELONERO

Departments of Periodontology and Microbiology
Schools of Dentistry and Medicine
State University of New York at Buffalo
Buffalo, New York 14214

A limited number of bacteria have been implicated in the etiology of dental caries.[5] Of these, *Streptococcus mutans* is of prime importance in the initiation of smooth surface and "pit and fissure" caries. To a lesser extent, other species of streptococci, lactobacilli, and filamentous bacteria also contribute to "pit and fissure" and root-surface caries. The identification of *S. mutans* as the principal pathogen makes immunization a feasible approach for the prevention of dental caries.[15] This has been supported by animal studies where stimulation of local secretory IgA by vaccination has resulted in a reduced caries incidence and/or decreased *S. mutans* colonization.[3,9,12,18]

Additional studies are necessary concerning the safety of *S. mutans* vaccines before they can be employed in humans. Of major concern is the potential pathology generated by streptococcal components that immunologically cross-react with mammalian tissues (CR antigens). Numerous studies have revealed that group A streptococci share antigens with cardiac muscle, skeletal muscle, smooth muscle, and kidney.[8,20] More recently, studies by van de Rijn *et al.*,[19] Ferretti *et al.*,[4] and Hughes *et al.*[7] have suggested similar CR antigens are shared by *S. mutans* and mammalian heart tissue. The purpose of the studies reported here was to evaluate *S. mutans* for CR antigens shared with mammalian brain, heart, kidney, liver, skeletal muscle, smooth muscle, and skin.

MATERIALS AND METHODS

Bacteria

The streptococci studied are listed in TABLE 1. A number of cultures were kindly supplied by Dr. Richard Evans, State University of New York at Buffalo, Buffalo, New York.

Media and Cultural Conditions

Bacteria were grown at 37°C in a dialysate of trypticase–soy broth supplemented with 0.1% yeast extract. Cultures in mid-logarithmic to early stationary phases of

[a]Supported by U.S. Public Health Service Contract DE-62488 and Grant DE-05696.

401

growth, as determined turbidimetrically, were harvested by centrifugation at 10,000 × g for 15 minutes at 4°C. The resulting pellet was washed twice with 0.01 M NaH_2PO_4–Na_2HPO_4, 0.15 M NaCl, pH 7.2 (PBS).

Protein Assay

Protein was measured by the method of Lowry et al.[10] using bovine serum albumin as a standard.

TABLE 1. Antiserum Titers to Microorganisms Studied

Microorganism	Strain	Bacterial Titer[a]
Streptococcus mutans, serotype a	AHT	640–1280
	E-49	160–640
Streptococcus mutans, serotype b	FA-1	640–5120
	BHT	640–1280
Streptococcus mutans, serotype c	10449	320–640
	GS5	80–640
Streptococcus mutans, serotype d	OMZ-176	320–10,240
	B-13	1280–2560
Streptococcus mutans, serotype e	LM-7	640–1280
	MT703	640–1280
Streptococcus mutans, serotype f	OMZ-175	640–1280
	MT557	160–5120
Streptococcus mutans, serotype g	OMZ-65	1280
	K-1	640
Streptococcus salivarius		
serotype I	ATCC9759	320–1280
serotype II	ATCC13419	640–1280
Streptococcus sanguis		
serotype I	ATCC10556	640–1280
serotype II	ATCC10557	1280–2560
Streptococcus pyogenes		
serotype 6	ATCC12348	640–1280
serotype 12	ATCC12351	640–2560

[a]Titers determined by indirect immunofluorescence on smear of the S. mutans immunizing strain.

Preparation of Antigens for Immunization and Absorption

Washed bacteria were disrupted in a Braun homogenizer (Bronwill Scientific, Rochester, NY) at 4°C as described by Bleiweis et al.[2] Cell walls, cytoplasmic membranes, and soluble components were then isolated following the procedures of Stinson et al.[17] and adjusted to 1, 2, and 5 mg protein per ml, respectively.

Tissue homogenates of monkey heart and kidney were prepared in a Sorvall omnimixer (DuPont Instruments, Norwalk, CT) at 4°C. The tissue was placed in PBS, blended for 1 minute and then cooled for 5 minutes. This was repeated four times, followed by filtration through cheesecloth for removal of connective tissue. The kidney homogenate was sedimented by centrifugation at 18,000 × g for 15 minutes and then resuspended in PBS to a final protein concentration of 35 mg protein per ml. The heart suspension was further homogenized by shaking with glass beads in a Braun homogenizer. The beads were removed by filtration and the tissue fragments sedimented at 100,000 × g for 1 hour. The pelleted heart material was then suspended in PBS to 11 mg protein per ml.

Antibody absorptions were performed by mixing 1 ml of antiserum with either 1 ml of bacterial suspension (5×10^9 cells) or 1 ml of one of the above antigen preparations. The mixture was then incubated at 37°C for 1 hour and 4°C for 18 hours. As a control, 1 ml of PBS was added to the antiserum instead of the antigen preparation. After clarification by centrifigation, the treated antiserum was tested for tissue reactivity by indirect immunofluorescence.

Preparation of Antisera to Streptococci

New Zealand white rabbits were immunized with streptococci using either intradermal or intravenous injections. For intradermal immunization, the antigens were diluted in PBS, pH 7.0, to 2 mg of protein per ml and emulsions were prepared with equal volumes of complete or incomplete Freund's adjuvant (Grand Island Biological Co., Grand Island, NY). Rabbits received weekly intradermal injections with 0.1 ml containing 0.1 mg protein of the appropriate vaccine. Immunization continued for 2 to 3 months until maximal antibacterial titers were achieved. For intravenous immunization, the method described by van de Rijn and Bleiweis[19] was employed. Rabbits were injected three times a week with 1 mg of antigen the first week, 2 mg of antigen the second week, and 3 mg of antigen the third week. During the fifth week, the animals were test-bled. This immunization regimen was repeated to obtain maximal antibody production.

Assay for Bacterial and Cross-reacting Antibody Titers

Rabbit anti-streptococcus sera were tested by indirect immunofluorescence on heat-fixed smears of the immunizing bacteria and on cryostat-cut frozen mammalian tissues. Brain, heart, kidney, liver, skeletal muscle, smooth muscle, and skin were tested from humans, Rhesus monkeys, and cynomologous monkeys. For these tests a fluorescein-labeled goat anti-rabbit IgG conjugate was prepared and characterized.[1] The conjugate had an F/P molar ratio of 2.5 and was diluted for tests to 1/4 unit of antibody /ml.

The indirect immunofluorescence tests were performed using serial twofold dilutions of preimmune and immune antisera.[14] Tests were read with a Leitz Ortholux II microscope equipped with an incident-light source. Tests were graded as to intensity of immunofluorescence and titer. Results were only considered positive when titers of the immune sera were at least two doublings dilutions greater than titers of the preimmune sera. As a control for the FITC-conjugated goat anti-rbbit IgG sera, PBS-BSA was substituted for rabbit anti-streptococcus serum. Antisera to other bacteria[16] reactive with brain, heart, and kidney were routinely titered and used as a positive control in subsequent assays.

RESULTS

Antibody titers to the streptococci as determined by indirect immunofluorescence are listed in TABLE 1. Titers to the homologous immunizing bacteria ranged from 160 to 10,240 with most falling in the range of 320 to 640.

TABLE 2. Indirect Immunofluorescence Tests with Antiserum to *S. mutans* Serotype g, Strain OMZ-65

Anti-*S. mutans*, Strain OMZ-65	Heart Titer[a]		Kidney Titer	
	Monkey	Human	Monkey	Human
Preimmune	—	—	—	—
Immune	16	16	32	—
absorbed with OMZ-65[b]				
washed cells	16	16	32	ND
cell walls	16	ND	< 2	ND
soluble fraction	< 4	ND	32	ND
cell membranes	< 4	ND	32	ND
absorbed with heart homogenate	16	16	ND	ND

[a]Corrected for dilution as a result of absorption.
[b]Three absorptions with indicated antigen.
[c]ND = not done. Dashes indicate titer <1.

The antisera were also tested for reactivity with tissue by indirect immunofluorescence. The majority of the antisera were nonreactive on mammalian tissue or reacted at the same titer as the preimmune sera and, therefore, were considered nonreactive. However, a number of anti–*S. mutans* sera reacted with components of heart, skeletal muscle, and/or kidney.

Antisera to *S. mutans* serotype g, strains OMZ-65 and K-1, reacted with cardiac muscle (TABLES 2 and 3). Antisera from two of three rabbits immunized with *S. mutans* strain OMZ-65 in incomplete Freund's adjuvant reacted with the striations of both human and monkey cardiac muscle (FIGURE 1). The antibody titers to heart were 8 and 32, compared to the preimmune sera, which were negative or had a titer of 4, respectively. TABLE 2 summarizes the results with the higher titered serum. Three absorptions with either a soluble or a cell membrane fraction of OMZ-65 reduced the anti-heart titer (corrected for dilution as a result of absorption) from 16 to ≤4. No change in titer resulted from three absorptions with strain OMZ-65 washed cells and cell walls or a monkey heart homogenate. Antisera from one of the three rabbits immunized with strain K-1 in incomplete Freund's adjuvant reacted with both monkey cardiac and skeletal muscle tissues at titers of 32 and 64 while the preimmume serum was negative (TABLE 3). The other two sera, as well as the one serum prepared by immunization with complete Freund's adjuvant, were non-reactive. Three absorptions with K-1 whole cells, cell walls, or cell membranes failed to reduce the titer.

TABLE 3. IIF Titers on Monkey Tissue with Antiserum to *S. mutans* Serotype g, Strain K-1

Anti-*S. mutans*, Strain K-1	Heart Titer[a]		Skeletal Muscle Titer[a]	
	Monkey	Human	Monkey	Human
Preimmune	—	—		16
Immune	32	—	64	128
absorbed with K-1[b]				
washed cells	32	ND[c]	64	128
cell walls	32	ND	64	128
cell membranes	32	ND	64	128

[a]Titers corrected for dilution as a result of absorption.
[b]Three absorptions with indicated antigen.
[c]ND = not done. Dashes indicate titer <1.

Cardiac muscle reactivity to a titer of 64 was also observed with anti-*S. pyogenes* type 6 sera. One of three rabbits immunized with incomplete Freund's adjuvant and one of three rabbits immunized with complete Freund's adjuvant reacted with heart components. These antisera were used as additional positive controls for heart-reactive antibodies during assays of anti-*S. mutans* sera.

In addition to cardiac muscle reactivity, antisera to *S. mutans* reacted with other tissues. Antisera to *S. mutans* strains OMZ-65, K-1, and MT703 also reacted with the glomeruli of mammalian kidney (FIGURE 2). One of the three rabbits immunized with *S. mutans* OMZ-65 in incomplete Freund's adjuvant had a titer of 32 on glomeruli while the preimmune serum was negative (TABLE 2). The antiserum prepared by immunization with *S. mutans* OMZ-65 in complete Freund's adjuvant was negative. Antibodies in the reactive serum bound to the glomerular basement membrane but not to the tubular basement membrane or to the basement membranes of skin or other

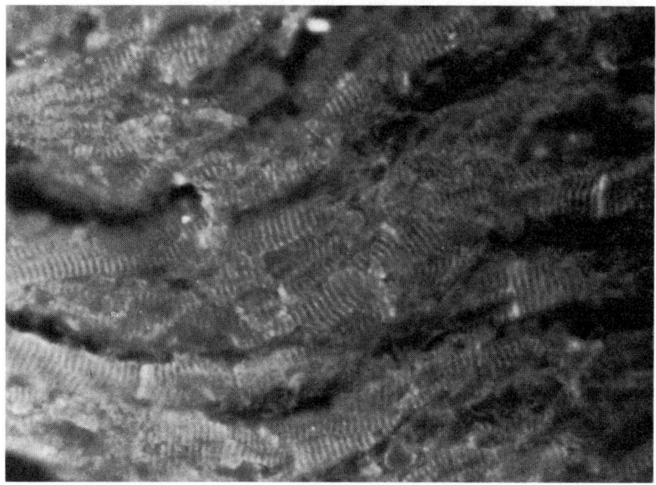

FIGURE 1. Indirect immunofluorescence staining of monkey heart muscle with a rabbit antiserum to *S. mutans*, strain OMZ-65 at a 1:4 dilution. ×400.

tissues. To determine antibody specificity, aliquots of the serum were absorbed three times with disrupted *S. mutans* OMZ-65 cells, isolated cell walls, or isolated cell membranes. Only the cell wall preparation caused a significant reduction in the quantity of kidney-reactive antibodies (titer ≤2). An antiserum to *S. mutans* MT703 also reacted with the glomerular basement membrane of monkey, human, and rabbit kidney (TABLE 4), while the preimmune serum was nonreactive. Three absorptions with either intact or disrupted *S. mutans* MT703 cells failed to change the antibody titer to glomeruli. Three absorptions with a homogenate of monkey kidney also failed to alter the antibody titer on monkey kidney.

Antisera from one of three rabbits immunized with *S. mutans* strain K-1 in incomplete Freund's adjuvant reacted with skeletal muscle (TABLE 3). This same antiserum gave comparable reactivity on cardiac muscle. Three absorptions with intact bacteria, cell walls, or cell membranes failed to change the antibody titer on either muscle tissue.

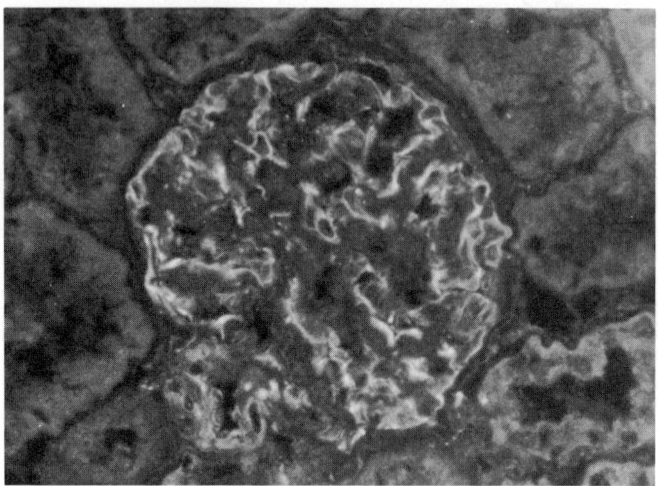

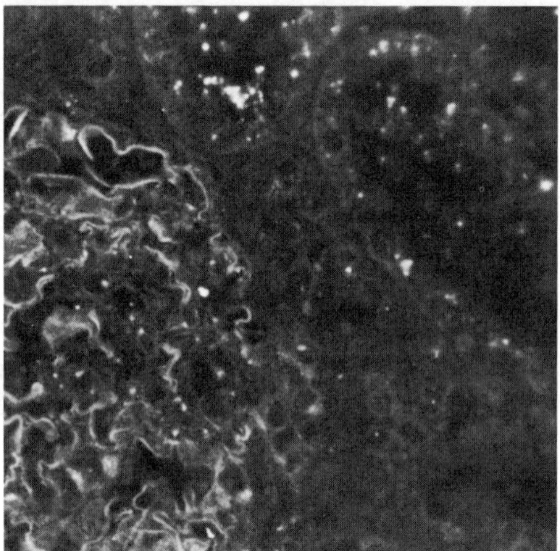

FIGURE 2. (Top) Indirect immunofluorescence staining of monkey kidney glomeruli with rabbit antiserum to *S. mutans,* strain OMZ-65 at a 1:8 dilution. ×400. (Bottom) Indirect immunofluorescence staining of monkey kidney glomeruli with rabbit antiserum to *S. mutans,* strain MT-703 at a 1:32 dilution. ×400.

DISCUSSION

The report by van de Rijn, Bleiweis, and Zabriskie[19] suggested that CR antigens, detected by indirect immunofluorescence, are shared among all serotypes of *S. mutans* and cardiac muscle tissue. The results reported here do not support this. Instead, antibody reactivity on cardiac muscle was observed only with antisera to *S. mutans* serotype g, strains K-1 and OMZ-65. These discrepancies may partly result from differences in the test procedures and interpretations placed on the observations. Van de Rijn *et al.* evaluated only 1:5 dilutions of the preimmune and immune rabbit sera for staining intensity.[19] Because the sera of many normal rabbits contained heart-reactive antibodies that were detectable at this dilution, these workers found it necessary to test each rabbit prior to immunization. Only those rabbits with negligible anti-cardiac muscle activity were used in their studies (van de Rijn, personal communication). This problem was highlighted by our own experience with cardiac muscle reactivity in non-immunized rabbits. In sera from 48 rabbits tested undiluted and in twofold doubling dilutions, 20 (41%) had titers less than or equal to 4. These titers ranged as high as 64. The presence of this reactivity requires further evaluation including nonspecific (non-immunologic) protein binding. To avoid these obvious complications, the present study was based on titrations. Tissue reactivity was only considered significant if the titer of the immune serum was at least two doubling dilutions greater than the preimmune serum.

More recently, Ferretti *et al.*[4] detected two CR antigens in heart extracts using concentrated rabbit antisera to *S. mutans,* serotype d, strain B13 in rocket and crossed immunoelectrophoresis. The antisera also reacted weakly with skeletal muscle but not with liver or kidney tissues. In similar studies Hughes *et al.*[7] reported that antisera to four of eight strains of *S. mutans* reacted with human cardiac muscle as measured by immunofluorescence. Antisera to *S. mutans,* Inbritt strain, cell walls, and cytoplasm also reacted with cardiac muscle suggesting that they contained the CR antigen. Hughes detected the CR antigens with several other techniques including immunoelectrophoresis, radioimmunoassay, and anaphylactic reactions. Both research groups reported that when heart-reactive antibodies were present in anti-streptococcal sera, they were in very low quantity. These findings are consistent with the low immunofluorescent titers reported here.

The previous studies of *S. mutans* CR antigens may be considered preliminary because of the small differences between the concentrations of heart-reactive antibodies in the preimmune sera and the immune sera. Future studies must include titrations and absorptions of antisera. Rigid criteria must be observed when analyzing rabbit sera for heart-reactive antibodies induced by immunization with bacteria. In addition to a difference of two doubling dilutions in titer between preimmune and immune sera, the antibody titers should be reduced by at least two doubling dilutions by absorption with bacteria and tissue fractions.

These criteria have not been completely fulfilled for the five tissue-reactive antibodies reported here. Up to three absorptions with immunizing or purified antigens did not always fully eliminate the tissue reactivity of the antisera. This may result from inappropriate antigen selection or the need for more highly purified antigens for absorption. Additionally, the CR antigen may be a minor antigenic component of the bacteria and tissue even though it may be highly immunogenic. An example can be seen in TABLE 2 where absorption with washed bacterial cells did not reduce the titer on kidney while absorption with purified cell membrane was successful.

TABLE 4. IIF Titers with Antiserum to *S. mutans* Serotype e, Strain MT-703

Subtrate	Preimmune Serum[a]	Immune Serum	No. Rabbits Pos./Total
S. mutans strain MT-703	—	640–1280	3/3
Human kidney	—	128	1/3
Monkey kidney	—	256–512	2/3
Rabbit kidney	—	32	1/3

[a]Negative with undiluted sera.

The location of the CR antigen in *S. mutans* strain OMZ-65 was indicated by specific absorption experiments. The CR antigen shared with heart was in the bacterial cell membranes and the soluble fraction but not in the cell wall. The CR antigen shared with kidney glomeruli appeared to be located in the bacterial cell wall.

The presence of CR antigens shared by kidney and bacteria have also been previously reported.[6,11] Group A streptococci and nephritogenic strains of *E. coli* had CR antigens demonstrable by agar-gel double-diffusion techniques. The species specificity of antiserum to *S. mutans* strain MT703 reactive with kidney glomeruli was investigated on human, monkey, and rabbit kidneys. The titers on human and monkey kidney sections were similar, while titers on rabbit kidney were two doubling dilutions less. The sharing of the CR antigen by several species suggests the possibility that this may be a heterophile antigen. The heart reactivity may also reflect a heterophile antibody. Nicholson *et al.*[13] reported several cardiac staining patterns in human sera from several diseases. Striational straining, the most common pattern observed in the present report was identified as a heart-specific rather than heterophile antibody. These human findings do not exclude the possibility that striational staining in immunized rabbits could be heterophile antibodies.

There are several explanations for tissue reactivity of sera with muscle and kidney glomeruli. Serum reactivity with tissue may follow a bacterial infection or vaccination as in the present experiments. Those antibodies, specific for CR antigens may then react with the CR antigens within the tissue. Alternatively, in immunized animals, bacterial antigens may first directly bind non-immunologically to tissues such as cardiac muscle,[11] followed by binding of the bacterial antibodies to the bacterial antigens. Similarly, circulating immune complexes of bacterial antigens may also bind non-immunologically to the tissue through the bacterial antigens. Naturally occurring tissue reactivity of antibodies may also follow heart damage as seen in human cardiotomy or polyclonal B cell activation as in SLE.

The possibility that CR antigens are shared by oral bacteria and mammalian tissue requires further study if immunization against dental caries is to become a reality. For any vaccine contemplated for human use, CR antigens must be identified and if possible, eliminated from the vaccine. The production of antibodies to CR antigens, however, may not necessarily lead to tissue damage. Tissue pathology would require *in vivo* reactivity with the tissue in sufficient quantity. This could be influenced by route of immunization, use of an adjuvant, and immunization regimen.

SUMMARY

The presence of cross-reacting antigens shared by oral streptococci and human and monkey tissue was investigated by indirect immunofluorescence. Rabbit antisera prepared against two strains of each of the seven serotypes of *Streptococcus mutans* were incubated on sections of cardiac muscle, skeletal muscle, kidney, brain, liver, and skin. Antisera to *S. mutans,* serotype g, strain OMZ-65 reacted with cardiac muscle components. This reactivity could be reduced by absorption with bacterial cell membranes. Antisera to *S. mutans* serotype g, strain K-1 reacted with both cardiac and skeletal muscle but could not be reduced by absorption with bacterial cells, cell walls, or cell membranes. Antisera to *S. mutans* serotype g, strain OMZ-65 reacted with kidney glomeruli and could be completely absorbed by a cell wall preparation. *S. mutans,* serotype e, strain MT703 also reacted with kidney glomeruli but could not be reduced by absorption with bacterial antigens.

REFERENCES

1. BEUTNER, E. H., M. R. SEPULVEDA & E. V. BARNETT. 1968. Quantitative studies of immunofluorescent staining. II. Relationships of characteristics of unabsorbed antihuman IgG conjugates to their specific and non-specific staining properties in an indirect test for antinuclear factors. WHO Bull. **39:** 587–606.
2. BLEIWEIS, A. S., W. W. KARAKAWA & R. M. KRAUSE. 1964. Improved technique for the preparation of streptococcal cell walls. J. Bacteriol. **88:** 1198–1200.
3. EVANS, T. R., F. G. EMMINGS & R. J. GENCO. 1975. Prevention of *Streptococcus mutans* infection of tooth surfaces by salivary antibody in iris monkeys (*Macaca fasciculoris*). Infect. Immun. **12:** 293–302.
4. FERRETTI, J. J., C. SHEA & M. W. HUMPHREY. 1980. Cross-reactivity of *Streptococcus mutans* antigens and human heart tissue. Infect. Immun. **30:** 69–73.
5. GIBBONS, R. J. & J. VAN HOUTE. 1975. Bacterial adherence in oral microbial ecology. Annu. Rev. Microbiol. **29:** 19–43.
6. HOLM, S. E., J. HOLMGREN & J. AHLMEN. 1972. Immunological cross-reactivity between human kidney and certain *E. coli* and streptococcal strains. Int. Arch. Allergy. **43:** 63–71.
7. HUGHES, M., S. M. MACHARDY, A. J. SHEPPARD & N. C. WOODS. 1980. Evidence for an immunological relationship between *Streptococcus mutans* and human cardiac tissue. Infect. Immun. **27:** 576–588.
8. KAPLAN, M. H. & M. MEYESERIAN. 1962. An immunological cross-reaction between group A streptococcal cells and human heart tissue. Lancet **1:** 706–710.
9. LEHNER, T., S. J. CHALLACOMBE & J. CALDWELL. 1975. An immunological investigation into the prevention of caries in deciduous teeth of rhesus monkeys. Archs. Oral Biol. **20:** 305–310.
10. LOWRY, O. H., N. J. ROSEBROUGH, A. L. FARR & R. J. RANDALL. 1951. Protein measurement with the folin phenol reagent. J. Biol. Chem. **193:** 265–275.
11. MARKOWITZ, A. S., H. ARMSTRONG & D. S. KUSHNER. 1960. Immunological relationships between the rat glomerulus and nephritogenic streptococci. Nature (London) **187:** 1095–1097.
12. MCGHEE, J. R., S. M. MICHALEK, J. WEBB, J. M. NAVIA, A. F. R. PAHMAN & D. W. LEGLER. 1975. Effective immunity to dental caries: Protection of gnotobiotic rats by local immunization with *Streptococcus mutans*. J. Immunol. **114:** 300–305.
13. NICHOLSON, G. C., R. L. DAWKINS, B. L. MCDONALD & J. D. WETHERALL. 1977. A classification of anti-heart antibodies: Differentiation between heart specific and heterophile antibodies. Clin. Immunol. Immunopath. **7:** 349–363.
14. NISENGARD, R. J. & E. H. BEUTNER. 1970. Immunologic studies of periodontal disease. V. IgG type antibodies and skin test responses to actinomyces and mixed oral flora. J. Periodontol. **41:** 149–152.
15. SCHERP, H. 1971. Dental caries: Prospects for prevention. Science **173:** 1199–1205.
16. STINSON, M. S., R. J. NISENGARD, L. PELONERO & A. J. DRINNAN. 1979. Immunological cross-reactivity of oral non-streptococcal bacteria with mammalian tissue. Infect. Immun. **24:** 532–538.
17. STINSON, M. W., R. J. NISENGARD & E. J. BERGEY. 1980. Binding of streptococcal antigens to muscle tissue in vitro. Infect. Immun. **27:** 604–613.
18. TAUBMAN, M. A. & D. J. SMITH. 1974. Effects of local immunization with *Streptococcus mutans* in induction of salivary immunoglobulin A antibody and experimental dental caries in rats. Infect. Immun. **9:** 1079–1091.
19. VAN DE RIJN, I., A. S. BLEIWEIS & J. B. ZABRISKIE. 1976. Antigens in *Streptoccus mutans* cross reactive with human heart tissue. J. Dent. Res. (Special Issue C):C59–C64.
20. ZABRISKIE, J. B. 1967. Mimetic relationships between group A streptococci and mammalian tissues. Adv. Immunol. **7:** 147–188.

Trends in the Localization of Bacterial Antigens by Immunoelectron Microscopy

P. D. WALKER AND J. E. BEESLEY

Department of Bacteriology
Research and Development
Wellcome Research Laboratories
Beckenham, Kent, England BR3 3BS

INTRODUCTION

Previous work on the location of bacterial antigens with electron microscopy has used antibodies labeled with ferritin[12,20–22] and enzymes[22,23] or, alternatively, the peroxidase-antiperoxidase (PAP) technique.[16,17] Both enzyme-labeled antibodies and the PAP technique have been used to locate antigens on ultrathin sections of fixed and embedded material. Unfortunately, due to its tendency to adhere nonspecifically to ultrathin sections, the use of ferritin has largely been confined to pre-embedding staining, as a result of which only surface antigens have been located. Location of internal antigens using ferritin-labeled antibodies can be determined following partial disintegration of the cell but the information is limited since the architecture of the cell is destroyed.[20]

Unlike ferritin, which produces a discrete opaque spot in the electron microscope, the end product of enzyme staining or the PAP technique tends to be somewhat diffuse and lack the precision of ferritin. Notwithstanding this the techniques have been used to locate a variety of antigens. For example, using bacterial sporulation and germination as a model, it has been shown that certain antigenic structures of the bacterial spore are synthesized in the cytoplasm of the vegetative cell, while others are synthesized within the spore core itself.[17]

In an attempt to find an alternative marker, attention has been focused on the protein A-gold technique (PAG).[13,14] The basis of this technique is that protein A obtained from staphylococci can be attached to gold particles of a specific size, which are visible in the electron microscope. Protein A attaches specifically to the IgG component of several animal species[4,6,7] and therefore will adhere to antibody molecules attached to antigens on ultrathin sections. Sections are treated with the appropriate specific antiserum, thoroughly washed to remove excess antibody, and then stained with PAG. Discrete particles of gold are seen at the antigen-antibody complex.

As routine fixation and embedding have been shown to decrease the sensitivity of immunohistochemical methods at both the light[10] and electron microscopic level,[3,8,9] an attempt has been made to compare the sensitivity of various staining and embedding procedures. The sensitivity of peroxidase-labeled antibodies, PAP and PAG have been compared using the bacterial spore and antibodies to the exosporium or outer spore layer as a model. In addition the effect of fixation with glutaraldehyde, glutaraldehyde followed by osmium, cryofixation with and without primary glutaraldehyde fixation, and embedding in methacrylate and Araldite have been compared using the PAG technique.

410

MATERIALS AND METHODS

Bacterial Spores and Antiserum

Spores of *Bacillus cereus* strain M8 spore and antisera were prepared as described previously.[1]

Fixation and Embedding

Spores were fixed in 2.5% glutaraldehyde in 0.1 M phosphate buffer pH 7.4 for 30 minutes and embedded in glycol methacrylate containing 3% *n*-butymethacrylate or Araldite. Additionally spores were fixed in glutaraldehyde (0.5 hour) followed by osmium (1 hour) and embedded in methacrylate.[15]

Gold-colored sections were cut with glass knives on an LKB Ultratrome III (LKB, Bromma, Sweden) and collected on Butvar-coated, 400-mesh gold grids.

Cryo-ultramicrotomy

Spore suspensions were either processed without chemical fixation or alternatively fixed for 18 hours in 2.5% glutaraldehyde in 0.1 M phosphate buffer pH 7.4 prior to freezing.

All specimens were mounted on silver pins and rapidly frozen in liquid nitrogen slush.

Ultrathin sections were cut on an LKB Ultratrome III fitted with a cryokit and cryotools (LKB, Bromma, Sweden) using a knife temperature of $-90°C$ and a specimen temperature of $-110°C$. Sections of unfixed material were collected dry whereas fixed material was collected on a drop of saturated sucrose. All sections were mounted on Butvar-coated, 400-mesh gold grids.

Reagents

PAP was obtained from Miles Laboratories (Slough, U.K.), donkey anti-rabbit from Wellcome Diagnostics (Dartford, U.K.) and horseradish peroxidase–labeled anti-rabbit IgG from Dako (Mercia Brocades, Weybridge, U.K.). The protein A-gold probe (PAG) was prepared by coupling staphylococcal protein A to 20 nm gold particles as previously described.[13,14]

Staining

PAP Technique

The PAP technique was carried out as previously described[16,18] using dilutions of serum shown in TABLE 1.

Enzyme-labeled Antibody Technique

For the enzyme-labeled antibody technique sections were incubated on drops of the appropriate dilution of rabbit antibody for 1 hour at room temperature followed by a

TABLE 1. A Comparison of the Peroxidase-Linked Antibody, Peroxidase-Antiperoxidase, and Protein A-Gold Techniques on Methacrylate Sections Using Bacterial Spores and Antisera Against the Exosporium or Outer Spore Layer

| | Serum Dilution | | | | |
Label	10^{-1}	10^{-2}	10^{-3}	10^{-4}	10^{-5}
Protein-A-gold	+	+	+	+	±
Peroxidase-labeled antibody	+	+	+	−	−
PAP	+	+	−	−	−

5-minute wash in running tap water. After washing, the sections were floated for 15 minutes on a ⅕ dilution of horseradish peroxidase–labeled anti-rabbit IgG. After a further 5-minute wash in running tap water the peroxidase was developed as described by Sternberger.[18] The sections were finally washed for 5 minutes in running tap water, dried, and stained for thirty minutes in osmium tetroxide vapor.

Protein A-Gold Technique

Fixed and embedded sections and fixed cryosections were incubated with the appropriate dilution of specific antiserum for 1 hr, washed for ten minutes in running tap water, and incubated for 1 hr on a drop of PAG. They were then washed thoroughly with tap water and viewed immediately.

Some of the methacrylate-embedded sections were contrasted with osmium vapor and likewise some of the glutaraldehyde-fixed cryosections were stained with 2% ammonium molybdate pH 6 before examination.

Unfixed cryosections were treated as above but after incubation with specific antiserum and PAG the sections were washed with Tris buffer. Finally they were fixed for 15 minutes in 1% osmium tetroxide in Tris buffer, rinsed in tap water, and viewed directly.

TABLE 2. A Comparison of the Protein A-Gold Technique on Methacrylate, Araldite, and Cryo-embedded Material Using Bacterial Spores and Antisera Against the Exosporium or Outer Spore Layer

| | Serum Dilutions | | | | | |
Embedding Medium	10^{-1}	10^{-2}	10^{-3}	10^{-4}	10^{-5}	10^{-6}
Methacrylate						
Glut. fixed	+	+	+	+	+	−
Glut./OsO₄ fixed	+	+	+	+	+	−
Araldite						
Unetched	+	−				
Etched 3 min 5% H₂O₂	+	+	+	−		
Etched 10 min 10% H₂O₂	+	+	+	−		
Cryo						
Glut.-fixed	+	+	+	+	+	−
Unfixed	+	+	+	+	+	−

Glut. = glutaraldehyde, + = antigen/antibody reaction, and − = negative, and Cryo = Cryofixation.

Etching

In view of the poor staining observed with sections embedded in Araldite, it was necessary to etch the specimen. This was carried out as previously described by Sternberger[18] (3 min 5% H_2O_2) and Roth[13,14] (10 min 10% H_2O_2).

RESULTS

A comparison of the sensitivity of the three methods of staining using glutaraldehyde-fixed methacrylate sections is shown in TABLE 1. It can be seen that the most sensitive technique was PAG followed by peroxidase-labeled antibodies and PAP.

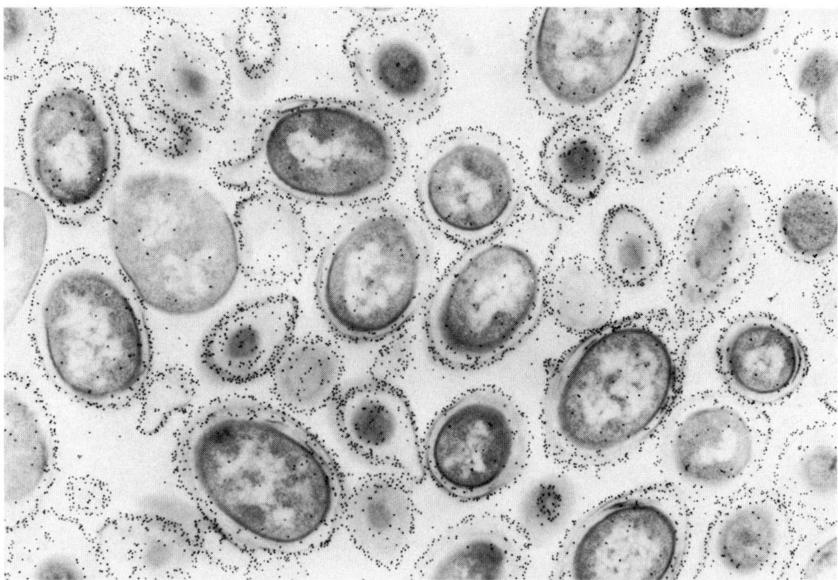

FIGURE 1. Ultrathin section of spores of *B. cereus* M8 fixed in glutaraldehyde and embedded in methacrylate after staining with specific antiserum 10^{-1} and protein A-gold (PAG). Note particles of gold surrounding the spore. ×9,500.

A comparison of different methods of fixation and embedding using PAG as label is shown in TABLE 2. It can be seen that methacrylate-embedded specimens when fixed with glutaraldehyde alone or glutaraldehyde followed by osmium show good sensitivity. On the other hand, embedding in Araldite considerably reduces the sensitivity although this was slightly increased by etching of the specimen. Cryofixation on its own or cryofixation combined with fixation with glutaraldehyde shows good sensitivity equivalent to that of methacrylate sections.

Examples of the titrations can be seen in FIGURES 1 to 12. FIGURES 1 to 3 show titration of PAG on methacrylate sections. At a dilution of 10^{-1} strong staining of the exosporium of the spore can be seen (FIGURES 1 and 2). At higher dilutions (10^{-4}), a small number of gold particles can be seen approaching the end point (FIGURE 3). FIGURE 4 shows staining with PAG after glutaraldehyde followed by osmium fixation. FIGURE 5 shows staining with enzyme-labeled antibodies, which is similar to that obtained with PAP. Strong specific staining of the exosporium is seen but the end product of the reaction is much more diffuse than PAG. At higher dilutions (FIGURE 6) it is often difficult to determine the small end-product reactions. FIGURE 7 shows staining with PAG after embedding in Araldite. This section is not etched and staining is poor compared to other methods. Some improvement was noted after etching. Prolonged etching did not increase sensitivity (TABLE 2).

Staining of cryofixed sections with and without prior glutaraldehyde fixation is shown in FIGURES 8 and 9. Good staining with colloidal gold is seen particularly within

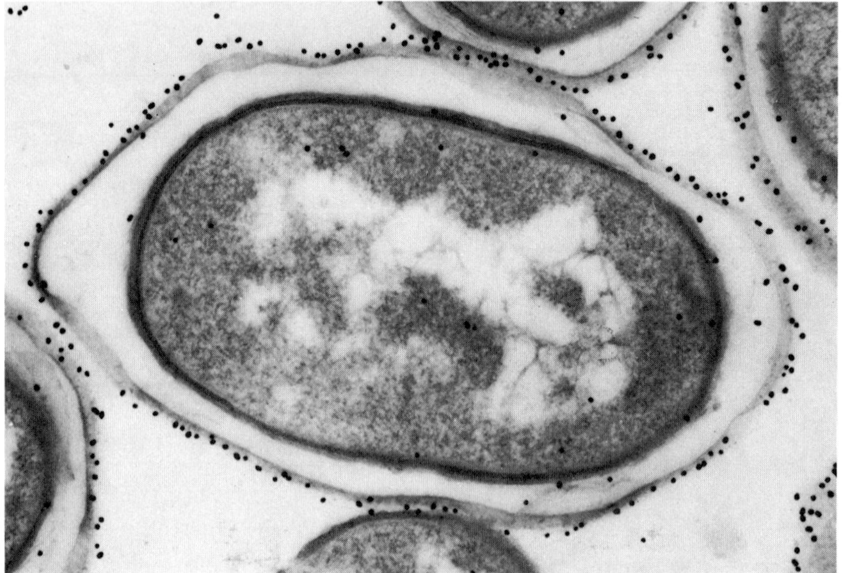

FIGURE 2. Higher magnification of FIGURE 1 showing particles of gold adhering to the exosporium. ×35,000.

the exosporium. FIGURES 10, 11, and 12 show staining of cryosections prefixed with glutaraldehyde and negative-stained with ammonium molybdate. FIGURE 10 shows the structure of spores obtained using this method and FIGURES 11 and 12 show staining with PAG. Good staining is observed at lower dilutions of antiserum (FIGURE 11) and this is readily detectable near the end point.

DISCUSSION

The results of staining with PAG produced a very definite end product that can be readily visualized and determined even when very small numbers of antigen-antibody combinations have taken place. The type of staining is more akin to that of ferritin

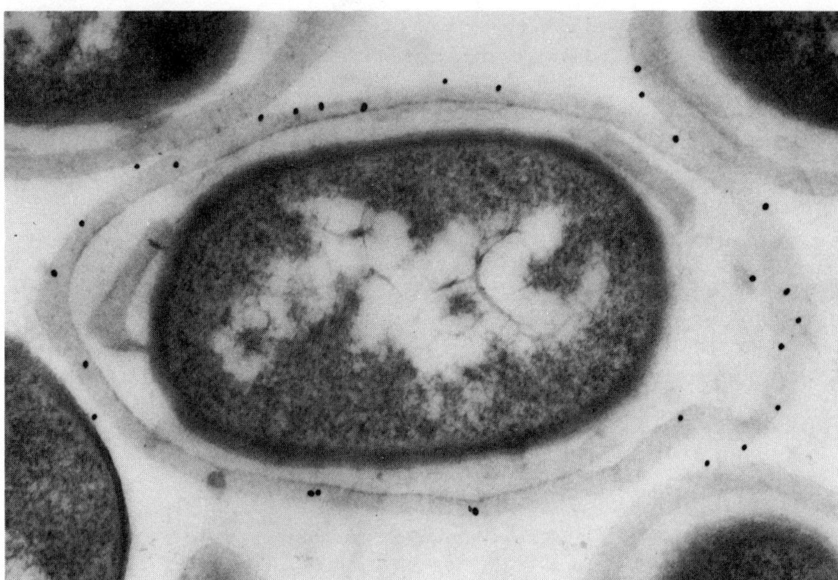

FIGURE 3. Spore of *B. cereus* M8 processed as FIGURE 1 but stained with specific antiserum at 10^{-4}. Small numbers of gold particles are discernable adherent to the exosporium approaching the end point of the titration. ×35,000.

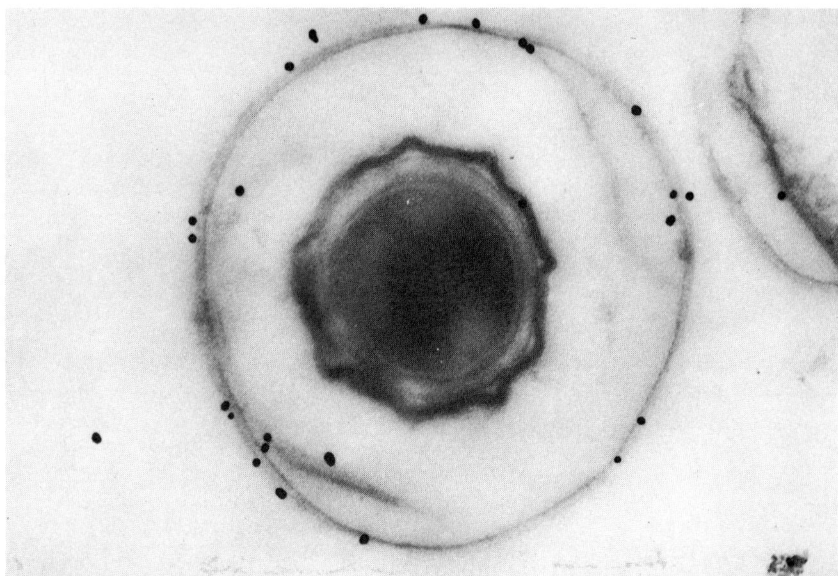

FIGURE 4. Spore of *B. cereus* M8 fixed in glutaraldehyde followed by osmium stained with specific antiserum 10^{-4} and PAG. Note similar numbers and distribution of particles to FIGURE 3. ×59,000.

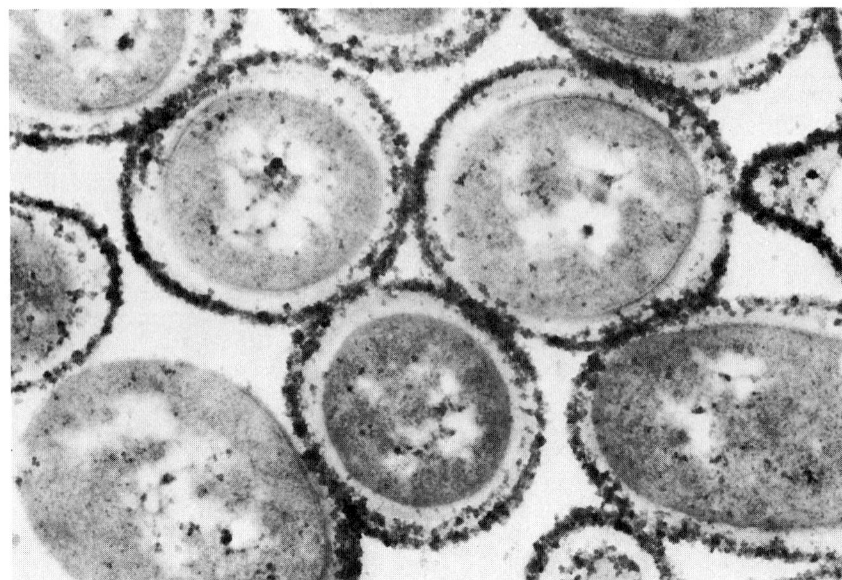

FIGURE 5. Spores of *B. cereus* M8 fixed in glutaraldehyde and embedded in methacrylate stained with specific antiserum 10^{-1} followed by horseradish peroxidase–labeled anti-rabbit antibody. Note strong specific staining of the exosporium by reaction product of 3,3-diaminobenzidine. ×19,400.

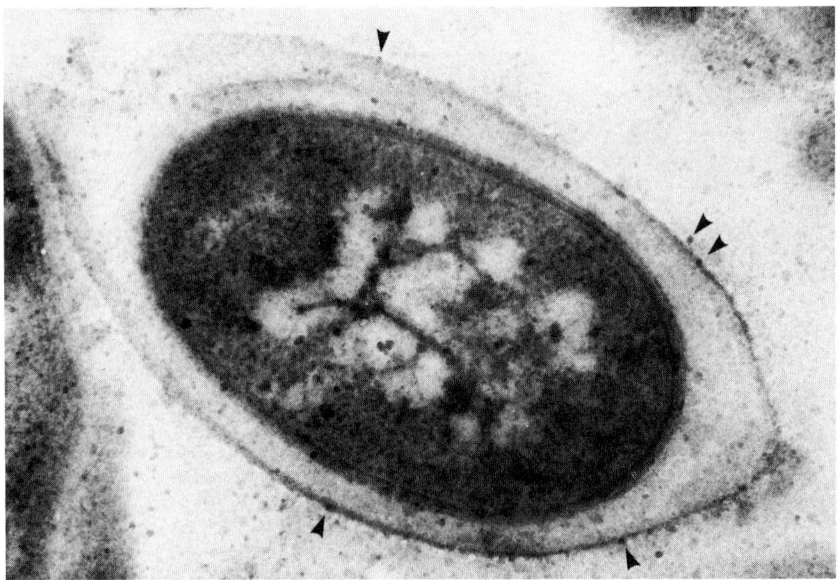

FIGURE 6. Spore of *B. cereus* M8 processed as FIGURE 5 but stained with dilution of specific antiserum 10^{-4}. Note small amount of reaction products on the exosporium (arrows) making determination of the end point difficult. ×41,000.

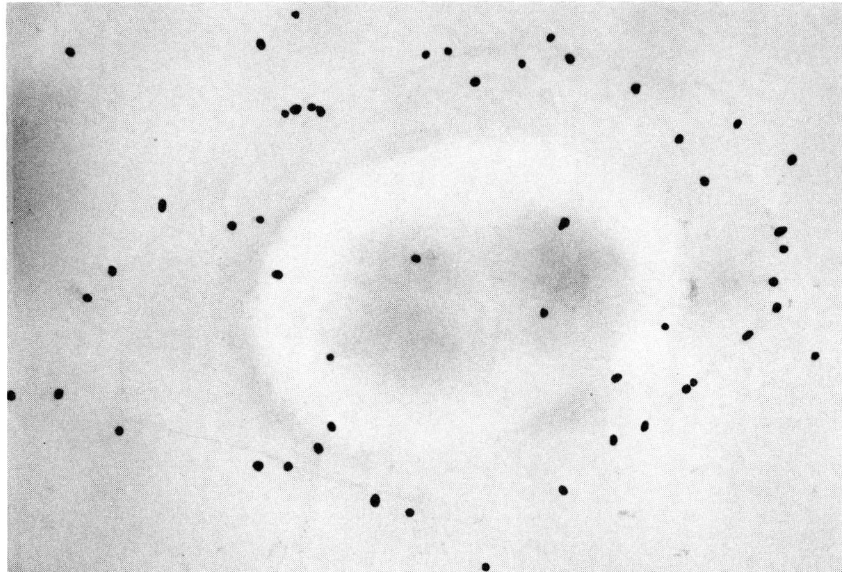

FIGURE 7. Ultrathin section of spore of *B. cereus* M8 fixed in glutaraldehyde and embedded in araldite stained with specific antiserum 10^{-1} and PAG. Note relatively poor staining compared to FIGURE 2. ×56,000.

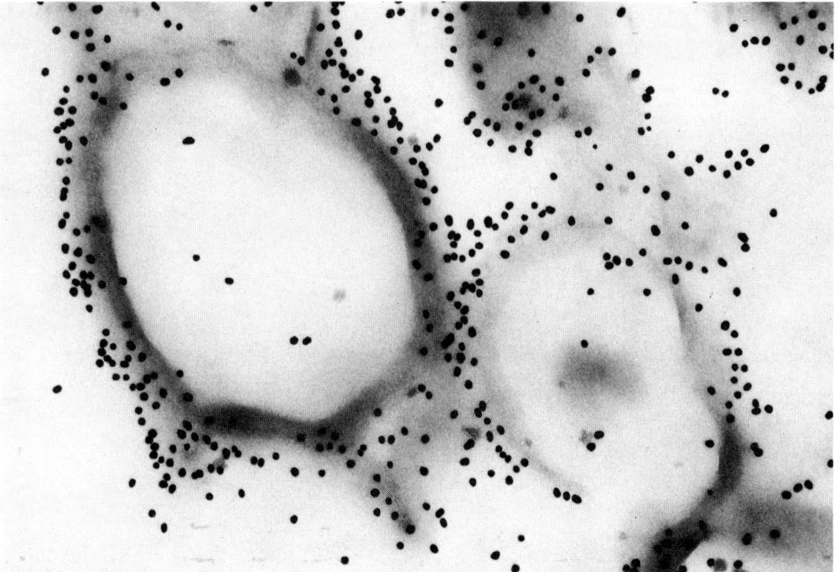

FIGURE 8. Cryofixed section of *B. cereus* M8 spores after staining with specific antiserum 10^{-1} and PAG. Note strong staining of the exosporium with gold particles particularly within the exosporium compared to FIGURE 2. ×55,000.

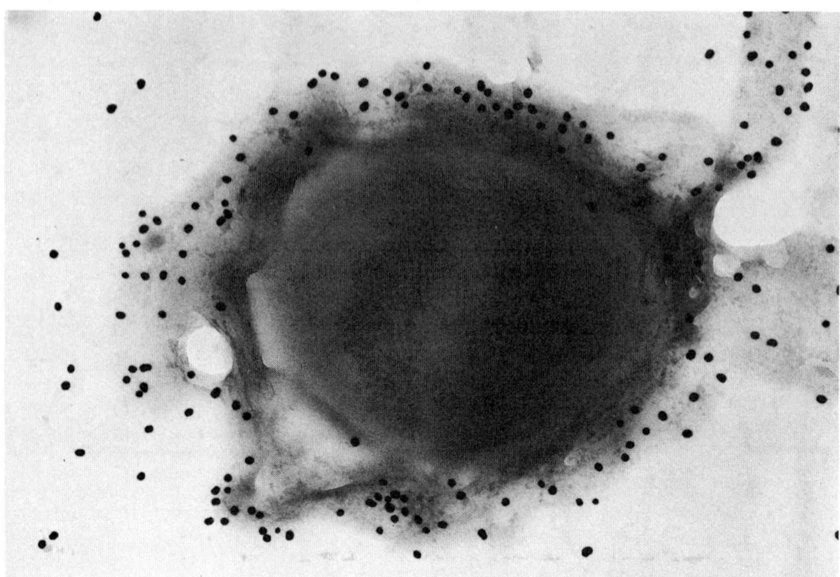

FIGURE 9. Cryofixed section of spore of *B. cereus* M8 previously fixed in glutaraldehyde and stained with specific antiserum 10^{-1} and PAG. Note strong staining with gold particles and increased contrast over FIGURE 8. ×47,000.

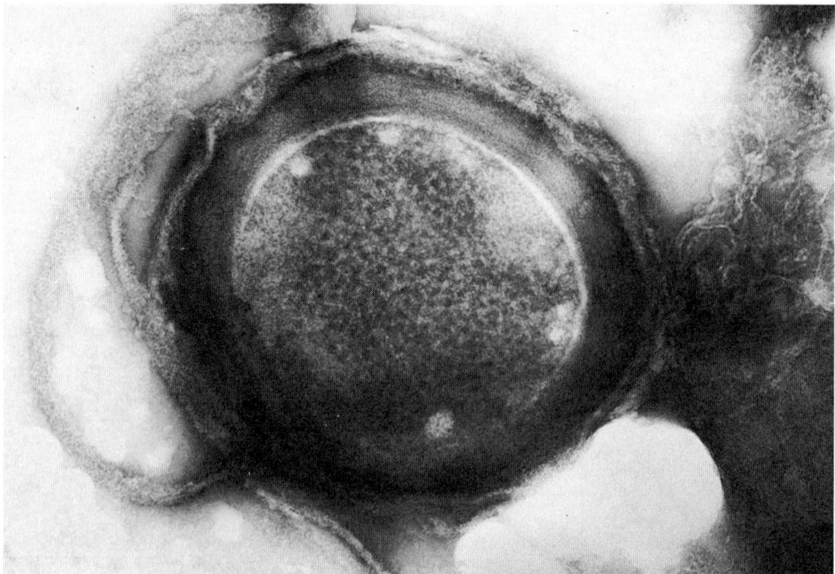

FIGURE 10. Cryofixed section of *B. cereus* M8 previously fixed with glutaraldehyde and stained with ammonium molybdate. Note increased resolution of ultrastructure compared to FIGURES 8 and 9. ×89,000.

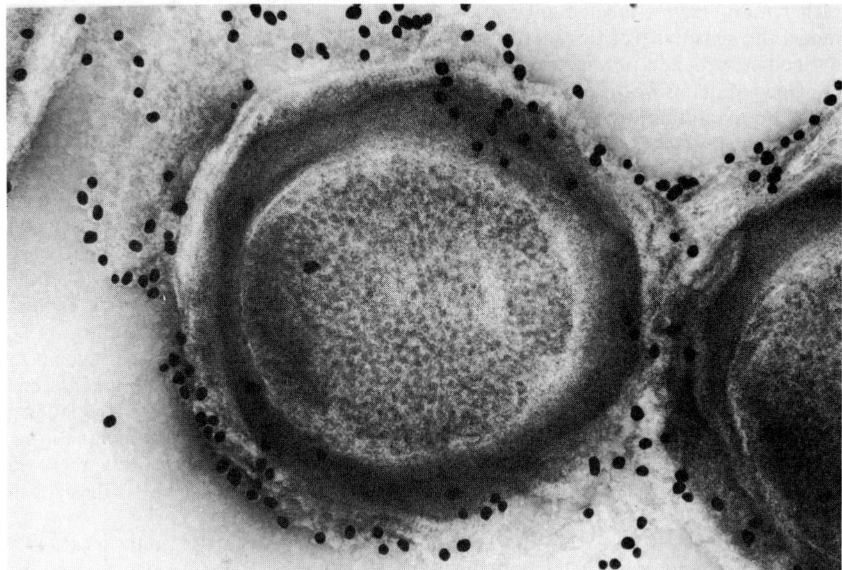

FIGURE 11. Cryofixed section of spore of *B. cereus* M8 processed as FIGURE 10 stained with specific antiserum 10^{-1} and PAG. Note strong staining of exosporium with gold particles together with details of the ultrastructure. ×82,500.

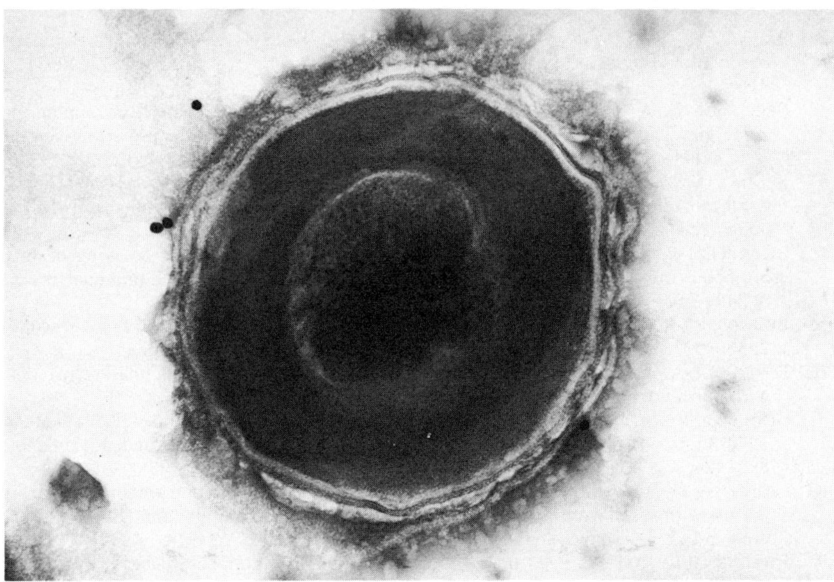

FIGURE 12. Cryofixed section of spore of *B. cereus* M8 processed as in FIGURE 11 except that specific antiserum dilution 10^{-5}. Note small numbers of gold particles present at the end point of the titration. ×73,000.

rather than enzyme-labeled antibodies or PAP. The results have shown that in this model the sensitivity of this method is somewhat higher than that of enzyme-labeled antibodies, which in turn is somewhat higher than PAP.

Good staining with PAG was obtained both on specimens fixed in glutaraldehyde alone or in glutaraldehyde plus osmium and embedded in methacrylate. Clearly, osmium in combination with glutaraldehyde does not inactivate these antigens. On the other hand, some of our previous work with Kellenberger fixation failed to demonstrate any staining. Similarly, equally good results were obtained on material prepared by cryofixation. Although cryostat sections have been extensively used in immunofluorescence studies, the introduction of ultrathin cryosections into electron microscope immunocytochemistry is fairly recent.[5,11,19] This latter method probably introduces the least artefact into the specimen and in combination with negative staining enables considerable ultrastructure to be observed in combination with specific staining.

By utilizing gold of various sizes it is possible to locate more than one antigen on a specimen. Specimens are stained with specific antibody and labeled with PAG of one diameter following which a second antiserum can be applied to the section and labeled with PAG of a different diameter. Preliminary observations with this technique using antisera to different spore components have confirmed results previously reported using the PAP[17] technique; namely that different antigen structures are synthesized in different parts of the bacterium.

The protein-A-gold system appears to have significant advantages in the location of bacterial antigens and confirms the results of Beesley *et al.*[2] on the location of the polysaccharide capsule of pasteurella.

REFERENCES

1. BAILLIE, A., R. O. THOMSON, I. BATTY & P. D. WALKER. 1967. Some preliminary observations on the location of esterases in *Bacillus cereus*. J. Appl. Bacteriol. **30:** 312–316.
2. BEESLEY, J. E., A. ORPIN & C. ADLAM. 1982. A comparison of immunoferritin, immunoenzyme and gold labelled protein A methods for the localization of capsular antigen on frozen thin sections of *Pasterurella haemolytica*. Histochem. J. (In press.)
3. BENDAYAN, M., J. ROTH, A. PERRELET & L. ORCI. 1980. Quantitative immunocytochemical localization of pancreatic secretory proteins in subcellular compartments of the rat acinar cell. J. Histochem. Cytochem. **28:** 149–160.
4. BIBERFELD, B., V. GHETIE & J. SJOQUIST. 1975. Demonstration and assaying of IgG antibodies in tissues and on cells by labelled staphylococcal protein A. J. Immunol. Meth. **6:** 249–259.
5. BRANDS, R., R. KONINKX-PEETERS, J. W. SLOT & H. J. GEUZE. 1980. Electron microscopy. **2:** 330–332. Seventh European Congr. of Electron Microscopy Foundation.
6. FORSGREN, A. & J. SJOQUIST. 1966. Protein A from *S. aureus:* I. Pseudoimmune reaction with human gamma globulin. J. Immunol. **97:** 822–827.
7. GOUDSWAARD, J., J. A. VAN DER DONK, J. A. NOORSZI, R. H. VAN DAM & J. P. VAERMAN. 1978. Protein A reactivity of various mammalian immunoglobulins. Scand. J. Immunol. **8:** 21–28.
8. GRIFFITHS, G. W. & B. M. JOCKUSCH. 1980. Antibody labelling of thin sections of skeletal muscle with specific antibodies: A comparison of bovine serum albumin (BSA) embedding and ultra-cryomicrotomy. J. Histochem. Cytochem. **28:** 969–978.
9. KRAEHENBUHL, J. P., L. RACINE & J. D. JAMIESON. 1977. Immunocytochemical localization of secretory proteins in bovine pancreatic exocrine cells. J. Cell Biol. **72:** 406–423.
10. MASON, D. Y. & P. BIBERFELD. 1980. Technical aspects of lymphoma immunohistology. J. Histochem. Cytochem. **28:** 731–745.
11. PAINTER, G., K. T. TOKUYASU & S. J. SINGER. 1973. Immunoferritin localization of

intracellular antigens: The use of ultra-cryotomy to obtain ultrathin sections suitable for direct immunoferritin staining. Proc. Natl. Acad. Sci. USA **70:** 1649.

12. ROLL, F. J., J. A. MADRI, J. ALBERT & H. FURTHMAYR. 1980. Codistribution of collagen types IV and AB_2 in basement membranes and mesagium of the kidney. An immunoferritin study of ultrathin frozen sections. J. Cell Biol. **85:** 597–616.

13. ROTH, J., M. BENDAYAN & L. ORCI. 1978. Ultrastructural localization of intracellular antigens by the use of protein A-gold complex. J. Histochem. Cytochem. **26:** 1074–1081.

14. ROTH, J. & M. BINDER. 1978. Colloidal gold, ferritin and peroxidase as markers for electron microscopic double labelling lectin techniques. J. Histochem. Cytochem. **26:** 163–169.

15. SABATINI, D. D., K. BENSCH & R. J. BARRNETT. 1963. The preservation of cellular ultrastructure and enzymatic activity by aldehyde fixation. J. Cell Biol. **17:** 19–58.

16. SHORT, J. A. & P. D. WALKER. 1975. The location of bacterial antigens on sections of *Bacillus cereus* by use of the soluble peroxidase-anti-peroxidase complex and unlabelled antibody. J. Gen. Microbiol. **89:** 93–101.

17. SHORT, J. A., P. D. WALKER, P. HINE & R. O. THOMSON. 1977. Immunocytochemical localization of spore specific antigens in ultrathin sections. J. Appl. Bacteriol. **43:** 75–82.

18. STERNBERGER, L. A. 1979. Immunochemistry. 2nd edit., p. 133. John Wiley and Sons. London.

19. TOKUYASU, K. E. & S. J. SINGER. 1976. Improved procedures for immunoferritin labeling of ultrathin frozen sections. J. Cell Biol. **71:** 894–906.

20. WALKER, P. D., A. BAILLIE, R. O. THOMSON & I. BATTY. 1966. The use of ferritin labelled antibodies in the location of spore and vegetative antigens of *Bacillus cereus*. J. Appl. Bacteriol. **29:** 512–518.

21. WALKER, P. D., R. O. THOMSON & A. BAILLIE. 1967. Use of ferritin labelled antibodies in the location of spore and vegetative antigens of *Bacillus subtilis*. J. Appl. Bacteriol. **30:** 317–320.

22. WALKER, P. D., I. BATTY & R. O. THOMSON. 1971. The localization of bacterial antigens by the use of fluorescent and ferritin labelled antibody techniques. *In* Methods in Microbiology. J. R. Norris & D. W. Ribbons, Eds.: 219–254. Academic Press. London.

23. WALKER, P. D., R. O. THOMSON & J. A. SHORT. 1971. Location of chemical components on ultrathin sections of bacteria. *In* Spore Research. A. N. Barker, G. W. Gould & J. Wolf, Eds.: 202–209. Academic Press. London.

The Use of the Solid Phase Indirect Immunofluorescent Assay (FIAX®) in the Serodiagnosis of Amebiasis

K. W. WALLS AND M. WILSON

Division of Parasitic Diseases
Centers for Disease Control
Public Health Service
U.S. Department of Health and Human Services
Atlanta, Georgia 30333

Very common worldwide, amebiasis remains one of the most prevalent parasitic infections in the United States. Although it is most commonly seen as an intestinal disease with diarrhea and abdominal discomfort as the major features, intestinal perforations frequently permit the amebae to attack other, more vital tissues. The most common extra-intestinal involvement is a liver abscess with less frequent penetration into the pleural cavity or extensive systemic spread to lung and brain. Although severe diarrhea can occur, the intestinal disease is usually relatively mild, and therapy is uncomplicated. Extra-intestinal amebiasis, however, is difficult to differentiate from other liver defects and requires an entirely different therapeutic approach than other liver defect conditions. Management of the patient frequently depends upon the identification of *Entamoeba histolytica* as the causative agent—or often, ruling out amebiasis so that other diagnoses can be entertained.

Diagnosis of intestinal disease is usually made by simple stool examination. In active diarrheal disease, the trophozoites are usually numerous and readily identified. Occasionally, parasites are rare and multiple stool samples are required to find the organism.

Extra-intestinal disease is more difficult to identify since there may be no recollectable history of diarrhea. Although not diagnostic, typical liver defects can be detected by X-ray, ultrasound, or CAT scan. If the cyst can be tapped, the trophozoites may or may not be seen in the "anchovy paste" exudate. Definitive clinical diagnosis is impossible without demonstration of amebae in patient material.

Many serological tests have been devised for amebiasis with varying amounts of success. Early tests included agar gel diffusion[4] and complement fixation.[3] Although each was useful, they both lacked sensitivity and specificity resulting from the use of impure antigens. The major obstacle preventing progress was the inability to grow *E. histolytica* without a bacterial concomitant. Finally in 1968, Diamond[1] developed the axenic medium in which *E. histolytica* could be maintained in pure culture. With the advent of the axenic culture technique, new tests using the "clean" antigen were possible.

In 1965, Kessel *et al.*[3] described an indirect hemagglutination test (IHA) that has

---

Abbreviations: IHA = indirect hemagglutination, CIE = counter immunoelectrophoresis, TTBS = 0.05 M Tris-buffered saline (pH 8.5) with 0.35% Tween 20, FSU = fluorescent signal units, STIQ = plastic paddle for FIAX®.[a]

since been further standardized and evaluated by Healy[2] to become the primary test for the serodiagnosis of amebiasis. He showed the specificity to exceed 90% and the sensitivity in extra-intestinal disease to be 92% and in intestinal disease to be only about 80%.

Recently, latex agglutination and counter-immunoelectrophoresis (CIE) tests have been introduced and are gaining popularity. Although both tests are useful, each has its shortcomings. The latex test, very rapid and simple to perform, is overly sensitive and tends to yield false positives. As such, it is an excellent screen test to rapidly rule out amebiasis from clinical consideration, but positives should be confirmed by other procedures. CIE, on the other hand, shows good specificity and bands correlate well with IHA at significant levels. However, it does not give quantitative results and consequently is diagnostically effective but cannot be used to monitor the progress of the disease or the effectiveness of therapy.

FIAX[®][a] is a solid-phase indirect immunofluorescent assay. Initially introduced for the measurement of immunoglobulin levels and antinuclear antibodies, it has now been used for a variety of infectious agents. Basically, the FIAX[®] procedure involves a plastic paddle (called a STIQ[®][a]) with a cellulose acetate disc, 8 mm in diameter, at the end on each side. Antigen is added to one side, and the other side acts as a control surface. The sensitized STIQ is then processed through serum, wash, conjugate, wash, and then read in a FIAX[®]-100 fluorometer. The resulting fluorescent signal unit (designated FSU) is transmitted to a Hewlett-Packard HP-85 microcomputer for analysis. Usually results are read against standards and quantitative results interpolated. Using this procedure, Taylor and Perez[5] described a qualitative test for amebiasis. Their procedure, however, only served to differentiate positive from negative patients and gave no measure of antibody changes.

We describe here a modification of the Taylor-Perez procedure based on an earlier technique used by us for toxoplasmosis,[6] which permits a quantitative measure of antibody levels against *Entamoeba histolytica* in a single serum dilution. Comparisons with IHA and evaluation against clinical data have shown the method superior to IHA in every respect.

METHODS AND MATERIALS

Sera

Four groups of sera were used for this study: (1) sera from patients with clinically diagnosed amebiasis with *E. histolytica* demonstrated either in stool or histological sections; (2) sera from patients with clinically diagnosed amebiasis and who responded to therapy but in whom amebae were not demonstrated; (3) patients with clinical symptomatology similar to amebiasis, but the final diagnosis was other than amebiasis; and (4) sera submitted routinely to the Centers for Disease Control for amebic serology. Four sera from Group 1 were selected that most closely fell on the correlation points of 128, 256, 512, and 1024. These sera were used throughout the remainder of the study as "calibrators" and served to establish the correlation curve for each test.

[a]FIAX and STIQ are registered trademarks of International Diagnostic Technology, Santa Clara, California. Use of trade names and commercial sources is for identification only and does not constitute endorsement by the Public Health Service or the U.S. Department of Health and Human Services.

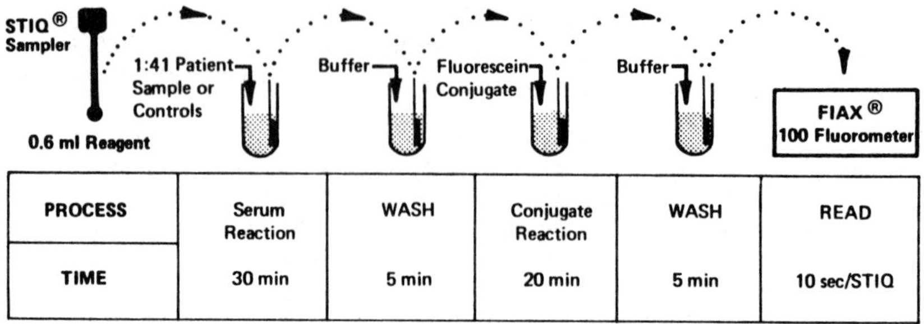

PROCESS	Serum Reaction	WASH	Conjugate Reaction	WASH	READ
TIME	30 min	5 min	20 min	5 min	10 sec/STIQ

FIGURE 1. Schematic diagram of FIAX procedure for amebiasis.

Antigen

All antigens were prepared from the HK-9 strain of *E. histolytica*. Parasites were harvested from axenic culture, washed with phosphate-buffered saline, and disrupted on a Heat Systems-Ultrasonica, Inc. sonicator (model W-375) pulsed on a 70% cycle at 20% power for 3 minutes using a standard probe. Three separate lots were prepared, keeping conditions as similar as possible to evaluate lot-to-lot effects on results.

Test Procedure

STIQs were sensitized by placing 20 μg of antigen diluted in distilled water to contain 200 μg/ml of protein on one side and allowing it to dry at room temperature for at least 2 hours. STIQs that were not used were stored with desiccant at 5°C for future tests. The test procedure is illustrated in FIGURE 1. Sensitized STIQs are placed in 0.6 ml of patient's serum diluted 1:41 in 0.05 M Tris-buffered saline (pH 8.5) with 0.35% Tween 20 (TTBS). After 30 minutes agitation at room temperature, the STIQs are transferred to 0.6 ml TTBS and washed for 5 minutes. They were then transferred to 0.6 ml fluorescein isothiocyanate–labeled anti-human immunoglobulin (Wellcome, Lot K7350, diluted 1:400 in TTBS) and agitated for 20 minutes. After final wash of 5 minutes in 0.6 ml TTBS, the blank and antigen slides of the STIQs were read and the value expressed as a Δ FSU.

TABLE 1. Within-Day and Day-to-Day Reproducibility of FIAX®

Spec. #	Day 1	Day 2	Day 3	Day 4	Day 5	$\overline{X}$	s.d.
1	307	355	433	216	188	304.6	78.2
	340	321	277				
2	632	506	472	472	467	517.1	62.4
	486	506	596				
3	1,440	1,878	1,195	1,081	1,134	1,317.5	268.3
	1,472	1,206	1,134				
4	1,082	834	879	684	713	838.4	158.6
5	2,523	2,375	2,749	2,048	2,013	2,341.6	313.9

Data Analysis

Quantitation was obtained by use of a correlation curve established daily from known calibrator sera. To maintain consistency, the highest calibrator is adjusted each day so that the antigen side of the STIQ reads 160 FSU. With this setting the Δ FSU for all four calibrators is determined and a standard curve established from which the value of the remaining STIQs is extrapolated. Since indirect hemagglutination (IHA), up to now, has been used as the reference serologic test, FIAX® titers were initially correlated with IHA titers. Except for TABLES 1 and 2, FIAX® data have been grouped into equivalent $\log_2$ titers to compare more directly with IHA.

RESULTS

FIGURE 2 illustrates the correlation between IHA and FIAX® that was obtained when sera from 66 proven cases of amebiasis having a range in IHA titer from

TABLE 2. Effect of Antigen Lot on FIAX® Reproducibility

	Antigen Lot						
Spec. #	A (Day 1)	A (Day 2)	A (Day 2)	B (Day 3)	C (Day 3)	$\bar{x}$	s.d.
1	10	37	19	6	10	16.4	12.4
2	33	123	58	10	54	55.6	42.3
3	99	225	170	37	68	119.8	76.7
4	129	311	197	352	452	288.2	127.5
5	489	967	996	523	444	683.8	273.4
6	678	924	698	485	370	631.0	213.3
7	1,150	1,741	1,501	1,029	529	1,190.0	465.0
8	1,941	2,208	2,252	909	820	1,625.8	705.7
9	2,585	3,589	3,671	1,831	1,575	2,650.2	968.9
10	5,981	6,913	6,803	2,580	2,393	4,934.0	2,264.0

1:128–1:1024 were used. Note the straight-line relationship. The standard deviation at 1:128 is deceiving since the Δ FSU at this point is very low (60), and minor variations are very apparent.

TABLE 1 illustrates the within-day and day-to-day reproducibility when five sera were tested on five separate days. Specimens 1, 2, and 3 were tested in duplicate on three days; all others were tested only once. Within-day variation was minimal, with specimen 1 on day 3 and specimen 2 on day 2 showing the most extreme variations; however, neither of these exceeded a single twofold variation, which would be acceptable in an IHA test. Day-to-day variation, with the exception of specimen 1, was comparable to within-day variation and is probably an expression of a constant procedural variation. The greatest variation is seen in the lowest titered specimen and still is less than half of what would be expected in a manual IHA test.

Test reproducibility is somewhat lower when different lots of antigens are used (TABLE 2). In these data it is apparent that lots B and C are less reactive than lot A. Although overall mean and standard deviations are good, obvious variants are apparent, specifically lot A on day 2 for specimen 3 and lots B and C for specimen 7.

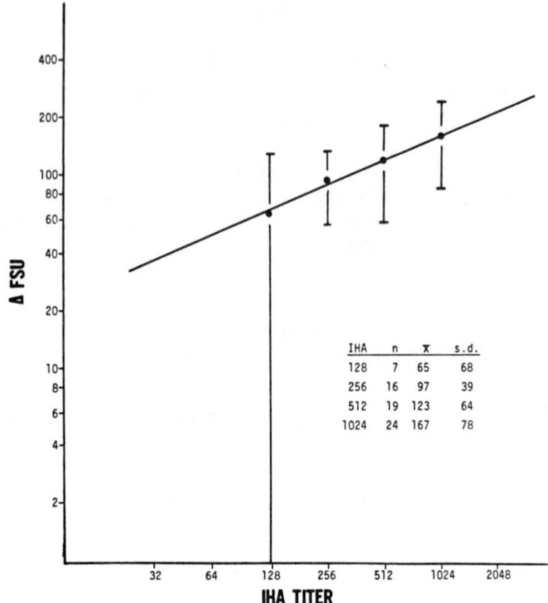

FIGURE 2. Reference curve for ameba FIAX®. (Derived from Proven Case Battery.)

Since these are single tests on one day, differences may reflect individual specimen error rather than antigen effect, but overall agreement is considerably less than that seen in TABLE 1.

To ascertain specificity, 123 sera from clinical cases resembling amebiasis but whose final diagnosis was other than amebiasis (such as pyogenic abscess, carcinoma, ulcerative colitis, and Crohn's disease) were tested. These results are seen in TABLE 3. The "cut-off" value for IHA in amebiasis was established some time ago as 1:128.[5] As can be seen, at this level IHA showed 18% positive reactors while FIAX®, when adjusted to comparable twofold titer levels, only gave 2% reactors. In fact, even at 1:64 FIAX® only had 5% reactors.

TABLE 3. Comparison of FIAX® and IHA in Cases with a Final Diagnosis not Amebiasis[a]

	IHA			
FIAX	≤32	64	128	256
≤16	57	33	17	1
32	2	4	1	1
64		1	3	
128		1		
256		1		
512		1		

FIAX® ≥ 128 = 2%

IHA ≥ 128 = 18%

[a]Liver abscess, ulcerative colitis, colitis, etc.

TABLE 4. Comparison of FIAX® and IHA in Invasive and Noninvasive Amebiasis[a]

FIAX®	<64	64	128	256	IHA 512	1,024	2,048	≥4,096
<64	32		6			1		
64	1			1				
128	2	1		1	1			
256	1	1		1		2		
512	2		2	2	1	2	1	1
1,024		1		2	4	1	1	6
2,048		1		3	9	4	2	11
≥4,096						5	7	16

FIAX® ≥ 128 = 78%

IHA ≥ 128 = 78%

[a]*E. histolytica* demonstrated in stool or biopsy specimens.

To determine sensitivity of FIAX®, sera from patients with *E. histolytica* identified in stool or biopsy were examined by both tests and the results are shown in TABLE 4. Each test detected 78% at titers ≥ 128 though not the same cases. By clinical history, the FIAX®-positive/IHA-negative cases were old cases. The IHA-positive/FIAX-negative cases were all but one at the minimum level. The clinical history on the one patient with a high titer cannot be obtained. Likewise, the FIAX-2048/IHA-64 remains unclassified. There is, however, strong agreement between the tests.

When the proven invasive amebiasis cases in which *E. histolytica* has been demonstrated were selected (TABLE 5), FIAX® detected more (89% versus 85%) than IHA. Again the greater sensitivity was primarily limited to the older cases.

In a more realistic clinical situation, cases may be identified by clinical presentation and response to therapy without necessarily demonstrating the parasite. Results of serologic tests on 200 such cases are shown in TABLE 6. Slightly higher percentage positive results (90%) are seen in these data than that seen in TABLE 5. The close agreement between these data and those cases in which *E. histolytica* was demonstrated certainly supports the clinical diagnosis.

TABLE 7 illustrates the data acquired through testing 291 successive specimens submitted for routine serodiagnosis for amebiasis. Clinical information is not available

TABLE 5. Comparison of FIAX® and IHA in Proven Cases of Invasive Amebiasis[a]

FIAX®	<64	64	128	256	IHA 512	1,024	2,048	≥4,096
<64	5		2					
64				1				
128	1			1	1			
256	1	1				1		
512	1		1	2			1	1
1,024			1	2	2		1	6
2,048		1			8	2	1	7
≥4,096						3	6	12

FIAX® ≥ 128 = 89%

IHA ≥ 128 = 85%

[a]*E. histolytica* isolated or histologically demonstrated.

TABLE 6. Comparison of FIAX® and IHA in Cases of Clinically Diagnosed Invasive Amebiasis[a]

FIAX®	<64	64	128	256	IHA 512	1,024	2,048	≥4,096
<64	11	1	3	1				
64			2	1				
128	1	2		3	4			
256	2	1		2	3	2		
512	1		1	4	2	3	3	1
1,024			3	5	8	2	4	13
2,048		1		3	14	7	9	26
≥4,096					1	7	12	31

FIAX® ≥ 128 = 90%

IHA ≥ 128 = 90%

[a]Cases identified as clinically compatible and/or therapy-responsive but *E. histolytica* not demonstrated in each case.

on most of the specimens, but it is assumed the sera were submitted either for diagnosis or rule-out. The data certainly describe such a group. Although there are many more negatives than seen in the previous tables, the agreement between the tests would indicate a large number of noncases and a good correlation between the two procedures in detecting positives.

The data in TABLES 3 and 4 are summarized on TABLE 8 in which it can be seen that FIAX® is slightly more sensitive in both invasive and intestinal amebiasis while being less reactive (and thus more specific) in other forms of intestinal disease. As might be expected from the type of infection, both IHA and FIAX® are insensitive in the intestinal disease.

DISCUSSION

We have demonstrated that FIAX®, a solid-phase, quantitative fluorescent immunoassay, is as effective as IHA in measuring antibodies to *E. histolytica*. Our data have shown that FIAX® is in fact more sensitive and specific than IHA. On the other hand, the data did indicate that FIAX® tends to react with sera from older cases thus introducing a degree of confusion in diagnosis. Generally, those cases tended to give

TABLE 7. Comparison of FIAX® and IHA in Sera Routinely Submitted for Amebiasis

FIAX®	<64	64	128	256	IHA 512	1,024	2,048	≥4,096
<64	41	1	7	1				
64	1	0	2	2				
128	2	3	1	4	5	1		
256	2	2	0	7	4	4	1	
512	2		2	5	3	7	4	1
1,024			3	5	12	3	4	13
2,048		1		6	15	9	10	30
≥4,096					1	10	13	41

minimal titers while acute, active cases gave much higher titers. At the same time, some cases gave minimal titers (1:128) with IHA and were negative ($\leq$1:32) by FIAX®. The specificity of FIAX® would indicate that a lower reactive level could be accepted without significant increase in nonspecificity, and most of the minimal IHA reactors would be detected. We have concluded, therefore, that the lowest useful serologic titer in the FIAX® test for amebiasis as we perform it is 1:64.

Reproducibility is an apparent problem. Well within the limits of IHA reproducibility, one would still hope for coefficients of variations to be less than 10% and in many cases in the data presented, the CV reached 20–25%. It is obvious in these studies as well as earlier work on toxoplasmosis that a major source of error is in the application of antigen to the STIQs. Although it is a soluble antigen, great care must be exercised in applying the 20 μl drop to the surface to assure even distribution across the surface. Uneven drying as well as uneven application seems to affect the results. In our hands, toxoplasma STIQs, which have the antigen applied and are sensitized by the manufacturer, have a higher degree of reproducibility than those that are prepared in our laboratory.

The simplicity of the procedure tends to create an atmosphere for error. We have

TABLE 8. Comparison of Sensitivity and Specificity of FIAX® and IHA Tests for Amebiasis

		Cumulative % Positive at						
		64	128	256	512	1,024	2,048	4,096
	Invas[a]	92	90	85	80	73	56	29
FIAX®	Intes	49	48	44	41	33	28	11
	Other	5	2	1	1	0		
	Invas	89	86	81	72	57	49	36
IHA	Intes	52	51	43	37	30	16	13
	Other	52	19	2	0			

[a]N: Invasive = 72, Intestinal = 63, Other = 123.

detected automatic pipetting errors resulting from hastily diluting large numbers of specimens and one has a tendency to remove the pipettor from the sample before the aspiration cycle is complete. Since the total volume withdrawn is only 15 μl, any error is great.

Cost comparisons between tests are very difficult. On a single specimen basis, IHA, which requires no special equipment, is undoubtedly less expensive. On the other hand, if one is testing numerous specimens by FIAX® (and they need not all be for the same test), FIAX® becomes highly cost-effective because of the speed and economy in labor.

When reagents become available, FIAX® for amebiasis appears to be an excellent procedure for the diagnosis, prognosis, and therapy evaluation.

REFERENCES

1. DIAMOND, L. S. 1968. Techniques of axenic cultivation of *Entamoeba histolytica Schaudinn,* 1903 and *E. histolytica*-like amebae. J. Parasit. **54:** 1047–1056.
2. HEALY, G. R. 1968. Use and limitations of the indirect hemagglutination test in the diagnosis of intestinal amebiasis. Health Lab. Sci. **5:** 174–179.

3. KESSEL, J. F., W. P. LEWIS, C. M. PASQUEL & J. A. TURNER. 1965. Indirect hemagglutination and complement fixation test in amebiasis. Am. J. Trop. Med. Hyg. **14:** 540–550.
4. MADDISON, S. E. 1965. Characterization of *Entamoeba histolytica* antigen-antibody reaction by gel diffusion. Exp. Parasit. **16:** 224–235.
5. TAYLOR, R. G. & T. R. PEREZ. 1978. Serology of amebiasis using the FIAX® system. Arch. Invest. Med. (Mex.) **9**(Suppl 1): 363–366.
6. WALLS, K. W. & E. R. BARNHART. 1978. Titration of human serum antibodies to *Toxoplasma gondii* with a simple fluorometric assay. J. Clin. Micro. **7:** 234–235.

Index of Contributors